EARTH
SCIENCE

EARTH SCIENCE

Cesare Emiliani
Professor and Chairperson,
Department of Geological Sciences
University of Miami
Coral Gables, Florida

Linda B. Knight
Earth Science Teacher
Paul Revere Middle School
Houston, Texas

Mark Handwerker
Instructional Support Teacher for Science
Temecula Middle School
Temecula, California

HBJ Harcourt Brace Jovanovich, Publishers
Orlando San Diego Chicago Dallas

Printed in the United States of America

ISBN 0-15-361451-X

ACKNOWLEDGMENTS

SENIOR EDITORIAL ADVISOR

Lee Suttner, Ph.D.
Professor of Geology
Indiana University
Bloomington, Indiana

CONTENT ADVISORS

George Greenstein, Ph.D.
Professor of Astronomy
Amherst College
Amherst, Massachusetts

Brian Skinner, Ph.D.
Professor of Geology and
  Geophysics
Yale University
New Haven, Connecticut

Carolyn Sumners, Ed.D.
Director of Astronomy and
  Physics
Houston Museum
  of Natural Science
Houston, Texas

CURRICULUM ADVISORS

Ronald E. Charlton, Ph.D.
Science Coordinator
Mt. Lebanon School District
Pittsburgh, Pennsylvania

Dee Drake *
Earth Science Teacher
Huron High School
Ann Arbor, Michigan

Larry Enochs, Ed.D.
Professor of Curriculum and
  Instruction
Kansas State University
Manhattan, Kansas

Robert Frank *
Earth Science and Geography
  Teacher, Science
  Coordinator K–12
Jefferson Junior High School
Caldwell, Idaho

Loistene Harrell
Earth Science Teacher,
  Department Chairperson
Lake Braddock Secondary
  School
Fairfax County, Virginia

Tom Howick *
Earth Science Teacher
Woodward Academy
College Park, Georgia

Nancy Hampton Johnson
Earth Science Teacher
Haggard Middle School
Plano, Texas

Bobbie G. Reed
Department Chairperson
Crestwood Middle School
Baton Rouge, Louisiana

David Sorenson
Science Teacher
Southwest High School
Green Bay, Wisconsin

Lou Travelsted
Science Teacher
Franklin-Simpson Middle
  School
Franklin, Kentucky

Doris Tucker *
Earth Science Teacher
Rutherford County Schools
Spindale, North Carolina

* Outstanding Earth Science Teacher,
  as awarded by the National
  Association of Geology Teachers

READING/LANGUAGE ADVISORS

Patricia S. Bowers, Ph.D.
Science Reading Coordinator
Division of Curriculum and
  Instruction
Chapel Hill–Carrboro City
  Schools
Chapel Hill, North Carolina

iv

Sue Porro
English Teacher
Dr. Phillips High School
Orlando, Florida

Karen Rugerio
Chairperson, Fine Arts
    Department
Dr. Phillips High School
Orlando, Florida

SERIES FIELD TEST
TEACHERS

John Benning
Morse Middle School
Milwaukee, Wisconsin

Jerry East
Nimitz Middle School
Tulsa, Oklahoma

Freddie Fight
Bartow Junior High School
Bartow, Florida

Corinne Fish
Morse Middle School
Milwaukee, Wisconsin

Larry French
Deland Junior High School
Deland, Florida

Rick Herbert
Mac Arthur Junior High School
Lawton, Oklahoma

Stan Hitomi
Monte Vista High School
Danville, California

Jennifer Jones
Coosa Middle School
Rome, Georgia

Ellen McCullough
Edwards Middle School
Conyers, Georgia

Carl Nebelsky
Bloomfield Junior High School
Bloomfield, Connecticut

Anna Rice
Bloomfield Junior High School
Bloomfield, Connecticut

John Richardson
Letha Raney Junior High
    School
Corona, California

Bernard Sanner
South Division High School
Milwaukee, Wisconsin

Mary Satterwaite
Morse Middle School
Milwaukee, Wisconsin

Charles Siebert
South Division High School
Milwaukee, Wisconsin

Peggy Stewart
Glasgow Middle School
Baton Rouge, Louisiana

CONTENTS

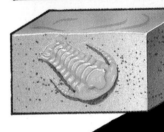

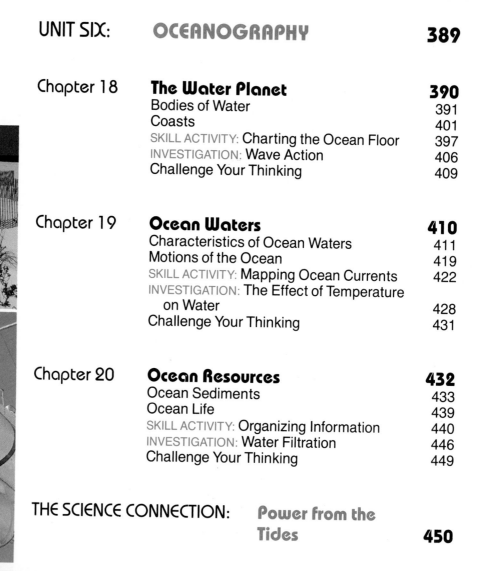

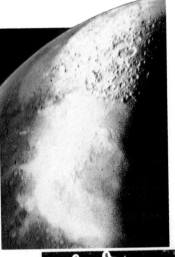

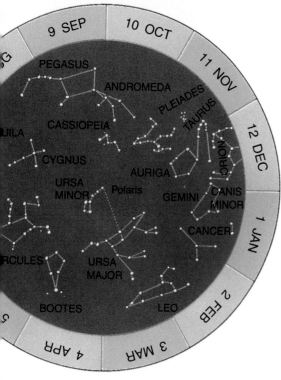

REFERENCE SECTION 547

SKILL ACTIVITIES

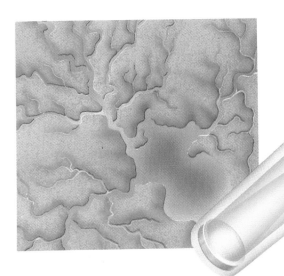

INVESTIGATIONS

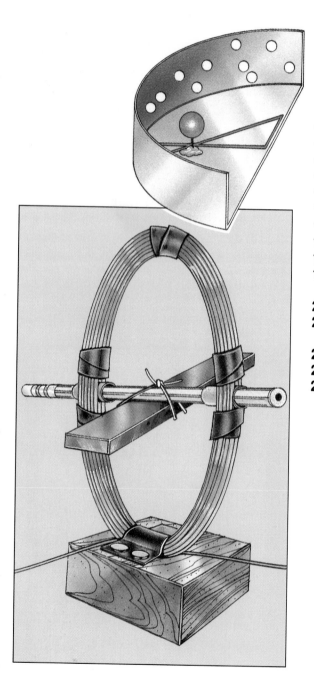

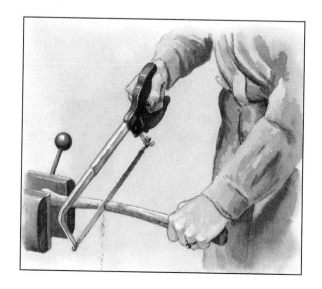

BIOGRAPHIES: THEN AND NOW

CAREERS

TECHNOLOGY

ACTIVITIES

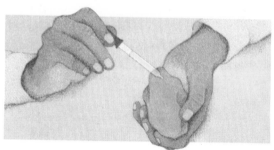

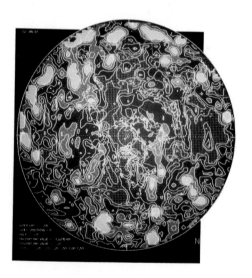

DISCOVERS

A Message to Students About
Earth Science

How often is your life touched by some aspect of earth science? You might quickly answer "never" or "not very often." However, if you stop and think for a few minutes, you might be surprised as to just how important earth science is to your life.

Do you know what the weather forecast is for today? If you do, you probably received your information from a newspaper or a radio or television broadcast. In turn, these media might have received their weather information from NOAA, the National Oceanic and Atmospheric Administration. NOAA is a department of the federal government that supplies weather data and forecasts for the United States. NOAA employs many meteorologists—earth scientists who specialize in the study of the weather.

On a clear night, have you ever looked at the moon or the stars through a telescope? Although the moon and the stars are not part of the earth, astronomy—the study of the universe—is a part of earth science. So, too, is space science. Many orbiting satellites have sensors that scan the earth and provide information about natural resources or soil erosion. Probes have landed on Venus and Mars and have sent data that may help increase understanding of Earth's processes.

Do you live on a mountain, on a plateau, or on a coastal plain? Have you ever seen a volcano, felt an earthquake, or heard thunder? Have you ever built a sand castle, tasted salt, or smelled rain? All these things are part of earth science.

You may be taking this course because you are interested in studying the earth or perhaps because it is required by your school. Whatever the reason you are taking this course, you should prepare yourself to make a discovery. In your study of *Earth Science,* you will learn many interesting and exciting things. Allow yourself to explore *Earth Science* and to discover the ideas, information, and beauty within its pages. You will find that learning about earth science is enjoyable as well as beneficial. You will learn many things that will help you understand the processes of the earth. You may even develop an interest in earth science as a career.

In addition to learning scientific concepts, with *Earth Science* you will develop other skills that will help you now and in the future. For example, you will learn how to improve your reading comprehension, how to make and read tables and graphs, and how to

understand diagrams and photographs. As you begin your study of *Earth Science,* you also will start developing your skills as a trained observer. Much of science is based on precise observations of nature. From their observations, scientists form questions and develop experiments to expand on their observations. From these close and precise observations, scientists try to answer their questions and form conclusions.

Answers to questions and records of observations have allowed earth scientists to create a body of knowledge that serves as the primary resource for engineers and technicians. This body of knowledge is the basis of technological growth. Computers, radiotelescopes, fusion reactors, submersibles, and the space shuttle are just a few applications of scientific knowledge. These devices, which are related to earth science, help improve our standard of living.

Technology is the method of putting science to work. For example, through the study of scientific principles, Thomas Edison discovered

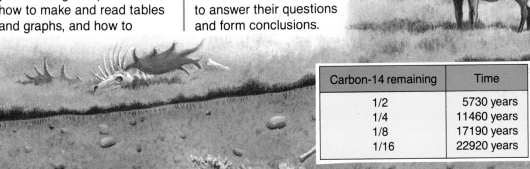

Carbon-14 remaining	Time
1/2	5730 years
1/4	11460 years
1/8	17190 years
1/16	22920 years

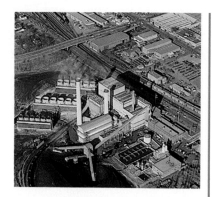

that electricity flowing through a thin wire would make the wire glow. As a result of this knowledge, Edison was able to create the light bulb. The invention of the light bulb allowed people to put electricity to work to improve their lives.

Further scientific discoveries created an understanding of how to generate the quantities of electricity needed to light every home. This information led to the development of fossil-fuel power plants that produce electricity by burning coal and oil. The need for large quantities of fossil fuels led to further studies of the earth and its resources. A lack of fossil fuels led to the discovery of alternative energy sources. In the 1950s, nuclear energy was first used to generate electricity commercially. Nuclear power may help overcome a shortage of fossil fuels in the immediate future, but the 1986 accident at Chernobyl, in the Soviet Union, has shown that nuclear power is not the final answer to our need for electricity. In the future, technology may make solar energy and fusion energy practical and safe alternatives.

In *Earth Science,* you will learn more about some topics with which you may already be familiar. For example, you will review the use of a scientific method for solving problems. You will also practice measuring, using SI. You will learn how maps are made and about the different kinds of maps an earth scientist uses.

The structure of the earth and the composition of the atmosphere are described in great detail in *Earth Science.*

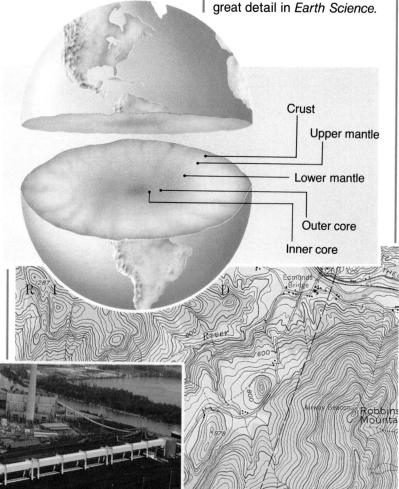

Crust

Upper mantle

Lower mantle

Outer core

Inner core

The processes that build and erode the earth are discussed. *Earth Science* presents the history of our planet and projects its future as well. Finally, in *Earth Science* you will learn about our great frontiers—the oceans and space.

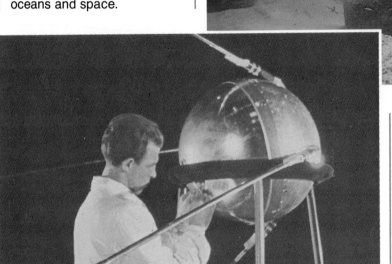

The study of *Earth Science* will give you a new perspective on why and how things happen around you. You will begin to ask your own questions and supply your own answers. *Earth Science* includes the ongoing processes of asking and answering questions about the earth. Curiosity and imagination are guiding lights for earth scientists. By virtue of this curiosity and imagination, earth scientists are able to create a never-ending list of questions about the universe in which we live.

Many of the questions asked by today's earth scientists will not be answered in their lifetimes. Their observations and research, however, will help future scientists achieve the answers. By pursuing a career in earth science, you may become a beneficiary of the knowledge, the dreams, and the aspirations of the scientists who lived before you. Your work in earth science may help answer your own questions as well as questions posed by the earth scientists of the past.

The authors of *Earth Science* hope that you will enjoy reading and learning about the earth, and that perhaps one day you may help others discover the importance of earth science in their lives.

INTRODUCTION TO EARTH SCIENCE

In 1972 earth science took a giant leap forward with the launch of the first Landsat satellite. In the years that followed, four more Landsat satellites were launched and are still circling the earth. These satellites have given scientists more than a million pictures of the earth. Each multi-colored photograph shows an area about 185 km wide. Details as small as a football field can be seen.

- Why do you think different colors are used in Landsat photographs?

- How has the exploration of space helped to solve some problems on Earth?

- How might a mapmaker use a tool like Landsat to create more accurate maps?

By reading the chapters in this unit, you will learn the answers to these questions. You will also develop an understanding of concepts that will allow you to answer many of your own questions about the study of the earth.

Satellite photographing Earth

CHAPTER

1

Studying the Earth

Few people can look at this photograph and not admire the beauty of Earth. The cloud patterns in the air contrast with the land masses below. The large bodies of water provide the rich blue color that makes Earth so different from other planets. Earth is a fascinating place, and its study reveals many wonders.

Earth, as it appears from space

1 What Is Earth Science?

SECTION OBJECTIVES

After completing this section, you should be able to:

- **Describe** the five scientific fields that are studied in earth science.
- **Explain** how the fields of biology, physics, and chemistry relate to earth science.

NEW SCIENCE TERMS

geology
hydrology
oceanography
meteorology
astronomy
science
fossils

1.1 The Earth Sciences

Have you ever taken something apart to see how it works? In order to really understand how something works, you must examine all its parts. Complicated things seem so much simpler if you can study one part at a time. The same is true when you study the earth. In order to understand how the earth works, you must look at it from the point of view of scientists who specialize in studying its parts.

Scientists are constantly striving to understand the earth and its secrets more fully. To do this, scientists observe, classify, and collect information about the earth.

Earth science can be divided into the areas of *geology, hydrology, oceanography, meteorology,* and *astronomy.* Although you can get a better understanding of parts of the earth through these specialties, you must study all of them to really understand how the earth works.

Figure 1–1. The oceans, the land, the clouds, and space are all subjects of study in earth science.

Geology

The study of the solid earth is called **geology** (jee AHL uh jee). A geologist studies minerals, rocks, and fossils, as well as the processes that change the earth. Studying volcanoes, mountain building, and erosion is also part of geology.

Hydrology and Oceanography

The study of all the water found on land is called **hydrology** (hy DRAHL uh jee). A hydrologist studies the water under the ground, as well as the water in streams and lakes. Hydrologists are concerned with the movement of ground water and the quality of fresh water supplies.

The oceans are studied in **oceanography** (oh shuh NAHG ruh fee). In this field, all aspects of the oceans are studied. Oceanographers study the shape and composition of the ocean floor, the chemical composition of ocean water, the movements of tides and currents, and the plants and animals that live in the oceans.

Figure 1–2. From a small rock sample, a geologist can learn many things about the earth. The presence of fossils in the sample may even help to determine the rock's age. What else might a geologist learn from a rock sample?

Figure 1–3. These oceanographers (right) are studying the organisms living in a 1-m² section of the ocean floor. The hydrologist (below) is determining the amount of water running off a bare field. She might compare this to the amount of water flowing from a planted field of the same size.

Meteorology

The study of the earth's atmosphere is called **meteorology** (meet ee uh RAHL uh jee). Some meteorologists analyze the composition of the gases that surround the earth. They also study the motions of the atmosphere and the interaction of the atmospheric gases with the liquid and solid surface of the earth. Such interactions affect weather and climate. Other meteorologists predict the weather for airlines, governments, newspapers, and radio and TV stations.

Figure 1–4. This weather forecaster (left) has studied current weather conditions, including satellite photographs, and is now predicting tomorrow's weather for his TV audience. The astronomer (right) is studying more distant sources of information—stars that are billions of kilometers from Earth.

Astronomy   The study of the planets, the sun, and the stars is called **astronomy** (uh STRAHN uh mee). Astronomers probe the universe trying to answer questions about the origin of the earth, the solar system, and the universe.

○ *What are the five specialties of earth science?*
○ *Why is the study of earth science divided into these parts?*

1.2 Related Sciences

Science is an organized body of knowledge that has developed through observation and experimentation. Earth science is just one branch of science. The five divisions of earth science use the same basic principles and methods as the other fields of science. The scientific fields of biology, chemistry, and physics are each important to an understanding of the earth. All fields of science are related to one another.

   Biology is the study of living things, their environment, and their behaviors within their environment. *Chemistry* is the study of the composition and characteristics of matter. The science of *physics* deals with the study of matter and energy. How do you think each of these areas of study relates to earth science?

   To see the relationships among the sciences, consider an example from geology—the search for oil. This search might actually begin with the study of fossils. **Fossils** are the remains of once-living organisms, preserved in the earth. *Paleontology* (pay lee uhn TAHL uh jee), the study of fossils, is an area in which scientists need a thorough knowledge of biology.

Figure 1–5. These fossils are all that remain of organisms that once were common on the earth. What can scientists learn about the earth's past by studying fossils?

Figure 1–6. The people working on this oil platform are trying to recover the remains of other fossil organisms. The oil that they seek is called a *fossil fuel.*

Sometimes paleontologists are hired by oil companies to help locate rock layers in which oil might be found. Paleontologists exploring for oil must work closely with chemists who analyze oil.

You can see from this example that earth science problems require the experience of specialists. No one scientist could have all the knowledge necessary to solve complicated problems.

○ *What is science?*
○ *How do biology, chemistry, and physics help in the study of earth science?*

Section Review

READING CRITICALLY

1. Which one of the fields of earth science would include the study of glaciers?
2. What type of work would a physicist do?

THINKING CRITICALLY

3. Why would a geologist studying the way in which ground water dissolves limestone rock need a strong background in chemistry?
4. After the discovery of an oil deposit, what science specialties might be needed to recover, transport, and refine the oil? Explain your answer.

SECTION
2 Methods of Scientific Study

SECTION OBJECTIVES

After completing this section, you should be able to:

- **Explain** the steps in a scientific method.
- **Compare** and **contrast** a hypothesis and a theory.
- **Discuss** the value of models to the study of science.

NEW SCIENCE TERMS

data
hypothesis
experiment
variable
control
conclusions
theory
law
scientific method
model

1.3 Scientific Method

What is your goal in life? Do you have any idea what you want to do after high school? Maybe you want to go to college or trade school, join the military, or begin working. Everyone needs to set goals for the completion of a task. Often it is much easier to decide what needs to be done when you have a goal to reach. In a similar manner, scientific problem solving requires goals.

Before scientists begin working on a project, they must determine the goal of the project. Whether the goal is to discover more oil or to predict the location of an earthquake, the whole scientific team must work toward that goal. What might happen if one person on a team were working toward a different goal from the rest of the team?

Making Observations   Often the goal that scientists hope to reach is the answer to a question or the solution to a problem. In order to reach their goal, scientists must make predictions and decisions based on data (DAY tuh). **Data** is the information gathered by observation or investigation. For instance, observing that it takes the earth about $365\frac{1}{4}$ days to orbit the sun provides data about the earth's orbit.

Figure 1–7. Just as a mining team must work together for mutual safety, scientists must also work as a team to solve problems.

Forming Hypotheses   Have you ever predicted that something was going to happen and had your prediction come true? Maybe your prediction was not just a wild guess, but based on careful observations over a long period.

Scientists often begin their investigations by making an educated guess about what they think will occur. This scientific guess is called a **hypothesis** (hy PAHTH uh sihs). Once a hypothesis is stated, it becomes the goal the scientists try to reach.

[plural, *hypotheses*]

Figure 1–8. These students are being careful not to introduce unwanted variables into their experiment.

Conducting Experiments   Scientists work carefully collecting data to test a hypothesis. This data is usually gathered through experiments. An **experiment** is a test designed to give a scientist information under carefully controlled conditions.

Scientists know that reliable data can come only from well-designed experiments. Any well-designed experiment tests a variable. In a scientific experiment, a **variable** is something that can be changed. Characteristics such as temperature, volume, time, and color are examples of the variables in an experiment. The part of an experiment that does not change is called the **control,** or the *controlled variable.*

In the experiment shown in Figure 1–8, the students try to find out if water will move more rapidly through gravel or sand. They think that the water will pass through the gravel faster. The variable in this experiment is time. In order to test their hypothesis, the students must control all other variables that might affect their experiment. For example, they must be sure they pour the same amount of water into each tube. What other variables do you think they need to control?

Drawing Conclusions   Scientists analyze the data they have collected to reach a conclusion. **Conclusions** are statements about the original hypothesis, based on all the information that has been gathered. Sometimes the conclusion confirms the hypothesis. In most cases, however, the conclusion creates more questions than it answers.

Scientists usually conduct many experiments to test a single hypothesis, so that they can be more sure about their conclusions. When new, conflicting information becomes available, the conclusion must be changed. Once scientists have collected data from many experiments, they may find that the data consistently supports their hypothesis. The hypothesis may then become a theory or a scientific law.

A **theory** is a statement that explains why things happen the way that they do. A scientific law is different from a theory. A **law** describes what happens in a given situation.

A scientific law applies to many situations and explains what has occurred. A law does not, however, explain why events occur. For example, the law of gravity allows you to predict that a book will hit the floor if you drop it from your desk. However, a theory would help you explain to your teacher why the book is on the floor.

Using a Scientific Method   The steps that scientists use to answer questions are called a **scientific method.** Scientists do not always use these steps in the same order. The order chosen depends on the goal of the scientists.

Scientists are not the only people who use a scientific method. A scientific method is also used by historians, police detectives, and even football coaches. Examine Figure 1–9 to see how a scientific method might be used to plan strategies for a high-school football game. Using the steps of a scientific method, try to think of labels for each of the drawings.

The football coach in the illustration found that his hypothesis did not lead to a successful game. Scientists experience similar frustrations. Often experiments fail to support the scientists' hypotheses. Scientists know that they must conduct many experiments. They may revise a hypothesis several times before they completely understand the problem they are studying. This may take years of work and experimentation.

○ *What is a hypothesis?*
○ *How is data usually obtained?*

DISCOVER

Using a Scientific Method
Think of something that you do every day, such as getting from your home to your school. Now pretend that you have never done this before. Write out the steps that you would use to solve this "problem" by using a scientific method.

Figure 1–9. How does this situation illustrate a scientific method of problem solving?

SKILL ACTIVITY: Reading for Meaning

BACKGROUND

Reading science books and magazines is often a challenge because each paragraph may contain both new concepts and new vocabulary terms. The chapters in this textbook are designed with special features to help you learn vocabulary and understand concepts easily and quickly.

PROCEDURE

1. VOCABULARY New vocabulary terms are highlighted in several ways in this textbook. A list of terms is presented at the beginning of each section as an introduction to vocabulary that might be new to you. When the term is defined in the chapter, it is in **boldface type.** The name of an unfamiliar object or process is in *italic type.*

   Sometimes there are signals that a definition is coming. Definitions are often written using defining verbs like *is* and *are*. Words such as *called, means,* and *explained as* are also good clues. The definition may appear before or after the term in the sentence. For example, the definition of geology might be written:

   A. "**Geology** is the study of the solid earth."

   or

   B. "The study of the solid earth is called **geology**."

   To help you even more, new science terms are listed at the beginning of each section and at the end of the chapter. The page number on which the term is defined follows each term. How many new terms are there in Section 1?

2. CONCEPTS Each chapter develops a group of related ideas, or concepts. You can use the headings in each chapter to help you understand the concepts presented. The headings are presented in outline form on the first page of each chapter.

   The technique of writing summary sentences for concepts may help you understand the material being presented. These summary sentences provide you with an expanded outline of the chapter. Often the statements in your expanded outline include definitions of new terms. Use the summary statements in the Chapter Review to check your outline.

APPLICATION

1. Find the definition of *geology* in Section 1. What clues were used to help identify the new term?
2. Locate three other definitions in Chapter 1. Copy these terms and their definitions. Then rewrite each sentence. Make sure each term is still defined.
3. Look at the first section of Chapter 1. An expanded outline of Section 1 follows.

A. What Is Earth Science?

   1. The Earth Sciences

      a. The study of the solid earth is called geology. Some of the topics geologists study are minerals, rocks, fossils, and processes that change the earth.
      b. Hydrology is the study of the water on land.
      c. Oceanography is the study of the oceans.
      d. Meteorology is the study of the earth's atmosphere.
      e. Astronomy is the study of the planets, the sun, and the stars.

   2. Related Sciences

      a. Biology is the study of living things.
      b. Chemistry is the study of the composition of matter.
      c. Physics is the study of matter and energy.

---

4. Practice outlining by completing the outline for Chapter 1.

Figure 1–10. Some models, unlike this ship, do not look like the real thing. In science, it is more important that models act like the real thing. Why do you think that this is so?

1.4 Models in Science

If you have ever built a model plane, boat, or car, you know that a lot of time is involved. You want your model to be perfect! Sometimes your goal is not just to have your model look like the real thing, but also to have it work in the same way as the real thing. A **model** is something scientists create to help them understand how things work. With these models, scientists can study objects and events that would be difficult to study otherwise.

Scientists often use a theory to create a model. Models show the relationships that scientists observe as they gather data. Such models are useful for explaining structures and processes, especially if an object is too large or too small to be observed directly. Models are continually refined and changed as more information is gathered. A good model allows scientists to predict what will happen in experiments that have not yet been performed. Today scientists use many computer models.

Consider how the model of the solar system has changed. In about 150 B.C., scientists constructed the first mathematical model of the solar system with the earth in the center and the sun orbiting around the earth. Since then, the model of the solar system has been refined many times as more information has been obtained.

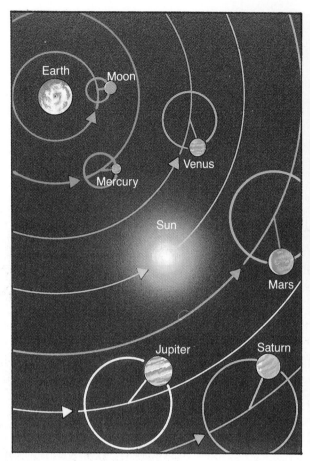

Figure 1–11. Early models of the solar system were Earth centered; that is, Earth was believed to be the center of the solar system. Because of the work of scientists during the sixteenth century, the earth-centered model was replaced by the sun-centered model of the solar system.

The present model of the solar system places the sun in the center with the planets revolving around the sun. Space travel has added more data to refine this model.

○ *Why are models used in science?*
○ *Why do models change?*

Section Review

READING CRITICALLY

1. What is the difference between a theory and a law?
2. Why is it necessary for scientists to do more than one experiment to form a valid conclusion?

THINKING CRITICALLY

3. Think of a problem, and tell how you might solve it using a scientific method.
4. List the control and variables for an experiment you might conduct to answer a question about temperature.

SECTION
3 Measurements in Science

SECTION OBJECTIVES

After completing this section, you should be able to:
- **Measure** length, volume, mass, and weight using SI.
- **Calculate** the density of an object.

NEW SCIENCE TERMS

volume
mass
weight
density

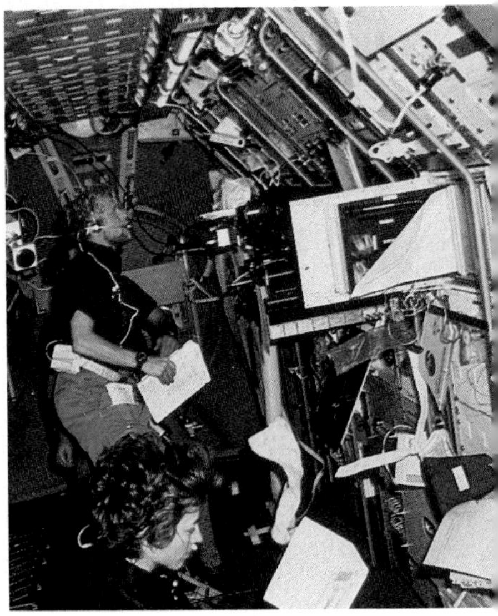

Figure 1–12. These scientists are collecting data from an experiment on board the space shuttle. They are recording the data in SI units. Why do scientists use SI units to record data?

1.5 The International System of Units

As scientists gather information in support of a hypothesis, they collect two types of data. The first type of data is descriptive. This data is based on observations made directly by the senses of sight, touch, smell, hearing, and taste.

The other type of data is measurable. As scientists study a problem, they may take measurements of distance, volume, time, or mass. The quality of any scientific effort depends on the correctness of both types of data.

Although there are several different systems of measurement, scientists around the world have agreed to use one system. This system is called the *International System of Units (SI)*. Using SI allows scientists to relate their research results in measurements understood by scientists everywhere. What do you think would happen if scientists in every country used a different system for measuring common objects? Some of the SI units are shown in Table 1–1.

TABLE 1–1: SOME SI UNITS

SI Unit	Symbol	Property Measured
meter	m	length or distance
liter	L	liquid volume
cubic meter	m^3	volume
kilogram	kg	mass
newton	N	weight
kelvin degrees Celsius	K °C	temperature

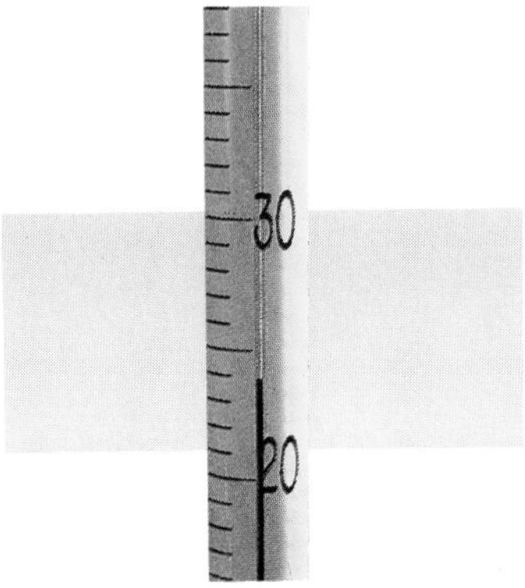

Figure 1–13. For most laboratory work, temperature is measured in degrees Celsius.

Figure 1–14. SI units are easier to imagine if you think of common objects. For example, a softball bat is about 1 m long and most soft drinks come in 2–L bottles.

The basic unit is not always appropriate for common measurements. For example, distance measurements in SI are in meters. A softball bat is nearly a meter long. However, many measurements must be made of distances much shorter than or much longer than a meter. To measure a distance shorter than a meter, you start with the basic unit, the meter. Then you add a prefix to indicate the fraction of the meter needed.

SI is based on the decimal system, or units of ten. Each prefix indicates some fraction or multiple of ten. Some common prefixes used in SI are shown in Table 1–2.

TABLE 1–2: SI UNIT PREFIXES				
Multiplication Factor	**Prefix**	**Symbol**	**Pronunciation**	**Term**
$1\ 000\ 000\ 000\ 000\ 000\ 000 = 10^{18}$	exa	E	EHKS uh	one quintillion
$1\ 000\ 000\ 000\ 000\ 000 = 10^{15}$	peta	P	PEHT uh	one quadrillion
$1\ 000\ 000\ 000\ 000 = 10^{12}$	tera	T	TEHR uh	one trillion
$1\ 000\ 000\ 000 = 10^{9}$	giga	G	JIHG uh	one billion
$1\ 000\ 000 = 10^{6}$	mega	M	MEHG uh	one million
$1\ 000 = 10^{3}$	kilo	k	KIHL uh	one thousand
$100 = 10^{2}$	hecto	h	HEHK toh	one hundred
$10 = 10^{1}$	deka	da	DEHK uh	ten
$0.1 = 10^{-1}$	deci	d	DEHS uh	one tenth
$0.01 = 10^{-2}$	centi	c	SEHN tuh	one hundredth
$0.001 = 10^{-3}$	milli	m	MIHL uh	one thousandth
$0.000\ 001 = 10^{-6}$	micro	μ	MY kroh	one millionth
$0.000\ 000\ 001 = 10^{-9}$	nano	n	NAN oh	one billionth
$0.000\ 000\ 000\ 001 = 10^{-12}$	pico	p	PEEK oh	one trillionth
$0.000\ 000\ 000\ 000\ 001 = 10^{-15}$	femto	f	FEHM toh	one quadrillionth
$0.000\ 000\ 000\ 000\ 000\ 001 = 10^{-18}$	atto	a	AT oh	one quintillionth

For instance, if you needed to measure the length of your little finger, you would certainly need a unit shorter than a meter. Even a decimeter, which is one-tenth of a meter, would be too long. You would probably need to use the unit for one-hundredth of a meter (0.01 m)—the centimeter (cm). What unit would you use to measure the total length of interstate highways in your state? Why?

These same prefixes are also used for measurements of volume and mass. For instance, you might need less than a liter of water to conduct an experiment. The best unit might be the milliliter—one-thousandth of a liter. What unit would you use to measure the mass of a blue whale? Why?

○ *Why do all scientists use the International System of Units?*
○ *What units would you use to measure your textbook? Why?*

1.6 Measuring Volume

All objects take up space. The total space that an object occupies is its **volume.** Consider the volume of a shoe box. A shoe box has height, length, and width. Since a shoe box is rectangular in shape, its volume could be calculated by multiplying its dimensions.

DISCOVER

Making SI Measurements

Pick out 10 objects to measure. For each object, pick a unit that you think would be appropriate and the tool that you would use to make the measurement. Now guess what you think the measurement will be. Then make the actual measurements, and check your estimates. Try the same thing with 10 new objects and see if you improve your skill at estimating.

Sample Problem

If you measured a shoe box and found the dimensions to be 30 cm, 15 cm, and 10 cm, what would the volume be?

Write the equation: Volume = length × width × height

$$V = l \times w \times h$$

Substitute the values of l, w, and h into the equation:
$$V = 30 \text{ cm} \times 15 \text{ cm} \times 10 \text{ cm}$$

Solve the equation: $V = 4500 \text{ cm}^3$

Notice that the unit for volume is cubic centimeters (cm^3). This is because the units are multiplied as well as the numbers.

The standard unit for measuring liquid volume is the liter. Since liters and cubic meters are both measurements of volume, their relationship can be calculated. A cube that is 1 cm × 1 cm × 1 cm (1 cm^3) will hold exactly one milliliter of liquid; therefore, one cubic centimeter equals one milliliter (1 cm^3 = 1 mL).

Figure 1–15. These two vessels, a flask (right) and a graduate (left), may be used to measure the volume of liquids.

Do you suppose the shoes that come in a box have the same volume as the box? How could you find the volume of the shoes? In order to calculate accurately the volume of an irregularly shaped object, like a pair of shoes or a rock, you use a method called *displacement.* The object is placed in a container with a known volume of water. The water level in the container will rise, or be displaced by the object. The amount of water displaced by the object will be equal to the volume of the object.

○ *What is the formula for finding the volume of a regularly shaped object, such as a box?*
○ *What method is used to determine the volume of an irregularly shaped object?*

ACTIVITY: Finding Volume

How can you measure the volume of an irregularly shaped solid?

MATERIALS (per group of 3 or 4)

graduate, 50 mL; water; string; galena; calcite; granite; cork; metric ruler

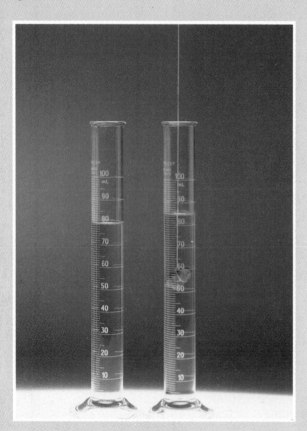

PROCEDURE

1. Copy the data table.

TABLE 1: VOLUME BY DISPLACEMENT

Object	Starting Volume	Final Volume	Volume of Object
Galena			
Calcite			
Granite			
Cork			

2. Fill the graduate about half full with water. Record the volume on your data table.
3. Tie a string around the galena and gently lower it into the water. Record the new volume of the water on your data table.
4. Subtract the starting volume from the final volume to find the volume of the galena.
5. Repeat steps 2, 3, and 4 using the other three specimens. Describe how you handled the cork.

CONCLUSIONS/APPLICATIONS

1. The faces of galena and calcite are nearly rectangular. Measure their dimensions with a ruler and calculate their volumes. How close are the measured volumes to the calculated volumes? Why don't they match exactly?
2. Explain how the displacement method works.

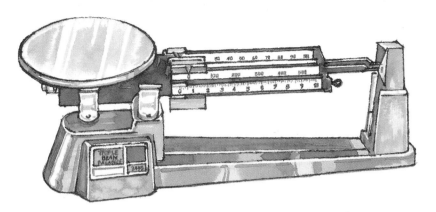

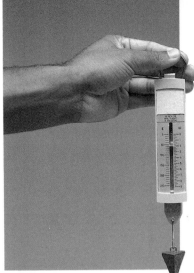

Figure 1–16. The balance (left) is used to measure mass, whereas the spring scale (right) is used to measure weight.

1.7 Comparing Mass, Weight, and Density

Have you ever heard of an object being described as "massive"? A mountain is a massive natural structure. The Pentagon in Washington, D.C., is a massive building. Jupiter is a massive planet.

Mass   An object is massive because it has a lot of matter. **Mass** is the amount of matter an object contains. All objects, whether they are solids, liquids, or gases, have mass. In SI, mass is measured in kilograms. A balance is the instrument scientists use to measure mass. Figure 1–16 shows a common laboratory balance.

Weight   Many people confuse mass with weight. **Weight** is the measure of the force, or pull, of gravity on the mass of an object. A spring scale, like the one shown in Figure 1–16, measures weight. The SI unit for weight is the newton. All forces are measured in newtons.

The earth's gravitational force decreases as you move away from the earth. Astronauts have traveled far enough from the earth to be almost weightless. On the moon the gravitational force is about one-sixth of the gravity on Earth. If an astronaut weighed 780 N on the earth, on the moon the same astronaut would weigh 130 N. Does this mean that the astronaut's body has lost mass? Explain your answer.

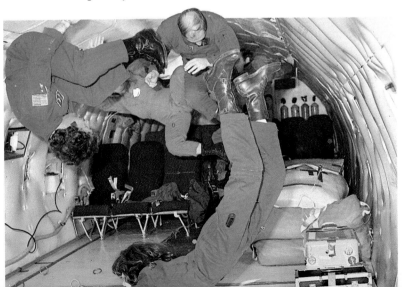

A MATTER OF FACT

The *standard mass* is a carefully machined one-kilogram cylinder of platinum-iridium alloy stored in a special vault in the International Laboratory of Weights and Measurements at Sèvres, near Paris, France.

Figure 1–17. These astronauts are experiencing free fall in a training airplane, while preparing for a space shuttle flight.

17

Density

Another property of matter related to mass is density. **Density** is the measure of the mass contained in a certain volume of matter. For example, if a brick and a block of wood are the same size, they would have the same volume. They would have different masses, however, so they would also have different densities.

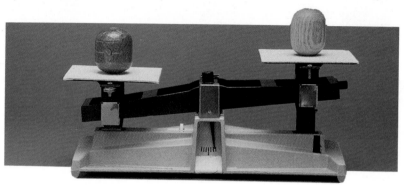

Figure 1–18. Both objects on the balance have the same volume; however, the lead has more mass than the wood.

The density of a substance does not change. A large piece of copper will have the same density as a small piece. Figure 1–18 shows that the densities of different substances vary greatly. Since density is always the same for a substance, it is a characteristic property that can be used to identify the substance.

The greater the mass of an object, the greater the pull of gravity on that object. Suppose you have two objects with the same volume. The object that is more dense will be heavier than the less dense object. A bowling ball, for example, is heavier than a melon of the same size because the bowling ball is denser. Mass, weight, and density are all related quantities.

○ *What SI unit is used to measure mass?*
○ *What is density?*

Section Review

READING CRITICALLY

1. A scientist recorded a measurement of 10 g. Was this a measurement of mass, weight, or density?
2. Explain how you would find the volume of an irregularly shaped rock.

THINKING CRITICALLY

3. Compare a 1-cm^3 sample of lead and a 1-cm^3 sample of wood. Do they have the same mass? Which sample has the greater density? Explain.
4. Why is it important for scientists to use the same system of measurements?

JAMES HUTTON (1726–1797)

James Hutton lived most of his life in Edinburgh, Scotland. His interest in science led him to investigate many parts of his environment. As a result, he made major contributions to the fields of geology, agriculture, chemistry, physics, and philosophy.

As a scientist, Hutton was a careful observer. He studied the land and drew conclusions about the origin of the rocks and the processes that changed them. Eventually he constructed his "theory of the earth." This theory proved to be so important to the development of the field of geology that Hutton is now called the founder of modern geology.

Hutton believed that the interior of the earth was composed of lava, with the crust acting as a container for the lava. Volcanoes acted as safety valves, allowing the molten rock to escape from the interior. According to Hutton, as this lava came to the surface, the overlying sedimentary rocks were tilted.

In his book *Theory of the Earth,* Hutton suggested a theory that the earth was much older than most scientists then thought. He believed that earth processes were continuous cycles. As support for this theory, he described the rock-forming processes that had occurred. He hypothesized that the same processes that can be observed today were working in the past and will continue to work in the future. Hutton's theory has been so thoroughly tested and so well supported by experimental data that it forms the basis for the science of geology.

LUIS ALVAREZ (1911–1988)

Luis Alvarez was an American physicist. He was awarded the 1968 Nobel Prize in physics. In 1980 he discovered that there was an unusually large amount of the element iridium in certain layers of rock in Italy. Iridium is very rare in rocks formed on Earth. This layer was deposited about 65 million years ago, at about the time of the disappearance of the dinosaurs.

Alvarez was surprised that such a concentration of iridium would appear in rocks of this age. Once he had observed this concentration in Italy, he hypothesized that there might be similar concentrations in rocks of the same age in other places. He discovered that there were other similar deposits.

This information led Alvarez to hypothesize that the dinosaurs, and other large animals, died because an asteroid collided with the earth. This event would have released so much dust into the atmosphere that the radiation from the sun would have been blocked for about three years. During this time, plant life would have been reduced and large animals would have starved. Although Alvarez's hypothesis seems reasonable, it is not accepted by all scientists.

An extension of the hypothesis proposed by Alvarez has recently been applied to the possibility that a similar situation migh develop if a large-scale nuclear war should occur. Dust and smoke produced by many nuclear explosions could block out the sun and lower the earth's temperature enough to cause the extinction of many plants and animals.

INVESTIGATION 1: Determining Density

PURPOSE

To calculate the density of several objects

MATERIALS (per group of 3 or 4)

Balance
Galena
Calcite
Granite
Bar magnet
Glass rod
Metric ruler
String
Graduate, 100 mL
Paper towels

PROCEDURE

1. Look at, but do not pick up, the objects to be tested. Write a hypothesis stating which object you think will have the greatest density. Record your hypothesis on your paper. *Which object do you think will have the lowest density?*
2. Make a data table like the one shown.
3. Using a balance, determine the mass of each object. Record the masses in the data table.

4. Determine the volume of each object. For some objects you may measure the dimensions and multiply. For others you will have to determine the volume by the displacement method described in the activity on page 16.
5. Calculate the density of each object. (Density = mass ÷ volume.) Your answer will be in g/cm^3 or g/mL. Since 1 mL equals 1 cm^3, convert all answers to g/cm^3 for easy comparison.

ANALYSES AND CONCLUSIONS

1. Which object has the lowest density?
2. Which object has the highest density?
3. Were your hypotheses correct? Explain. If you were incorrect, discuss what properties might have misled you.
4. If you doubled the size of the galena sample, would you double the density? Explain.
5. Explain how this activity proves that the density of water is less than that of calcite, granite, and galena.

APPLICATION

Look at the two photographs. Using your knowledge of density, explain why the ice floats in a glass of water, but sinks in a glass of oil.

TABLE 1: FINDING DENSITY			
Object	Mass (g)	Volume (cm^3)	Density (g/cm^3)
Galena			
Calcite			
Granite			
Bar magnet			
Glass rod			

GOING FURTHER

Make a list of five common materials of different densities found in your home or classroom. Before finding their densities, try to arrange them in order of increasing density. Then determine the density of each to see if you listed them in the correct order.

SUMMARY

- The major fields studied in earth science are geology, hydrology, oceanography, meteorology, and astronomy. (1.1)

- Geology is the study of minerals and rocks, and of the processes that shape the earth. (1.1)

- Hydrology is the study of the underground and surface waters of lakes and streams. (1.1)

- Oceanography is the study of the oceans. (1.1)

- Meteorology is the study of the atmosphere and the weather of the earth. (1.1)

- Astronomy is the study of the planets, the sun, and the stars. (1.1)

- The earth sciences apply principles from the basic scientific fields of biology, physics, and chemistry. (1.2)

- Biology is the study of living organisms. (1.2)

- Chemistry is the study of the composition of matter. (1.2)

- Physics is the study of matter and energy. (1.2)

- Problems can be studied by using the investigative procedure called a scientific method. (1.3)

- Scientists often present theories in the form of models. (1.4)

- Scientific work uses the International System of Units. (1.5)

- The space that an object occupies is the object's volume. (1.6)

- The volume of an irregularly shaped object is measured by displacement. (1.6)

- Mass is the amount of material or matter in an object. (1.7)

- Gravitational attraction affects an object's weight but not its mass. (1.7)

- The density of an object depends on the relationship between the object's mass and volume. (1.7)

Write all answers on a separate sheet of paper.

SCIENCE TERMS

Correctly use each of the following terms in a sentence.

astronomy **(5)**
conclusions **(8)**
control **(8)**
data **(7)**
density **(18)**
experiment **(8)**
fossils **(5)**
geology **(4)**
hydrology **(4)**
hypothesis **(8)**
law **(8)**
mass **(17)**
meteorology **(4)**
model **(11)**
oceanography **(4)**
science **(5)**
scientific method **(9)**
theory **(8)**
variable **(8)**
volume **(15)**
weight **(17)**

SCIENCE QUIZ

Modified True-False

Mark each statement *true* or *false.* If the statement is false, change the underlined term to make the statement true.

1. The water in lakes is studied in the field of geology.

2. Scientists use a theory with more confidence than they use a hypothesis.

3. A kilogram equals 1000 liters.

4. The density of a sample of iron ore will be less on the moon than it is on Earth.

5. The International System of Units is based upon the decimal system.

continues

Multiple Choice

Write the letter of the term that best answers the question or completes the statement.

6. A length that is 1/100 of a meter is a
   a) centigram.   b) millimeter.
   c) centimeter.   d) decimeter.

7. A kilogram is a measure of
   a) length.   b) mass.
   c) weight.   d) volume.

8. In order to compute the volume of an object, you must obtain measurements of
   a) length, height, and weight.
   b) mass and weight.
   c) mass, height, and length.
   d) length, width, and height.

9. The formation of clouds is studied in the field of
   a) meteorology.   b) geology.
   c) astronomy.   d) hydrology.

10. One liter is equal to
    a) 10 cm^3.   b) 100 cm^3.
    c) 1000 cm^3.   d) 10 000 cm^3.

Completion

Complete each statement by supplying the correct term.

11. Matter and energy are studied in the science of _____.

12. Usually, the step in a scientific method that follows experimentation is _____.

13. The cubic meter is a measure of _____.

14. The _____ of 1 cm^3 of water is 1 g.

15. A spring scale is used to measure _____.

Short Answer

16. Select a single object in your classroom and make five observations about it. Find out if your classmates can identify the object based on your list of observations.

17. Explain the procedure that you would follow to measure the volume of a heart-shaped eraser.

18. A block of metal 4 cm × 15 cm × 10 cm has a mass of 1620 g. What is the density of the block?

Writing Critically

19. Explain the difference between a scientific theory and a scientific law.

20. Why would a meteorologist need a strong background in physics?

EXTENSION

1. Design and build a model to demonstrate that the density of a substance will not change even if the amount of the substance is reduced.

2. Write a short story about a historian who has decided to research the life of a Mississippi River pirate who lived in the 1800s. Demonstrate in your story how the historian uses a scientific method in his or her effort to obtain material for the book he or she plans to write.

3. Visit a building-supply store and make a list of 20 items that are not measured by the use of SI. Visit a grocery store and note 20 products that use SI measurements. What difficulties might a person find in trying to convert the SI units to the system used in the United States?

4. Go to your school or public library and find out about the conference of international scientists that decided on the standard units of measurement for scientific research. In a written report, describe the basic units they decided on, and explain what the standard for each unit is and where it is located.

APPLICATION/CRITICAL THINKING

1. Many people are "afraid" to change to SI because they feel the conversion back and forth between SI and conventional measurements will be very difficult. Is this really a problem? Explain.

2. Your uncle drives his restored 1957 car across the border into Canada. He immediately notices that the speed limits are in kilometers/hour (km/h). You know that there are 0.6 km/mile. Devise a simple chart that shows him the SI equivalents of driving 15 mph, 25 mph, 35 mph, 45 mph, 55 mph, and 65 mph.

FOR FURTHER READING

Harrison, J., ed. *Science Now.* New York: Arco Publishing, 1984. Sixty-five illustrated articles describe the way scientists have affected our lives and are shaping our future.

National Bureau of Standards, U.S. Department of Commerce. *The International System of Units (SI).* Special Publication 330, 1981. This publication provides descriptions and examples of using SI in everyday situations.

Trefil, J. *Meditations at Sunset.* New York: Scribner, 1987. In this book, the author explains that the laws of nature, on which scientists work for countless years in laboratories, can be seen all around us in the natural world. This book specifically explores the wonders of the atmosphere and explains many of the objects and phenomena seen in the night sky.

Challenge Your Thinking

This airplane is being built by a consortium, or group, of manufacturers from Britain, France, Germany, and Spain. How do you suppose they get all the parts to work together if they are made in four different countries? What problems in getting parts to fit might develop if the United States were part of this consortium?

Earth Science Models

For centuries, scientists and explorers have been gathering information about the earth's surface. Early civilizations considered the exploration of the earth's surface an exciting challenge. Even today, there are still unexplored regions of the earth that hold mysteries yet to be discovered.

The southern California coast

SECTION

1 Models of the Earth

SECTION OBJECTIVES

After completing this section, you should be able to:

- **Describe** some maps used by early civilizations.
- **Explain** how a map is a model.
- **Identify** some modern techniques used in mapmaking.

NEW SCIENCE TERMS

cartography
bench mark
topographic maps
geologic maps
hydrologic maps

2.1 Early Models

Try to imagine yourself as a Babylonian (bab uh LOH nee uhn) explorer in 750 B.C. You think that the earth is a flat disk completely surrounded by an ocean called *Bitter Waters*. You also believe that outside the disk are seven islands that link the world to an outer *Heavenly Ocean,* which is where the gods live.

The ancient map shown in Figure 2–1 is an example of one of the oldest surviving maps of the Mediterranean area. Even older maps were used to show land boundaries and personal property. These maps date back to 2200 B.C.

Early civilizations used a variety of mapmaking techniques. Inuits (Eskimos), for instance, cut shapes of coastal islands out of dark-colored animal skins. The shapes were then sewn onto a light-colored skin that represented the ocean. The Egyptians engraved maps on gold, silver, and copper plates to identify the

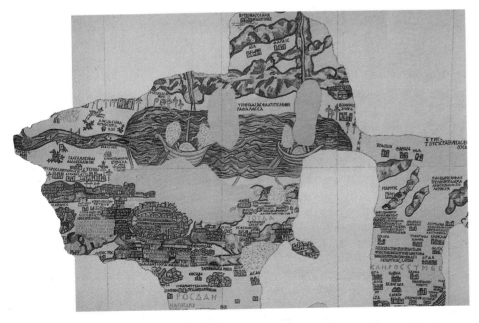

Figure 2–1. This is a photograph of one of the oldest known maps of the Mediterranean Sea.

Section 1   Models of the Earth   **25**

Figure 2–2. The people of Micronesia made maps such as this to help them locate neighboring islands.

locations of gold mines and other valuable properties. The people of the Pacific Ocean islands made maps similar to the one in Figure 2–2 to indicate the location of nearby islands.

These ancient maps led to the development of **cartography** (kahr TAHG ruh fee), the science of mapmaking. In ancient times cartography was more of an art than a science. Mapmaking was based more on imagination, theory, and travelers' tales than on precise measurements. Ancient cartographers constructed the best models they could, based on the small amount of information available. Each time an adventurer returned from a trip, new information was added to the data already collected. Maps were revised to include this new information.

During the second century A.D., mapmaking became more precise. Ptolemy (TAHL uh mee), an Egyptian astronomer and geographer, made cartography into a science. Many of his contributions still influence mapmaking today.

Ptolemy was the first mapmaker to put north at the top of a map. Through his work in astronomy, he improved the understanding of the actual size of the earth and the distances between land masses by designing his maps based on a round Earth.

Ptolemy's maps were so good that they were still being used in the 1400s. At that time Johann Gutenberg, a German printer, printed copies of Ptolemy's maps, making them readily available. About the same time, interest in geography blossomed as adventurers explored unknown parts of the world. With this new exploration, more was learned about the sizes and positions of the continents.

○ *Of what materials were some early maps made?*
○ *Who was Ptolemy?*

Figure 2–3. Ancient cartographers used the descriptions of explorers to make models of the earth.

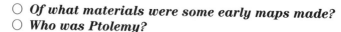

2.2 Modern Maps

Scientists are still involved in making maps. Since 1807 the United States Coast and Geodetic Survey (now called the *National Ocean Survey*) has been surveying much of the surface of the earth to improve the accuracy of maps.

Within the United States, each specific place that is surveyed is marked with a bronze marker called a **bench mark.** Perhaps you have noticed such markers while walking along a trail, in a field, or even in a city. Engraved on the bench mark are the point's elevation and exact location. Bench marks enable cartographers to show exact points on their maps. These points help land surveyors to set property boundaries and engineers to plan bridges and highways. In addition, bench marks allow construction crews to locate sites properly for buildings and dams.

During an average year, surveys covering about 64 000 km² of land are completed. These surveys establish about 3000 exact geographic positions. As a result of this effort, large parts of the United States have now been mapped. Greater accuracy and speed are now possible through the use of satellite and aerial photographs. However, it will still take many years to finish the job of accurately mapping the entire country.

Figure 2–4. A bench mark shows the exact location and elevation of this spot.

Figure 2–5. Satellite photographs help cartographers draw accurate representations of coastal areas.

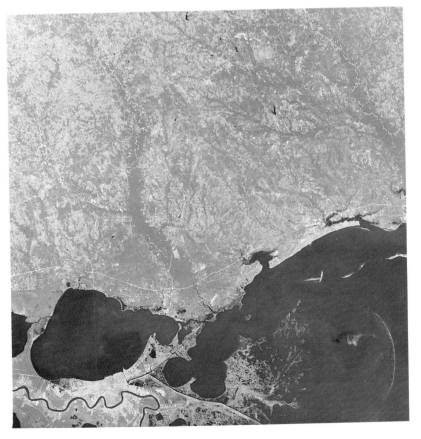

A MATTER OF FACT

Colorado is the highest state, with an average elevation of about 2073 m.

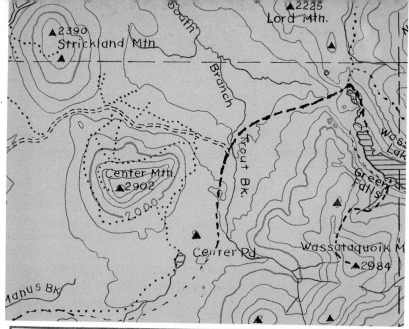

Figure 2–6. Shown here are examples of a topographic map (top), a hydrologic map (bottom), and a geologic profile map (center).

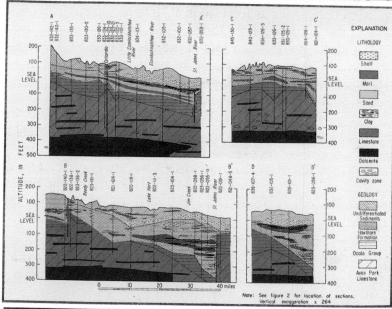

Making Maps

Make a map of your classroom using a scale of 1 cm = 0.1 m. Draw the map with north at the top and measure to accurately locate doors, windows, desks, and other permanent features.

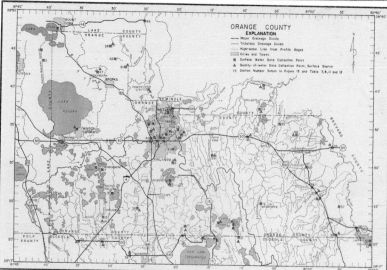

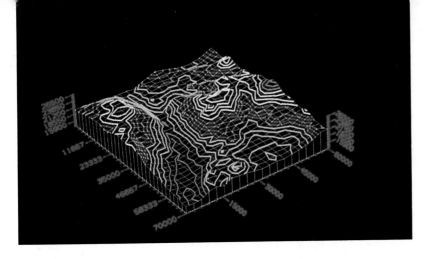

The survey work of the National Ocean Survey is very important to mapmaking efforts in the United States. However, the actual development of many maps is done by another agency of the government, called the *United States Geological Survey* (USGS). The USGS was founded in 1879 to prepare maps as well as to perform other, related tasks.

Maps are an important tool of the earth scientist. There are three basic types of maps used in earth science: *topographic maps*, *geologic maps*, and *hydrologic maps*.

Topographic maps show the shape of the land. These maps use fine lines drawn in patterns to indicate features such as mountains, valleys, and rivers. **Geologic maps** show different layers and types of rock by the use of different colors. Geologists use the patterns of colors to represent the age and structure of rock layers. **Hydrologic maps** show where surface and underground water are located. They show drainage patterns and areas where water has produced special landforms. Figure 2–6 shows examples of different types of maps.

○ *What government agency is responsible for surveying the United States?*
○ *What are three types of maps used by earth scientists?*

Section Review

READING CRITICALLY

1. Why are the works of Ptolemy so important to modern cartography?
2. How is data gathered for modern mapmaking?

THINKING CRITICALLY

3. How are modern maps similar to maps used by ancient civilizations? How are they different?
4. Why is the ancient map of the world in Figure 2–1 really a model of the world?

② World Maps

SECTION OBJECTIVES

After completing this section, you should be able to:

- **Compare** and **contrast** latitude and longitude.
- **Explain** the advantages and disadvantages of various map projections.

2.3 Latitude and Longitude

Imagine again that you are an early explorer about to set sail for some distant port. You have a map to help guide you from your home to your destination, but the map has few points of reference. You cannot tell which way to sail to reach your destination.

In order to locate places accurately on a map, a system of north-south and east-west lines was developed. The imaginary lines that run east-west around the earth are called lines of **latitude** (LAT uh tood). Latitude lines are parallel—they never cross. As you can see in Figure 2–8, the zero latitude line is the equator. All other lines of latitude are parallel to the equator and are measured from 0° to 90° from the equator. Another name for lines of latitude is *parallels*. The equator divides the earth into two equal halves—the Northern Hemisphere and the Southern Hemisphere. Therefore, a compass direction of north or south must always be included with the degrees from the equator when describing latitude—for example, 23°N or 45°S.

The imaginary lines that run from the North Pole to the South Pole are called lines of **longitude** (LAHN juh tood). Longitude lines are not parallel. They touch at the poles and are farthest apart at the equator. Lines of longitude cross parallels at right angles.

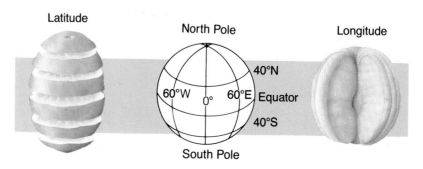

Figure 2–8. Lines of latitude are like slices of an orange (left). Lines of longitude are like sections of an orange (right).

In 1884 a zero-degree longitude line was established by international agreement. This longitude line, called the *prime meridian*, passes through Greenwich, England. Lines of longitude, also called *meridians*, are measured, in degrees, either east or west of the prime meridian. As Figure 2–9 shows, the prime meridian and the 180° meridian divide the earth into two equal halves. The halves are called the *Eastern Hemisphere* and the *Western Hemisphere*. All longitude readings must include both the degrees and direction from the prime meridian—for example, 29°E or 41°W.

Any position on the earth can be located with latitude and longitude. Latitude and longitude can also be used to measure distance. A single degree of latitude is equal to about 111 km. Sometimes, however, maps must show distances of less than one degree. Each degree can be divided into 60 minutes (60′). Each minute is equal to 1.85 km and can be divided into 60 seconds (60″). Each second is equal to 0.03 km, or 30 m.

Meridians can serve as a measurement of time. The earth rotates through 15° of longitude each hour. One 24-hour period is a complete circle. The 180° meridian separates consecutive days—it is one day earlier east of this meridian. This meridian is called the *international date line*. If you stood with one leg on each side of it, each half of you would be in a different day.

○ **What are lines of latitude and longitude?**
○ **In addition to location, for what can latitude and longitude be used?**

DISCOVER

Finding Your Home Town
On a world map, find your home city or the city nearest to you and indicate its latitude and longitude.

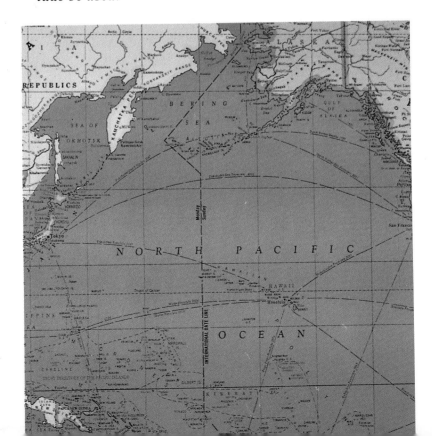

Figure 2–9. The international date line is the same as the 180° meridian, except in cases where the date line would split a nation.

How can you use latitude and longitude to locate places on a map?

MATERIALS

A world map

PROCEDURE

1. Review the definitions of latitude and longitude lines.
2. Familiarize yourself with the map by locating the equator and the prime meridian. Determine the number of degrees between the latitude and longitude lines.
3. Locate the 30°N latitude line on the east and west borders of the map. List the continents that 30°N passes through.
4. Locate the 80°W longitude line on the north and south borders of the map. List the continents that 80°W passes through.

CONCLUSIONS/APPLICATIONS

1. What city is located at 42°N, 87°W?
2. What city is located at 34°S, 18°E?
3. What city is located at 22°N, 88°E?
4. What are the latitude and longitude of Houston, Texas?
5. What are the latitude and longitude of New York City?
6. What are the latitude and longitude of Hong Kong?
7. What are the latitude and longitude of Sydney, Australia?
8. What are the largest latitude and longitude readings possible?

2.4 Map Projections

Maps have many shortcomings as models of the earth. Maps are two-dimensional models of the earth's three-dimensional surface. Making a flat representation of a spherical surface causes distortions in the appearance of many of the earth's features. As a result, maps do not show totally accurate shapes of the continents and oceans.

In the 1500s, Gerardus Mercator (juhr AHR duhs muhr KAYT uhr), a Flemish cartographer, made a breakthrough in the art of mapmaking. The map that Mercator made showed all parallels and meridians at right angles to each other. This type of map is known as a *Mercator projection.*

The Mercator projection made it much easier to navigate using a compass and a map. For example, if you were sailing from New York to London, you would simply draw a straight line between the two cities. Then you would determine the angle of the line from any meridian and sail at that angle until reaching London. The Mercator projection widens and lengthens the areas at high latitudes. Greenland, for instance, appears almost as large as Africa. In fact, Africa is 15 times larger than Greenland.

A MATTER OF FACT

The earth is not perfectly round. The distance around the equator is over 100 km greater than the distance around the earth through the poles.

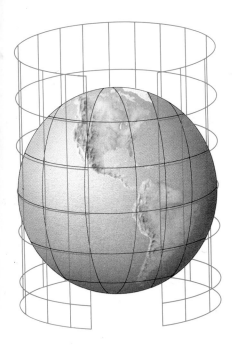

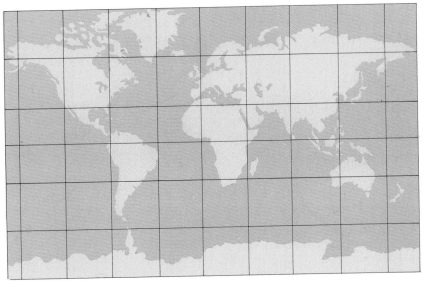

Figure 2–10. This illustration shows the development of a Mercator projection. Notice the distortion of the land areas at high latitudes.

There are other map projections that can be used to study specific parts of the world. If you wanted to study the countries around the Arctic Ocean, you might use a *polar projection,* such as the one shown in Figure 2–11. Polar projection maps are made as if the observer were looking down on the world from the North or South Pole. The polar projection is circular rather than rectangular like the Mercator projection. The pole is at the center of the map, and the equator forms the outer boundary. Land areas near the pole are shown in true proportions, but areas near the equator are greatly distorted.

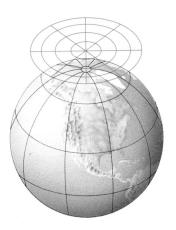

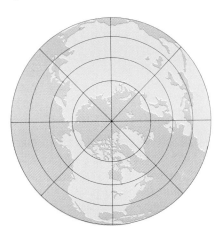

Figure 2–11. Polar projections, such as this, show areas at high latitudes without much distortion.

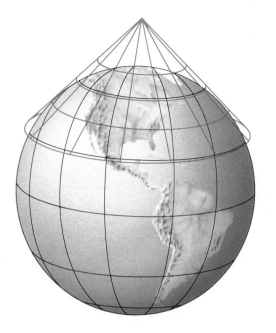

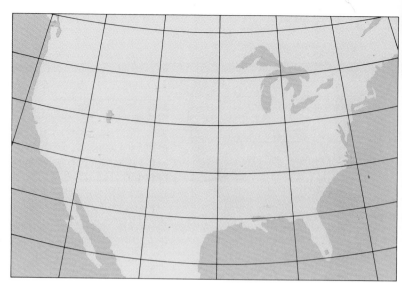

Figure 2–12. Polyconic projections are often used to make small area maps because such projections do not distort the shape of the land.

The *polyconic projection,* shown in Figure 2–12, shows small areas of the world without distortion of any land masses. As you can see, however, these maps would not be very good for navigation because many different maps would have to be used. Polyconic projections are often used for road and topographic maps.

○ *Why is it hard to produce an accurate world map?*
○ *What one problem do most map projections share?*

Section Review

READING CRITICALLY

1. Describe how and why a Mercator projection changes the shape or size of land masses that are located near the poles.
2. Write out 23°15′30″W, using words instead of symbols. Is this a longitude reading or a latitude reading? Why?

THINKING CRITICALLY

3. Explain why 95°30′15″N and 190°30′15″E are impossible locations on a map of the world.
4. On Mercator and polar projections, extend the edge of a sheet of paper from New York City to Moscow. Why do these maps show totally different routes for travel between the same cities?

CAREERS

SURVEYOR

In the original American colonies, most boundaries were determined by natural features, such as rivers. Early *surveyors* were responsible for determining the boundaries of most of the states that were not part of the original 13 colonies. Surveyors today determine ownership of private property from fixed points. such as USGS bench marks.

Many surveyors work for state and local governments, where they establish right-of-ways for roads, waterlines, and government buildings.

To do the work of a surveyor, you must have a high school diploma.

For Additional Information
American Congress on
    Surveying and Mapping
210 Little Falls Road
Falls Church, VA 22046

GEOLOGIST

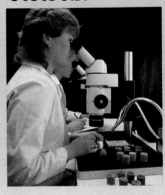

Studying rocks and the processes that change the surface of the earth is the job of a *geologist*. Most geologists have majored in the sciences and mathematics in college and have obtained an advanced degree in a specialized field of geology.

Some geologists who specialize in earth materials are economic geologists. Petroleum geologists specialize in the search for oil and natural gas. Other geologists study the earth's more violent changes, such as volcanoes, that sculpture the earth's surface.

For Additional Information
American Geological
    Institute
5205 Leesburg Pike
Falls Church, VA 22041

CARTOGRAPHER

Mapmaking, a combination of science and art, is the job of a *cartographer*. Cartographers use data from field geologists, surveyors, and other sources to make maps. This work takes imagination and a high mechanical-drawing aptitude. Computers are playing an increasingly important part in this field.

Part of cartography is the actual drafting of the maps from computer interpretation of data. This position requires two years of training after high school and some on-the-job training.

For Additional Information
American Congress on
    Surveying and Mapping
210 Little Falls Road
Falls Church, VA 22046

3 Topographic Maps

NEW SCIENCE TERMS

scale
topography

SECTION OBJECTIVES

After completing this section, you should be able to:

- **Relate** map scales to actual distances.
- **Interpret** contour lines on a topographic map.

2.5 Map Scales

If you were planning a trip from your home to your state capital, you probably would not find a world map very helpful. Many times cartographers need to show more details than can be shown on a world map. To show more details of an area, cartographers change the scale of a map. The **scale** of a map is the relationship between a distance on the map and a distance on the earth. For example, a scale of 1 cm = 100 km means that 1 cm on the map represents 100 km on the earth's surface. What would 2 cm on the map represent with a scale of 1 cm = 1 km?

The example of 1 cm = 1 km is called a *verbal scale*. This type of scale equates two different units, centimeters and kilometers in this case. This type of scale is often found on road maps, because it is easy for people to understand.

Figure 2–13. Changing the scale of a map allows for greater detail (right) of a small portion of a larger area map (below).

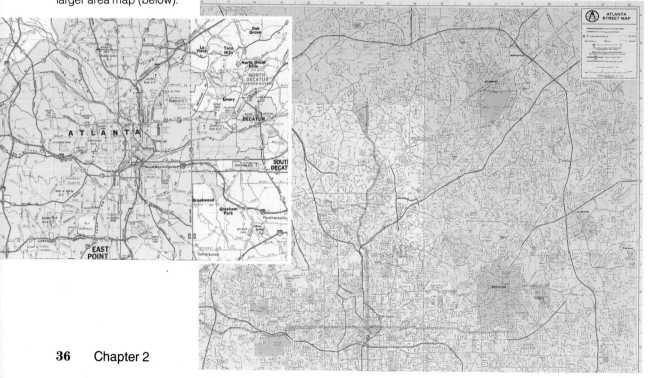

Map scales are usually written as ratios. Such a scale is called a *representative fraction (RF)*. The RF shows the ratio of a represented distance on a map to the actual distance on the earth's surface. The two numbers must be in the same units. For instance, a verbal scale of 1 cm = 1 km would be written as 1:100 000. Since a ratio is not really a measurement, it is never written with units.

In addition to or in place of an RF, many maps show a *graphic scale*. A graphic scale consists of a measured line marked off in specific distances. To show the scale of 1:100 000, the line would be drawn 1 cm long. One end of the line would be marked 0 cm, while the other end would be marked 100 000 cm or 1 km. Look at Figure 2–14 to see an example of a graphic scale.

Although scientists have agreed to use SI measurements for all scientific work, most of the maps available in the United States are still drawn using customary units. This system of measurement uses inches and miles. Many of the maps available to you will use verbal scales of 1 in. = 1 mi., which is the same as the RF 1:62 500 (63 360).

○ *What is a map scale?*
○ *What are representative fractions and graphic scales?*

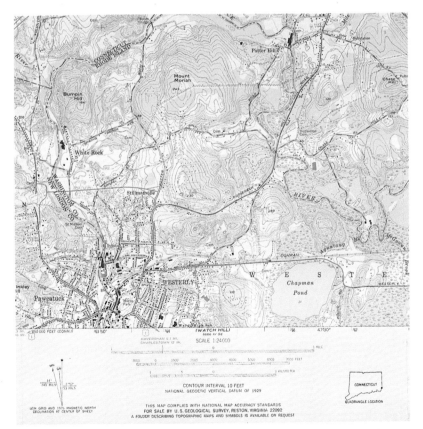

Figure 2–14. The key of this topographic map shows a graphic scale and a representative fraction.

BACKGROUND

Since maps are scale representations of actual locations, any interpretation of a map depends on an accurate understanding of the scale of the map. Although there are three different types of map scales, only the graphic scale and the representative fraction usually appear on the map. When people discuss the scale of a map, they frequently use the verbal scale.

Map scales can be presented in either the SI or the customary units of measurement. Most of the maps produced in the United States have been based on the customary units. These maps are still in use today, even though the scientific community works in SI units. As a student of maps, you will need to be able to interpret scales in both systems.

PROCEDURE

The graphic scale is especially handy to use because you do not need to convert from one unit to another. You can use the graphic scale of a map even if you do not have a ruler.

To use the graphic scale to measure a distance on a map, hold the edge of a sheet of paper along the distance you wish to measure. Mark both ends of the distance on your paper. Then hold the marked paper along the graphic scale of the map to read the distance.

APPLICATION

Using the Wet Fish Island map, complete the following:

1. Hold the side of a sheet of paper along an imaginary north-south line between the two coastlines.
2. Mark both ends of the island on the paper.
3. Place the paper along the graphic scale on the map, lining up one of the marks with the zero mark on the scale.
4. By comparing where the second mark on your paper falls along the scale, you can determine the distance between your two marks.

USING WHAT YOU HAVE LEARNED

Use the graphic scale to determine the following distances on the Wet Fish Island map.

1. What is the length of High Island?
2. What is the width of High Island?
3. What is the width of the map?
4. What is the width of Wet Fish Island?

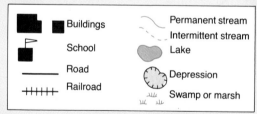

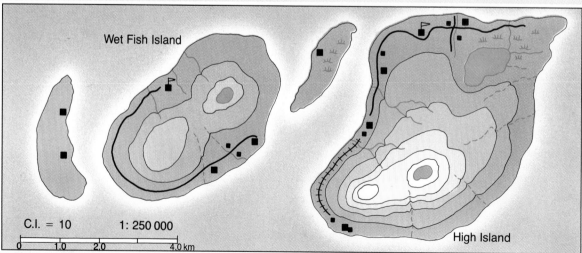

Wet Fish Island

C.I. = 10   1: 250 000

0   1.0   2.0   4.0 km

High Island

2.6 **Contour Lines**

If you were planning a hike through mountainous country, you would probably want to know the elevation of various places along your path. You might also want to know how steep certain trails were, so you could pick the easiest one to reach your destination. One of the most useful maps for these purposes is the topographic map. On a topographic map, a pattern of lines is used to show elevation. Streams, lakes, and other natural features are also represented on topographic maps, as well as land boundaries, towns, roads, and other structures.

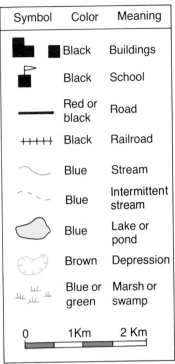

Symbol	Color	Meaning
▉	Black	Buildings
⚑	Black	School
—	Red or black	Road
+++++	Black	Railroad
∿	Blue	Stream
- - -	Blue	Intermittent stream
🌫	Blue	Lake or pond
⌇	Brown	Depression
�	Blue or green	Marsh or swamp

0   1Km   2 Km

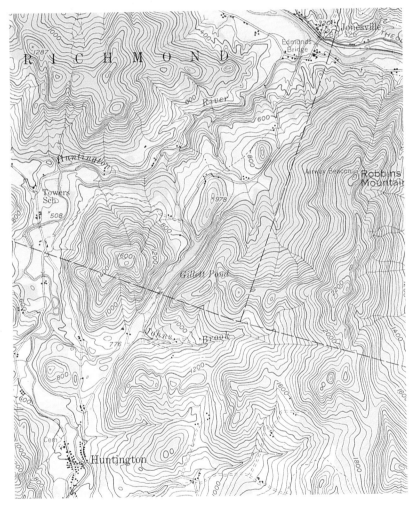

Figure 2–15. Contour lines connect places with the same elevation. Notice that the contour lines on this map are brown, and that some of them are labeled with the elevation. The labeled contour lines are called *index contours*. Special map features can be identified from the key above.

Topography is the contour, or shape, of the land surface. On a topographic map, topography is shown by using *contour lines*. Contour lines are lines on a map that connect points of equal height above sea level. A zero contour line is at sea level. Follow one of the contour lines in Figure 2–15. Notice that you stay at the same elevation.

DISCOVER

Drawing Contours

Take a large lump of modeling clay and form it into a hill-shaped mass. On a plain sheet of paper, draw contour lines to represent this hill. Use a contour interval of 1 cm and an RF of 1:1.

Look at the pattern on the map in Figure 2–16. This pattern represents a mountain. Anytime you see a pattern similar to this one on a topographic map, it represents a mountain. The difference in elevation between two consecutive contour lines is called the *contour interval*. For example, if the contour interval were 10 feet, the lines on either side of a line representing the 400-foot contour would represent 390 feet and 410 feet.

Figure 2–16. Shown here is a topographic map (right) and an aerial photograph (left) of Mount St. Helens, an active volcano in the state of Washington.

Many times a topographic map has so many contour lines that the elevation of a single line is difficult to determine. As an aid in reading the topographic map, every fifth contour line is drawn darker and thicker so that it stands out. Each of these darker lines, called *index contours*, is marked with the elevation. Locate an index contour line in Figure 2–16. What is its elevation?

The topographic map shown in Figure 2–16 is a map of the area also pictured in Figure 2–16. Notice the contour lines that show the hill. This map also shows special patterns for steep slopes, gentle slopes, and valleys. What is the pattern for each of these features?

Compare the topography in Figures 2–16 and 2–17. Why would you want to use a different contour interval for each area?

A MATTER OF FACT

The average elevation of the United States (excluding Alaska and Hawaii) is about 762 m above sea level.

By decreasing the contour intervals, mapmakers are able to show more details of flat surfaces. Larger contour intervals are used to show the features of mountainous areas.

Table 2–1 explains a few simple rules about contour lines. These rules should help you interpret topographic maps.

○ *What is topography?*
○ *How is elevation represented on a topographic map?*

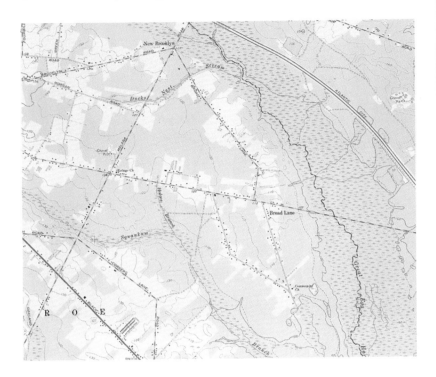

**TABLE 2–1:
INTERPRETING
CONTOUR LINES**

1. A contour line connects points of equal elevation.
2. A contour line is an endless line; it either closes upon itself on the map or at some point outside the map area.
3. Contour lines never branch or fork.
4. Contour lines never cross one another.
5. Closely spaced contour lines represent steep slopes; widely spaced contour lines represent gentle slopes.
6. Circular depressions in the earth's surface are shown by closed contours that have marks on the inside called *hachure marks.*
7. Contour lines point upstream when crossing a stream, river, or valley.

Figure 2–17. This topographic map is of a relatively flat coastal area of New England. Using the steps in Table 2–1 above, study the features of this map.

Section Review

READING CRITICALLY

1. Maps always use symbols to represent features. Study some of the features shown on the symbol sheet in Figure 2–15. Which ones can also be found in the map shown in Figure 2–17?
2. What is a topographic map?

THINKING CRITICALLY

3. Write a representative fraction and draw a graphic scale for the verbal scale of 1 cm = 100 km.
4. Examine the two road maps shown in Figure 2–13. They show some of the same places, although the scales of the two maps are different. What are some of the advantages of a small scale, such as 1:625 000, compared to a larger scale, such as 1:62 500? What are some of the advantages of the smaller scale?

INVESTIGATION 2: Interpreting Topographic Maps

PURPOSE

To identify landforms and elevations from contour-line patterns

MATERIALS (per group of 3 or 4)

Topographic map of Boothbay quadrangle on page 581 of the Reference Section

PROCEDURE

Answer the following questions by using this map.

1. What is the maximum elevation of each of the following locations?
   a) Pumpkin Ledges   b) Pumpkin Island
   c) The Cuckolds

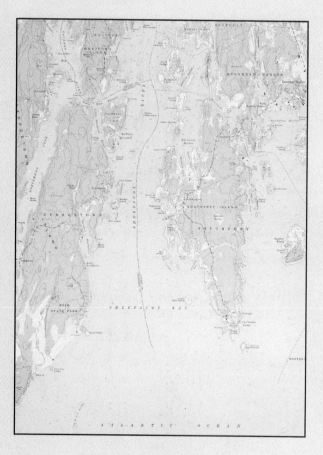

2. Name and locate (using latitude and longitude readings) two islands below 20 feet in elevation.
3. Name and locate one example of an island for each of the following elevations:
   a) More than 20 feet, less than 40 feet
   b) More than 40 feet, less than 60 feet
   c) More than 60 feet, less than 80 feet
4. Of the following elevations, which ones would you expect to be indicated with a contour line on this map? 20, 59, 120, 200, 40, 75, 150, 180, 60, 105, 260
5. What is the elevation of the house on Ram Island?
6. What is the elevation of Adams Pond?

ANALYSES AND CONCLUSIONS

1. Why do you think there are so many swamps in the southern part of Georgetown Island?
2. Locate Cushman Hill in the NW quarter of the map. Which side of the hill would prove to be the most difficult hike to the top? Explain how you reached your conclusion.
3. Describe the way High Island would look if you were approaching it from a distance. Sketch or describe its profile, or side view.
4. Locate the small stream that flows along the western side of Whaleback Ridge. In which direction does this stream flow? Explain how you reached your conclusion.
5. Explain how you can identify an island below 20 feet in elevation.
6. Locate Riggs Hill along the western boundary of the map. Its exact elevation is identified as 217 feet. Why was the number written on the map instead of just being indicated with the contour lines?

APPLICATION

On a topographic map of your own area, find your home, your school, or a nearby park, and several other natural features.

SUMMARY

- As long ago as 2000 B.C., people constructed maps and models of the earth. (2.1)

- Ptolemy was the first mapmaker to put north at the top of a map. (2.1)

- Since 1807 the United States Coast and Geodetic Survey (now called the *National Ocean Survey*) has been making accurate maps. (2.2)

- Bench marks are points where the exact elevation and location are known. (2.2)

- Latitude and longitude lines are used to indicate on maps the location of points on the earth's surface. (2.3)

- Lines of latitude are also known as parallels and lines of longitude are known as meridians. (2.3)

- Map projections show a spherical Earth on a flat page. (2.4)

- Many map projections show parts of the world in a distorted view. (2.4)

- The representative fraction scale, the verbal scale, and the graphic scale are three ways to present the relationship between map distance and represented distance on the earth's surface. (2.5)

- Although scientific work is in SI, many topographic maps are drawn with customary units of measurement. (2.5)

- Contour lines show places of equal elevation above sea level. (2.6)

- Topography is the shape of the earth's surface. (2.6)

Write all answers on a separate sheet of paper.

SCIENCE TERMS

Correctly use each of the following terms in a sentence.

bench mark **(27)**
cartography **(26)**
geologic maps **(29)**
hydrologic maps **(29)**
latitude **(30)**
longitude **(30)**
scale **(36)**
topographic maps **(29)**
topography **(39)**

SCIENCE QUIZ

Modified True-False

Mark each statement *true* or *false*. If a statement is false, change the underlined term to make the statement true.

1. The scale 1:100 000 is an example of a <u>verbal scale</u>.

2. The closer the contour lines, the <u>steeper</u> the slope that they represent.

3. The Mercator projection distorts landforms in the area around the <u>equator</u>.

4. A <u>verbal</u> scale never appears on a map.

5. All contour lines are measured in height above <u>sea level</u>.

6. A <u>polar projection</u> shows areas near the poles without distortion.

7. <u>Topographic</u> maps show rock structure.

8. <u>Bench marks</u> show exact elevation.

9. <u>Contour lines</u> connect points of equal elevation.

10. Most small area maps are <u>polyconic</u> projections.

continues

Multiple Choice

Write the letter of the choice that best answers the question or completes the statement. All of the questions refer to the topographic map of Clark's Falls.

11. In which direction does Green Fall River flow?
a) north
b) south
c) east
d) west

12. What is the contour interval of this map?
a) 5 feet
b) 10 feet
c) 15 feet
d) 20 feet

13. Which side of the hill northwest of Clark's Falls is the steepest?
a) north
b) south
c) east
d) west

14. What are the elevations of the two bench marks?
a) 90 and 98
b) 150 and 190
c) 199 and 255
d) 90 and 199

15. What is the elevation of Point D?
a) 130 feet
b) 240 feet
c) 200 feet
d) 170 feet

Completion

Complete each statement by supplying the correct term.

16. A person who makes maps is called a _____.

17. The zero latitude line is the _____.

18. The _____ projection is best when showing land at high latitudes.

19. Longitude lines are measured east and west of the _____.

20. A map that indicates elevation by the use of contour lines is called a _____ map.

Short Answers

21. Explain why even during the earliest civilizations, people needed to have maps.

22. Using the world map on page 592, determine what date and time it would be in Atlanta if it is 5:00 P.M., October 15, in Sydney, Australia. Explain why different time zones are necessary.

23. Describe how you could use latitude and longitude to locate any place on Earth to within 30 m.

Writing Critically

24. If you visited Los Angeles, you would be as far from the equator as what major city in the Southern Hemisphere?

25. If a city had a latitude reading of 25°S and a longitude reading of 130°E, in what hemisphere and on what continent would it be located? Explain how you arrived at your answer.

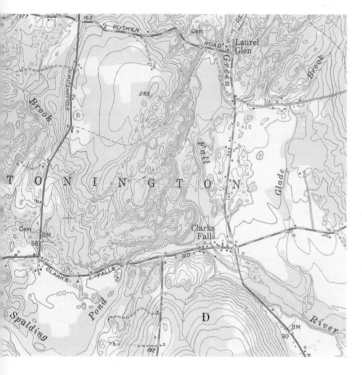

EXTENSION

1. Build a clay model of the Clark's Falls map. Using the point of a pencil or a straightened paper clip, carve the contour lines into the clay model.

2. Design your own topographic map showing an oval hill that is steepest on the west side and has an eastward-flowing stream with one branch, a lake, and a depression. The contour lines for each of these landforms must form a continuous sequence.

3. Research the development of Landsat maps of North America. Write a report describing the process for changing the satellite images into maps.

APPLICATION/CRITICAL THINKING

1. When comparing distances in Alaska with distances in Brazil, why would degrees of latitude be used instead of degrees of longitude to express the relationship between distance and degrees?

2. Explain why the international date line is not a straight line like all other meridians.

FOR FURTHER READING

Carey, H. H. *How to Use Maps and Globes.* New York: Franklin Watts, 1983. This book tells how to get information from maps and globes and how to draw maps.

Nordenskiold, A. E. *Facsimile Atlas to the Early History of Cartography.* Mineola, New York: Dover Publications, Inc., 1973. This atlas reproduces 169 of the most important maps printed before 1600.

Challenge Your Thinking

These photographs show a very complex system of levees, locks, and dams that protect much of the country of Holland. The components of these structures have been tested on a working scale-model. Of what value is a working model for predicting the consequences of various happenings?

UNIT 1 THE SCIENCE CONNECTION:

Landsat Imaging

Landsat satellites circle at altitudes as high as 920 km. They scan the earth with electronic sensors that find parts of the world not seen by the human eye.

The Landsat satellites gather information through two sets of sensors. A multispectral scanner (MSS) senses objects on the earth in great detail. The scanner can see in the visible light spectrum as well as in the infrared, or heat, spectrum.

The second type of sensor, the thematic mapper (TM), can record radiation from the surface of the earth. The TM sensor is responsible for the colors in the Landsat pictures. The TM records seven different types of radiation, or bands, as different colors. Colors are assigned to the bands to make it easier to analyze a region to be mapped. These computer-produced pictures are called *false-color images.*

The data for each of the seven bands can be made into separate black and white images. A computer combines the bands to make a colored image like that of the Texas coast shown below.

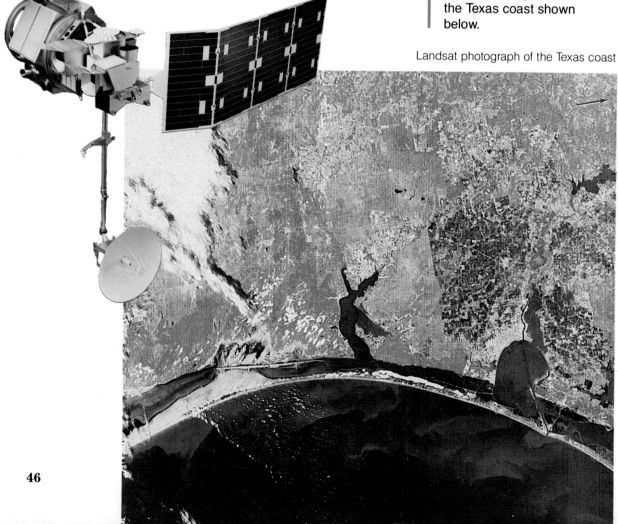

Landsat photograph of the Texas coast

Preparing maps from a Landsat image

Information provided by Landsat images has caused earth scientists to change some of their ideas about the earth and its resources. Through analysis of Landsat images, cartographers have surveyed regions in less developed countries where accurate maps have never been made. Using Landsat images, hydrologists have been able to locate unmapped lakes and other waterways. Landsat images have aided geologists in the discovery of oil in the Sudan, tin in Brazil, and copper in Mexico. The Landsat images have also been used to find uranium, zinc, copper, and nickel deposits in the United States.

A combination of bands can be selected to highlight certain features, such as a forest or river.

There are filters in the sensors that isolate the colors blue, green, and red. In addition, several types of radiation that cannot be seen by the human eye can be isolated.

Each band shows something different. The blue band shows areas where there are plants. This allows scientists to distinguish between bare soil and planted fields. The blue band can also detect clear, shallow water and show its depth. The green band shows healthy vegetation and can detect cloudy water. The red band separates different crops and shows other features such as cities and highways. Some of the other bands highlight fast-growing plants, show surface temperatures, and reveal sources of heat.

Testing a Landsat before launch

GEOLOGY

The formation of diamonds has been studied for a long time, but much of the process remains a mystery. Scientists are relatively certain that these pure-carbon gems begin to form at depths of about 150 km beneath the earth's surface. These diamond pipes begin in the upper mantle. In this region, there are many molten, or plasticlike, rocks.

- Is a diamond a mineral or a rock?
- What is fool's gold, and how is it different from real gold?
- Why are the rocks of the continental crust lighter than the rocks of the oceanic crust?
- What causes some of the rocks of the earth's interior to be plasticlike?

By reading the chapters in this unit, you will learn the answers to these questions. You will also develop an understanding of concepts that will allow you to answer many of your own questions about the study of the earth.

Cut diamonds

Minerals

How many different objects are pictured here? Do you think it is possible that all these objects are the same? Although their colors are different, these objects have many characteristics in common. What similar characteristics can you see? What other characteristics about these objects might help you decide if they are the same?

Various crystal minerals

1 Atomic Structure

SECTION OBJECTIVES

After completing this section, you should be able to:

- **Identify** the parts of the atom.
- **Compare** atoms, isotopes, and ions.

NEW SCIENCE TERMS

atoms
elements
protons
electrons
neutrons
nucleus
isotopes
ion

3.1 Atoms

Look around your classroom. How many different objects do you see? What do you think these objects have in common? People, tables, chairs, books, and even the air are all made of very small particles. These particles are called atoms. **Atoms** are the building blocks of matter.

Each kind of atom forms an element. **Elements** are substances made of only one kind of atom. There are 90 different elements that occur naturally on Earth. In addition to the 90 natural elements, 19 other elements are known. If they do not occur naturally, where do you think they come from?

The tiniest speck of dust and the largest mountain are both made of atoms. A speck of dust is very small, but it contains thousands of atoms. Try to imagine how small an atom is. Take a piece of paper and tear it in half. Now tear the half in half, and continue this until you cannot tear the paper anymore. You now have the smallest part of a piece of paper that still has all the characteristics of paper, but it is still far bigger than an atom. Imagine taking that smallest piece of paper and dividing it 1000 more times. You would now be approaching something close to the size of an atom.

There are many different kinds of atoms. Figure 3–1 shows a substance made of two kinds of atoms: fluorine (F) and calcium (Ca). Atoms cannot be broken down into smaller parts without changing their characteristics.

Fluorine and calcium combine to form *fluorite*. Different atoms can combine to make many different substances. Fluorite can be separated into atoms of fluorine and calcium.

Atoms are small, but they are not the smallest particles that scientists know about. Atoms are made up of subatomic particles called *protons*, *electrons*, and *neutrons*. These subatomic particles do not have the characteristics of the elements from which they come. The protons, neutrons, and electrons of fluorine are the same as the protons, neutrons, and electrons of calcium.

Figure 3–1. Calcium is a solid and fluorine is a gas. They combine to form the mineral fluorite, shown here.

Making Atomic Models

Cut out circles of colored paper to represent protons, electrons, and neutrons. Try to create flat models of atoms with one, five, and ten protons. In real atoms, the first energy level can have only one or two electrons, and the second level no more than eight electrons. Why do you think there are limits to the number of electrons in each level?

You may find it confusing that the same subatomic particles can form so many different kinds of atoms. Think of a common substance like wood. How many different things can be made of wood? Name at least ten. If wood is combined with two other substances, such as steel and concrete, how many things can be made? You can probably see now that making 109 different atoms from protons, neutrons, and electrons is not too difficult.

Protons are subatomic particles that have a positive electrical charge. **Electrons** are subatomic particles that have a negative charge. These opposite charges attract each other and hold atoms together, just as the opposite ends of two magnets attract and hold the magnets together. Protons repel protons, and electrons repel electrons. If you try holding like ends of two magnets together, they will repel each other.

Neutrons are subatomic particles that have no electrical charge. Protons and neutrons are clustered together to form the core of an atom. This core is called the **nucleus.** The neutrons keep the protons away from each other.

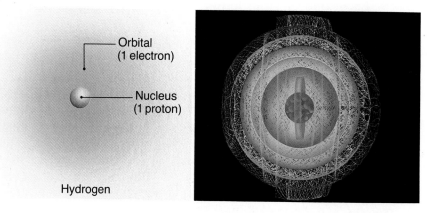

Orbital
(1 electron)

Nucleus
(1 proton)

Hydrogen

Figure 3–2. On the left is a model of a hydrogen atom and on the right is a computer-generated model of an atom. Notice the difference between the electron cloud of the hydrogen model and the electron orbit of the computer-generated model.

For hundreds of years, scientists have been trying to decide what atoms would look like if they could be seen. As a result, many models have been developed to help explain the structure of atoms. One of these models was developed by the Danish physicist Niels Bohr. The model Bohr proposed looks very much like a model of the solar system. At the center of the atomic model is the nucleus, with the electrons orbiting it — much as planets orbit the sun. The orbits in which the electrons move are called *energy levels.* The modern model of an atom is somewhat different from Bohr's model. The electron orbits are shown as a cloud surrounding the nucleus. Bohr's model is still used, however, because it shows the nucleus and electrons in a simplified manner.

○ *What are the three subatomic particles that make up atoms?*

○ *What is the electrical charge of each of these particles?*

3.2 Isotopes and Ions

When there are equal numbers of electrons and protons, an atom is electrically neutral. The number of neutrons, however, may or may not be the same as the number of protons and electrons.

Suppose you have three different chocolate chip cookies. One has six chips, another has seven chips, and the third has eight chips. You know that they are not all exactly the same; however, they are all chocolate chip cookies. Similarly, every carbon atom has six protons and six electrons, but a carbon atom may have six, seven, or even eight neutrons. They are all carbon atoms, but they are not exactly the same. Atoms of an element that have different numbers of neutrons are called **isotopes.**

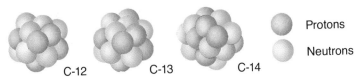

Figure 3–3. Although these forms of carbon appear to be very different, both the graphite and the diamond are composed mostly of Carbon-12. The other common carbon isotopes are shown at left.

Not all atoms are electrically neutral. Atoms can gain or lose electrons. When an atom gains or loses electrons, the electrical charge is no longer neutral. If an atom gains electrons, it has more electrons than protons. Therefore, the atom has a negative charge. If an atom loses electrons, it then has more protons than electrons. What kind of charge does it have? An atom that has gained or lost electrons and therefore has an electrical charge is called an **ion.** The opposite electrical charges of some ions pull them together. When ions are attracted to each other, they may combine, forming new substances.

○ *How many neutrons do different carbon isotopes have?*
○ *Why are ions attracted to each other?*

Section Review

READING CRITICALLY

1. An atom of the element hydrogen has one proton. How many electrons would there be in a positive hydrogen ion?
2. Since carbon has six protons, how many electrons would you expect to find in an electrically neutral carbon atom? Why?

THINKING CRITICALLY

3. If two fluorine atoms were to lose one electron each, would they likely be attracted to each other? Explain your answer.
4. Suppose that there are two isotopes of an element with eight protons and eight electrons. How many neutrons might each of these isotopes have?

② Chemical Compounds

NEW SCIENCE TERMS

molecule
compound
ionic bond
covalent bond
atomic number

SECTION OBJECTIVES

After completing this section, you should be able to:

■ **Describe** two kinds of chemical bonds.
■ **Identify** an element's properties by studying the periodic table.

3.3 **Molecules and Bonds**

You probably have friends that you like to spend time with. Maybe you spend time together at the library, in the lunchroom, or at the mall. Your group might even have a special name. Similarly, atoms are usually found together in groups called molecules. A **molecule** is a group of two or more atoms chemically combined. The atoms may be alike or they may be different.

Some molecules have only two atoms and are so small that millions of them could fit on the head of a pin. Some large molecules also exist. Many of the molecules in your body contain thousands of atoms; only a few hundred of these would fit on that same pinhead.

Atoms of different elements combine, or bond, to form compounds. A **compound** is a combination of two or more different kinds of elements.

There are two basic types of chemical bonds: *ionic* and *covalent*. An **ionic bond** forms when electrons move from one atom to another. Table salt, sodium chloride, has ionic bonds. When a sodium atom gives one of its electrons to a chlorine atom, the sodium becomes a positive ion, and the chlorine a negative ion. What do you think happens to these two ions?

Figure 3–4. These halite crystals, shown in both the photograph and the drawing, are composed of sodium (Na) and chlorine (Cl) atoms, which are held together by ionic bonds.

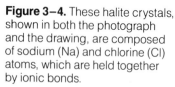

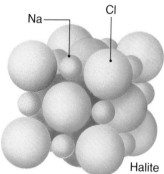

Halite

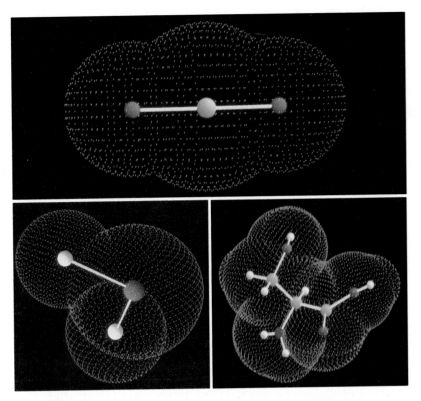

Figure 3–5. The computer-generated models show molecules that contain covalent bonds. The chemical formulas for carbon dioxiode (top), water (left), and serine (right) are shown in Table 3–1.

TABLE 3–1: COVALENT COMPOUNDS	
Covalent Compound	**Chemical Formula**
Water	H_2O
Carbon dioxide	CO_2
Serine	$CH_2CHOHNH_2COOH$

A **covalent bond** is formed when atoms share electrons. Water molecules form covalent bonds. In a single water molecule, two hydrogen atoms and one oxygen atom share electrons. Sharing electrons produces strong bonds between atoms, just as sharing experiences produces strong bonds between friends. Table 3–1 shows the chemical and structural formulas of several covalent compounds.

○ *What is the difference between an ionic bond and a covalent bond?*
○ *Give an example of each type of bond.*

3.4 The Periodic Table

In addition to the 90 naturally occurring elements, scientists have produced many other elements in the laboratory. To remember the characteristics of all these elements would be very difficult. To help scientists and students organize information about each element, all the elements have been placed in a table. This table, called the *periodic table,* may be found in the Reference Section on pages 558–559. Shown in Figure 3–6 are the 15 elements commonly found on Earth. The ten shown in red are common in the solid earth, and the five shown in blue are common in water and air.

The order of the elements in the periodic table is based on the properties of each element. The elements with similar properties are listed in groups. Each group is in order of increasing number of protons. The number of protons an atom contains is its **atomic number.** For example, helium has two protons and therefore an atomic number of two. Oxygen has eight protons. What is the atomic number of oxygen?

Figure 3–6. Highlighted in this version of the periodic table are the most common elements found on Earth.

PERIODIC TABLE OF ELEMENTS

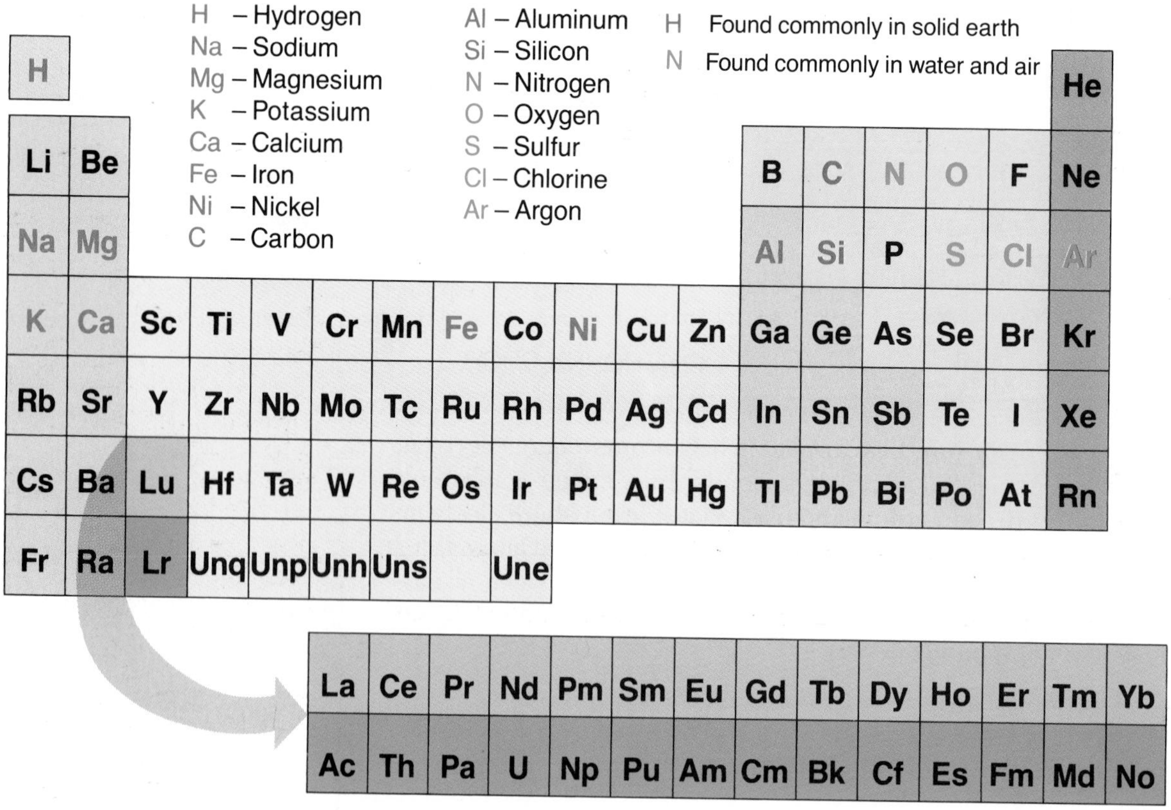

H – Hydrogen
Na – Sodium
Mg – Magnesium
K – Potassium
Ca – Calcium
Fe – Iron
Ni – Nickel
C – Carbon

Al – Aluminum
Si – Silicon
N – Nitrogen
O – Oxygen
S – Sulfur
Cl – Chlorine
Ar – Argon

H   Found commonly in solid earth
N   Found commonly in water and air

H																	He
Li	Be											B	C	N	O	F	Ne
Na	Mg											Al	Si	P	S	Cl	Ar
K	Ca	Sc	Ti	V	Cr	Mn	Fe	Co	Ni	Cu	Zn	Ga	Ge	As	Se	Br	Kr
Rb	Sr	Y	Zr	Nb	Mo	Tc	Ru	Rh	Pd	Ag	Cd	In	Sn	Sb	Te	I	Xe
Cs	Ba	Lu	Hf	Ta	W	Re	Os	Ir	Pt	Au	Hg	Tl	Pb	Bi	Po	At	Rn
Fr	Ra	Lr	Unq	Unp	Unh	Uns		Une									

La	Ce	Pr	Nd	Pm	Sm	Eu	Gd	Tb	Dy	Ho	Er	Tm	Yb
Ac	Th	Pa	U	Np	Pu	Am	Cm	Bk	Cf	Es	Fm	Md	No

As the atomic number increases, atoms may become less stable. Notice that the most common elements on Earth are elements with lighter nuclei.

The members of your family probably have many characteristics in common. Maybe all of you have dark hair and brown eyes. Each family of elements also has similar characteristics. Elements found in the same vertical column in the periodic table make up a family, or group. For example, the elements in Group 1 are all soft metals that are easily melted.

Figure 3–7. The two minerals shown at left are feldspar. The difference in color is due to the presence of potassium in one sample and sodium in the other sample. Sodium and potassium are shown above.

Elements in the same group sometimes substitute for each other in the formation of compounds. For instance, *feldspars*, a group of minerals, may contain potassium, sodium, or calcium. Although the metals affect the mineral's color, its other characteristics are not changed.

○ *What is meant by the atomic number of an atom?*
○ *What is the periodic table?*

Section Review

READING CRITICALLY

1. One neutral silicon atom has 14 protons and 14 neutrons. How many electrons does this silicon atom have? What is its atomic number?

2. Which elements are most commonly found in solid earth? Which are most common in water and air?

THINKING CRITICALLY

3. What elements could probably replace carbon in chemical compounds?

4. If there were an electrically neutral element with the atomic number of 110, how many protons and electrons would it have? How many neutrons might it have?

DISCOVER

Naming Simple Compounds

Look at the periodic table on pages 558–559. All the elements in Group 1 easily lose one electron, and all the elements in Group 17 easily gain electrons. See how many different compounds you can create by combining elements from these two columns. Using "sodium chloride" as an example, try naming the compounds as you form them.

CAREERS

LAPIDARY

A *lapidary* is a person who cuts rough mineral samples into beautiful gemstones. A single cut can be the difference between a gem worth thousands of dollars and one that is worthless. Lapidaries must understand the optical effects of light and the structure of minerals.

People who would like to learn the art of gem cutting usually begin as apprentices.

For Additional Information
Gemological Institute of America
1660 Stewart Street
Santa Monica, CA 90406

MINING GEOLOGIST

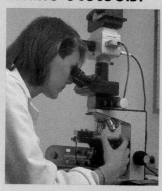

Mining geologists play a major role in finding minerals and getting them out of the ground. They are also involved in the important task of restoring the environment in mined areas.

Most jobs in mining geology are through private employers in the industry. A bachelor's degree is required, and an advanced degree is often recommended. People who have gained experience as technical assistants to mining geologists can obtain positions by passing a professional licensing test.

For Additional Information
American Geological Institute
5205 Leesburg Pike
Falls Church, VA 22041

STONEMASON

Although most large buildings are now built with steel, *stonemasons* still play an important part in construction work. Many buildings have stone decorations, window sills, door frames, and exteriors. These stones must be cut, fitted, and secured with precision. Most stonemasons learn their skills through apprenticeship programs. Sometimes they take classes in general construction skills.

For Additional Information
Associated General
    Contractors of America, Inc.
1957 E Street, NW
Washington, DC 20006

SECTION
3 Kinds of Minerals

SECTION OBJECTIVES

After completing this section, you should be able to:

- **Define** the term *mineral*.
- **Classify** minerals into families based on their chemical formulas.
- **Compare** native elements to other mineral families.

NEW SCIENCE TERMS

mineral
tetrahedron

3.5 Silicates, Carbonates, and Sulfates

Have you seen the commercials on TV for products that contain a whole day's supply of vitamins and minerals? What is a day's supply of minerals? Where do these minerals come from?

A **mineral** is a naturally occurring substance consisting of a single element or compound. Most minerals come from nonliving things, such as the earth itself. In addition, most minerals are solids, formed from common elements.

Silicates   Silicon and oxygen are two of the most common elements in the earth's crust. Therefore, it is not surprising that many minerals contain these two elements. Minerals that contain silicon and oxygen are called *silicates* (SIHL uh kayts). The silicate minerals make up about 92 percent of the earth's crust.

One common silicate mineral is formed from a single silicon atom bonded to four oxygen atoms. These atoms form a network called a **tetrahedron,** or a four-sided structure. Silicon-oxygen tetrahedrons can bond together to form chains or sheets. These chains and sheets form a three-dimensional network that produces a very hard mineral called *quartz* (KWAHRTS).

Figure 3–8. This beautiful mineral is one form of the element sulfur.

Figure 3–9. Quartz is a silicate mineral formed of silicon-oxygen tetrahedrons.

Feldspars are the largest group of silicate minerals. They are similar to quartz in that they are also silicates. Feldspars, however, contain calcium, potassium, and sodium atoms in addition to silicon atoms.

Just as paint may contain different-colored pigments, feldspars may contain other elements, such as iron and manganese. These different elements produce different-colored feldspars. However, they are all feldspars, just as all colors of paint have the basic characteristics of paint.

Carbonates Only eight percent of the earth's crust is composed of nonsilicate minerals. One important nonsilicate family is the *carbonates.* Carbonate ions contain three oxygen atoms joined to a single carbon atom. A carbonate ion bonded to a calcium ion forms a mineral called *calcite.* If magnesium is substituted for half of the calcium, a different mineral, called *dolomite,* is formed. Dolomite is often used as a mineral supplement for people. It helps provide calcium and magnesium for preventing some bone diseases.

Sulfates Another family of nonsilicate minerals is the *sulfates.* Sulfates have four oxygen atoms combined with one sulfur atom. If a sulfate ion bonds with calcium and water, a mineral called *gypsum* is formed. Gypsum, the most common sulfate mineral, is used to make drywall for buildings.

○ *What is the most common family of minerals?*
○ *What elements do sulfates contain?*

Figure 3–10. Many carbonate minerals, such as calcite (above), have commercial value. Gypsum, a sulfate mineral, can be made into drywall (right), a type of plasterboard used in the construction of many homes.

3.6 Oxides, Halides, Sulfides, and Native Elements

There are several other mineral families. Although these mineral families make up only a small percentage of the earth's crust, they are important as sources from which metals may be obtained.

Oxides   In the *oxides*, oxygen is bonded to another element, such as iron or aluminum. Iron oxides include several different minerals; all are mined as sources of iron. The most common source of aluminum is also an oxide.

Halides and Sulfides   Two other important mineral families are the *halides* and the *sulfides*. These families of minerals contain no oxygen. Sodium chloride is a halide called *halite. Pyrite*, also known as "fool's gold," is a sulfide of iron. Pyrite is known as fool's gold because it looks much like real gold, although it is nearly worthless. Look at Figure 3–12. Can you tell the fool's gold from the real thing? Some sulfides also have value as sources of metals such as lead, copper, and zinc.

Native Elements   Some minerals contain only one kind of element. These rare minerals are called *native elements*. Many of the minerals in this group are valuable metals. Some examples are gold, silver, and copper. Uncombined nickel and iron are believed to exist in large amounts deep within the interior of the earth. However, uncombined nickel and iron are very rare in the earth's crust. Diamond and graphite are also native elements; they are pure carbon. They differ only in the way the carbon atoms are arranged.

○ *Why are oxides important?*
○ *Name two native elements.*

Figure 3–11. Oxides such as those shown above are commercial sources of iron, aluminum, and other metals.

Figure 3–12. Gold is a native element, while "fool's gold" is a compound called *iron pyrite*. The real gold is on the right in this sample.

Section Review

READING CRITICALLY

1. What are the major groups of minerals that make up the earth's crust?
2. What is the difference between native elements and other minerals?

THINKING CRITICALLY

3. Why does the silicate family of minerals make up such a large part of the earth's crust?
4. Why can magnesium replace calcium in the mineral dolomite?

4 Physical Properties of Minerals

NEW SCIENCE TERMS

streak
luster
cleavage
fracture
crystal
gem

SECTION OBJECTIVES

After completing this section, you should be able to:
- **Describe** the five major mineral tests.
- **List** a mineral example for each of the crystal forms.
- **Explain** what qualities make a mineral a gem.

3.7 Simple Mineral Properties

Do you collect stamps, coins, or baseball cards? If you have a lot of them, you probably have them classified in some way, such as by postmark, date, or team. The minerals found on Earth can be classified, too.

A detailed analysis of minerals can be done only in a laboratory with special equipment. However, there are some simple physical properties that you can use to quickly classify a mineral sample.

Color Probably the first characteristic that you will notice about a mineral is its color. Unfortunately, this is one of the least dependable characteristics you can use to classify a mineral. Color may change with even a small amount of impurity. For instance, pure quartz is clear, but with impurities it can be pink, tan, red, black, or purple.

The surfaces of some minerals change color when exposed to air. The mineral *chalcopyrite* (kal koh PY ryt), which contains copper, iron, and sulfur, can tarnish. The copper in the mineral reacts with the air just as the copper in a penny does, making it dark and dull.

A MATTER OF FACT

Wayne F. Downey, Jr., a high-school student in Harrisburg, Pennsylvania, discovered a new mineral in a burning coal bed. This oxide of selenium was named Downeyite to honor his discovery.

Figure 3–13. These minerals are all samples of quartz. Pure quartz is colorless; the colors of these samples are due to a variety of impurities.

Streak   If you rub a piece of chalk on a sidewalk, it leaves a mark on the cement. The color of the powder that is left when a mineral is rubbed on a rough surface is called the mineral's **streak.** Sometimes the color of the streak is different from the color of the mineral sample. For instance, the streak of golden-colored pyrite is greenish black, and the streak of silver-colored hematite is brick red.

Figure 3–14. The girl on the left is preparing a mineral streak. Two mineral streaks are shown on the right. Streak is not influenced by such things as tarnishing and impurities. Therefore, streak is a better characteristic for identifying minerals than color is.

Luster   Have you ever admired the appearance of a shiny new car? What you are admiring is the luster of the surface. The **luster** of a mineral refers to the appearance of its surface in reflected light. There are two types of luster: metallic and non-metallic. If the mineral shines like gold, copper, or silver, then the luster is metallic. If the mineral does not shine like a metal, then its luster is nonmetallic. A nonmetallic luster can be further described using terms such as waxy, pearly, glassy, dull, or brilliant. What common objects can be considered as examples of each of these lusters?

Figure 3–15. Metals, such as gold and silver, have a luster that can be easily recognized. Other types of mineral luster are named for common substances, such as pearls and earth (soil).

63

Hardness   List ten different objects and rank them from the softest to the hardest. If you included a diamond, then it should be at the bottom of your list, because a diamond is the hardest substance on Earth. At the other extreme is the mineral *talc*, from which talcum powder is made. All minerals fall somewhere between these two extremes.

A set of ten standard minerals is used to make a scale to measure the hardness of all minerals. This scale, called *Mohs' Scale of Mineral Hardness*, is shown in Table 3–2. The number 1 is assigned to talc, and the number 10 is assigned to diamond. A mineral that can scratch calcite (hardness of 3) but cannot scratch fluorite (hardness of 4) would have a hardness between 3 and 4.

Frequently the minerals of the Mohs' Scale are not available for comparison, so geologists have created a set of more common objects that can be used. For instance, if a certain mineral can scratch a penny, its hardness is greater than 3. If a steel nail can scratch that same mineral, you know its hardness is less than 4.5, or somewhere between 3 and 4.5.

ACTIVITY: Testing a Mineral's Hardness

How can you use common materials to determine the hardness of mineral specimens?

MATERIALS (per group of 3 or 4)

penny, steel nail, glass plate,
mineral samples (5)

TABLE 1: A SCALE OF HARDNESS				
	Softer than			
Mineral	Finger-nail	Penny	Steel nail	Glass plate	Mineral hardness
1				
2				
3				
4				
5				

PROCEDURE

1. Make a data table like the one shown.
2. Assign numbers to each of your minerals, using the following guide: 1 = pink or tan color, 2 = gold color, 3 = silver color, 4 = white, pearly color, 5 = clear to white color, and glassy.
3. Scratch mineral 1 with your fingernail, the penny, the steel nail, and the glass plate. Record your results in the data table.
4. Test the other samples in the same way. Record your results in the data table.

CONCLUSIONS/APPLICATIONS

1. Which mineral was softer than your fingernail?
2. Which mineral was harder than your fingernail but softer than the penny?
3. Which mineral was harder than the penny but softer than the steel nail?
4. Which minerals were harder than the glass?
5. Review the hardness of the glass, nail, and penny from page 65, then assign a hardness to each mineral.

TABLE 3–2: MOHS' SCALE OF MINERAL HARDNESS

	Hardness — mineral	Common material — hardness
	1 — Talc	
	2 — Gypsum	
	3 — Calcite	Fingernail—2.5
	4 — Fluorite	Penny—3.5
	5 — Apatite	Steel nail—4.5
	6 — Feldspar	Glass plate—5.5
	7 — Quartz	Steel file—6.5
	8 — Topaz	
	9 — Corundum	
	10 — Diamond	

Figure 3–16. Halite (top left) will always break into cubes, while the cleavage of calcite (top center) is diagonal. Mica (top right) forms thin, flat sheets. The glass (below top) and asbestos (below bottom) fracture when they break.

Cleavage and Fracture   The terms *cleavage* and *fracture* describe how minerals break. Minerals that always break along flat surfaces have **cleavage.** Halite, for example, always breaks into little cubes. A mineral with cleavage will break where its chemical bonds are weakest.

If a mineral breaks along curved or irregular surfaces, it has **fracture.** Minerals can show fracture in several different ways. Some fractures have a curved surface similar to the inside of an eggshell. Another type of fracture produces long splinters or fibers. Fractures can also have jagged surfaces or sharp edges.

Special Properties   Some minerals show special properties that can be used to identify them. Lodestone, for example, is a type of iron oxide that is magnetic. A piece of lodestone will attract iron filings just as a magnet does.

Calcite can be identified by its chemical reaction with weak acids. If a weak solution of hydrochloric acid is dropped on calcite, the acid begins to bubble. Many other minerals can be identified by special properties as well.

○ *What is meant by a mineral's streak?*
○ *What scale is used to rate the hardness of minerals?*

Figure 3–17. The lodestone shown here is magnetic and will attract a variety of iron objects.

SKILL ACTIVITY: Using a Classification Key

BACKGROUND

Classifying things into groups in order to identify them is a skill that scientists find very useful. There are thousands of different minerals on the earth. To speed up the identification process, the testing information is often organized into a *classification key*. At each step in the key, one property is identified, and minerals not fitting the property are eliminated. This continues until only one mineral remains.

PROCEDURE

A mineral classification key is shown on page 566 in the Reference Section. Look at the first category of the key. *What property is tested?* Follow the sample that is given, and at each step, see if you can decide why that choice was made. Copy the table and fill it in as you classify your mineral samples.

1. Obtain a sample of galena from your teacher. Try to determine its luster. You should con-clude that it fits in Category I because of its metallic luster.
2. Determine its streak. Galena's black streak puts it into Category A.
3. Test its hardness. Galena can be scratched by a penny, so its hardness is less than 3.5. It belongs in Group 1.

APPLICATION

Obtain four unknown samples of minerals from your teacher and classify them down to their group, using the key. Check your answers with your teacher when you have classified the samples.

USING WHAT YOU HAVE LEARNED

1. Did you classify your minerals correctly? If not, what problems did you have?
2. What are the advantages and disadvantages of this type of classification key?
3. Why is color an unreliable characteristic for classification?

TABLE 1: MINERAL CLASSIFICATION

Sample	Luster	Streak	Hardness	Color
1. Galena				
2.				
3.				
4.				
5.				

TABLE 3—3: CRYSTAL SHAPES

Mineral	Geometry	Axes
Galena	Cubic	Three axes — all equal length and all right angles
Chalcopyrite	Tetragonal	Three axes—two equal length and all right angles
Quartz	Hexagonal	Four axes—three equal length and one at right angle to the others
Olivine	Orthorhombic	Three axes—all different lengths and all at right angles
Gypsum	Monoclinic	Three axes—length varies and two at right angles
Microcline	Triclinic	Three axes—all different lengths and no right angles

3.8 Crystal Structure

You may have seen the beauty of minerals in a jewelry store or a museum. Such minerals are not only beautiful, they usually exhibit interesting geometric patterns as well. These naturally shaped minerals are crystals.

A **crystal** is a three-dimensional structure with each face, or surface, having a definite shape and orientation. When minerals have enough space to grow, they form crystals. Each mineral has a characteristic crystal form that is useful in its identification. The basic crystal forms are shown in Table 3–3. Crystal forms should not be confused with cleavage or fracture surfaces. When a mineral fractures, the crystal form is destroyed. For instance, a quartz crystal is a beautiful six-sided column, topped by a pyramid. When quartz breaks, it forms surfaces with curved fractures.

Some minerals are classified as *gems*. A **gem** is a mineral that has beauty, durability, and value. The beauty of a gem is based on its richness of color, purity of composition, and brilliance of sparkle. Because gems have durability, they last when worn as jewelry. Since gems are rare and unusual, they usually have high monetary value. Sometimes jewelry is made from solids, such as glass or opal, which have no crystal structure. These stones are called *amorphous*, which means without regular form.

In some cases, only certain varieties of a mineral are considered to be gems. For instance, a ruby is a rare, dark-red variety of the mineral *corundum*. Emeralds and aquamarines are special varieties of *beryl*. These rare-colored gems are produced by impurities in the minerals.

○ *What is a crystal?*
○ *Why are gems valuable?*

Figure 3–18. These cut minerals have qualities that make them attractive as gems.

DISCOVER

Breaking Crystals

Pour some ordinary table salt onto a dark piece of paper. Using a magnifying glass, look at the shape of the crystals. What shape are the crystals?

With the point of a sharp pencil, try to break some of the crystals. Is the shape of the broken crystals the same as the shape of the unbroken crystals? Explain.

Section Review

READING CRITICALLY

1. Describe what is meant by the properties of streak and luster.
2. Explain why geologists might have to scratch or break a mineral to discover its true color.

THINKING CRITICALLY

3. Amethyst, a variety of quartz, is considered to be a semiprecious gem. Why do you think it is not as highly valued as emeralds and rubies are?
4. What do calcite and dolomite have in common that make them react to hydrochloric acid?

INVESTIGATION 3: Properties of Minerals

PURPOSE

To use the properties of hardness, streak, color, and luster to identify minerals

MATERIALS (per group of 3 or 4)

Penny
Steel nail
Glass plate
Streak plate
Minerals: pyrite, chalcopyrite, sphalerite, galena, magnetite, quartz, hematite, talc, calcite, limonite

TABLE 1: CHARACTERISTICS OF COMMON MINERALS

Mineral	Hardness	Color	Streak	Luster
Pyrite				
Chalcopyrite				
Galena				
Magnetite				
Talc				
Quartz				
Calcite				
Sphalerite				
Limonite				
Hematite				

PROCEDURE

1. Make a data table like the one shown. Fill in the table as you complete the investigation.
2. Using the penny, nail, and glass plate, test the hardness of each mineral. Record your results in the data table.
3. Record the color of each mineral in the data table.
4. Using the streak plate, determine the streak of each mineral. Record your results in the data table.
5. Describe the luster of each mineral using the terms *metallic, glassy, pearly,* or *waxy.* Record your results in the data table.

ANALYSES AND CONCLUSIONS

1. Which two properties would be most helpful in distinguishing between galena and chalcopyrite?
2. Which one property would be helpful in telling calcite from quartz?
3. Which two properties would be helpful in telling pyrite from chalcopyrite?
4. Which three properties would be helpful in telling sphalerite from pyrite?
5. Which one property would be helpful in telling calcite from talc?
6. Which two properties would be helpful in telling hematite from limonite?
7. Compare the streaks of the minerals with metallic luster to those with nonmetallic luster. What general statement can you make about the relationship between luster and streak?

APPLICATION

If you had an unknown mineral sample, what tests would you use to determine the mineral family to which the unknown sample belongs?

SUMMARY

- All things are made of atoms. Atoms are made of protons, electrons, and neutrons. (3.1)

- There are 90 different kinds of atoms, called elements, that occur naturally. (3.1)

- Isotopes of an element have the same number of protons but a different number of neutrons in their nuclei. (3.2)

- An ion is an atom with an electrical charge. (3.2)

- Atoms are often joined together as molecules. (3.3)

- Compounds are held together by two types of bonds: ionic and covalent. (3.3)

- To help scientists and students organize information about each element, all the elements have been placed in a periodic table. (3.4)

- The elements in the periodic table are listed in order of increasing proton number. The number of protons in an element is its atomic number. (3.4)

- Minerals are grouped into families based on their chemical composition. (3.5)

- Families of minerals include silicates, carbonates, sulfates, oxides, halides, and sulfides. (3.5, 3.6)

- Minerals formed of only one element are native elements. (3.6)

- Minerals can be identified by common physical properties. (3.7)

- Some minerals have special properties that aid in their identification. (3.7)

- Most minerals build crystal formations. (3.8)

- A gem is a rare mineral with the special characteristics of beauty, durability, and value. (3.8)

Write all answers on a separate sheet of paper.

SCIENCE TERMS

Correctly use each of the following terms in a sentence.

atomic number **(56)**
atoms **(51)**
cleavage **(66)**
compound **(54)**
covalent bond **(55)**
crystal **(69)**
electrons **(52)**
elements **(51)**
fracture **(66)**
gem **(69)**
ion **(53)**
ionic bond **(54)**
isotopes **(53)**
luster **(63)**
mineral **(59)**
molecule **(54)**
neutrons **(52)**
nucleus **(52)**
protons **(52)**
streak **(63)**
tetrahedron **(59)**

SCIENCE QUIZ

Modified True-False

Mark each statement *true* or *false.* If a statement is false, change the underlined term to make the statement true.

1. A silicate is composed of silicon and <u>calcium</u>.

2. Compounds can have either ionic bonds or <u>electrical</u> bonds.

3. A carbon isotope is an atom of carbon that has the number of protons characteristic of carbon but a different number of <u>neutrons</u>.

4. Of all the properties of minerals, <u>color</u> is the least dependable to use for identification.

continues

5. Gold and silver are <u>native</u> elements.

6. Elements in the periodic table are listed in order by increasing number of <u>protons</u>.

Multiple Choice

Write the letter of the choice that best answers the question or completes the statement.

7. The smallest particle of an element is
   a) a neutron.
   b) an atom.
   c) an electron.
   d) a proton.

8. The atomic number of an element is equal to its number of
   a) protons.
   b) electrons.
   c) neutrons.
   d) atomic particles.

9. Quartz belongs to the group of minerals called
   a) silicates.
   b) feldspars.
   c) carbonates.
   d) sulfides.

10. All of the following occur as native elements except
    a) graphite.
    b) silver.
    c) gold.
    d) quartz.

11. Fluorite has a mineral hardness of
    a) 1.
    b) 2.
    c) 3.
    d) 4.

12. Feldspar is a
    a) carbonate.
    b) sulfate.
    c) halide.
    d) silicate.

Completion

Complete each statement by supplying the correct term or phrase.

13. There are _____ naturally occurring elements.

14. Two of the most common elements in the earth's crust are _____.

15. The _____ is the basic structure of quartz.

16. The softest mineral on Mohs' Scale of Mineral Hardness is _____.

17. Diamonds and emeralds are minerals that are called _____.

18. The appearance of the surface of a mineral is known as its _____.

Short Answer

19. Compare and contrast the properties of fracture and cleavage.

20. Explain why some mineral specimens are considered to be gems.

21. Describe the steps to estimate the hardness of a mineral if Mohs' Scale of Mineral Hardness is not available.

22. Describe the placement of elements in the periodic table, and explain the relationship between atomic number and proton number.

Writing Critically

23. Considering the definition of a mineral, discuss why ice should or should not be considered a mineral.

24. Which two mineral identification tests do you think are the most useful? Explain why you chose these two.

25. Compare and contrast an atom, an element, a molecule, and a compound. Explain the fact that most minerals are compounds, while others are elements. Also explain why it is possible to have a single atom of some minerals, while for most minerals a molecule is the smallest unit possible.

EXTENSION

1. Gold is not used just in making jewelry. It has been and is still being used in the treatment of some diseases. Write a report about the medical uses of gold in both the past and the present.

2. Research the story of the Hope diamond. Be sure to include information about the people who have owned this diamond.

APPLICATION/CRITICAL THINKING

1. Suppose you found a beautiful white mineral on the roadside while on vacation in Colorado, and you wanted to identify it. What tests would you perform on it? If the mineral you found was soft and showed three directions of cleavage, what might it be? Explain your answer.

2. Special organisms that live in the sea can take minerals from the water to build structures called reefs. The reefs are similar in composition to certain types of materials used for building human structures. Explain how artificial reefs might be constructed from building wastes. Describe the problems that might occur from using waste materials.

FOR FURTHER READING

Bains, R. *Rocks and Minerals.* Mahwah, New Jersey: Troll, 1985. This book simplifies the description of mineral composition for anyone looking for a clear explanation.

Cheney, G. A. *Mineral Resources.* New York: Watts, 1985. This book is a survey of America's most important minerals. It includes information on their composition, location, mining, and use.

Cotterill, R. *The Cambridge Guide to the Material World.* New York: Cambridge University Press, 1985. The book is well illustrated, and topics include crystals, minerals, metals, ceramics, and glass.

O'Neil, P. *Gemstones. Time-Life Planet Earth Series.* Alexandria, Virginia: Time-Life, 1983. This book presents extensive information on sources and processing of gems.

Challenge Your Thinking

This person is standing inside a special mineral formation known as a *geode.* Most geodes are not this large, but they do represent some of Earth's largest mineral structures. A geode is a hollow ball with mineral crystals growing on the inside. How do you think a geode might form?

Rocks

Earth is a special planet. It has air, water, and a solid surface. This solid surface, or crust, is composed of rocks. As this picture shows, rock formations are often spectacular. As well as being spectacular, rocks are also important scientifically and economically. The rocks of Earth contain the history and most of the riches of our planet.

Long columns of igneous rocks

1 Studying Rocks

SECTION OBJECTIVES

After completing this section, you should be able to:

- **List** the three rock families and the general characteristics used to classify rocks into these families.
- **Describe** the rock cycle and **relate** it to the formation of the three families of rocks.

NEW SCIENCE TERMS

rocks
magma
lava
igneous rocks
sedimentary rocks
metamorphic rocks
rock cycle

4.1 Rock Families

What makes you a person? Is it your hair, or your face, or perhaps the fact that you have two arms and two legs? Whatever it is, you have definite characteristics that make you recognizable as a person. Still, you are different from every other person. You are unique.

Rocks have definite characteristics also. Look at the photographs in Figure 4–1. You can probably tell that the objects in the photographs are *rocks*. All rocks share some characteristics that make them rocks, but they also have many differences that make them unique. For example, the rocks in the photographs are different colors. One rock has large crystals, and the other appears to have no crystals at all. Rocks are like people; they are alike, yet each is unique.

In Chapter 3, minerals are defined as naturally occurring substances made of specific elements or compounds. **Rocks** are naturally occurring combinations of minerals. A few rocks consist of only one type of mineral. However, most rocks consist of two or more different minerals. Although hundreds of different kinds of minerals are found in rocks, only about 20 minerals are common. In Chapter 3, you practiced classifying minerals; rocks can be classified in much the same way.

Geologists classify rocks in several ways. One method of classification is based on the origin of the rocks; that is, on how

Figure 4–1. Although these rocks look different, they belong to the same rock family. As with brothers and sisters of human families, members of the same rock family share many characteristics.

they were formed. All rocks may be classified into one of three families according to the way in which they were formed.

Understanding how rocks form can be somewhat confusing. You can observe similar processes by using simple materials. Suppose your teacher takes some old candles and melts them in a pan, and then carefully pours the melted wax into a mold for you. If you allow the wax to cool completely, it will harden. You now have a solid object formed from liquid, or *molten*, material.

Rocks are formed in a similar manner inside the earth. Where temperature is very high, rocks melt. Molten rock within the earth is called **magma.** Many pockets of magma are found in the earth's interior.

Molten rock that reaches the surface of the earth is called **lava.** You may have heard the term *lava* used while listening to news reports about volcanoes. As magma or lava cools, it hardens, and solid rocks form. The rocks that form from magma or lava are known as **igneous** (IHG nee uhs) **rocks.** The word *igneous* comes from the Latin word *ignis,* meaning "fire."

Now suppose you take your wax block and carefully scrape off some of the wax with the edge of a knife. When you have made a small pile of wax shavings, you cover it with a piece of paper and a heavy book and press it as hard as you can, the wax shavings stick together. A second type of rock forms from small rock fragments in much the same way.

Rocks exposed to the weather break into small pieces, which form a sediment (SEHD uh muhnt). Sediments can be transported by water or wind to new locations. Eventually the sediments are deposited in layers. As layers of sediments are compressed and cemented, they stick together. Compressed sediments form **sedimentary** (sehd uh MEHN tuhr ee) **rocks.** The word *sedimentary* also comes from Latin, from the word *sedimentum,* meaning "that which has settled."

Figure 4–2. Molten rocks from within the earth often come to the surface as lava from a volcano. As the lava cools, igneous rocks form.

Figure 4–3. The mud of this dry lake bed may eventually become sedimentary rock. Some sedimentary rocks show mud cracks such as these or ripple marks from shallow water.

Sandstone

Quartzite

Granite

Gneiss

Limestone

Marble

Figure 4–4. Shown here are three examples of metamorphic rocks. In each case the sample on the right is the metamorphic rock, and the sample on the left is the parent rock from which it formed.

Finally, if you warm the remaining wax slightly, you should be able to bend it and squeeze it without having to melt it. In a similar fashion, heat and pressure change igneous and sedimentary rocks.

Heat and pressure cause the minerals in rocks to form new combinations, or new crystals. These changes form rocks called **metamorphic** (meht uh MOR fihk) **rocks.** The word *metamorphic* comes from a Greek word, *metamorphosis*, which means "change."

◎ *Name the three rock families.*
◎ *How are rock families named?*

4.2 The Rock Cycle

People who don't know you may still be able to tell that you belong to a certain family because you have characteristics that identify you with that family. The family to which a rock belongs can usually be determined by studying certain characteristics of the rock. For instance, rocks can be identified by their mineral composition.

Some minerals occur in only a few types of rocks. The mineral calcite, for example, is rarely found in igneous rocks, but it is common in sedimentary rocks. Mica and feldspar usually occur in igneous rocks. Another characteristic of rocks is the layering of minerals, which often occurs in sedimentary and metamorphic rocks.

In our society, families change because of death, divorce, adoption, and marriage. The rock families change because of weather, heat, and pressure. New rocks are constantly being formed from old rock material. This continuous process of

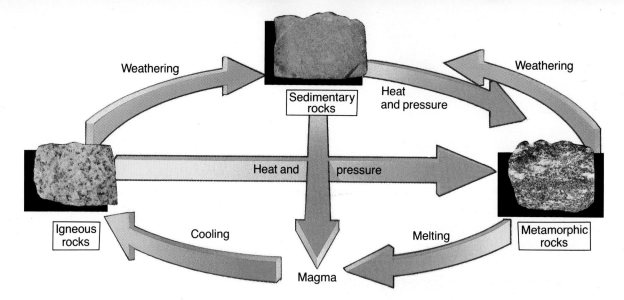

Figure 4–5. The rock cycle is not a one-way process. Rocks may enter the cycle at any place and be changed into any other type of rock.

change within rock families is called the **rock cycle.** A simplified version of the rock cycle is shown in Figure 4–5.

All rocks go through this cycle. For example, as magma cools it may form *granite,* an igneous rock. Granite can be broken by sea waves into sand-sized sediments. The sand can then be deposited and compacted to become *sandstone,* a sedimentary rock. If the sandstone is exposed to heat and pressure, it can change to *quartzite,* a metamorphic rock. Quartzite may later be forced deep within the earth and remelted into magma. The cycle may repeat itself many times.

The rock cycle is not a one-way process. While it is possible for igneous rocks to be changed into sedimentary rocks, it is also possible for metamorphic rocks to be changed into sedimentary rocks. Granite may be changed into sandstone by weather and water. In fact, any rock may change from any one of the rock families to any one of the others.

○ *What is the rock cycle?*
○ *Into what kinds of rocks can igneous rocks change?*

DISCOVER

Studying the Rock Cycle

Look at samples of granite, sandstone, and quartzite. Try to decide which rock belongs in each family. Make a sketch of the rock cycle with these three samples. Which of the three rocks probably formed first? Why?

Section Review

READING CRITICALLY

1. What are the origins of each of the rock families?
2. Explain why the rock cycle is not a one-way process.

THINKING CRITICALLY

3. What kind of rocks must have formed first on Earth? Explain your answer.
4. Starting with sandstone, a sedimentary rock, give several possible steps of a rock cycle that would end with granite, an igneous rock.

2 Igneous Rocks

SECTION OBJECTIVES

After completing this section, you should be able to:

- **Discuss** the process that forms igneous rocks.
- **Classify** common igneous rocks by their characteristics.
- **List** several uses for igneous rocks.

NEW SCIENCE TERMS

texture
intrusive rocks
extrusive rocks

4.3 Formation of Igneous Rocks

What is the most common family name on Earth? What is the most common name in North America, or in your school? Which of your classmates has the largest family? You could ask similar types of questions about rocks.

Igneous rocks are by far the most common rock family in the earth's crust. Igneous rocks make up about two-thirds of the earth's crust. The type of igneous rock that forms depends mainly on two conditions: the chemicals within the magma or lava and the rate at which the magma or lava cools. If, for example, a large amount of a dark element such as iron is present in magma, dark-colored rocks, such as *basalt* (buh SAWLT), form.

As magma cools, ions bond to form mineral crystals. Magma that cools slowly forms visible crystals. These crystals join in tightly interlocking units, producing very dense rocks. Remember that lava is found on the earth's surface, so lava cools much more quickly than magma. This rapid cooling prevents the formation of large crystals.

Does slow cooling form large or small crystals in rocks? Explain your answer.

4.4 Classification of Igneous Rocks

Igneous rocks are classified by texture and mineral content. **Texture** is the size of the crystals in the rocks. Mineral content determines a rock's density and color.

Texture Geologists divide igneous rocks into two main groups: those with fine texture, formed from lava, and those with coarse texture, formed from magma. How does the texture of a rock relate to the size of the mineral crystals in the rock?

Rocks with large mineral crystals are called **intrusive** (ihn TROO sihv) **rocks.** Individual crystals can easily be seen in intrusive rocks.

Figure 4–6. Shown here are basalt (top) and porphyry (bottom). The porphyry cools unevenly so small and large mineral crystals form.

Figure 4–7. This is rhyolite, an extrusive rock. Notice the fine texture of this rock.

Smaller, invisible mineral crystals form in lava. Lava quickly loses its heat to the atmosphere, preventing the formation of large crystals. The fine-grained rocks formed from lava are called **extrusive** (ehk STROOS ihv) **rocks.** *Rhyolite*, shown in Figure 4–7, is an example of extrusive rock.

Mineral Content   The rate of cooling determines the mineral content of rocks. The first crystals to form are those of minerals that are stable at high temperatures. Many of these minerals are rich in iron and magnesium. Two types of high-temperature minerals are olivine and pyroxene. These dark crystals sink into the lower parts of the magma because of their high density. This leaves the remaining magma with a high percentage of less dense minerals.

Igneous rocks that contain dense minerals are called *mafic* (MAF ihk) *rocks.* The magma that forms mafic rocks has a temperature of 900°C to 1200°C. Mafic rocks are usually black or dark green. Most of the rocks of the oceanic crust and of the Pacific islands are mafic rocks. Two examples of mafic rocks are basalt and *gabbro.*

Igneous rocks containing low-density minerals are called *felsic* (FEHL sihk) *rocks.* The minerals in felsic rocks are light-colored crystals of quartz and some types of feldspar. The temperature of the magma that forms felsic rocks is about 800°C. Rhyolite and *andesite* are felsic rocks.

Special Igneous Rocks   Some igneous rocks do not fit neatly into either the intrusive or extrusive category. These rocks usually have no visible crystals at all.

Figure 4–8. The dark rocks of the Pacific islands (left) are mafic rocks, while the light rocks of the continent (right) are felsic rocks.

One special type is volcanic glass, or *obsidian*, shown in Figure 4–9. Obsidian is an igneous rock that does not have crystalline structure. The rock hardens so fast that no crystals form. Obsidian contains many of the same minerals as rhyolite, but it looks very different. Arrowheads of early Native Americans were often made from obsidian.

Pumice, shown in Figure 4–10, is a low-density, glassy rock. Pumice cools so rapidly that gases in the lava are trapped inside the rock. The gases escape after the rock cools, leaving many spaces, or pores. There are so many of these pores in pumice that it can float on water. *Scoria* also has pores, but it is too dense to float.

Figure 4–9. Obsidian, or volcanic glass, cools so quickly that no crystals form.

○ *How do intrusive rocks form?*
○ *Which igneous rocks have light-colored minerals?*

Figure 4–10. Pumice (left) is so porous that it will float. Although scoria (right) is also porous, it will not float.

ACTIVITY: Classifying Igneous Rocks

How do you classify igneous rocks?

MATERIALS (per group of 3 or 4)

pencil, paper, igneous rock set, hand lens

TABLE 1: ROCK COMPOSITION AND TEXTURE			
TEXTURE	**ROCK COMPOSITION (COLOR)**		
	Felsic (Light)	Intermediate (Medium Light)	Mafic (Dark)
Glassy			
Fine			
Coarse			

PROCEDURE

1. Make a table like the one shown. Fill in the chart as you complete the activity. You may not have a rock for each box in the chart.
2. Divide the igneous rocks in your set into three groups based on their composition.
3. Separate each group by texture. Place the names of the rocks in your chart to indicate the proper texture and composition for each rock. You may need to review the chapter for help in matching the rock name to the proper texture.

CONCLUSIONS/APPLICATIONS

1. How would you distinguish granite from rhyolite?
2. Why would you have a problem placing pumice on this simple classification chart?
3. How were the holes formed in scoria?
4. Write a thorough description of granite. Explain what the texture of granite tells you about its origin.

Figure 4–11. The black granite used to make this memorial will probably last for centuries.

4.5 Uses of Igneous Rocks

Many igneous rocks are used in the construction industry because of their strength. These rocks can also be polished to shiny, smooth surfaces. Carvings and inscriptions in granite can withstand the effects of weather for centuries. The Vietnam Veterans' Memorial in Washington, D.C., is made of 150 panels of granite from Barre, Vermont.

Pumice is a valuable polishing material. Sometimes it is ground up and used as a scouring powder. Pumice is also used in a lump form called a *pumice stone*.

Igneous rocks are valuable sources for some minerals and metals. These rocks are often mined for uranium, iron, copper, and other metals.

○ *For what purpose is pumice used?*
○ *Why are igneous rocks used in monuments?*

Section Review

READING CRITICALLY

1. What are the two characteristics used to classify igneous rocks?
2. Explain the difference between felsic and mafic rocks and give examples of each.
3. What relationship exists between high-temperature stable crystals and density?

THINKING CRITICALLY

4. Give a possible origin of a rock that has large crystals of feldspar and very fine crystals of quartz.
5. What kinds of elements might be present in mafic rocks that make them dark and dense?

SKILL ACTIVITY: Designing Tables

BACKGROUND

As scientists study a particular topic, they gather large amounts of data. In order to make sense of all the information, they often organize the data into a table.

PROCEDURE

When you design a table, you must give the table a title. List all the information or relationships that you wish to record in your table. These are your headings. Headings may be listed either horizontally or vertically in the table.

For example, the activity on page 81 shows a data table for igneous rocks. In the table, the two major characteristics to be recorded are color and texture. Each of these characteristics is divided into smaller groups.

A different format is used for the table in Investigation 4, which is on page 94. One column lists the numbers of the rocks to be studied.

The other columns list categories of information about each rock.

Still another format can be seen in Table 1. This table has illustrations to show what the particles in different types of igneous rocks look like.

APPLICATION

Design a data table that shows the relationship between the degree of metamorphism and the metamorphic rocks formed from shale.

USING WHAT YOU HAVE LEARNED

1. Consider the characteristics of sedimentary rocks that are described in this chapter. Design a table to show the difference in sediment size of six sedimentary rocks.
2. Consider the origin of various metamorphic rocks and design a table to show the parent material of four different metamorphic rocks.

Rock Formed	Temperature of Magma	
	800°C	900–1200°C
Felsic rocks		
Mafic rocks		

TABLE 1: IGNEOUS ROCK FORMATION

NEW SCIENCE TERMS

cementation
clastic sediments
chemical sediments
organic sediments
evaporites
precipitates

SECTION OBJECTIVES

After completing this section, you should be able to:
- **Describe** three types of sedimentary rocks.
- **Trace** the formation of compact limestone.
- **Identify** common sedimentary rocks by their characteristics.

4.6 Formation of Sedimentary Rocks

Even though igneous rocks make up most of the earth's crust, the rocks found on the earth's surface are mostly sedimentary rocks. You are probably unaware that sedimentary rocks are forming all around you. Fragments, or pieces of broken rocks, leaves falling from trees, and streams full of mud are all sources of sediments for new sedimentary rocks.

Recall the model sedimentary rock made by pressing together wax shavings. Fine sediments, such as clay and silt, also stick together if pressure is applied. They easily form solid rocks. Coarse sand and pebbles do not stick together unless something is present to hold them together.

The water that soaks through the earth carries with it natural cements. Some of these natural cements are silica (from quartz), calcium carbonate (from limestone), and iron oxides. They coat sediments, such as sand, and bind them together. This process is called **cementation.** The weight of additional sediments is usually enough to complete the formation of sedimentary rocks.

Natural cementation is really very similar to the process of making concrete. To make concrete for a driveway or sidewalk, you need three things: sand, water, and cement. The cement you

Figure 4–12. The fallen leaves on this pond are one source of sediment for the formation of sedimentary rocks.

Figure 4–13. The natural cement of sedimentary rocks is similar to the cement in the concrete these workers are pouring.

use in making concrete has the same function as the natural cement in sedimentary rocks; it holds the sand together. The water ensures even distribution of sand and cement.

Sand is only one kind of sediment that forms rocks. There are three types of materials that form sedimentary rocks: *clastic, chemical,* and *organic.*

Clastic sediments are pieces of other rocks. These sediments are produced as rocks are exposed to heat, plants, water, and air. For example, some rocks are broken into fragments when water collects in tiny cracks. The water freezes, and the ice expands and fractures the rock. The small rock pieces are then available for the formation of clastic sedimentary rocks.

Chemical sediments come from minerals dissolved in water. The salts in sea water can become chemical sediments. If the sea water evaporates, the chemicals collect on the bottom and form chemical sediments. Compaction turns the sediments into sedimentary rocks.

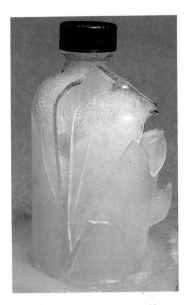

Figure 4–14. As the water in a bottle freezes, it expands and breaks the bottle. Water freezing in the cracks of a rock can break the rock.

Figure 4–15. The salt being loaded into these trucks is a chemical sediment.

Organic sediments are the hard remains of once-living organisms, such as shells or the skeletons of corals. Many seafloor sediments are organic. In some places these remains accumulate in thick layers, forming organic sedimentary rocks.

○ *Name three kinds of sediments.*
○ *What kind of sediments are the remains of once-living organisms?*

Figure 4–16. Organic sedimentary rocks often contain fossils of the organisms from which they were formed.

Figure 4–17. Coquina rock is a sedimentary rock formed from shell fragments.

Figure 4–18. Conglomerate (left), breccia (center), and sandstone (right) are all clastic sedimentary rocks.

4.7 Classification of Sedimentary Rocks

Remember that the classification of igneous rocks is based on whether the rock solidified from magma or lava. Sedimentary rocks are classified by the type of sediments present in the rocks. Therefore, the types of sedimentary rocks that form are the same as the types of sediments.

Clastic Sedimentary Rocks

Clastic sedimentary rocks are formed from fragments of other rocks. Clastic rocks may be further classified by texture.

Coarse rocks are *conglomerates* (kuhn GLAHM uh rayts) and *breccias* (BREHSH ee uhs). Both of these rocks contain pebble-sized pieces with diameters of 2 mm or more. The difference between conglomerates and breccias is the shape of the rock pieces. Conglomerates have rounded pieces and breccias have sharp-edged pieces.

Sandstone is formed of sand-sized grains that are usually between 0.06 mm and 2.0 mm in diameter. The grains are usually quartz.

Clastic rocks formed from smaller sediments are *siltstone* (formed of grains 0.004 mm to 0.06 mm in diameter) and *shale* (formed of particles smaller than 0.004 mm). The minerals found in siltstone are quartz and feldspar, while those found in shale are clays.

Figure 4–19. Shale (left) and siltstone (right) are composed of nearly microscopic particles.

Clastic sedimentary rocks are deposited in layers. For example, in slow-moving water, only clay-sized sediments are carried by the water; larger, sand-sized sediments are deposited. If the water stops flowing altogether, even the clay-sized sediments are deposited. Two different types of sedimentary rocks are formed as a result—sandstone and shale. Each type of rock is in a separate layer.

Clastic sedimentary rocks can be identified by the layers they contain. Other types of sedimentary rocks do not show clearly defined layers.

Chemical Sedimentary Rocks

Chemical sedimentary rocks are made of substances dissolved in water. For chemical rocks to form, the dissolved substances must somehow be separated from the water.

Two chemical sedimentary rocks, evaporites and precipitates, are formed by different processes. **Evaporites** (ih VAP uh rytes) are chemical rocks that form when substances dissolved in water are left as the water evaporates. The two most common evaporites, shown in Figure 4–20, are *rock salt* and *rock gypsum*.

Precipitates (pruh SIHP uh tayts) are chemical rocks that form when dissolved substances precipitate, or fall out of solution. Precipitates are usually deposited on the sea floor. The most common precipitate is *compact limestone*, shown in Figure 4–21, which forms when calcite precipitates from sea water. Silica also precipitates out of sea water, forming some types of *chert*. Chert is a very dense rock formed of quartz.

Figure 4–20. Rock salt (top) and rock gypsum (bottom) are chemical sedimentary rocks.

Figure 4–21. Compact limestone (left) and chert (right) are both precipitated from sea water.

A MATTER OF FACT

Rock salt is mined from huge deposits found in several locations in the United States. For example, the salt beds near Detroit, Michigan, are about 300 m thick. The openings in the salt mines of Louisiana are more than 35 m high.

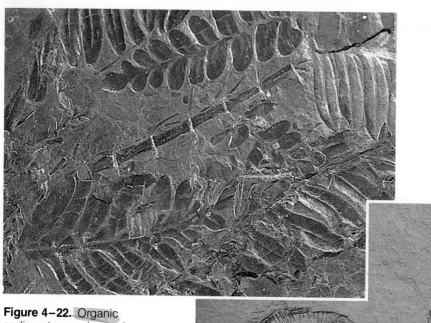

Figure 4–22. Organic sedimentary rocks, such as coal (above), often contain fossils of plants. The organic limestone (right) contains the fossils of tiny fish.

Organic Sedimentary Rocks Organic sedimentary rocks form from the remains of once-living organisms. Although limestone is a chemical rock, it can also be an organic rock. Organic limestone forms from the hard remains of animals such as corals. Organic limestone often contains fossils of animals trapped in the sediments, as shown in Figure 4–22. These fossils provide clues about life during the time the organisms lived.

Organic sedimentary rocks also form from compressed plant material. In a swamp, plants decay very slowly to form *peat*. Over years the buried peat becomes *lignite*, sometimes called brown coal. Brown coal is compacted, but it is not as hard as other coals. With more pressure and time, a harder coal called *bituminous* (bih TOO muh nuhs) *coal* forms. Bituminous coal is used primarily in the production of electricity. Millions of years are required for bituminous coal to be produced from plant material. Many good plant fossils are found in layers of bituminous coal, as Figure 4–22 shows. Coal often provides the only clues about ancient plant life, for soft plant parts do not easily form fossils.

○ *Give an example of an organic rock.*
○ *What are two types of chemical sedimentary rocks?*

4.8 Uses of Sedimentary Rocks

Many sedimentary rocks are used in industrial processes and construction projects. Shale and siltstone are used for manufacturing sewer pipe, drain tile, and chimney linings.

Many buildings are constructed of sandstone. Huge blocks are sawed out of thick sandstone beds. These blocks are then cut into smaller rectangular blocks to meet the size requirements of the construction industry. The White House in Washington, D.C., is built mostly of sandstone, painted white to hide the damage of a fire set by British soldiers during the War of 1812.

Rock gypsum is used by the building industry in the production of drywall and plaster. The quartz in sandstone is the basic mineral component of glass and porcelain. Porcelain—used in fine china, bathtubs, sinks, and toilets—is a combination of clay, feldspar, and quartz. Fiberglass insulation and the silicon chips used in computers are also made from quartz.

○ *Why is the sandstone of the White House painted?*
○ *What is porcelain?*

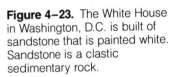

Figure 4–23. The White House in Washington, D.C. is built of sandstone that is painted white. Sandstone is a clastic sedimentary rock.

Section Review

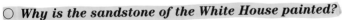

READING CRITICALLY

1. What kind of rocks form from rounded pebble-sized sediments?
2. Why is rock salt considered to be an evaporite?

THINKING CRITICALLY

3. Why would silt be deposited by slow-moving water but not by more rapidly moving water?
4. Why might you expect to find fossils in organic limestone but not in chemical limestone?

BIOGRAPHIES: Then and Now

RENÉ JUST HAÜY (1743–1822)

René Haüy (ah YOO ee) was a French priest who developed an interest in science. While visiting a friend, he accidentally dropped a calcite specimen from his friend's mineral collection. The mineral broke into many small pieces. He was very embarrassed, of course, but as he picked up the mineral fragments, he noticed that all the broken pieces formed the same geometric shape. The shape was that of a rhombohedron, a slanted box. Later he tested more calcite and verified that calcite would always break into rhombohedral shapes.

Haüy then hypothesized that each mineral is built up of layers or sections of a simple geometric shape. He proposed that this geometric shape is related to the internal crystalline form of the mineral. That is, the form is based upon the molecular makeup of the mineral.

For example, crystals of halite are always cubic in shape. Crystals of quartz are always hexagonal (six-sided) with a pyramidal top.

This was the beginning of the science of crystallography. Haüy became a professor of mineralogy at the Museum of Natural History in Paris during the reign of Napoleon. While at the museum, he wrote the first important textbook on crystallography. Haüy's work with crystals has led to the development of many consumer products based on the characteristics of quartz.

RONALD B. PARKER (1932–)

Ronald Parker earned a Ph.D. in geology from the University of California at Berkeley. He has been teaching geology at the University of Wyoming for over seventeen years. Dr. Parker has written many articles and several books on various topics in geology. Through witty stories and simple language, he leads his readers into the complex study of the earth.

Parker has also done much research in geology and has proposed a hypothesis to explain why some geologic events occur as great bursts of energy rather than in gradual steps. He calls this the Buffer Hypothesis. In this hypothesis, Parker argues that many earth features, such as volcanoes, glaciers, mountain slopes, and geysers, store energy as a reservoir stores water. Eventually, so much energy is stored that the system reaches its energy capacity. At that point, the system is so out of balance that the energy is released in the form of a dramatic event such as a landslide, volcanic eruption, or burst of water from a geyser.

Although not proven, this hypothesis may explain why earthquakes and volcanic eruptions are so hard to predict. So far, scientists have not been able to measure the amount of energy that is built up in the earth. If a way could be found to measure this energy buildup, it might be possible to predict these natural disasters.

4 Metamorphic Rocks

SECTION OBJECTIVES

After completing this section, you should be able to:

- **Explain** how metamorphic rocks form.
- **Define** the terms *foliated rocks* and *nonfoliated rocks*.
- **Identify** some common metamorphic rocks by their characteristics.

NEW SCIENCE TERMS

foliated rocks
nonfoliated rocks

4.9 Formation of Metamorphic Rocks

Many animals go through a metamorphosis, or change, in their lives. Frogs, for example, start life as fishlike tadpoles. They gradually change their body form, by growing legs and lungs. Their life style changes also, as they become adults. Instead of living in water, they can live on land. Other animals, such as butterflies and moths, also undergo great changes in form.

Many rocks change physically, as well. Metamorphic rocks form when heat and pressure cause chemical and structural changes in existing rocks. During a structural change, heat and pressure rearrange the minerals within the rock. This change results in a more compact mineral pattern or in a different form of crystal. When a chemical change occurs, elements recombine to form different minerals from those that were in the original rocks.

The pressure that forms metamorphic rocks can come from mountain-building processes. Mountains, which are discussed more fully in Chapter 6, are built up as forces within the earth push and fold huge areas of land. A source of heat is large bodies

Figure 4–24. Heat and pressure can change the minerals in igneous and sedimentary rocks, forming metamorphic rocks.

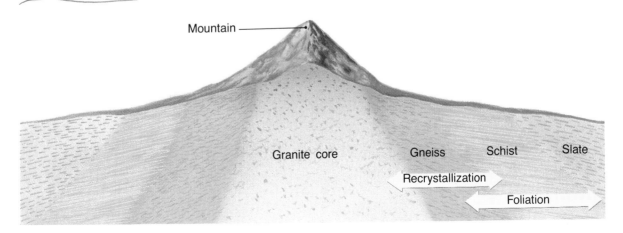

Mountain

Granite core   Gneiss   Schist   Slate

Recrystallization

Foliation

Observing Metamorphic Rocks

Study pieces of marble, slate, and quartzite. What sedimentary rocks might each of these have come from? On what observations did you base your answers?

of magma within the earth. A body of magma can cause changes in rocks for hundreds of kilometers around it.

○ *How might heat and pressure affect the minerals in a rock?*

○ *Name two sources of heat for the formation of metamorphic rocks.*

4.10 Classification of Metamorphic Rocks

Metamorphic rocks are divided into two groups: foliated and nonfoliated. **Foliated rocks** have obvious layers. Foliated rocks are formed by pressure from one direction. The layers often appear shiny, due to the new mineral crystals that have formed. Some foliated rocks show a pattern of bands. The amount of foliation is used to classify metamorphic rocks. For example, *slate, phyllite, schist,* and *gneiss* (NYS) show increasing amounts of foliation. **Nonfoliated rocks** have no visible layers. This is because pressure during formation comes from several directions. *Marble,* for instance, is nonfoliated.

Identifying the minerals in the original rocks helps to classify metamorphic rocks. For example, when limestone is changed to a metamorphic rock, the original limestone becomes more compact. However, the mineral can still be identified as calcite. The resulting rock is marble. Several kinds of metamorphic rocks are shown in Figure 4–25.

Another example of the metamorphic process occurs when bituminous coal is put under intense heat and pressure. This soft coal becomes a harder, more compact form of coal called *anthracite.*

○ *How can foliated rocks be identified?*
○ *What is a nonfoliated rock?*

Figure 4–25. Marble (below), quartzite (bottom left), and slate (bottom right) are all metamorphic rocks.

4.11 Uses of Metamorphic Rocks

Metamorphic rocks have many common uses. Marble, for example, is extensively used as a building material because its surface polishes beautifully, displaying interesting patterns. Marble is also used for monuments and sculptures. The Lincoln Memorial in Washington, D.C., is made of polished marble.

Slate is used as roofing tile in some areas, because it easily breaks into thin slabs and it is waterproof. Slate was once widely used for gravestones. The inscriptions on these gravestones, some over 100 years old, are still readable. What characteristic of slate allows it to last so long?

Anthracite coal is very hard and has few of the sulfur impurities found in bituminous coal. For these reasons, anthracite coal is a very good fuel for heating, but it is also very rare and expensive.

○ *Why is marble used for buildings and monuments?*
○ *What is slate used for?*

Figure 4–26. Because of its beauty and durability, many monuments, such as the Lincoln Memorial, are made of marble.

Section Review

READING CRITICALLY

1. What are metamorphic rocks?
2. What characteristic of a metamorphic rock determines whether it is classified as foliated or nonfoliated?

THINKING CRITICALLY

3. How can the alignment of crystals in a metamorphic rock provide information about the direction in which pressure was applied to the rock?
4. Why are mineral crystals not foliated in marble?

A MATTER OF FACT

Very pure, white marble is found in the mountains around Carrara, Italy. Mining of this marble has been going on for at least 2000 years. The interior of the Pantheon in Rome was built of this marble. It can be found in such different places as the Leaning Tower of Pisa, the pavement of St. Peter's Basilica in Rome, and in the Kennedy Center in Washington, D.C.

INVESTIGATION 4: Classifying Rocks

PURPOSE

To classify igneous, sedimentary, and metamorphic rocks, using a rock classification key

MATERIALS (per group of 3 or 4)

Safety goggles
Laboratory apron
Unlabeled rock specimens
Rock Classification Key, page 568
Hydrochloric acid, 0.5 M
Hand lens

PROCEDURE

1. **CAUTION: Put on your safety goggles and laboratory apron and leave them on for the entire investigation.**
2. Make a table like the one shown. Fill in the chart as you complete the investigation.
3. Select one of the rock specimens. Classify it by using the key on page 568. This key is organized to lead you step by step, eliminating various possible rock types as you move through the key. The first category is sedimentary rocks. If a fresh surface of your rock specimen looks like fragments cemented together, then you should further explore the subdivisions in Category I.

4. If your rock does not look like a sedimentary rock, you should move through the key to Category II.
5. If Category II does not match, you should continue through the key until you find the description that matches your specimen.
6. When you decide the name of the rock, you should fill in all the information in your Rock Identification Table.
7. Follow this same procedure for all the specimens.

ANALYSES AND CONCLUSIONS

1. Which rock did you find to be the most difficult to classify? Explain why it was difficult.
2. Do you think the range of mineral colors would be less useful for classifying rock specimens than for classifying mineral specimens? Explain.
3. Of all the rock specimens discussed in this chapter, which one is the easiest to identify? Why?

APPLICATION

If you owned some property with a lot of rocks, how would you find out if the rocks were limestone and, therefore, good for building?

TABLE 1: ROCK IDENTIFICATION				
Rock	Name	Mineral Content	Rock Family	Key Classification Characteristics
1				
2				
3				
4				
5				
6				
7				
8				

SUMMARY

- Rocks are naturally occurring minerals or combinations of minerals. (4.1)

- Rocks are classified into three families: igneous, sedimentary, and metamorphic. (4.1)

- Rocks are classified by their origin and the types of minerals they contain. (4.1)

- The continuous process of change that affects all rocks is called the rock cycle. (4.2)

- The rock cycle is not one way. Any rock may be changed into any other type of rock. (4.2)

- Igneous rocks are formed as magma or lava cools and crystallizes. (4.3)

- The crystal size in igneous rocks is determined by the rate of cooling. (4.4)

- Many igneous rocks are used in buildings because of their strength. (4.5)

- Sedimentary rocks are formed from the cemented fragments of other rocks. (4.6)

- Sedimentary rocks, which are classified by the way the sediments form, are called clastic, chemical, or organic. (4.7)

- Sedimentary rocks are used in many industrial processes. (4.8)

- Metamorphic rocks are formed when heat or pressure is applied to other rocks. (4.9)

- Metamorphic rocks are classified as foliated or nonfoliated depending on whether or not mineral layers are visible. (4.10)

- Many metamorphic rocks are useful as building materials. (4.11)

Write all answers on a separate sheet of paper.

SCIENCE TERMS

Correctly use each of the following terms in a sentence.

cementation **(84)**
chemical sediments **(85)**
clastic sediments **(85)**
evaporites **(87)**
extrusive rocks **(80)**
foliated rocks **(92)**
igneous rocks **(76)**
intrusive rocks **(79)**
lava **(76)**
magma **(76)**
metamorphic rocks **(77)**
nonfoliated rocks **(92)**
organic sediments **(85)**
precipitates **(87)**
rock cycle **(78)**
rocks **(75)**
sedimentary rocks **(76)**
texture **(79)**

SCIENCE QUIZ

Modified True-False

Mark each statement *true* or *false*. If a statement is false, change the underlined term to make the statement true.

1. Granite is an example of an <u>igneous</u> rock.

2. When heat and pressure are applied to igneous rocks, <u>sedimentary</u> rocks are formed.

3. Rocks are classified by the properties of <u>density</u> and origin.

4. Rock salt is an example of an <u>evaporite.</u>

5. The term <u>foliated</u> refers to the layers of mineral crystals in metamorphic rocks.

6. The three main types of sedimentary rocks are <u>clastic</u>, organic, and chemical.

continues

Multiple Choice

Write the letter of the choice that best answers the question or completes the statement.

7. Granite consists of feldspar, mica, and
   a) quartz.   b) olivine.
   c) shale.   d) magnetite.

8. What is the main difference between felsic and mafic rocks?
   a) the rate of cooling of the molten rock
   b) the degree of metamorphosis
   c) the color of the minerals in the rocks
   d) the depth of magma source

9. Bituminous coal is considered a sedimentary rock because it
   a) is formed from clastic material.
   b) hardens under great heat and pressure.
   c) has been compressed and made more compact.
   d) solidified from a molten body.

10. Which choice lists only metamorphic rocks?
    a) shale, gneiss, slate, schist
    b) slate, shale, granite, gneiss
    c) basalt, schist, slate, phyllite
    d) slate, phyllite, schist, gneiss

11. A rock formed from the cementing of fine clay minerals is
    a) sandstone.   b) shale.
    c) limestone.   d) marble.

12. Which of the following is not a type of clastic sedimentary rock?
    a) limestone   b) breccia
    c) sandstone   d) conglomerate

Completion

Complete each statement by supplying the correct term.

13. Rocks are classified into _____ families.

14. The term clastic refers to particles in _____ rocks.

15. The term texture refers to the overall appearance and feel of a rock due to the _____.

16. The _____ cooling of magma results in the formation of large mineral crystals.

17. When heat and pressure are applied to existing rocks, _____ rocks are formed.

18. Sedimentary rocks formed from the compressed, cemented fragments of shells are called _____.

Short Answer

19. Compare and contrast intrusive and extrusive rocks.

20. Describe how anthracite is formed.

21. Explain how rocks from each of the three rock families differ.

22. Explain why the rock cycle is not a one-way process.

Writing Critically

23. Would it be possible for fossils to be preserved in igneous rocks? Explain your answer.

24. Explain the rock cycle. Include changes that occur to at least three different rocks as they move through the rock cycle.

25. Why do the layers of sediment that are clearly visible in sandstone seem to be entirely absent in quartzite?

EXTENSION

1. Research the process that results in a rock formation called a *pegmatite*. Locate areas in the continental United States where pegmatites are found. Why are pegmatites of economic interest?

2. Find out what type of rocks are in your region of the country. Make a chart to classify the rocks.

3. Go to the library and read about *geodes*. Explain how scientists think geodes are formed and draw a map to show locations where geodes are found.

APPLICATION/CRITICAL THINKING

1. Oil is found in certain rocks. In what type of rocks could oil be found? Explain your answer.

2. Coquina is a rock made of pieces of broken shells. Describe the conditions in the environment that would allow such a rock to form.

3. Stone fireplaces are almost never constructed of sandstone because of the danger of fire spreading beyond the fireplace. What is there about the structure of sandstone that might cause sandstone fireplaces to be dangerous?

FOR FURTHER READING

MacFall, R. *Rock Hunters' Guide.* New York: Crowell, 1980. This book is a good introductory guide to finding and identifying collectible rocks as a hobby.

McGowen, T. *Album of Rocks and Minerals.* New York: Rand McNally, 1981. This book explains that rocks are more interesting than you might think. Their uses range from the practical to the destructive.

Selsam, M. and J. Hunt. *A First Look at Rocks.* New York: Walker and Company, 1984. This book relates some of the pleasures of looking for and at rocks.

Weiner, J. *Planet Earth.* New York: Bantam, 1986. This book provides the reader with many facts about the earth, including resources of the earth.

Challenge Your Thinking

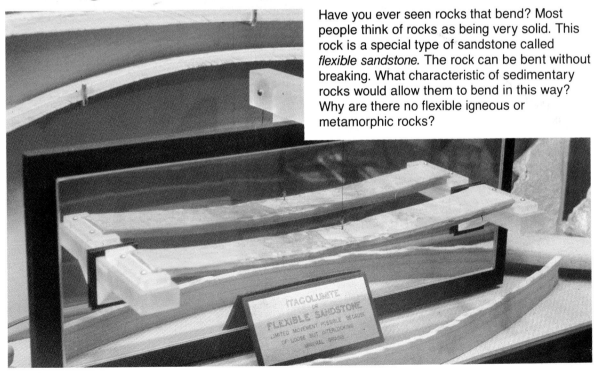

Have you ever seen rocks that bend? Most people think of rocks as being very solid. This rock is a special type of sandstone called *flexible sandstone.* The rock can be bent without breaking. What characteristic of sedimentary rocks would allow them to bend in this way? Why are there no flexible igneous or metamorphic rocks?

CHAPTER 5

Earth's Resources

What does this photograph show? You may think that this is an aerial view of a modern city. However, it is really a photograph of a computer chip from a personal computer. The chip is made from quartz. Quartz, a resource from the earth, has many unique properties that allow it to be used in a number of important industries.

A computer chip

1 Metals and Nonmetals

SECTION OBJECTIVES

After completing this section, you should be able to:

- **List** the metals used by early humans.
- **Discuss** the process of smelting metals.
- **Explain** how iron and steel are produced.
- **Describe** the uses of quartz.

NEW SCIENCE TERMS

ores
smelting
alloy
forge

5.1 Metals

You probably know that gold is one of the most valuable metals in the world. Part of its value is due to the fact that gold occurs as a native element—that is, it is not combined with other elements. Occasionally silver will occur as a native element, but it is found mainly as a compound with sulfur or chlorine. Most metals occur combined with other substances in economically important mineral compounds called **ores.**

Since gold occurs as a native element, it was probably the first metal used by early humans. Gold is soft and easily worked, so it is likely that it was used for making ornaments and jewelry. Gold has also been formed into coins and used as money since about 2000 B.C.

Copper, like silver, sometimes occurs as a native element also. Native copper was used 10 000 years ago to make ornaments, jewelry, and—since it is stronger than gold and silver—simple cutting tools. Unfortunately, copper is not often found as a native element. However, humans soon found that heating copper ore in air produced copper metal. The separation of metals from their ores by the use of heat is called **smelting.**

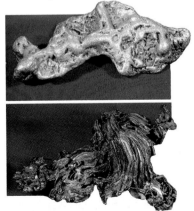

Figure 5–1. These coins (left) were made from native elements, such as gold and silver (below).

Figure 5–2. Copper (below) is separated from its ore in huge smelters (right).

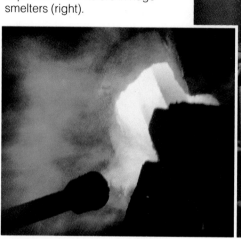

About 3000 B.C. it was discovered that copper could be made stronger by adding tin. This new material was an alloy called *bronze*. An **alloy** can be either a mixture of two or more metals or a mixture of metals and nonmetals.

For 2000 years bronze was commonly used to make knives, swords, helmets, shields, mirrors, combs, and measuring tools. Bronze was so important that this period of history is called the *Bronze Age*.

○ *What is an ore?*
○ *How is copper smelted from its ore?*

Figure 5–3. Shown here are some tools of the Bronze Age.

SKILL ACTIVITY: Writing a Report

BACKGROUND

Humans first began using metals over 10 000 years ago. From copper and bronze to iron and steel, civilization has progressed through the use of metal alloys.

In the 19th Century, an English engineer, Henry Bessemer (1813–1898), invented an inexpensive way of making steel from iron. The method involved blowing air through the molten metal in a container called a Bessemer converter. This was done in order to remove impurities from the molten iron.

PROCEDURE

1. Use the card catalog of your school or local library to research the steel making process of Henry Bessemer. The words underlined in the background paragraphs should help you to locate the information you need.

2. After locating several sources of information, make notes on the process on index cards or in a notebook. Include sketches of the Bessemer converter in your notes.

3. Write a two-page report, complete with diagrams, that describes the Bessemer process of making steel. Be sure to include several paragraphs on how the process transforms iron into steel. You might also want to include some information about Henry Bessemer, and mention some of the advantages that steel has when compared to iron.

USING WHAT YOU HAVE LEARNED

After you have finished reading the chapter, pick another topic, perhaps the formation of coal. Use the card catalog to locate library sources, and write another report, including diagrams, on your chosen topic.

Figure 5–4. Iron for these tools (right) was made in a forge and then hammered into shape by a blacksmith (above).

Figure 5–5. Steel, shown here being poured into a mold, has replaced iron as the most important metal in our society.

5.2 Iron and Steel

About 1100 B.C. a process for smelting iron was developed in Persia and Eastern Europe. The process for smelting iron caused a dramatic change in the ways in which humans worked and defended themselves.

Iron occurs most often as the iron oxides hematite (Fe_2O_3) and magnetite (Fe_3O_4). Producing metallic iron from these minerals is not as easy as producing copper from its ore.

To make iron, a forge is needed. A **forge** is a furnace through which a stream of air is forced. This produces a very hot flame—hot enough to force the iron oxide to react with the carbon in the charcoal or coal that is used for fuel. The reaction in the forge produces carbon dioxide and *bloom*, a spongy, gray mixture of metallic iron and iron carbide (Fe_3C). Blacksmiths, as iron workers were called, found that hammering red-hot bloom would remove most of the iron carbide, leaving fairly pure iron.

Blacksmiths also discovered that hot iron plunged into cold water hardened rapidly. The hardened iron would maintain a sharp edge longer than any known metal, even bronze. Good iron was so difficult to make that an iron sword or plow was considered a valuable possession. The Bronze Age had been replaced by the *Iron Age*.

Today most iron is made into steel. Although iron can be hardened, it remains brittle and can be broken easily. Iron is also hard to shape except when it is very hot. Steel, an alloy made of iron and about one percent carbon, is as hard as iron, but it is less brittle and can be shaped when cold.

Figure 5–6. Stainless steel has many uses because of its resistance to oxidation. The body of this sports car should last indefinitely.

You may recall from Chapter 3 that there are many metals in the center of the periodic table. Some of these metals, such as chromium, manganese, cobalt, and molybdenum, are added to steel. The addition of each of these metals improves certain, specific qualities of steel, producing alloys called *specialty steels*. One of the most important metals used in the production of specialty steels is nickel. Nickel occurs in igneous rocks and in some soils. Nickel gives steel added strength and improved electrical, thermal, and antirust properties.

The United States is the world's largest producer of specialty steels and the largest consumer of the metals needed for specialty steels. However, only iron and nickel occur in significant quantities in North America; the other metals must be imported from around the world.

○ *What is an alloy?*
○ *What two metals are alloyed most often in the production of steel?*

A MATTER OF FACT

Tin neither rusts nor corrodes; therefore, it is ideal for coating the inside of steel food cans. Over 50 billion "tin cans" are made each year.

5.3 Quartz

Quartz, one of the most common nonmetallic minerals, is also one of the most important to the modern world. Quartz is used extensively in the production of common glass, fine crystal, high-temperature glass, precision optical glass, and electronic circuits.

Figure 5–7. Glass, manufactured from quartz sand, can be made into many products.

Figure 5–8. Although most glass products are now made in molds, the art of glassblowing is still practiced.

Have you ever seen anyone blowing glass as the person in Figure 5–8 is doing? At one time all glass was made this way. Today common glass is manufactured by heating quartz sand, sodium carbonate (Na_2CO_3), and calcium carbonate ($CaCO_3$) to a temperature of nearly 1000°C. At this temperature the minerals fuse, and molten glass is produced. The soft glass is removed from the furnace and poured into molds to produce the desired shape. Some fine crystal is still blown by hand.

TABLE 5–1: MATERIALS USED TO MAKE COLORED GLASS

Addition to Glass	Resulting Color
Gold	Red
Selenium	Red
Ferric oxide	Brown
Silver	Yellow
Uranium oxide	Yellow
Ferrous oxide	Green
Cobalt oxide	Blue
Manganese oxide	Violet

Specialty glass is similar to specialty steel in that various elements and compounds are added to the molten glass to give it special properties. Table 5–1 shows the colors that result from the addition of metals and minerals to molten glass.

Most glass contains impurities. Glass made of pure silica (SiO_2) has a very high melting point (1713°C) and is used only for special purposes that require high temperatures.

Just as sugar can be spun into fibers of cotton candy, glass can be spun into *fiberglass*. In a loose condition, fiberglass can be used as a filtering material or as insulation for buildings. Mixed with chemicals called *resins*, fiberglass is used wherever a strong, lightweight substitute for metal is needed, such as in boat hulls or sports car bodies.

Figure 5–9. Many specialty products, such as high-temperature glass (left) and fiberglass (right), can also be produced from quartz glass.

You have probably seen advertisements for quartz watches; in fact, you may even have one. You know that the watch is not made of quartz, but do you know why it is called a quartz watch? The reason is that these watches are controlled by a crystal of quartz. Quartz has a unique property that allows it to vibrate at a constant frequency, making a quartz watch very accurate.

Crystals of quartz can transform vibrations into electrical voltages, or conversely, transform electrical voltages into vibrations. Quartz has the property of *piezoelectricity* (pee AY zoh ih lehk TRIS uh tee)—that is, it can produce voltages. When pressure is applied to a quartz crystal, positive and negative charges develop on opposite sides of the crystal. If the pressure is alternately applied and relaxed, a current of electricity flows.

If subjected to a vibrating electric field, a quartz crystal will vibrate at a constant frequency. A typical quartz crystal may vibrate at a constant rate of 100 000 Hz, resulting in an extremely accurate timing device. Quartz watches work on this principle.

Silicon chips, made from quartz, have special properties that make them useful in electronic components. Miniature electronic circuits can be printed on them in much the same way that a diagram is printed on paper. This property makes silicon chips important in the production of microcircuits for computers and many other electronic devices.

○ *State five uses of quartz.*
○ *What is piezoelectricity?*

Figure 5–10. A quartz watch, such as the one shown here, is very accurate because of the special properties of the quartz crystal it contains.

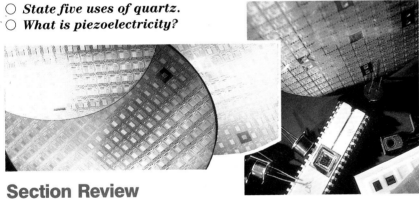

Figure 5–11. Most electronic products can now be made more economically because of silicon chips such as these.

Section Review

READING CRITICALLY

1. Why is iron more difficult to smelt than copper?
2. How is glass made?

THINKING CRITICALLY

3. Why was the production of bronze important to the development of civilization?
4. If quartz were as rare and expensive as gold, how would the electronics industry be affected?

All computers rely on circuits to control the flow of electric current. In the late 1950s, a computer chip could hold only one circuit. Today the number of circuits molded onto a single chip can be greater than 1 million.

A printed circuit

Computer chips are made from a thin slice of silicon called a *wafer.* A single wafer may be used to make hundreds of separate computer chips.

Why is silicon used to make these chips? Silicon has several advantages that make it ideal for use in microcircuitry. For example, it is a semiconductor. That is, sometimes it can conduct electricity and sometimes it cannot. Silicon's properties as a semiconductor remain intact over a wide range of temperatures. Since silicon chips have a natural resistance to heat, they are more efficient semiconductors than other types of chips.

In most atoms the electrons in the outer energy level are ready to react with other atoms. In silicon, however, the outer electrons are bound tightly to each other. This situation can be changed by inserting materials that can "donate" or "steal" electrons from silicon atoms. The insertion of impurities can change the electrical conductivity of the chip. Doing this to the chip is called *doping the chip.* After doping, the flow of electric current in the wafer can be precisely controlled.

A data processing center

The manufacture of computer chips requires great precision. A procedure called *optical lithography* is used to "print" computer chips in much the same way that photographic plates are made for printing magazines. Computer chips are made under microscopes, since the circuits are microscopic in size.

With improved manufacturing technology, the size of computer chips continues to get smaller and smaller. In the 1960s the smallest circuits were about 25 micrometers across. During the 1980s, that size has been reduced to 2.5 micrometers. Some scientists predict that by the year 2000, the size could be 0.25 micrometers.

Of course, these advances in miniaturization will depend on the chemical and structural characteristics of the raw materials used and the photographic processes and materials used to make the chips.

A personal computer

SECTION
2 Fossil Fuels

SECTION OBJECTIVES

After completing this section, you should be able to:
- **Describe** the chemical structure of simple hydrocarbons.
- **Explain** how hydrocarbons form.
- **List** the stages of coal development.

NEW SCIENCE TERM

hydrocarbons

5.4 Hydrocarbons

What do you think is the single most important resource in the world? Do you think it might be gold or diamonds or iron? Would it help you to know that in the mid-1970s the United States' economy suffered severely as a result of less than a 20 percent reduction in the availability of this resource? You have probably guessed by now that the resource that is so important to the modern way of life is oil. No other resource is so directly involved in the way we work and play.

Oil belongs to a group of compounds known as hydrocarbons. **Hydrocarbons** are compounds containing only hydrogen and carbon. The simplest hydrocarbon, methane (CH_4), consists of four hydrogen atoms bonded to a single carbon atom.

Natural hydrocarbons are fossils that were once part of living organisms. Most hydrocarbons burn easily, so they are often used as sources of energy. For these reasons, hydrocarbons are called *fossil fuels.*

○ *What are hydrocarbons?*
○ *What are fossil fuels?*

Figure 5–12. In the mid-1970s, many oil refineries (left) had to close because of a lack of crude oil. This created a shortage of gasoline that resulted in many problems for motorists (above).

5.5 Oil and Gas

A liquid mixture of hydrocarbons is called *oil* or *petroleum*, while a mixture of gaseous hydrocarbons is called *natural gas*. Solid hydrocarbons are *asphalts*.

Much oil and gas is found in rocks that are less than 100 million years old, and in rocks between 200 and 300 million years old. However, under favorable conditions, the physical and chemical processes that change organic matter into hydrocarbons may take as little as 1 million years.

Petroleum and natural gas form in fine-grained, sedimentary rocks called *source rocks*. Under the pressure of overlying rocks, the oil and gas are squeezed out of their source rocks and move up through whatever space exists. If the oil or gas reaches the surface, it is oxidized and lost. However, if on its way to the surface it becomes trapped beneath a layer of impermeable rock, the oil or gas forms a deposit.

Figure 5–13. As this diagram (right) shows, oil is produced in sedimentary deposits. To remove the oil, it is often necessary to drill into those deposits (above).

Figure 5–14. Oil under pressure often produces a gusher when the pressure is suddenly released.

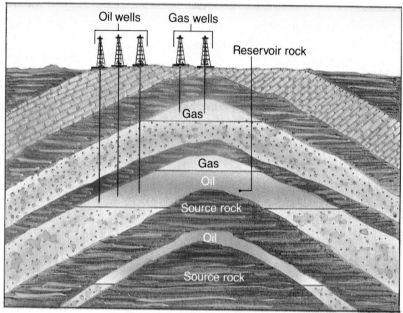

The rocks that contain large deposits of oil and gas are called *reservoir rocks*. Nearly 60 percent of the world's petroleum is found in sandstone reservoirs. The remainder is in limestone and other rocks. Petroleum is obtained by drilling into reservoir rocks and pumping out the oil; gas rushes out by itself.

Sometimes the oil is not concentrated enough to be pumped out directly, so water or steam is forced into the reservoir rock through a nearby well. On rare occasions the oil will be under such tremendous pressure that it rushes to the surface by itself, forming a gusher. Oil gushers waste much valuable oil, so modern wells are capped to prevent this from happening.

North America has large deposits of petroleum and natural gas along the Gulf and California coasts and on the north slope of Alaska and Canada. In spite of this, the United States still imports about 20 percent of its oil from the Middle East, South America, and Africa.

Many petroleum and natural gas deposits are found offshore, deep under ocean water. To drill at these locations, derricks are mounted on platforms, which are set on the sea floor. The largest platform—over 300 m tall—operates in the North Sea, between Britain and Scandinavia.

○ *What are source rocks and reservoir rocks?*

Figure 5–15. Oil deposited under the oceans requires the construction of special drilling platforms. Some of these platforms are constructed on the ocean bottom (above), while others are constructed on land, then towed to the drilling site (left).

ACTIVITY: Determining the Oil-Bearing Capacity of Rocks

How can you determine the oil-bearing capacity of different rocks?

MATERIALS (per group of 3 or 4)

rock samples (sandstone, limestone, and shale); plastic Petri dishes (3); mineral oil; medicine dropper

PROCEDURE

1. Place one of the rock samples into a Petri dish.
2. Repeat the procedure with the other samples, placing each sample in a separate dish. Be sure that the grain sizes of each sample are noted.
3. Fill the medicine dropper with mineral oil and place a drop of oil on top of each sample.

4. Observe and record the time required for the oil to be soaked up by each sample.

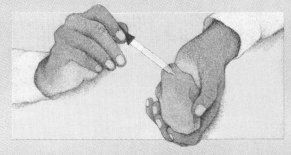

CONCLUSIONS/APPLICATIONS

1. Which sample soaked up the oil the fastest?
2. Do you think oil deposits can be found in most any rock? If not, which rocks could be oil reservoirs?

Section 2   Fossil Fuels   **109**

5.6 Coal

Although the United States must import some percentage of the petroleum it uses, it has vast supplies of coal. Coal is no longer used much as a heating fuel, but it is very important for the production of electricity. Carbon from coal is also used in the steel industry.

Coal forms from decayed plants in areas that were once swamps. Fresh plant matter is about 80 percent water and 20 percent carbon. When plants die, they become waterlogged and sink to the bottom. In the first stage of coal development, the activity of bacteria and fungi transforms the plants into peat. Peat is a dark brown, mushy substance in which the carbon content is about 60 percent. Peat is now forming in many swamps, including the Dismal Swamp of Virginia and North Carolina and the Okefenokee Swamp of southeastern Georgia.

Figure 5–16. Coal is used extensively for producing electricity (right). In many places, such as the Okefenokee Swamp in Georgia (above), coal is still being formed. However, reserves of coal are being used much faster than new deposits are forming.

DISCOVER

Conserving Fossil Fuels

Ask your parents or guardians to carefully record the distance they drive and the amount of gasoline they use in a week. Determine how much fuel could be saved in a year if they cut their driving by 10 percent.

The second step in coal formation occurs as peat is buried under sediments and layers of additional peat. The pressure increases and the temperature rises. Water is expelled from the peat, and any oxygen present in the organic matter combines with carbon to form carbon dioxide. Next, the peat gradually becomes lignite, which is nearly 70 percent carbon. If the process continues, lignite becomes subbituminous coal, then bituminous coal, and eventually anthracite. Anthracite is over 90 percent carbon.

The major coal fields of the United States are in the Appalachian Mountains, extending from Pennsylvania to Alabama, and under the prairies of the Midwest. Coal is also found under the western plains from New Mexico to North Dakota.

Figure 5–17. In some areas, coal is mined by stripping away the overlying rocks and soil (left). In other areas, coal is removed from deep underground mines (right).

Coal beds range in thickness from a few centimeters to nearly 50 m in the Powder River Basin of Wyoming. Most Eastern coal is deep-mined from tunnels dug directly into the coal seams or stripped from the ground by removing the overlying rock layers. Western coal is almost exclusively stripped.

○ *What is peat?*
○ *Where are the major coal fields of the United States?*

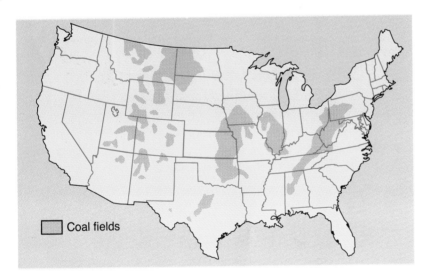

Coal fields

Figure 5–18. This map shows the location of major deposits of coal in the United States.

Section Review

READING CRITICALLY

1. How are hydrocarbons formed?
2. Why is coal important to the steel industry?

THINKING CRITICALLY

3. What are the dangers of the United States being dependent on imported oil?
4. How might the use of coal be expanded to take advantage of the vast reserves in the United States?

INVESTIGATION 5: Crystal Formation

PURPOSE

To observe how crystals form

MATERIALS (per group of 3 or 4)

Cotton string (30 cm)
Petri dishes (3)
Water
Hot plate
Copper sulfate
Ferrous sulfate
Sodium chloride
Beakers, 100 mL (3)
Glass stirring rod
Ice-filled tray

PROCEDURE

1. Cut the string into three 10-cm lengths and lay each piece of string across a Petri dish.
2. Prepare a saturated solution of copper sulfate by heating water in the beaker on a hot plate and slowly adding copper sulfate until no more will dissolve.
3. Soak the string from the first petri dish with the copper sulfate solution.
4. Prepare saturated solutions of ferrous sulfate and sodium chloride in the same manner and soak the other strings in those solutions.
5. As the water evaporates from each string, observe the formation of crystals on the string.
6. When the strings are nearly dry, place the beakers into the ice-filled tray and then dip each string into the beaker containing its dissolved salt. Observe the results.

ANALYSES AND CONCLUSIONS

1. What happened when the "crystal-covered strings" were placed into the cooling solution of the same salt? Explain why this occurred.
2. Can you devise a method of making rock candy using a similar procedure? Explain exactly what you would do.

APPLICATION

How might a knowledge of the way that crystals grow help to produce large, pure quartz crystals for the manufacture of silicon chips?

5 REVIEW

SUMMARY

- Gold was probably the first metal used by humans. (5.1)

- Copper was the first metal to be smelted from its ore. (5.1)

- Bronze is an alloy of copper and tin. (5.1)

- Iron is obtained by smelting iron ore in a forge. (5.2)

- Steel is an alloy of iron and one percent carbon. Nickel and other metals may be added to produce specialty steels. (5.2)

- Quartz, one of the most plentiful minerals, has many uses in the modern world. (5.3)

- Colored glass is made by adding small amounts of specific metals or compounds to the molten glass. (5.3)

- The property of piezoelectricity makes quartz important in the electronics industry. (5.3)

- Hydrocarbons are chemical compounds of carbon and hydrogen. (5.4)

- Most hydrocarbons burn easily. Since they are good sources of energy, they are known as fossil fuels. (5.4)

- Petroleum is a mixture of liquid hydrocarbons; natural gas is a mixture of gaseous hydrocarbons; and asphalts are solid hydrocarbons. (5.5)

- Oil and gas are produced in rock layers called *source rocks*. (5.5)

- A drill is used to release oil and gas from reservoir rocks. (5.5)

- North America has large deposits of hydrocarbons along the Pacific and Gulf coasts, and on the north slope of Alaska and Canada. (5.5)

- Coal forms from plant matter that accumulates at the bottom of swamps. (5.6)

- The United States has many deposits of coal in the eastern mountains and the western plains. (5.6)

Write all answers on a separate sheet of paper.

SCIENCE TERMS

Correctly use each of the following terms in a sentence.

alloy **(100)**
forge **(102)**
hydrocarbons **(107)**
ores **(99)**
smelting **(99)**

SCIENCE QUIZ

Modified True-False

Mark each statement *true* or *false*. If a statement is false, change the underlined term to make the statement true.

1. One of the first metals used by humans was <u>gold</u>.

2. Native <u>copper</u> was used 10 000 years ago.

3. Steel is an <u>ore</u>.

4. The process of obtaining metals from ore is called <u>forging</u>.

5. <u>Petroleum</u> is a liquid, natural hydrocarbon.

6. <u>Coal</u> is a solid fossil fuel.

Multiple Choice

Write the letter of the choice that best answers the question or completes the statement.

7. Early metal tools were probably not made of
   a) gold.　　　　　　b) copper.
   c) bronze.　　　　　d) iron.

8. Early blacksmiths hammered bloom into
   a) iron.　　　　　　b) steel.
   c) iron carbide.　　d) lead.

9. Some steel alloys contain
   a) nickel.　　　　　b) iron oxide.
   c) bloom.　　　　　d) carbon dioxide.

continues

10. Which material can transform electrical voltage charges into vibrations?
    a) iron ore
    b) quartz
    c) steel
    d) copper

11. What are the elements in hydrocarbons?
    a) carbon and oxygen
    b) hydrogen and oxygen
    c) carbon and hydrogen
    d) methane and quartz

12. Hydrocarbons include all of the following except
    a) coal.
    b) oil.
    c) gas.
    d) asphalts.

Completion

Complete each statement by supplying the correct term or phrase.

13. Sedimentary rocks in which hydrocarbons form are called _____.

14. Combinations of either two or more metals or of metals and nonmetals are called _____.

15. Nickel is found in igneous rocks and some _____.

16. The _____ process is used to remove metals from ores.

17. Peat will become _____, and then coal.

18. Computer chips are made of silicon, which can be obtained from the mineral _____.

Short Answer

19. What is the procedure for making iron in a forge?

20. Describe the process for obtaining oil from reservoir rocks.

21. Which areas of the United States have large deposits of coal?

22. What must be added to molten glass to make it red? To make it yellow? To make it blue? To make it brown? To make it green?

Writing Critically

23. Explain how the production of specialty steels could be interrupted by international disagreements.

24. In what ways are the making of specialty glasses similar to the making of specialty steels?

25. Why are peat, lignite, and coal considered fossil fuels, but not hydrocarbons?

EXTENSION

1. Write a report that traces the evolution of the transistor from vacuum tube to silicon chip. Be sure to describe the materials that were, and currently are, used in the manufacture of each device.

2. Research and report to the class on the technological and environmental problems associated with the surface mining of coal.

3. Construct a model of an oil rig from simple materials, such as popsicle sticks and plastic straws, and then use your model to explain the drilling process to your class.

APPLICATION/CRITICAL THINKING

1. During the time of Napoleon, the element aluminum was considered more valuable than gold. Today aluminum is considerably less expensive than gold. What advances might have led to the decline in the value of aluminum?

2. How would your life be different if all fossil fuels suddenly disappeared?

3. This photograph shows one result of the energy crisis of the mid-1970s. What other changes resulted from this drive to conserve energy?

FOR FURTHER READING

Bailey, D., and L. Castoro. *Careers in Computers.* New York: Julian Messner, 1985. If you are interested in a career working with computers, this book is a good guide to the various opportunities available in the field. It describes types of computer-related jobs in industry, business, government, education, and research.

Bass, G. "Splendors of the Bronze Age." *National Geographic* 172 (September 1987): 693. This article describes improvements in underwater research equipment which is expanding the science of marine archaeology. The article includes illustrations and photographs of the exploration of ancient sea-going vessels and their Bronze Age treasures.

Bell, P. and D. Wright. *Rocks and Minerals.* New York: Macmillan, 1985. This field book contains colorful photos and illustrations accompanied by fact-filled descriptions of the earth's many mineral resources.

Cotterill, R. *The Cambridge Guide to the Material World.* New York: Cambridge University Press, 1985. This is a beautifully illustrated and comprehensive description of many different types of materials, including metals, polymers, ceramics, and glass, as well as all the naturally occurring minerals.

Ward, F. "Jade: Stone of Heaven." *National Geographic* 172 (September 1987): 282. This article gives a fascinating history of jade, one of civilization's most prized precious stones.

Challenge Your Thinking

This is a photograph of one of the first electronic computers. This computer filled an entire room, cost hundreds of thousands of dollars, used a tremendous amount of electricity, and constantly broke down because of the heat it produced. Yet it had less computing power than a modern pocket computer. What changes would there be in your life style if this tremendous miniaturization in electronics had not taken place?

Structural Geology

How was this beautiful landform created? What forces could have caused these rocks to be positioned like this? The formation of geologic features that may have taken millions of years to develop is difficult to imagine. Often only small pieces of landforms remain as evidence of past processes. An even greater challenge is trying to understand the internal processes of the earth.

Folded layers of rock

1 Inside the Earth

SECTION OBJECTIVES

After completing this section, you should be able to:

- **Identify** the layers of the earth.
- **Locate** the major discontinuities between layers.
- **Explain** how convection currents may cause movements of the earth's crust.

NEW SCIENCE TERMS

crust
mantle
core
lithosphere
asthenosphere
convection

6.1 The Interior of the Earth

Scientists know much about the surface of the earth, but no one has been able to take samples of the center of the earth for study. Using current technology, geologists can drill only about 13 km into the earth; the center of the earth is about 6400 km below the surface. However, information about the earth's center is available from the study of earthquakes. This information has helped scientists determine the structure and composition of the earth's interior.

The Crust   The earth is divided into three layers: the *crust,* the *mantle,* and the *core.* The **crust** is the thin, rocky layer on the surface of the earth.

Like the crust of a pie, the earth's crust is not the same thickness throughout. Under the oceans the crust is only about 7 km thick, and it is composed of dense mafic rock such as basalt. A thin layer of sediments lies on top of the basalt. The continental crust, which is much thicker than the oceanic crust, ranges from 25 km at the edges of the oceans to 70 km near the center of the

Figure 6–1. The continental crust, composed mostly of granite and sedimentary rocks, is much lighter and thicker than the oceanic crust, which is composed mostly of basalt and sediments. Therefore, the continental crust sits higher than the oceanic crust.

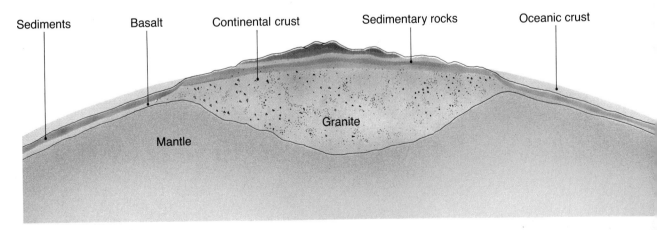

Sediments   Basalt   Continental crust   Sedimentary rocks   Oceanic crust

Granite

Mantle

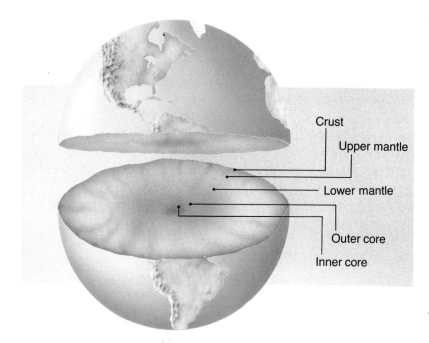

Crust

Upper mantle

Lower mantle

Outer core

Inner core

Figure 6–2. If the earth could be cut open, you would see several distinct layers. Some of the layers are solid, while others are pliable or even liquid.

continents. Continents are made of various igneous, sedimentary, and metamorphic rocks. Because of this mix of rock types, the continental crust is less dense than the oceanic crust. The continental crust is also much older than the oceanic crust. Some continental rocks may be more than 20 times older than the oldest rocks on the ocean floor.

There is a definite boundary to the earth's crust. Andrija Mohorovičić (ahn DREE jah moh hoh ROH vuh chihch), a Yugoslavian geologist, discovered the boundary between the crust and the rest of the interior by analyzing earthquake data. Mohorovičić noted that the waves produced by earthquakes did not pass directly through the earth. Instead, the waves were deflected at a point about 40 km below the surface. This boundary, or discontinuity, is called the *Mohorovičić discontinuity*, or *Moho*, in his honor.

The Mantle   The Moho is about 7 km below the ocean floor and, on the average, about 35 km below the surface of the continents. The Moho separates the crust from the next layer. The **mantle** is the layer immediately beneath the Moho. The mantle, which is about 2900 km thick, is divided into two sections. The upper mantle and the lower mantle are separated by an unnamed boundary. This boundary marks the place where the rocks of the mantle become soft and pliable. The upper mantle is composed of rocks even more dense than the rocks of the oceanic crust. The lower mantle is made of similar rocks, but due to the great pressure, the rocks are even denser.

The Core   Below the mantle lies the innermost layer of the earth, called the **core**. The core is separated from the mantle by a boundary called the *Gutenberg discontinuity.* This boundary is about 2900 km beneath the earth's surface.

Like the mantle, the core is also divided into two parts. The outer core is liquid and probably consists of iron. The inner core is solid and is probably composed of iron and nickel. Scientists believe the liquid outer core may be responsible for producing the magnetic fields of the earth.

The extreme pressure at this depth, due to the weight of the overlying rocks, probably keeps the inner core solid. Most elements expand when they melt, but because the pressure of the overlying layers is so great, the elements cannot expand or melt.

○ *Name the layers of the earth's interior.*
○ *What is a discontinuity?*

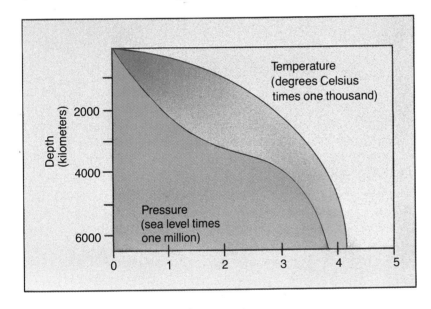

Figure 6–3. Temperature and pressure increase as you go deeper into the earth. At one point in the mantle, there is enough heat to overcome the pressure and to partially melt some of the rocks.

6.2 Movement Within the Mantle

The layers of the earth may also be grouped according to their consistency. The solid **lithosphere** (LIHTH uh sfihr), composed of crust and upper mantle, floats on the asthenosphere (as THEHN uh sfihr).

The **asthenosphere** is a 100-km-thick layer near the top of the mantle. In the asthenosphere high temperature and pressure combine to cause a small amount of melting of the rocks. This makes the asthenosphere a bit softer than the rest of the mantle.

Some of the heat still lingers from the earth's formation. However, scientists believe that most of the heat comes from the decay of radioactive elements. Radioactive decay is discussed in Chapter 15.

Figure 6–4. The lithosphere, composed of the crust and upper mantle, floats on the plasticlike asthenosphere.

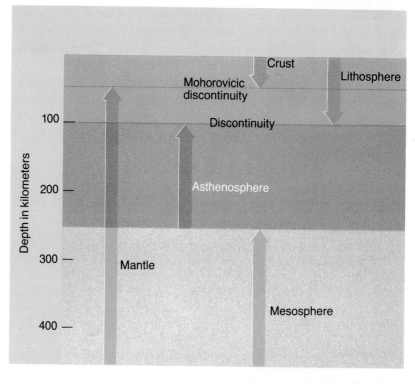

When rocks are heated, they become less dense and tend to rise through the mantle. Currents within the mantle carry warmer rocks up and cooler rocks down. These currents are very slow, moving the rocks only a few centimeters per year. The transfer of heat by the circulation or movement of solids, liquids, or gases is called **convection**. These currents create pockets of circulation, known as *convection cells,* throughout the mantle. Convection cells can be observed in a laboratory by carefully heating a thick, colored liquid in a beaker. However, movement within the earth is far more complicated than movement in a laboratory beaker. Convection cells in the earth may have eddies, or currents within currents.

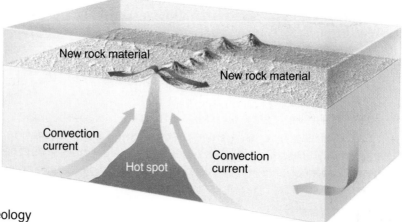

Figure 6–5. Convection cells bring molten rock material to the earth's surface where new rock is added to the crust.

Since the lithosphere is floating on the asthenosphere, the movements within convection cells influence the entire lithosphere. There is much evidence that the lithosphere does not remain stationary. In fact, some evidence suggests that entire continents may be drifting on these convection currents.

○ *What are convection cells?*
○ *Where does the heat to produce convection cells come from?*

Figure 6–6. These photographs show evidence of movement of the earth's crust. For example, the Himalaya Mountains (right) grow several centimeters each year, the rift valley in Iceland (upper left) continues to expand, and these cracks in the earth continue to widen (lower left).

Section Review

READING CRITICALLY

1. Diagram the layers and discontinuities of the earth's interior.
2. Explain the difference between the lithosphere and the asthenosphere.

THINKING CRITICALLY

3. How might the study of earthquakes have been used to gain knowledge about the earth's interior?
4. Explain how convection cells could cause movement of the lithosphere.

② Crustal Adjustments

NEW SCIENCE TERMS

fold
orogeny
anticline
syncline
dip
strike
fault
tension
compression
shear
continental drift

SECTION OBJECTIVES

After completing this section, you should be able to:

- **Describe** the process that causes the crust to fold.
- **Classify** faults.
- **List** four kinds of evidence that support the theory of continental drift.

6.3 Folds

Studying a map may give you the idea that the earth's surface is fixed, and cannot change. However, because of forces within the earth, the surface is slowly but constantly changing. Look at Figure 6–7. This is not a trick; these rocks really are positioned like this. Although sediments are deposited in horizontal layers, sedimentary rocks are not always horizontal. Some rock layers may even be vertical. Great forces within the earth have pushed these rocks into this position.

Sometimes these forces build up slowly, causing rocks to warp or fold. A **fold** is a bend in rocks caused by force. You can make a fold by pushing gently on one end of a sheet of paper while holding the other end in place. What happens to the paper? In the same way, slow, continuous force can build mountain ranges. This process of mountain building is called **orogeny** (oh RAHJ uh nee).

Figure 6–7. Forces within the earth can cause rocks, such as those shown here, to bend or break.

Mountain ranges, such as the Rocky Mountains, that are formed in this way are called *orogenic belts.* In addition, large areas can be lifted without breaking the rocks in the formation. Areas formed in this way are called *uplifts.* The Colorado Plateau, shown in Figure 6–8, is an example of a large-scale uplift.

Two factors that cause rock layers to bend are pressure and temperature. On or near the earth's surface, rocks fold if steady pressure is applied over a long time. Deep within the earth, high temperatures change rocks to a softer, plasticlike material that folds more easily.

Interesting landforms develop where folded rocks are exposed on the earth's surface. The landforms are most easily studied from a geologic map of the area. On a geologic map, each rock layer is represented by a color and a number. Folds often appear as U-shaped patterns on geologic maps. When formations are folded upward, an **anticline** is formed. A downward fold forms a **syncline**. Anticlines and synclines are shown in Figure 6–9.

Figure 6–8. The city of Denver, just east of the Rocky Mountains, is on a plateau, or an uplifted area.

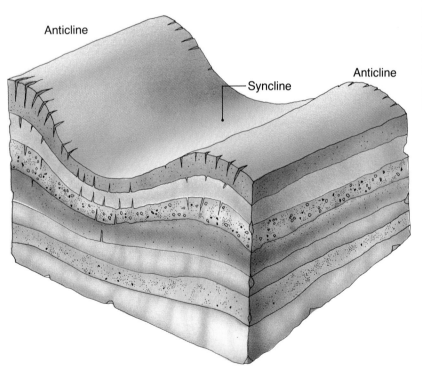

Anticline

Syncline

Anticline

Figure 6–9. Anticlines and synclines form as rock layers are folded. Anticlines often form parallel ridges, with a syncline forming a valley between them.

Geologists have developed a system to describe synclines and anticlines and to show them on geologic maps. This system is called *strike and dip.* **Dip** is the angle at which the rock layer is tilted from the horizontal. Dip also includes the direction in

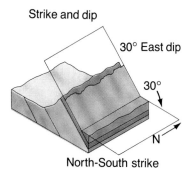

Strike and dip

30° East dip

30°

N

North-South strike

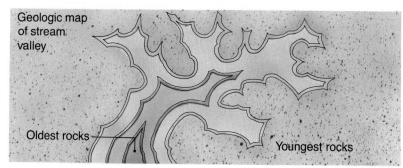

Geologic map of stream valley

Oldest rocks

Youngest rocks

Figure 6–10. Geologic maps show rock layers, as well as special features such as strike and dip (left). Rock layers are usually shown in colors (right), while special features are shown with symbols.

which the layer is tilted. Thus, dip is described by both an angle and a compass direction. The angle tells to what degree the layer is tilted, and the compass reading tells the direction in which the layer is tilted. The rock layer shown in Figure 6–10 is dipping 30° east.

As the formation dips, it forms a straight line called the **strike** along the earth's surface. The direction of this line is always at right angles to the dip. If the dip is east, then the strike will be north-south. Therefore, the rock layer shown in Figure 6–10 would strike north-south and dip 30° east. Strike and dip are shown on geologic maps, along with individual formations.

○ *What are folds?*
○ *How are folds shown on a geologic map?*

6.4 Fractures and Faults

Sometimes the forces within the earth are strong enough to break, or fracture, rocks instead of folding them. An example of fracturing can be seen in the quartzite cliffs near Devil's Lake, Wisconsin.

Figure 6–11. Forces within the earth are sometimes strong enough to cause rock layers to fracture. The fractured rocks shown here are at Devil's Lake, Wisconsin.

Figure 6–12. These rock joints are caused by fractures in the rocks. Where joints occur there is no movement along the fracture.

If there is no movement along the fracture, the fracture is called a *joint*. Joints can also form as igneous rocks cool and contract. These joints often form long columns in rocks, as shown in Figure 6–12.

If the forces are great enough, the fractured rocks may move, forming a fault. A **fault** is a rock fracture along which there is movement. All faults share some characteristics. For instance, all faults have two sides, the *footwall* and the *hanging wall,* as shown in Figure 6–13. All faults show movement between the walls, as well. Different types of faults result from movement in different directions. Earthquakes, which are discussed in the next chapter, often occur along faults.

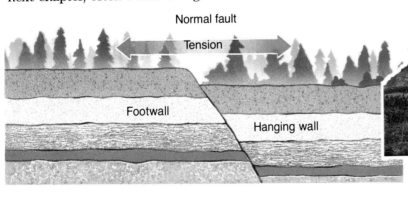

Normal fault

Tension

Footwall

Hanging wall

Figure 6–13. Fault-block mountains result from movement along normal faults. The fault-block mountains shown here are part of the Rocky Mountains.

Normal Faults Movement along faults is caused by three different kinds of force—tension, compression, and shear. **Tension** occurs when the earth's crust is pulled apart. Take a block of modeling clay and pull it apart. Tension separates rocks in a similar manner, and the hanging wall moves down. The type of fault formed by tension is called a *normal fault.* Normal faults are seen in the Basin and Range area of the southern Rocky Mountains. In this area pieces of the earth's crust may move many kilometers. As crustal blocks move and tilt, they form *fault-block mountains.*

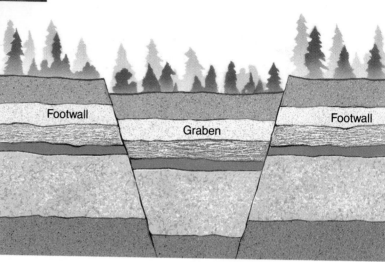

Figure 6–14. The Great Rift Valley of Africa (above) is an example of a graben, which results as a hanging wall drops between two normal faults (right).

Sometimes a large block falls between two normal faults. The sides of this block are the hanging walls of the two faults. This sunken area is called a *graben* (GRAH behn). The Great Rift Valley of Africa, shown in Figure 6–14, is an example of a graben.

Reverse Faults   Instead of being pulled apart, sections of the crust may be pushed together. This pushing together of the crust is called **compression,** which is the second type of force. A fault formed by compression is called a *reverse fault.* In a reverse fault, the hanging wall moves up in relation to the footwall, as shown in Figure 6–15. If the angle of the reverse fault is very low, the fault is called a *thrust fault.* In some thrust faults, the hanging wall can slide hundreds of kilometers over the footwall.

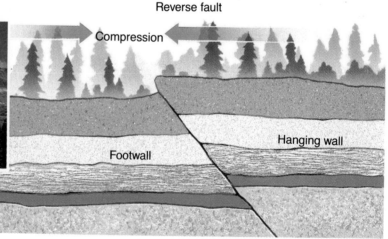

Figure 6–15. These mountains are the result of a thrust, or reverse, fault. A diagram of a reverse fault is shown on the right.

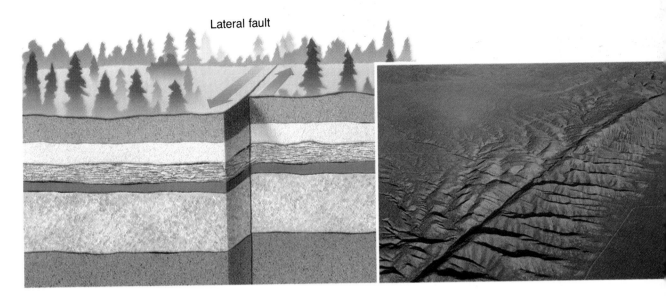

Lateral fault

Lateral Faults The third type of force on pieces of the crust is parallel to the pieces. This force, called **shear,** occurs when pieces of the crust drag against each other in opposite directions. Try sliding two sheets of sandpaper past each other. The sheets of sandpaper resist sliding, but if enough force is applied, they may suddenly move.

This sudden movement, which is shear, produces a third type of fault, called a *lateral fault.* In a lateral fault the hanging wall moves parallel to the footwall but in the opposite direction. The famous San Andreas fault in California, shown in Figure 6–16, is an example of a lateral fault.

Like a fold, a faulted area also forms interesting patterns on geologic maps. A geologic map of a faulted area often indicates rock layers that are repeated or that are missing. These repeated or missing layers are caused by movements along faults. A formation may be displaced several hundred kilometers by faulting. Figure 6–17 shows a geologic map of an area that has been both folded and faulted.

○ *What is a fault?*
○ *Name the three forces that produce faults, and the type of fault each produces.*

Figure 6–16. The famous San Andreas fault in California (above) is a lateral fault. A diagram of a lateral fault is shown on the left.

Figure 6–17. A geologic map of an area that has been both folded and faulted can be very complex.

127

SKILL ACTIVITY: Interpreting Geologic Maps

BACKGROUND

Interpreting a geologic map is often the first stage in understanding the geologic history of an area. Just as topographic maps represent landscapes with symbols, geologic maps use symbols to show where different rock units appear on the earth's surface.

The rocks shown on a geologic map are called formations. On most geologic maps, graphic symbols are used to indicate more common rock types. Letters are also used to indicate the geologic age of a formation. Other symbols are used on geologic maps to show strike and dip, horizontal beds, and the axes of synclines and anticlines.

PROCEDURE

Study the geologic map symbols and the area of the map along the line marked AB. Along the line AB, locate different types of rock layers, anticlines and synclines, and locate the oldest and youngest rocks along the line.

APPLICATION

1. Along the line marked AC, name the different types of rocks.
2. How many anticlines and synclines are there along the line AC?
3. Name the oldest and youngest rock layers along the line AC.

USING WHAT YOU HAVE LEARNED

At the bottom of the map is a cross section of the area along the line AB. Study it to determine how it was constructed, and then try drawing a similar cross section for the area along the line AC.

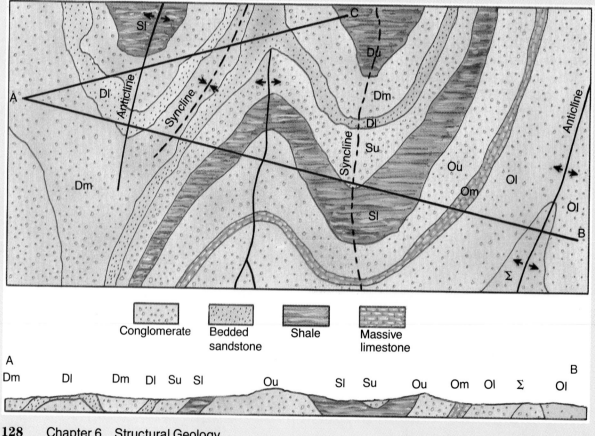

Conglomerate   Bedded sandstone   Shale   Massive limestone

6.5 Continental Drift

The evidence presented so far has shown that small areas of the earth's crust move. There is, however, evidence that large portions of the crust, entire continents in fact, are also moving. This movement of large portions of the earth's crust is called **continental drift.**

In the early 1900s, a young German scientist, Alfred Wegener (VAY guh nuhr), became fascinated by the matching shapes of the coastlines of Africa and South America. In his book, *The Origin of Continents and Oceans,* published in 1915, Wegener claimed that these coastlines appeared similar because at one time they were joined. Wegener further proposed that the continents, because they are made of lighter rocks than the ocean basins, float on top of the oceanic crust.

When Wegener proposed his theory of continental drift, most scientists disagreed with him; he was even ridiculed by the scientific community. Like Wegener, many scientists had noticed that there is an amazing match in the coastlines of Africa and South America; however, most dismissed the similarity as mere coincidence. Careful study of the rocks along the edges of these two continents provided support for Wegener's theory. The rocks match in structure and type, indicating that the continents were once joined.

Other matches of similar rocks have been discovered. The Appalachian Mountain range, in the eastern United States, seems to disappear into the northern Atlantic Ocean. The rocks in this mountain range, however, are very similar to those found in Scotland's Grampian Mountains. This geologic evidence indicates that North America and Europe were once joined as well.

DISCOVER

Solving a Puzzle

On the back of an old jigsaw puzzle, sketch the rough outlines of Africa and South America joined along a common line. Now separate the puzzle pieces, and see if a classmate can reconstruct the joined supercontinent. How is solving this puzzle similar to what Wegener must have done in developing his theory of continental drift?

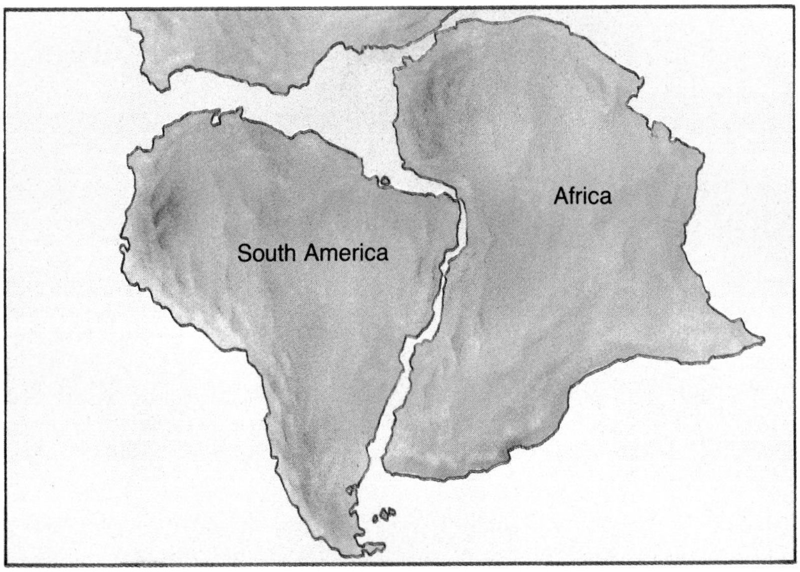

Figure 6–18. This diagram shows how the continents on either side of the Atlantic Ocean may have been arranged before they began drifting apart. The similarities of the coasts of Africa and South America support the theory of continental drift.

Figure 6–19. Fossils found on several continents, provide evidence for the theory of continental drift.

More evidence supporting the theory of continental drift comes from the comparison of fossils. Fossils of the same mammal-like reptiles were found in Antarctica, Africa, South America, and India. These ancient animals could not swim. Scientists wondered how these animals could have crossed hundreds of miles of open ocean to reach Antarctica from Africa or India. Each new bit of evidence added support to the theory of continental drift. However, more evidence was needed to convince scientists who found it difficult to believe that solid rock could flow.

○ *What is the theory of continental drift?*
○ *What evidence exists to prove this theory?*

Section Review

READING CRITICALLY

1. What are the three types of force and the faults they produce?
2. What conditions are needed to fold rocks?

THINKING CRITICALLY

3. Why would it be dangerous to construct a building on or near a fault?
4. If the theory of continental drift is correct, on what layer of the earth's interior could the continents actually be moving?

ACTIVITY: Studying Ancient Glaciers

How can data from ancient glaciers support the theory of continental drift?

MATERIALS (per group of 3 or 4)

pencils, tracing paper, textbook, scissors, colored pencils, glue, legal-size paper

PROCEDURE

1. Trace the continents of South America, Africa, India, and Australia from the world map in the reference section of pages 592–593. Cut out the tracings of the continents.
2. Using a colored pencil, color the areas where ancient glaciers might have been found. Add arrows to indicate the likely direction of movement of the glacier.

3. Position the tracings so that the glaciers appear to move from a central location outward onto the various continents. Glue the tracings into place on the legal-size paper to show the single continent.

CONCLUSIONS/APPLICATIONS

1. Why do the continents not fit together exactly?
2. If you consider the shape of the shallow shelves just offshore, does the fit improve?
3. If you assume that all the continents were together when this glacier covered parts of these continents, where on the globe would this continent have been? What type of climate existed in North America at that time?

CAREERS

SEISMOLOGIST

A seismologist studies waves in the earth produced by earthquakes. Seismologists not only record and study earthquake data, but they use the data to predict earthquakes.

Seismologists also use earthquake data to analyze the rock formations of an area. This is sometimes useful in the search for natural resources such as oil and gas.

Some seismologists work in laboratories interpreting seismic records, while others work in the field using lasers, computers, and other sophisticated equipment. A career in seismology requires a college degree in geology or physics.

For Additional Information
American Petroleum Institute
1220 L Street, N.W.
Washington, DC 20005

SCIENTIFIC TRANSLATOR

Research and scientific work are being done by scientists in many countries. There is a great need for qualified translators of scientific writing, both into English and from English into other languages.

Translators are often employed by international agencies like the World Health Organization. Scientific journals also use the services of translators.

There is no standard training program for translators. A basic understanding of science is necessary as well as reading and writing experience in one or more foreign languages.

For Additional Information
Contact your local
community college
or university.

OIL-PLATFORM WORKER

Much of the oil and gas needed to run an industrialized world is found deep below the earth's surface. Some of this oil is below the surface of the sea, many kilometers from shore.

The process of obtaining oil and gas from undersea wells requires both scientists and technicians. Most of the jobs on an oil platform are exciting and dangerous. Many require an ability to work with tools and machinery.

The technician's jobs require at least a high-school diploma; some require specialized training such as a trade school or technical school would offer.

The wages are high and there is opportunity for travel and advancement. However, the hours are long and the conditions in places such as the North Sea are unpleasant.

For Additional Information
American Geological Institute
5205 Leesburg Pike
Falls Church, VA 22041

INVESTIGATION 6: Analyzing Structural History

PURPOSE

To analyze the history of an area using information from rock formations

MATERIALS

Paper and pencil

PROCEDURE

1. Study the cross section of rock layers shown. Examine it for rock type and structural deformations.
2. How is a syncline illustrated in the cross section? Identify a syncline in this cross section by listing the letters that represent the formation.
3. Locate an anticline in the cross section. List the letters of the formations involved.
4. Classify any faults found in this cross section. List the letters of the formations involved.

ANALYSES AND CONCLUSIONS

1. Which of the three structures in procedure steps 1, 2, and 3 was formed last? Explain your reasoning.

2. Which layer on the footwall side of the fault is the same as layer A?
3. In which direction does rock layer X dip?
4. In which direction does the fault dip?
5. Name the youngest rock layer.
6. Name the oldest rock layer.
7. Name the type of fault shown on this cross section.
8. List the rock layers that make up the footwall of the fault.
9. Match the rock layers of the footwall to the same layers in the hanging wall.

APPLICATION

What do you think the area east of the cross-section might look like? Draw an extension of the original cross-section showing how you think it might appear.

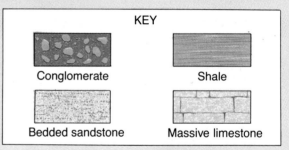

KEY

Conglomerate

Shale

Bedded sandstone

Massive limestone

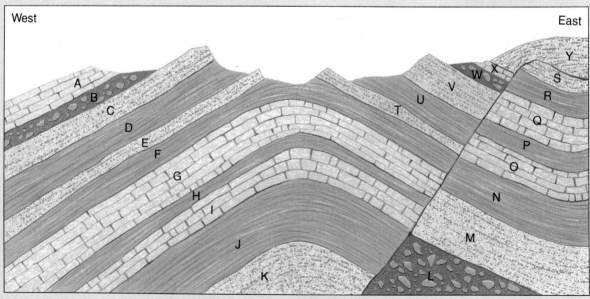

SUMMARY

- The earth is made of several layers that are distinct in composition, density, and thickness. (6.1) *The crust, mantle, + the core.*

- Beneath the crust are the mantle and the core of the earth. (6.1)

- The heated material within the earth moves by convection. Most of the heat in the interior of the earth is supplied by the decay of radioactive elements. (6.2)

- Forces within the earth's crust cause rocks to fold and fracture. Rocks fold into anticlines and synclines. (6.3)

- Large-scale folding produces mountain ranges by the process of orogeny. Plateaus may be formed by large-scale uplifting without folding. (6.3)

- Landforms created by folding are often studied by using geologic maps. (6.3)

- The position of rock formations can be described using strike and dip. (6.3)

- Faults form because of tension, shear, or compression. (6.4)

- Tension produces normal faults. Fault-block mountains may result from normal faults. Reverse faults are the result of compression. A low-angle reverse fault is a thrust fault. Lateral faults, like the San Andreas fault, are caused by shear. (6.4)

- In 1915 Alfred Wegener proposed the theory of continental drift. (6.5)

- The match of continental shapes, geological features, and fossil evidence are used to support the theory of continental drift. (6.5)

Write all answers on a separate sheet of paper.

SCIENCE TERMS

Correctly use each of the following terms in a sentence.

anticline **(123)**
asthenosphere **(119)**
compression **(126)**
continental drift **(129)**
convection **(120)**
core **(119)**
crust **(117)**
dip **(123)**
fault **(125)**
fold **(122)**
lithosphere **(119)**
mantle **(118)**
orogeny **(122)**
shear **(127)**
strike **(124)**
syncline **(123)**
tension **(125)**

SCIENCE QUIZ

Modified True-False

Mark each statement *true* or *false*. If a statement is false, change the underlined term to make the statement true.

1. A lateral fault forms when the crust pulls apart.

2. According to the theory of continental drift, continents can move.

3. Hot molten material rises in the mantle due to convection currents.

4. Grabens are formed by compression.

5. Mountains may occur as a result of folds in the crust.

continues

Multiple Choice

Write the letter of the choice that best answers the question or completes the statement.

6. The only portion of the earth thought to be entirely molten is
   a) the upper mantle.
   b) the inner core.
   c) the lower mantle.
   d) the outer core.

7. Pressure of overlying rocks affects the rocks below by
   a) lowering the melting point.
   b) altering the mineral content.
   c) raising the melting point.
   d) causing convection currents.

8. The lithosphere includes
   a) the crust and the mantle.
   b) the mantle and the upper core.
   c) the asthenosphere and the crust.
   d) the upper mantle and the crust.

9. Which type of fault is produced by shear?
   a) lateral
   b) normal
   c) reverse
   d) thrust

10. The evidence for continental drift comes from
   a) lateral faults.
   b) anticlines.
   c) matching rocks.
   d) radioactive material.

Completion

Complete each statement by supplying the correct term or phrase.

11. Wegener began a great scientific argument by proposing the _____.

12. The boundary between the core and the mantle is _____.

13. Fractures along which there is no movement are called _____.

14. Fossils of a land-dwelling African animal were found in _____.

15. The boundary between the crust and the mantle is called the _____.

Short Answer

16. Explain why the discovery of the same types of fossils in Africa and India was important in the development of the theory of continental drift.

17. What is the relationship between different forces and different types of faults?

18. How are the Appalachian Mountains similar to the Grampian Mountains? Why is this geologically important?

19. Draw a cross section of the earth's interior and label the discontinuities.

Writing Critically

20. What causes the discontinuities between the crust and the mantle, and between the mantle and the core?

21. Explain why the continents float higher than the ocean basin does.

22. Explain the relationship between spreading continents and convection cells.

EXTENSION

1. Build clay models of rock structures such as anticlines or fault blocks. Show the different landforms that are created and write a brief report identifying at least one location on Earth where such a landform exists.
2. Research *Lystrosaurus* fossils. Find out how they were preserved. Write a report about the type of environmental conditions that must have existed where they lived.
3. Find out what is unusual about Chief Mountain, Montana. Explain how it resulted from a thrust fault.

APPLICATION/CRITICAL THINKING

1. Design your own geologic map to show an anticline, a syncline, and a thrust fault.
2. Some geologists have said that the section of California west of the San Andreas fault will end up near Alaska someday. Using your knowledge of fault movement, explain why you agree or disagree.
3. Imagine two people walking out of a California travel agency. One character turns to the other and says, "The only way we'll ever get to Hawaii is by continental drift." Explain whether or not this would be possible.

FOR FURTHER READING

Lane, Frank W. *The Violent Earth.* Topsfield, Mass., 1986. This book offers an interesting discussion of violent actions on Earth, covering everything from earthquakes to avalanches.

McConnell, Anita. *The World Beneath Us.* New York: Facts on File, 1985. The author provides very interesting descriptions of the geologic time table, fossils, plate tectonics, earthquake prediction, and volcano activity.

Miller, R. *Continents in Collision.* Time-Life Books, 1983. This book traces the possible positions of the continents through the history of the earth and suggests possible movement in the future.

"Seafloor Spreading and Sinking." *Science News* 107 (March 22, 1986): 183. This article describes some of the sediments deposited in regions where there are plate boundaries.

Challenge Your Thinking

These two circular structures are related to synclines and anticlines. One is called a *basin* because of its bowl-like shape, and the other is called a *dome* because of its rounded shape. Basins and domes are circular, and both are folded formations. Using what you have learned in this chapter, explain why basins and domes might be considered as special types of synclines and anticlines.

CHAPTER

7

The Dynamic Earth

Erupting volcanoes are some of the most dramatic geologic events on Earth. They have always fascinated humans. Early civilizations explained these events by telling stories of angry gods and of animals with supernatural powers. You may find that the truth is even more fascinating! Geologists now know that volcanoes are produced as heated rocks are forced to the surface.

An active volcano erupting

1 Plate Tectonics

SECTION OBJECTIVES

After completing this section, you should be able to:
- **Explain** the theory of plate tectonics.
- **Identify** the three types of crustal boundaries and **describe** their associated landforms.
- **Predict** how the positions of the continents may change in the future.

NEW SCIENCE TERMS

theory of plate tectonics
earthquakes
volcanoes
divergent boundary
convergent boundary
transform fault

7.1 Crustal Plates in Motion

Imagine all the land areas of the world joined together in one supercontinent. This supercontinent would, of course, be surrounded by one superocean. Look at the shape of the continents on the world map in the Reference Section on page 592. Imagine cutting around the continents and joining the pieces together where there seems to be a natural fit. One fit, for example, would be the east coast of South America with the west coast of Africa.

Alfred Wegener hypothesized that all the continents were once joined together in one huge continent called *Pangaea* (pan GEE ah). Since the time of Pangaea, Wegener said, the continents have been drifting apart. Some scientists listened to Wegener's theory of continental drift, but many rejected his idea of a supercontinent. They could not believe that continents could drift through a solid crust.

Scientists now believe that about 550 million years ago two continental masses were drifting on the mantle. These early continents are called *Laurasia* and *Gondwana*. Evidence supports the belief that, about 300 million years ago, these two continents collided, forming Pangaea. Since that time the continents have been drifting apart, as Figure 7–1 shows.

Figure 7–1. This series of computer-generated diagrams shows the break-up of Pangaea about 200 million years ago, and the movement of the continents into their present positions.

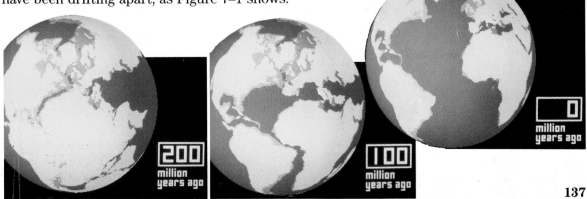

Shifting Plates

Using Figure 7–2, sketch the shape of several adjoining plates on a sheet of paper. Cut out the outlines, separate the plates, and transfer the patterns to thin sheets of polystyrene. Cut out the polystyrene plates, and float them in water in a glass baking dish. Gently warm the dish on a hot plate, and note the movement of the plates as convection cells begin to develop.

These conclusions led to the formulation of a new theory—that of plate tectonics. The **theory of plate tectonics** states that the lithosphere is divided into plates, each moving independently of the others.

Some of these plates consist of oceanic crust only, while other plates contain both continental and oceanic crust. These plates have an average thickness of about 70 km and extend to the top of the asthenosphere. These plates slide or float on the asthenosphere, driven by convection cells in the mantle.

The plates that form the lithosphere are moving at different speeds and in different directions. For example, the North American plate is moving toward the southwest at about 2 cm per year. The adjoining Pacific plate is moving to the northwest at about 13 cm per year. What might eventually happen if these plates continue to move as they do now?

○ *What is the theory of plate tectonics?*
○ *What causes the individual plates to move?*

Figure 7–2. This map shows the location of the major plates of the world.

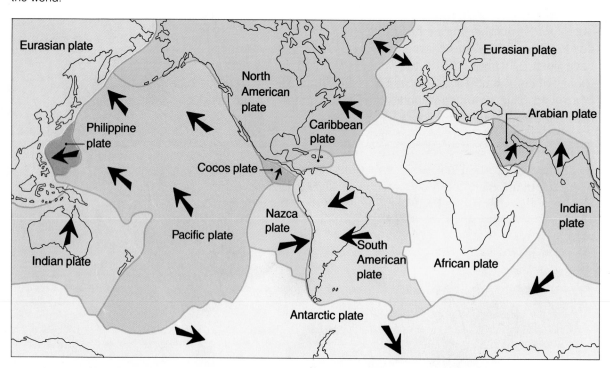

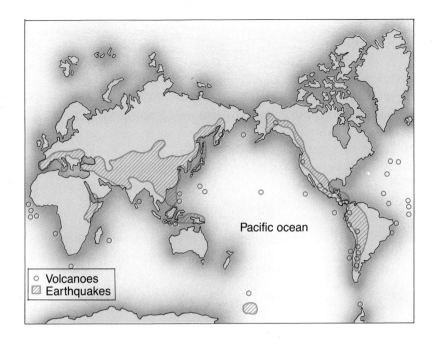

Pacific ocean

○ Volcanoes
▨ Earthquakes

7.2 Plate Boundaries

Exciting geologic events occur along the boundaries of plates. Although there are exceptions, most earthquakes and volcanoes occur near the edges of plates. **Earthquakes** are violent shakings of the earth's interior. **Volcanoes** are openings in the crust through which magma reaches the earth's surface.

So many earthquakes and volcanoes occur around the edge of the Pacific Ocean that this area is called the *Ring of Fire*. Earthquakes, volcanoes, mountains, and even ocean trenches occur near the boundaries between crustal plates. These earthquake- and volcano-prone areas are shown in Figure 7–3. You can see that these areas form a ring.

Divergent Boundaries   Three different types of plate boundaries have been identified. Approximately in the middle of the Atlantic Ocean is an area where molten rock from the mantle rises to the earth's surface through cracks, or *rifts*, in the ocean floor. Here newly formed crust is adding width to an underwater mountain range called the *Mid-Atlantic Ridge*. An area of the earth where the crust is spreading is called a **divergent boundary.** The word *diverge* means "to move apart."

> **A MATTER OF FACT**
>
> In 1883 there was a volcanic eruption on Krakatau in Indonesia. It was one of the world's worst disasters, killing nearly 36 000 people.

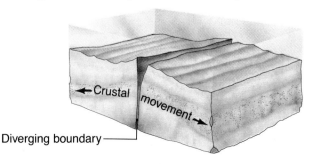

Diverging boundary

← Crustal movement →

Figure 7–4. Along divergent plate boundaries (left) new material is added to the crust and the plates spread apart.

Figure 7–5. The Arabian Sea is widening because of diverging plates. If the divergence continues, one day this sea may be known as the Arabian Ocean.

Divergent boundaries are also found in a few places on land, such as the Great Rift Valley of Africa. This huge valley, or rift zone, extends for over 4000 km, from the southern end of the Red Sea to Mozambique. The wide, flat valley is bordered by steep cliffs, over 600 m high in some places. Frequent earthquakes and volcanic eruptions also occur along the valley.

Africa is probably splitting apart along this rift. As the plates continue to separate, the valley floor drops. Many scientists believe that in the next few million years, the valley will become a sea, with the cliffs forming the coastlines of two separate continents.

Convergent Boundaries   Where two plates press together a **convergent boundary** is formed. *Converge* means "to come together." Converging plates produce mountains, volcanoes, and ocean trenches.

If an oceanic plate collides with a continental plate, the denser oceanic plate is forced below the continental plate. An area where one plate is forced under another is called a *subduction zone*. As the oceanic plate is pushed under the continental plate and into the mantle, the rocks heat up and melt. Some of this molten rock rises again, forming volcanic mountains along the continental edge of the boundary. Volcanoes are discussed in Section 2.

During subduction the oceanic plate pushes downward and a trench forms between the two plates. Trenches occur in many places, although most are located along the edges of the Pacific Ocean. The greatest ocean depths are found in these trenches. The *Mariana Trench*, located off the Mariana Islands, is the deepest trench at about 11 000 m.

Figure 7–6. Along converging plate boundaries (right), part of the crust is subducted into the mantle. There it melts and forms pools of magma. This magma may again reach the surface as lava (left).

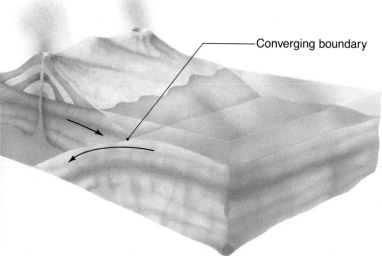

Converging boundary

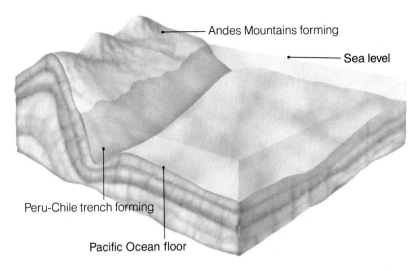

Andes Mountains forming

Sea level

Peru-Chile trench forming

Pacific Ocean floor

Figure 7–7. Where the Nazca Plate was forced under the South American Plate, a deep ocean trench (the Peru-Chile Trench) and a tall mountain range (the Andes Mountains) eventually formed.

On the western edge of the South American plate is another subduction zone and trench. Here, the Pacific plate is being forced below South America, forming a deep trench called the *Peru-Chile Trench*. The Andes Mountains are a result of this movement. Several trenches in the Pacific Ocean and in the Atlantic Ocean border chains of islands called *island arcs*. Look at the map on page 138. Where else might you find subduction zones?

When two continental plates converge, neither plate can be forced under the other. Instead, great mountain ranges are formed. The rocks fold and fracture, often rising to great heights. The Himalaya Mountains were formed when the Indian plate collided with the Eurasian plate. Mountains in this range continue to rise by about 5 cm each year. In the Western Hemisphere, the westward movement of North and South America formed the Cordilleras. This mountain system extends from Alaska to the southern tip of South America and includes the Rocky Mountains of the United States and Canada.

Figure 7–8. The Himalaya Mountains (left) were formed by the collision of continental plates. The Rocky Mountains (right) are fault-block mountains.

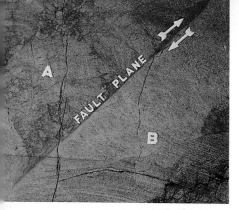

Figure 7–9. Transform faults occur where adjoining plates slide past each other in parallel but opposite directions.

Transform Faults When two plates slide past each other in parallel but opposite directions, a type of lateral fault called a **transform fault** forms. The San Andreas fault in California is caused by the North American plate grinding past the Pacific plate. Although this type of boundary does not form mountains or trenches, there are many earthquakes associated with this movement.

○ *What are the three types of plate boundaries?*
○ *Which type of boundary creates folded mountain ranges?*

Section Review

READING CRITICALLY

1. Compare and contrast the three types of plate boundaries.
2. How does the Ring of Fire provide evidence for the theory of plate tectonics?

THINKING CRITICALLY

3. Explain how the theory of plate tectonics is different from the theory of continental drift.
4. Describe the possible position of the continents fifty million years in the future. Support your answer with facts about present plate movement.

ACTIVITY: Examining Crustal Plates

How can you analyze information about crustal plates?

MATERIALS (per group of 3 or 4)

pencil, paper

PROCEDURE

1. Make a table similar to the one below, but allow room for 20 rows.
2. Carefully examine each plate shown on the map in Figure 7–2. On your data table, record the name of each plate, the name of any land-masses included on the plate, the type of boundary, and the direction of movement.

CONCLUSIONS/APPLICATIONS

1. How many different plates did you find?
2. How many plates contained landmasses?
3. Do you predict that any of the plates are going to eventually subduct under other plates? Explain your prediction.
4. Geologists have stated that it is relatively easy to identify a plate but often quite difficult to locate the plate boundaries exactly. Explain why this might be so.

TABLE 1: CRUSTAL PLATES				
Plate Name	**Landmasses on Plate**	**Types of Plate Boundaries**	**Direction of Movement**	**Other Features**

② Volcanoes

SECTION OBJECTIVES
After completing this section, you should be able to:
- **Explain** why volcanoes usually form near plate boundaries.
- **Classify** types of volcanoes.
- **Describe** the relationship between hot spots in the crust and volcanic island chains.

NEW SCIENCE TERMS
batholith
viscosity
shield volcanoes
cinder-cone volcano
composite volcano
caldera
hot spot
seamount

7.3 Lava Flows

One of the most dramatic effects of plate movement is the formation of volcanoes. Most volcanic activity occurs near the boundary of crustal plates. When plates collide, crustal material is subducted, or moves, into the asthenosphere. There an increase in pressure raises the temperature of the rocks. Eventually the rocks melt, forming magma. Where plates diverge, pressure is released, rocks melt, and magma rises through rifts, as in Iceland.

Often, rising magma does not reach the surface. When a body of magma cools beneath the surface, it forms an intrusion called a **batholith.** Smaller intrusive bodies are called *stocks.* The cores of many mountains are formed from batholiths. Sometimes thin fingers of magma extend from these intrusions. When magma squeezes between layers of rock, a *sill* is formed. When magma cuts across rock layers, it forms a *dike.*

If magma reaches the surface, it is called *lava.* Some magma rises slowly to the surface, creating lava flows that may stream great distances across the earth's surface. If the magma rises

> ### A MATTER OF FACT
>
> In 1815 the Tambora volcano in Indonesia killed 12 000 people when it erupted. The eruption sent over 80 km³ of ash into the atmosphere, blocking out so much of the sun that there was a general cooling of the entire earth, making 1816 the "year without a summer."

Figure 7–10. A lava flow, such as that shown, allows molten rock from within the earth to reach the surface.

Figure 7–11. Pahoehoe (left) is smooth, ropy lava. Aa (center) is rough and blocky. Pillow lava (right) forms rounded lumps.

quickly to the surface, lava is thrown into the atmosphere in semisolid chunks. Most of this lava hardens as it falls to the ground, forming lava masses called *bombs* and *lapilli*. At times the force of a lava eruption is so violent that large chunks of old, hardened lava blocks, or cinders, are thrown into the air. Ash—particles less than 2 mm in diameter—is also produced by violent eruptions. Ash can form great clouds which, when they fall to Earth, cover everything.

Lava flow is affected by viscosity (vihs KAHS uh tee). **Viscosity** is a liquid's resistance to flow. For example, honey has high viscosity, so it flows very slowly. Water has low viscosity, so it flows easily. The viscosity of lava is determined by the amount of water and silica in the magma. Low-viscosity lava has little silica but a lot of water. High-viscosity lava is rich in silica but has little water. Low-viscosity lava moves rapidly, forming smooth ropy flows of basalt called *pahoehoe* (pah HOH ee hoh ee). High-viscosity lava moves more slowly and forms rough, blocky flows of andesite called *aa* (AH ah).

○ *What is viscosity?*
○ *What are the two types of lava flows?*

7.4 The Structure of Volcanoes

Many volcanoes begin with relatively peaceful lava flows and then erupt explosively. In other cases the first eruption is the most explosive. A volcano's shape depends on the type of lava and the force of the eruption. There are three kinds of volcanoes, which are classified by shape: *shield volcanoes, cinder-cone volcanoes,* and *composite volcanoes.*

Shield volcanoes are formed by quiet eruptions of basalt lava that has a low silica content. A shield volcano has a cone with gentle slopes and a base that covers a wide area. Shield volcanoes make up most of the Hawaiian Islands.

Silica-rich magma produces viscous lava. This magma traps gases inside the volcano until enough pressure builds up to push the magma out of the earth. If there is much water present, it

Figure 7–12. A shield volcano has gentle slopes and a wide base. Most of the volcanoes of the Pacific islands are shield volcanoes.

quickly turns to steam. The pressure of the steam causes an explosion that may shoot gases, bombs, and ash several kilometers into the atmosphere. In contrast to shield volcanoes, this kind of eruption builds a steep-sided **cinder-cone volcano.** Cinder-cone volcanoes consist mainly of ash and *tuff.* Tuff is formed from compressed ash, cinders, and lapilli. Many of the volcanoes in Mexico and Central America are cinder cones.

If the lava flow changes, the nature of the eruption may change as well. A third type of volcano, formed from a series of alternating eruptions of different lavas, is a **composite volcano.** A composite volcano consists of alternating layers of ash, tuff, and andesite lava. The slope of a composite volcano is steeper than that of a shield volcano but not as steep as that of a cinder-cone volcano. Mount Fuji, in Japan, is a composite volcano.

Figure 7–13. A steep-sloped cinder-cone volcano (left) often erupts explosively (top). A great deal of ash is produced from a cinder-cone volcano (bottom).

A MATTER OF FACT

In 1902 Mount Pelee in Martinique erupted, leaving only two survivors in the town of St. Pierre. One of these was Auguste Ciparis, a criminal, who survived because he was protected by his thick-walled jail cell.

Figure 7–14. Japan's Mount Fuji, one of the world's most photographed mountains, is a composite volcano.

Figure 7–15. Old calderas often fill with water, forming beautiful lakes such as Crater Lake in Oregon.

An explosive eruption of an old composite volcano may blow away entire sections of the cone wall. Large sections of the wall then collapse into the hollow center, forming a *caldera*. A **caldera** is a wide basin, usually more than a kilometer across, in the center of a volcanic cone. Crater Lake in Oregon is in an old caldera, while Mount St. Helens, a volcano in the state of Washington, formed a new caldera when it erupted in 1980.

○ *What causes some volcanic eruptions to be explosive?*
○ *Describe the three shapes of volcanoes.*

7.5 **Hot Spots**

Although most volcanoes are formed near plate boundaries, some seem to develop in the middle of plates. These mid-plate volcanoes often occur in long, straight island chains, such as the Hawaiian Islands. At one end of the island chain, the volcanoes are old and extinct. At the other end of the chain, the volcanoes are young and active, such as those on the island of Hawaii.

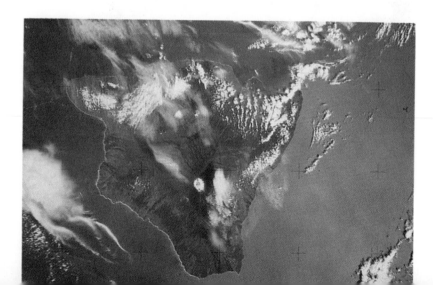

Figure 7–16. Active volcanoes on the island of Hawaii indicate that this island is located over a hot spot in the Pacific plate.

The exact cause of these mid-plate volcanoes is unknown; however, geologists believe these volcanoes develop from columns of magma rising from hot spots in the mantle. A **hot spot** is an area of concentrated heat deep within the earth. Hot spots melt the rocks of the mantle with which they come in contact. The molten rock creates a column of magma that rises toward the surface. The magma then erupts through the lithosphere. As plates move across the rising column of magma, volcanoes are produced. Geologists have identified nearly 120 hot spots on the earth.

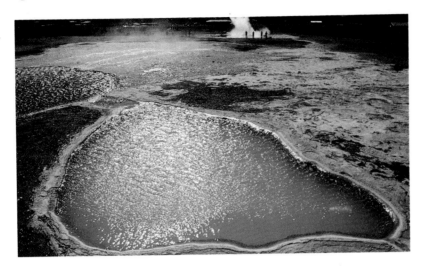

Figure 7–17. This steam field in Iceland (left) and Old Faithful geyser in Yellowstone National Park, Wyoming (below), are both located over hot spots.

Sometimes hot magma heats underground water, turning the water into steam. This steam is released through cracks in the earth's surface, producing steam vents called *geysers* (GY zuhrz). Yellowstone National Park, with its famous "Old Faithful" geyser, sits over a hot spot. Geologists predict that areas to the northeast of Yellowstone may soon begin showing signs of geyser activity, as well. Why would new geysers form to the northeast of Yellowstone National Park?

○ *What are hot spots?*
○ *What are geysers?*

7.6 Undersea Volcanoes

Volcanoes form not only on land but also under the oceans. In shallow water, volcanoes erupt violently, forming clouds of ash and steam. An underwater volcano is called a **seamount.** Seamounts look much like cinder-cone volcanoes on land.

There are more than 10 000 seamounts on the Pacific Ocean floor. However, many seamounts never reach the surface of the ocean. The Hawaiian Islands, for example, are a small part of an

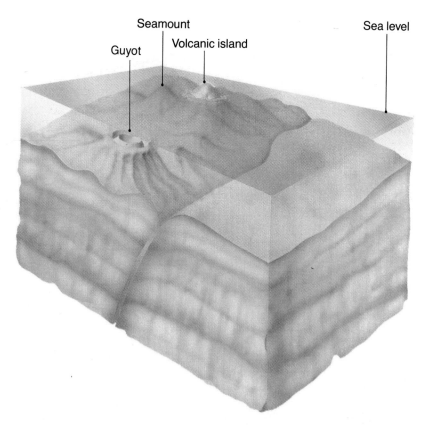

Figure 7–18. Seamounts are volcanoes that form under water. Guyots form as waves flatten the tops of seamounts near the water's surface (right). If the guyot is in a tropical ocean, a coral atoll may result (above).

underwater volcanic-mountain chain. Just eight of these volcanoes have become islands, while dozens remain submerged.

Another type of undersea volcano is called a *guyot* (gee OH). A guyot is an undersea volcano with a flat top. Geologists hypothesize that guyots were once seamounts that rose above the ocean's surface. There the tops of the volcanoes were flattened by weathering and wave action.

○ *What are seamounts?*

Section Review

READING CRITICALLY

1. Explain how a chain of volcanoes can form in the middle of a plate.
2. Why is most volcanic activity located near plate boundaries?

THINKING CRITICALLY

3. If guyots were flattened by weathering and the action of waves, why are most of them found far below the surface?
4. Where would you expect any new Hawaiian Island volcanoes to form? Explain.

SKILL ACTIVITY: Making a Graph

BACKGROUND

Scientists often present information in the form of a table. Sometimes, however, this format may not be the best way to present the information. Often a better interpretation can be obtained by placing the data on a graph. The procedures for placing any type of data on a graph are the same.

PROCEDURE

1. Title your graph. Write the title on the top of your paper.
2. Draw two axes on a sheet of graph paper. The horizontal, or *X,* axis should be near the bottom of the paper. The vertical, or *Y,* axis should be near the left edge of the paper. Label both axes with the type of data you are going to plot.
3. Choose the scale for each axis. The scale should be such that the graph covers as much information as possible and is spaced evenly over the entire sheet of graph paper.
4. Plot the data on your graph.

APPLICATION

Below is a table giving the ages of several of the Hawaiian Islands. Use the information in the table to make a graph. On one axis, plot the longitude of each island, starting with the island of Hawaii and moving to the west. On the other axis, plot the age of each island from east to west, starting with Hawaii.

USING WHAT YOU HAVE LEARNED

Use your graph to answer the following questions:
1. Which is the youngest island?
2. Which is the oldest island?
3. What is the relationship between the nearness of the islands to Hawaii and their relative age?
4. Using the graph and a map, predict the longitude of the next volcanic island to form in this chain.
5. The age of one of the islands does not seem to fit the pattern formed by the others. Explain.

TABLE 1: THE HAWAIIAN ISLANDS AND SEAMOUNTS

Island (or reef)	Approximate Age (in millions of years)	Longitude
Hawaii	0	155°30′ W
Kanum	39.0	170° E
Kauai	4.1	159°30′ W
Maui	0.6	156°15′ W
Midway Island	18.0	177°30′ W
Molokai	1.8	157° W
Necker Reef	10.1	164°30′ W
Nihoa	no data	162° W
Oahu	3.1	158° W
Pearl Reef	20.1	176° W
Yuryaku	42.3	168°30′ E

3 Earthquakes

focus
epicenter
P waves
S waves
surface waves
seismograph
Richter scale

SECTION OBJECTIVES

After completing this section, you should be able to:

■ **Compare** and **contrast** the motions of P, S, and surface waves.
■ **Explain** how the epicenter of an earthquake is found.
■ **List** some factors that may be helpful in predicting earthquakes.

A MATTER OF FACT

Although earthquakes are less likely to occur in central North America, there have been a few very strong ones. Earthquakes at New Madrid, Missouri, which occurred in 1811 and 1812, were felt as far away as Canada.

7.7 The Origin of Earthquakes

You may recall from Section 1 that strong convection cells within the mantle are constantly moving the earth's plates. At the plate boundaries, pressure builds as the friction between the plates stops their movement. Finally, the plates slip, and the pressure is released in a series of waves. These waves are similar to the ripples that move away from a pebble dropped into a pond of water. However, instead of moving through water, the waves move through the solid earth. This movement shakes the earth, causing an earthquake.

Many of the earthquakes in the United States occur where the Pacific plate scrapes past the North American plate, along the San Andreas fault. Some sections along the fault move smoothly past each other, causing frequent, but minor, earthquakes. In other sections the plates stick to each other as they try to move. These stuck-together sections, called *seismic gaps*, are areas where minor earthquakes do not occur. However, seismic gaps have a high probability of producing major earthquakes. Both San Francisco and Los Angeles are located on seismic gaps of the San Andreas fault.

Figure 7–19. The San Francisco earthquake of 1906 occurred along a seismic gap of the San Andreas fault.

Just as ripples in a pond move out from a falling pebble, earthquake waves move away from their source. The source of an earthquake is called the **focus**. The waves travel in all directions from the focus, which is located below the earth's surface. The area on the earth's surface directly above the focus is called the **epicenter** of the earthquake.

○ *What causes earthquakes?*
○ *What is the epicenter of an earthquake?*

7.8 Earthquake Waves

There are two types of waves that originate from the earthquake focus. The fastest-moving wave is a longitudinal wave. A longitudinal wave travels by compressing earth material in front of it and stretching material behind it. You can create longitudinal waves by moving a coil of wire back and forth.

Longitudinal waves are the fastest waves produced during an earthquake and, therefore, are the first waves that reach an earthquake recording station. For this reason, longitudinal waves are called *primary waves*, or **P waves.** P waves can travel through any type of material, even the dense center of the earth. A diagram of how P waves travel is shown in Figure 7–21.

The second type of earthquake waves is transverse waves, which move more slowly than P waves. These slower waves are called *secondary waves*, or **S waves.** They are similar to the movement of a rope shaken from side to side.

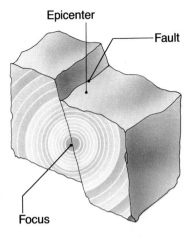

Figure 7–20. As this diagram shows, the focus of an earthquake is often far below the epicenter.

Epicenter

Fault

Focus

A MATTER OF FACT

Along the west coast of Mexico, the Cocos plate dives beneath the North American plate in a subduction zone. On September 19, 1985, the built-up pressure caused the Cocos plate to move 20 m below the surface. This caused a tremendous earthquake, releasing energy a thousand times greater than that of the atomic bomb dropped on Hiroshima, Japan, during World War II.

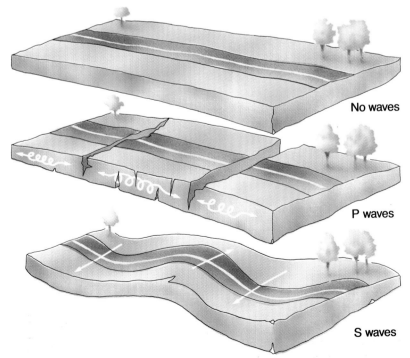

No waves

P waves

S waves

Figure 7–21. P waves and S waves, which originate at an earthquake's focus, cause only minor damage.

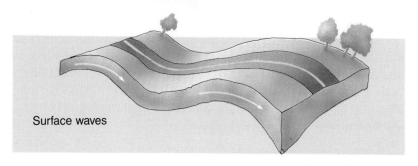

Surface waves

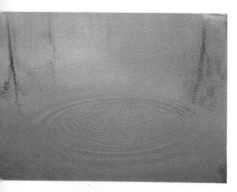

Figure 7–22. Surface waves, which originate at an earthquake's epicenter like ripples on a pond, cause severe damage because of their rolling action.

The interaction of P and S waves produces **surface waves** that cause the surface of the earth to shake and roll like a ship in a storm. These surface waves originate from the epicenter, not the focus, and cause more destruction than either P waves or S waves. P waves and S waves shake buildings back and forth or up and down, but the rolling action of surface waves causes many buildings to collapse.

The wave motion of earthquakes is very complex. As earthquake waves travel through the different layers of the earth, they tend to change characteristics because of the density differences of the layers. For example, P waves bend sharply as they enter the core. S waves, however, cannot pass through the liquid outer core. This creates an area on the earth's surface that is shielded from the effects of P and S waves. This area, known as the *shadow zone*, is shown in Figure 7–23.

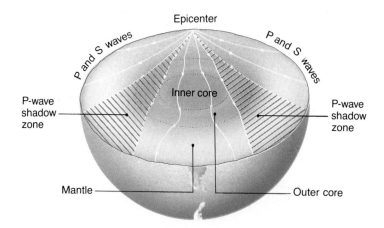

Figure 7–23. The shadow zone is an area of the earth protected from the effects of P and S waves. However, this area is not protected from the effects of surface waves.

The instrument used to record earthquake waves is called a **seismograph** (SYZ muh graf). A seismograph consists of a rotating drum wrapped with paper, and a pen attached to a suspended weight. The pen presses gently against the paper-wrapped drum. The structure that supports the drum is fastened to solid rock, so it will move up and down only if the rock moves. Any vibration of the rock will produce a zigzag line on the paper. The recording from a seismograph is called a *seismogram*.

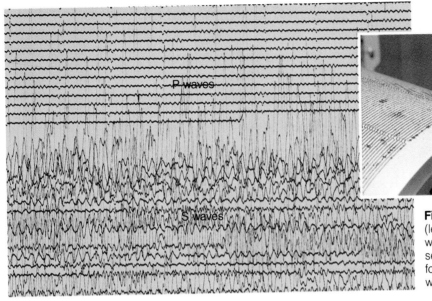

Figure 7–24. The seismogram (left) is a record of earthquake waves. P waves arrive at the seismograph (right) first, followed by S waves. Surface waves arrive last.

P waves travel fastest and are the first waves recorded on the seismogram. The P waves are recorded as a series of small zigzag lines. The S waves arrive after the P waves and appear as larger, more ragged lines. The surface waves arrive last and make the largest lines. The difference in travel time between the waves indicates the distance between the recording station and the epicenter. However, this is not enough information to locate the epicenter. The distance from at least three seismograph stations must be measured before the exact location of an earthquake epicenter can be determined.

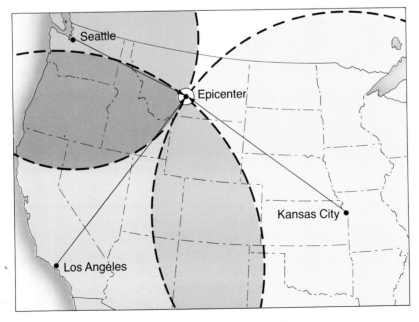

Figure 7–25. Triangulation, diagrammed here, is the process used to determine the location of an earthquake from the recordings of three seismic reporting stations.

Section 3   Earthquakes   **153**

The size, or *magnitude*, of an earthquake is reported using a special scale called the **Richter** (RIHK tur) **scale.** This scale measures the amount of energy released by an earthquake. Each increase of one unit on the Richter scale represents a tenfold increase in earthquake strength. For example, an earthquake magnitude of eight is ten times stronger than an earthquake magnitude of seven, and 100 times greater than a magnitude of six. Table 7–1 shows the magnitudes of some strong earthquakes rated on the Richter scale.

TABLE 7–1: MAJOR RECORDED EARTHQUAKES			
Year	Location	Magnitude	Deaths
1556	China	unknown	830 000[1]
1906	San Francisco	8.3	700
1906	Colombia/Equador	8.9	unknown[2]
1923	Tokyo	8.3	100 000
1976	China	8.0	750 000
1976	Guatemala	8.1	22 000
1985	Mexico City	8.1	5 000

[1]Most deadly earthquake
[2]Strongest earthquake

○ *What is a seismograph?*
○ *What is an earthquake shadow zone?*

Section Review

READING CRITICALLY

1. Compare and contrast P waves, S waves, and surface waves.
2. Why are areas of seismic gaps likely to have stronger, but less frequent, earthquakes than other sections of a fault?

THINKING CRITICALLY

3. Explain why three different seismic readings are needed to find the epicenter of an earthquake.
4. Why is it easier to locate a potential earthquake area than to predict when an earthquake will occur?

TECHNOLOGY: Seismic Records

In the eighteenth century, an Italian geologist developed a pendulum-based instrument that recorded the strength of an earthquake. This instrument works because a heavy, suspended weight resists movement and remains in a fixed position while the earth moves below it.

Scientist at the National Earthquake Center

Later seismographs used a suspended pen, which lightly rests on a rotating, paper-covered drum. The drum is anchored in rock, so that any movement of the earth will cause the pen to make zigzag lines on the drum. This seismograph records vertical movement. In a second type of seismograph, the pen is attached to an arm that swings freely. This seismograph records horizontal movements. Both types are needed to accurately measure earthquake magnitude.

In the 1930s Victor Hugo Benioff designed a seismic instrument that was similar to a telephone. A continuous electrical current was passed through the instrument. When the earth vibrated in an earthquake, this instrument showed tiny variations in current, which could be recorded. Modern instru-

ments are based on this same principle.

Technological advances in the past 20 years have greatly improved the scientist's ability to interpret seismograms. The orientation of the fault plane, the type of movement, and even the direction of movement along the fault can be determined. Computers have also improved the speed and the accuracy of recording seismographic data.

Seismic recording stations are located around the world. All these stations report their data to the National Earthquake Information Service in Golden, Colorado. Each month approximately

A computer-enhanced seismogram

60 000 seismic records are fed into the computers at the center. This information is analyzed by scientists to obtain a better understanding of earthquakes and to aid in developing a method for predicting earthquakes in the future.

A remote seismic station

INVESTIGATION 7: Plotting Earthquake Data

PURPOSE

To use earthquake data to classify a plate boundary

MATERIALS (per student)

Textbook
Graph paper
Pencil

TABLE 1: EARTHQUAKE DATA

Latitude	Longitude	Depth of Quake	Magnitude of Quake
15° N	130° E	10 km	8.0
15° N	130° E	10 km	7.8
15° N	129° E	10 km	8.2
15° N	129° E	50 km	7.7
15° N	128° E	95 km	7.0
15° N	128° E	165 km	7.1
15° N	128° E	180 km	7.6
15° N	128° E	155 km	7.3
15° N	127° E	200 km	7.1
15° N	127° E	305 km	7.2
15° N	127° E	225 km	6.9
15° N	126° E	310 km	7.0
15° N	126° E	275 km	7.5
15° N	125° E	375 km	6.1
15° N	124° E	350 km	6.5
15° N	124° E	445 km	6.4
15° N	123° E	570 km	6.7
15° N	121° E	550 km	6.1
15° N	121° E	510 km	6.9
15° N	120° E	690 km	6.2

PROCEDURE

1. Examine the data presented in the Earthquake Data table. Note that the data was recorded along a single latitude line but at different longitude positions. Refer to the world map on page 592 of the Reference Section to locate the general area represented by this data.
2. Graph the information in the table. Since latitude is constant, graph longitude versus the depth of the earthquake. Longitude should be on the X, or horizontal, axis. Place the reading of 120° E at the intersection of the axes. Space your graph to fill the entire sheet of graph paper.
3. The Y, or vertical, axis should represent the depth of the earthquake. Place 700 km at the top and 0 km at the bottom.
4. Plot the earthquake data on the graph, using the following symbols to indicate the magnitude of the earthquake.

$$+ \ = 6.0 - 6.9$$
$$\times \ = 7.0 - 7.7$$
$$O \ = 7.8 - 8.5$$

ANALYSES AND CONCLUSIONS

1. Sketch a map of the region represented by the earthquake data. On your sketch show where the earthquakes occurred.
2. Describe the pattern on your graph made by the different earthquake depths.

APPLICATION

1. What type of plate boundary do you think the earthquake data indicates? Explain what you think is happening at this boundary by discussing the plates involved and their direction of movement.
2. There are several mountains on the landmasses in this area. Based on your analysis of the type of plate boundary, what type of mountains would you expect to find? Explain how these mountains are related to the plate boundary.

CHAPTER 7 REVIEW

SUMMARY

- Alfred Wegener hypothesized that all the continents were once joined as the supercontinent Pangaea. (7.1)

- The theory of plate tectonics explains the movement of the earth's crust. (7.1)

- There are three types of plate boundaries: divergent, convergent, and transform fault. (7.2)

- Many geologic processes, such as subduction, folding, and faulting, occur along plate boundaries. (7.2)

- Magma sometimes intrudes into other rock layers where it forms batholiths. Fingerlike projections between or across rock layers form sills or dikes. (7.3)

- Lava flows have many characteristics that depend on the viscosity of the lava. (7.3)

- Volcanoes are classified as shield volcanoes, cinder-cone volcanoes, and composite volcanoes. (7.4)

- Hot spots form volcanic island chains in the middle of plates. Geysers form near hot magma. (7.5)

- Seamounts and guyots are undersea volcanoes. (7.6)

- The sudden release of energy along faults causes earthquakes. (7.7)

- Earthquake waves are classified as P waves, S waves, and surface waves. (7.8)

- Earthquake waves travel at different speeds. (7.8)

- Earthquake magnitude is measured on a scale called the Richter scale. (7.8)

Write all answers on a separate sheet of paper.

SCIENCE TERMS

Correctly use each of the following terms in a sentence.

batholith **(143)**
caldera **(146)**
cinder-cone volcano **(145)**
composite volcano **(145)**
convergent boundary **(140)**
divergent boundary **(139)**
earthquakes **(139)**
epicenter **(151)**
focus **(151)**
hot spot **(147)**
P waves **(151)**
Richter scale **(154)**
seamount **(147)**
seismograph **(152)**
shield volcanoes **(144)**
S waves **(151)**
surface waves **(152)**
theory of plate tectonics **(138)**
transform fault **(142)**
viscosity **(144)**
volcanoes **(139)**

SCIENCE QUIZ

Modified True-False

Mark each statement *true* or *false.* If a statement is false, change the underlined term to make the statement true.

1. Aa is a type of <u>lava</u> with high viscosity.

2. Most earthquake activity occurs along <u>the center</u> of a plate.

3. <u>Hot spots</u> melt small areas of the mantle.

4. The distance between a recording station and an earthquake source is calculated by determining the difference in <u>arrival time</u> between different types of waves.

5. The <u>Ring of Fire</u> provides evidence in support of the theory of plate tectonics.

6. <u>Batholiths</u> are igneous intrusions.

continues

Multiple Choice

Write the letter of the choice that best answers the question or completes the statement.

7. The lava that forms a volcano comes from a _____ in the earth's crust.
   a) volcanic cone
   b) seamount
   c) rift
   d) caldera

8. Large, streamlined masses of lava are called
   a) tuff.
   b) pahoehoe.
   c) bombs.
   d) aa.

9. Which of the following is not a type of intrusion?
   a) sill
   b) rift
   c) batholith
   d) dike

10. Cinder cones form from lava that is
    a) low in water and low in silica content.
    b) high in water and low in silica content.
    c) low in water and high in silica content.
    d) high in water and high in silica content.

11. A convergent boundary may be identified by
    a) an ocean trench.
    b) volcanic activity.
    c) a mountain range.
    d) All choices are correct.

12. The magnitude of an earthquake depends on
    a) the speed of the P waves.
    b) the amount of energy that is released by the earth.
    c) the material at the epicenter.
    d) All choices are correct.

Completion

Complete each statement by supplying the correct term.

13. Guyots are _____ volcanoes.

14. The earthquake wave that is a longitudinal wave is a _____.

15. The boundary between the North American plate and the Pacific plate is a _____.

16. Areas where faults stick together are _____.

17. Lava that has a low silica content is likely to form a _____ volcano.

18. Earthquake magnitude is measured on the _____ scale.

Short Answer

19. How can scientists use seismic gaps to predict the location of future earthquakes?

20. Why do scientists anchor a seismograph into solid rock?

21. What is the difference between a convergent and a divergent plate boundary?

22. Explain the process of triangulation and relate it to determining the location of earthquake epicenters.

Writing Critically

23. Explain why scientists predict that any new Hawaiian islands that form will do so to the southeast of the present chain.

24. How do viscosity, silica content, and steam affect the eruption of volcanoes?

25. Describe the motions of P waves, S waves, and surface waves. Explain the effects that each one would have on buildings constructed of wood, stone, and steel. Why is the motion of surface waves more destructive than either P waves or S waves?

EXTENSION

1. Study the effects of the 1985 earthquake in Mexico City. Write a report describing how the ancient lake bed increased the effects of the wave motion.

2. Research the way in which geothermal energy is used as a source of electricity in certain areas of the world. Relate geothermal energy to magma. Present your findings to the class.

3. Find out where the closest seismic recording station is to your home. If possible arrange to visit the station.

APPLICATION/CRITICAL THINKING

1. Explain how a lava flow, such as the one that formed the Columbia River Plateau, might be related to the volcanic activity and geysers of a hot spot below Yellowstone National Park.

2. What clues would you look for in trying to determine whether a lava flow was composed of aa or pahoehoe?

3. After reviewing Table 7–1 on page 154, explain why the strongest earthquakes do not necessarily cause the most deaths. Also explain why an early warning system for strong earthquakes would be vital for heavily populated areas.

FOR FURTHER READING

Aylesworth, T. G., and V. L. Aylesworth. *The Mount St. Helens Disaster: What We've Learned.* Danbury, CT: Franklin Watts, 1983. This book offers a complete account of the 1980 eruption of Mount St. Helens.

McConnell, A. *The World Beneath Us.* New York: Facts on File, 1985. The author provides very interesting descriptions of the geologic time table, fossils, plate tectonics, earthquake prediction, and volcano activity.

Walker, B., ed. *Earthquake.* New York: Time-Life Books, 1982. This book is an outstanding reference on all aspects of earthquakes.

Challenge Your Thinking

These buildings were all in good condition before the 1985 earthquake in Mexico City. One building (right) was a solid old structure built with bricks and cement, while another (far left) was constructed of steel and glass. The newer building was built to withstand the effects of an earthquake, while the older buildings obviously were not. What kinds of things might have been built into the newer building to make it earthquake resistant?

UNIT 2 THE SCIENCE CONNECTION:

Diamonds

From laboratory experiments, scientists have concluded that diamonds are formed under very special circumstances. The temperature must be at least 1400°C and the pressure about 40 000 times the surface pressure on the earth in order to cause atoms of carbon to crystallize into diamonds. Diamonds appear to form inside structures called *diamond pipes,* which are slender, carrot-shaped holes filled with igneous rock.

The diamond-bearing rocks probably reached the surface through these pipes when ancient concentrations of carbon dioxide gas exploded. Apparently, no new diamond pipes have formed in the past 15 million years.

When the igneous rocks of the diamond pipes moved toward the surface, the magma mixed with bits of rock from the surrounding area. As it moved upward, the water in the ground also interacted with the magma. The resulting rock, called *kimberlite,* is a jumbled mixture of minerals, including diamonds. Kimberlite is found only in the oldest and most stable continental regions. Although kimberlites have been found in Africa, Brazil, India, Australia, North America, and Siberia, those in Africa provide 95 percent of the world's total diamond supply.

You may think of diamonds only as expensive gemstones. Diamonds are also highly valued in industry for their special properties. Diamonds are the hardest known substance on Earth. Many gems are cut using a paper-thin disk made of hardened bronze coated with diamond dust.

Diamond pipe formation

Sorting diamonds

Grinding industrial diamonds

Diamonds also have the highest thermal conductivity of any known substance. This means that they do not get hot when used as a cutting instrument. The heat that develops from friction is rapidly conducted through the diamond so that the cutting edge stays cool.

Surgeons use diamond-tipped scalpels for delicate operations such as eye surgery. Diamonds are also used in high-quality record-player needles and as heat conductors in miniature transmitters. The Pioneer space probe of Venus used a diamond wafer as a tiny window. The diamond wafer was used instead of glass or plastic because diamonds can withstand the extremes of heat and cold in outer space.

Diamonds are also used in drilling operations. Crystals of diamonds are used as an abrasive called *bort.* Bort is mounted on the end of a cylinder called a *bit.* When the rotating bit is forced against a rock, it cuts through the rock. The mining industry makes extensive use of the bort drill.

The greatest use of industrial diamonds is in the production of diamond grinding wheels. These wheels are especially hard because they have diamond dust embedded in them. Diamond wheels are able to shape and sharpen the new ultrahard tools that are being produced by modern industry.

Jewelry quality diamonds

METEOROLOGY

A tornado can toss a train as if it were a toy or drive a board through a concrete wall. Most people flee from this awesome power, but some scientists actually chase tornadoes. They measure a storm's power with a special package of instruments named TOTO, after Dorothy's dog in the movie *The Wizard of Oz*. With TOTO, scientists are learning more about what causes tornadoes and how they can be predicted.

- What kinds of weather conditions create tornadoes and other violent storms?

- How does weather radar show storms?

- Why does weather occur only in the lower layer of the atmosphere?

- How do the TOTO weather chasers know where to find tornadoes?

By reading the chapters in this unit, you will learn the answers to these questions. You will also develop an understanding of concepts that will allow you to answer many of your own questions about the study of weather and climate.

A rapidly moving tornado

The Atmosphere

If you live in the northern part of the United States, you may have seen lights like these in the sky. These lights are caused by charged particles from the sun interacting with the atmosphere. In the Northern Hemisphere, these lights are known as the *aurora borealis* and in the Southern Hemisphere as the *aurora australis*.

The aurora borealis, or northern lights

1 Composition of the Atmosphere

SECTION OBJECTIVES

After completing this section, you should be able to:

■ **Explain** why the earth has an atmosphere.
■ **Compare** the earth's early atmosphere with the present atmosphere.
■ **Describe** the earth's atmosphere and **explain** why it does not disappear into space.

NEW SCIENCE TERMS

atmosphere
solar wind
photosynthesis
sunspots
ozone

8.1 The Primitive Atmosphere

Air is just about the only thing that is free. Most people take for granted an unlimited supply of air. Why does Earth have air? Where did this air come from?

The **atmosphere** is the layer of gases that surrounds the earth. Earth's original atmosphere probably formed along with the planet. Scientists theorize that the sun and all the planets formed from a cloud of gas and dust about 4.5 billion years ago. The cloud collapsed into a central body, which formed the sun and ten rings. The rings eventually became the nine planets. One ring did not form a planet.

It is likely that high-energy radiation from the young sun, called the **solar wind,** scattered into space most of the gases that had originally collected around the planets closest to the sun, including Earth. Some of the more distant planets, such as Jupiter and Saturn, have retained their first atmospheres.

After the solar wind scattered Earth's original atmosphere, a new atmosphere formed from gases such as hydrogen, methane, and water vapor that were trapped inside the earth when it formed. Most of these gases escaped during volcanic eruptions. In fact, escaping gases are still adding to the atmosphere today.

Figure 8–1. Many of the gases of the primitive earth's atmosphere probably came from volcanoes.

165

One abundant gas of this primitive atmosphere was water vapor. This water vapor produced rains that lasted for millions of years. These rains filled in the low places on the earth and formed the primitive ocean. Escaping hydrogen was lost into space. Other gases, such as carbon dioxide and sulfur dioxide, were created by chemical reaction and remained in the new atmosphere. This primitive atmosphere was very different from Earth's present atmosphere, as shown in Table 8–1.

TABLE 8–1: COMPARISON OF THE PRIMITIVE ATMOSPHERE AND THE PRESENT ATMOSPHERE

Gas	Percent of Molecules	
	Primitive Atmosphere	Present Atmosphere
Carbon dioxide (CO_2)	92.2	0.03
Nitrogen (N_2)	5.1	78.1
Sulfur dioxide (SO_2)	2.3	0.0
Hydrogen sulfide (H_2S)	0.2	0.0
Methane (CH_4)	0.1	0.0
Ammonia (NH_3)	0.1	0.0
Oxygen (O_2)	0.0	20.9
Argon (Ar)	0.0	0.9

The living organisms that evolved on Earth were probably responsible for the change from the primitive atmosphere to the present atmosphere. Bacteria, among the first living organisms on Earth, developed two processes that were fundamental to life. At first, they used ammonia from the primitive atmosphere to make the chemical compounds necessary for life. Later, the bacteria used sunlight as a source of energy to make their own food. These bacteria, called *cyanobacteria*, made their food through the process of photosynthesis (foht oh SIHN thuh sihs). During **photosynthesis,** water and carbon dioxide are combined chemically to make sugar.

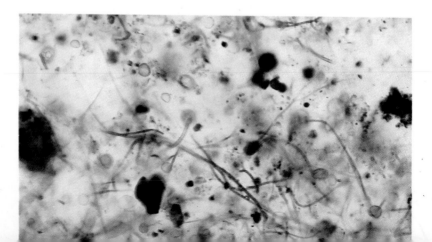

Figure 8–2. Shown here are fossils of cyanobacteria and algae. Cyanobacteria were the likely source of oxygen for the primitive earth's atmosphere.

Photosynthesis also releases oxygen as a byproduct. Other life forms needed this oxygen to survive. Today green plants, rather than cyanobacteria, produce most of the oxygen in the atmosphere.

Of the gases that were released into the primitive atmosphere, some, especially oxygen and carbon dioxide, dissolved in the ocean. In water, carbon dioxide reacts with rocks, forming calcium carbonate. By about 600 million years ago, enough oxygen had dissolved in the ocean to allow animal life to evolve there. Many of these early life forms had shells of calcium carbonate. Most of the carbon dioxide in the earth's atmosphere was used by these animals to form their shells. Only a trace of carbon dioxide remains in the atmosphere today.

Figure 8–3. Early corals and worms, from which these fossils formed, used much of the carbon dioxide dissolved in ocean water to make their skeletons and tubes.

With the removal of carbon dioxide from the primitive atmosphere, nitrogen became the most abundant gas. Oxygen, from photosynthesis, soon became the second most abundant gas. As oxygen increased, the sulfur dioxide in the primitive atmosphere was oxidized to sulfuric acid. The sulfuric acid then washed out of the atmosphere as "acid rain." The acid rain you read about today contains sulfuric and other acids, but these are from industrial pollution.

○ *What were the most common gases of the primitive atmosphere?*
○ *What caused the primitive atmosphere to change?*

8.2 **The Present Atmosphere**

You may wonder why the atmosphere does not disappear into space. What keeps it stuck to the earth? The answer is gravity.

Some gases, such as hydrogen and helium, the lightest gases, do escape because they tend to float to the top of the atmosphere. There they are scattered by the solar wind and escape into space.

DISCOVER

Making Oxygen

Light a small candle and cover it with a glass jar. **CAUTION: Be careful with the flame.** Time how long the candle will burn before it goes out. Why does the candle go out?

Now place a small green plant next to the candle and light the candle again. Cover both the plant and the candle with a glass jar. Time how long the candle burns before it goes out. Does the candle burn for the same time? Explain any difference in time.

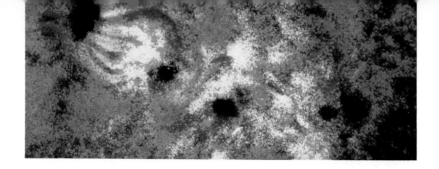

Figure 8–4. Sunspots create strong magnetic fields that accelerate the solar wind.

Figure 8–5. Scientists have discovered a hole in the ozone layer of the upper atmosphere over Antarctica. If similar holes should develop over populated areas, large amounts of dangerous ultraviolet radiation could cause an increase in genetic disorders and skin cancers.

The particles of the solar wind, mostly protons, are accelerated by strong magnetic fields created by sunspots. **Sunspots** are enormous whorls of gases on the surface of the sun. The number of sunspots changes in a cycle that repeats itself every 11 years. When there are few sunspots, the solar wind is weak; when there are many sunspots, the solar wind is strong. The solar wind increases the speed of all gas atoms in the upper atmosphere.

In addition to the elements and compounds shown in Table 8–1, the present atmosphere also contains variable amounts of water vapor. The amount of water vapor ranges from almost zero over the deserts to 2.5 percent over the tropic regions near the equator.

Another small but very important component of the atmosphere is ozone. An **ozone** molecule consists of three atoms of oxygen. Ozone is extremely important to life on Earth because it absorbs dangerous ultraviolet radiation from the sun. Ultraviolet radiation causes sunburn, skin cancer, and possible genetic damage inside cells.

The ozone layer is in danger of being destroyed by chemical pollution, particularly from substances added to the atmosphere by industrial processes and by the propellants in aerosol cans. In order to save the ozone layer, laws have been passed to limit this kind of pollution.

○ *What are sunspots?*
○ *Why is the presence of ozone in the atmosphere important?*

Section Review

READING CRITICALLY

1. How did the early atmosphere differ from today's atmosphere?
2. Why do hydrogen and helium escape from the atmosphere?

THINKING CRITICALLY

3. Why has the level of oxygen in the atmosphere not continued to increase beyond its present percentage?
4. What might happen to the number of cases of skin cancer if the ozone level of the atmosphere is reduced?

SKILL ACTIVITY: Reading for Meaning

BACKGROUND

In order for you to learn from textbooks, magazines, and newspapers, you must be able to understand what you read. One way to understand is to answer questions about what you have just read.

PROCEDURE

In this textbook, questions have been provided after sections to help you see if you have understood what you have read. When you read something that does not provide questions, you can ask yourself questions. Questioning is a very important technique in learning.

Of the questions you will answer about what you have read, some can be answered directly from the material and some you have to think about. If you can answer the ones you have to think about, you have really understood what you have read.

APPLICATION

Read the following selection and answer the questions that follow.

Atmospheric Gases

The many gases of the atmosphere are mixed in a delicate balance. Each gas, such as nitrogen or oxygen, is found in the atmosphere in very specific amounts. Each plays an important role in the maintenance of life on our planet.

The atoms and compounds that make up the gases circulate throughout the environment in very specific chemical pathways called *cycles*. There is a water cycle, an oxygen cycle, a carbon cycle, and a nitrogen cycle, which is shown below. Upsetting the flow of any of these cycles can have drastic consequences for our environment and the plants and animals that live in it.

Air pollution created by modern industries disturbs the balance of atmospheric substances. Industrial pollutants poured into the atmosphere can disturb natural chemical reactions that keep the cycles going. Although this problem has existed for several centuries, scientists have only, in the past few decades, begun to examine it very closely.

1. What is a cycle?
2. Why are cycles important?
3. Why would upsetting the way a cycle works cause problems for living organisms?
4. How does air pollution affect the cycles?

USING WHAT YOU HAVE LEARNED

Go to the library and find a newspaper or magazine article about air pollution. Write a brief report that answers the following questions:

1. What kind of pollution does the article discuss?
2. What specific pollutants are involved?
3. Which industry is mostly responsible for the pollution?

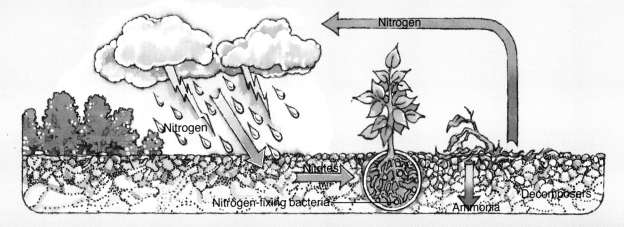

Nitrogen

Nitrogen

Nitrates

Nitrogen-fixing bacteria

Ammonia

Decomposers

② Solar Radiation

NEW SCIENCE TERMS

photons
electromagnetic waves
radiation balance
greenhouse effect

SECTION OBJECTIVES

After completing this section, you should be able to:

■ **Explain** the way in which Earth receives energy from the sun.
■ **Describe** the relationship of solar radiation to radiation balance.
■ **Discuss** how the addition of carbon dioxide to the atmosphere affects the radiation balance of the earth.

8.3 Electromagnetic Waves

You might think that the use of solar energy is something new. The fact is, practically all the energy on Earth is received as *solar radiation*, or energy from the sun.

Light consists of particles called **photons** (FOH tahnz) that travel from the sun to the earth in 8.3 minutes. Photons travel as electromagnetic waves. **Electromagnetic waves** are energy waves that travel at the speed of light.

Figure 8–6. Volcanic dust in the atmosphere causes more red light waves to reach the earth (left). The structure of waves is shown in the diagram (right).

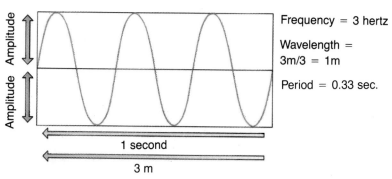

Frequency = 3 hertz

Wavelength = 3m/3 = 1m

Period = 0.33 sec.

Electromagnetic waves vibrate like the strings of a guitar. The number of vibrations that occur in one second is the *frequency* of the wave. Frequency is measured in SI units called *hertz* (Hz). One hertz is equal to one vibration, or one cycle, per second.

Consider, for example, the human heart. Your heart beats about 60 times per minute. The frequency of this heartbeat would be one hertz, or one beat per second.

Each wave has a high point, called a *crest*, and a low point, called a *trough*. The distance from crest to crest or from trough to trough is called the *wavelength*. The height of the wave, from the midpoint to the trough or from the midpoint to the crest, is the *amplitude*.

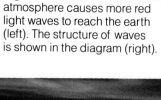

A MATTER OF FACT

The speed of photons in empty space is constant at 299 792 458 m/s and is called the *speed of light*. It would take 125 days to travel that distance in a car going 100 km/h.

Electromagnetic waves range in frequency from less than 1 Hz to 100 000 billion billion (10^{23}) Hz. The higher the frequency of the wave, the shorter the wavelength. Figure 8–7 shows different wavelengths in the *electromagnetic spectrum.*

Human senses can detect only a portion of the electromagnetic spectrum. However, humans may be affected by wavelengths they cannot detect. Some waves that affect humans are ultraviolet rays, X rays, and gamma rays.

Notice that the range of visible light is a very small part of the electromagnetic spectrum. The sun emits light of all colors. The human eye sees the combination of these colors as white.

○ **What are electromagnetic waves?**
○ **What is frequency?**

Figure 8–7. White light contains all the colors of the spectrum (below). The electromagnetic spectrum is shown in the diagram (left).

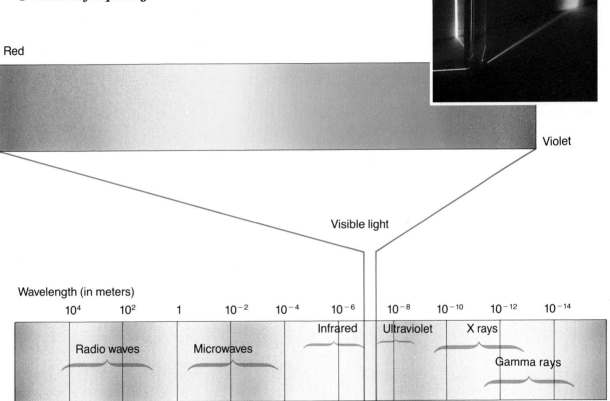

Red

Violet

Visible light

Wavelength (in meters)

| 10^4 | 10^2 | 1 | 10^{-2} | 10^{-4} | 10^{-6} | 10^{-8} | 10^{-10} | 10^{-12} | 10^{-14} |

Infrared   Ultraviolet   X rays

Radio waves   Microwaves   Gamma rays

8.4 Radiation Balance

Have you ever noticed that you feel warmer on a sunny day if you are wearing something dark in color? This is because different surfaces absorb different amounts of solar radiation. A black surface absorbs nearly all the solar radiation that falls on it. Consequently, any dark-colored object set out in the sun becomes very hot. A white surface reflects almost all the solar radiation that falls on it. For this reason, white objects stay cooler in the sun than objects of any other color.

Most atmospheric gases let solar radiation through without reflecting or absorbing much of it. Clouds absorb or reflect about 25 percent of the solar radiation entering the atmosphere. About 30 percent of the solar radiation that does reach the earth is in the form of heat.

For the earth to remain livable, there must be a balance between the amount of solar radiation that is absorbed and the amount that is reflected into space. **Radiation balance** is the balance between solar radiation coming into Earth's atmosphere and radiation going out from Earth. Most of this outgoing radiation is in the form of heat. Heat radiation is in the infrared part of the electromagnetic spectrum.

Infrared radiation is reflected and radiated into space from the ground and from cloud tops. If too much heat is reflected into space, the atmosphere cools down. If too little heat escapes, the atmosphere heats up. Either case could upset the weather conditions that now exist on Earth. Radiation balance is maintained in delicate adjustment by many physical and environmental factors.

Figure 8–8. This diagram shows how the radiation balance of the earth is maintained.

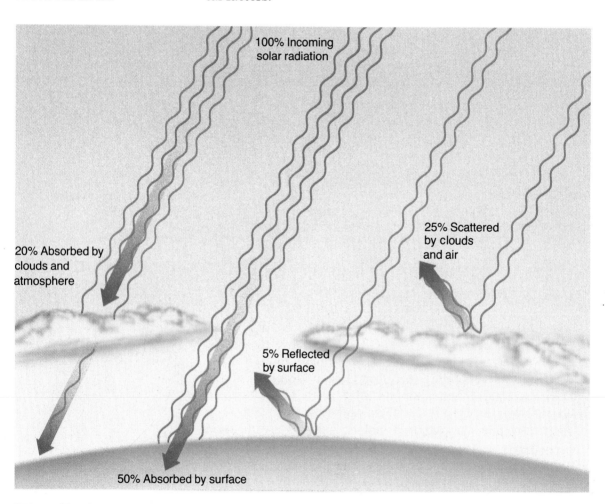

100% Incoming solar radiation

25% Scattered by clouds and air

20% Absorbed by clouds and atmosphere

5% Reflected by surface

50% Absorbed by surface

Figure 8–9. If the earth receives more heat than it loses, a warm climate results. If the earth receives less heat than it loses, a cold climate results.

The radiation balance of the earth has shifted many times throughout the earth's history. One extreme in atmospheric temperature occurs during times of extreme cold, when more of the earth's surface is covered with snow and ice than is now covered. These conditions cause more of the sun's radiation to be reflected into space. This higher rate of reflection causes the earth to be colder. Why would a covering of snow and ice cause the earth to reflect radiation?

The other extreme occurs when there is little or no ice on Earth, and the oceans extend over broad portions of the land. More solar energy is absorbed by the earth, and the earth is warmer. From one extreme to the other, however, the average temperature of the earth changes by only a few degrees Celsius. Temperature changes have never been large enough to endanger all life on the earth. However, some species, such as dinosaurs, may have become extinct because of temperature changes.

Unfortunately, activities of people worldwide threaten to change dangerously the radiation balance of the earth. Aerosols, chlorine-containing compounds, and carbon dioxide are being added continuously to the atmosphere. These substances may change the radiation balance enough to affect life on the planet.

Carbon dioxide, for example, traps infrared radiation that would normally be radiated back into space. This same effect can be seen in a greenhouse or in a car with the windows closed. Like carbon dioxide, glass lets sunlight through but traps the returning infrared radiation. The inside of the greenhouse or car warms up. The temperature rise caused by the addition of carbon dioxide to the atmosphere produces a **greenhouse effect.**

Some scientists believe that human activities may change the radiation balance of the earth enough to trigger another ice age or to melt the ice that now covers Antarctica and Greenland. What problems do you foresee if a new ice age were to begin? What would happen if all the polar ice were to melt and the level of the oceans were to rise?

○ *What is radiation balance?*
○ *What causes the greenhouse effect?*

Figure 8–10. A greenhouse traps heat, so the temperature remains high even on a cold day. The greenhouse effect would cause the average temperature of the earth to increase, possibly melting the polar ice.

Section Review

READING CRITICALLY

1. What is the effect of carbon dioxide on the radiation balance of the earth?
2. Why is the radiation balance of the earth important?

THINKING CRITICALLY

3. How could the fact that most of the sunlight received on Earth is in the blue-green wavelength affect the way life developed on Earth?
4. How can you explain the fact that some people think the earth's temperature may increase, while others think the temperature may decrease?

SECTION OBJECTIVES

After completing this section, you should be able to:

- **Describe** the cause of air pressure.
- **List** the layers of the atmosphere and **discuss** their characteristics.

NEW SCIENCE TERMS

atmospheric pressure
barometer
Van Allen belts

8.5 Atmospheric Pressure

Have you ever noticed how hard it is to open a new jar of pickles? Many food products are vacuum packed—that is, packed without air. The atmosphere pushes down on the lid and makes it very difficult to open.

If you could measure a column of air 1 cm² across that reaches from sea level to outer space, you would find that the air has a total mass of 1.033 kg. This much mass exerts a considerable force, or weight, on objects on the earth. **Atmospheric pressure** is the weight of the atmosphere. Atmospheric pressure is determined by measuring the force the atmosphere exerts on a column of mercury. Mercury is 13.6 times denser than water, so atmospheric pressure that would raise a column of mercury 760 mm high would raise a column of water 10.33 m high.

The pressure exerted by the atmosphere decreases with altitude. Mount Everest is one of the highest mountains on Earth. Atmospheric pressure at the top of Mount Everest is less than one-third of the pressure at sea level. Mountain climbers trying to reach the top have to carry oxygen tanks and breathe through regulators. This is necessary because high on the mountain the atmosphere becomes so thin that there is not enough oxygen for the climbers to do the strenuous work of climbing.

One instrument used to measure atmospheric pressure is called a **barometer** (buh RAHM uh tuhr). The most commonly used type is the *aneroid* (AN uhr oihd) *barometer,* shown in Figure 8–11. An aneroid barometer consists of a thin-walled metal chamber from which some of the air has been removed. As atmospheric pressure changes, the shape of the chamber changes. The change in shape activates a pointer that indicates the atmospheric pressure on a gauge.

○ *Why does the atmosphere exert pressure?*
○ *What is atmospheric pressure?*

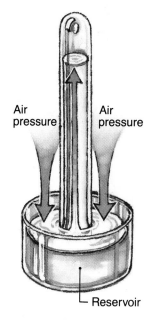

Figure 8–11. An early barometer (top) was made by an Italian scientist, Evangelista Toricelli. A modern aneroid barometer is also shown (bottom).

CAREERS

AUTOMOTIVE TECHNICIAN

Since the 1960s, *automotive technicians* have played a very important role in trying to keep our air free of pollution. They are skilled at adjusting features of gasoline engines that improve their efficiency.

Most automotive technicians work at neighborhood garages. Some own or manage businesses that specialize in automobile maintenance, such as tune-ups. Others are employed by large companies that manufacture cars. All automotive technicians must have at least a high-school diploma, and many must pass state and local examinations that test their knowledge of engines and laws governing pollution-control standards.

For Additional Information
Society of Automotive Engineers
400 Commonwealth Drive
Warrendale, PA 15096

ENVIRONMENTAL PROTECTION AGENT

The Environmental Protection Agency is a government agency that works to improve the environment. *Environmental protection agents* are dedicated to the control of pollution caused by sewage, agricultural pesticides, factory noise and emissions, automobile emissions, and more.

The agency employs people of many educational backgrounds. However, most have college degrees. You do not need a degree in science to work for the EPA, but researchers and field agents usually have one.

For Additional Information
The Environmental Protection
    Agency
401 M Street, S.W.
Washington, DC 20460

ATMOSPHERIC CHEMIST

Atmospheric chemists conduct research on atmospheric pollution. They study the effects of modern industry on rain, snow, and cloud chemistry.

A new and exciting field for atmospheric chemists has grown with the success of the space program, that is, studying the atmospheres of other planets.

Atmospheric chemists must have an advanced degree in chemistry. They work for government agencies like the EPA or the National Aeronautics and Space Administration (NASA) or do research for universities.

For Additional Information
National Academy of Sciences
Office of Public Affairs
2101 Constitution Avenue
Washington, DC 20418

ACTIVITY: Demonstrating Atmospheric Pressure

Can the fact that the atmosphere exerts pressure be shown?

MATERIALS (per group of 3 or 4)

safety goggles, laboratory aprons, gloves, water, empty soda can, bowl, Bunsen burner or alcohol lamp, tripod, wire gauze

PROCEDURE

1. **CAUTION: Put on safety goggles, a laboratory apron, and gloves before you begin this experiment.**
2. Pour about 2 mL of water into the soda can.
3. Fill the bowl nearly to the brim with COLD water.
4. Put the burner or lamp on a low flame.
5. Set the can on the wire gauze atop the tripod over the Bunsen burner or lamp.
6. Count for 30 seconds after the water in the can gives off steam and boils steadily.
7. Remove the flame from under the can.
8. Quickly turn the can upside down into the bowl of water. Record your observations.

CONCLUSIONS / APPLICATIONS

1. What happened to the can when it was placed in the cold water? Explain.
2. What might have happened to the can above? Explain.
3. Draw a conclusion about the surrounding atmosphere based on your observations.

8.6 Layers of the Atmosphere

Imagine a wedding cake four layers high. This is a very unusual wedding cake, for the guests all have different tastes. The bottom, or first layer, is marbled. There is cool, dense, dark chocolate swirled with warm, light, milk chocolate.

The second layer is yellow cake, but since it has just come out of the oven, it is cooling at the bottom and still very warm at the top. The third layer is dull, plain white cake, warm at the bottom where it touches the second layer, but much cooler at the top.

The fourth layer is the most exciting to see. Near the bottom of this layer, the cake is greenish with touches of red, and it seems to glow with a light of its own. The top of this layer is angel food, all light and thin; it does not really end, it just sort of trails off into nothing.

The earth's atmosphere is much like this layered cake; each layer has its own characteristics of temperature and composition.

> **A MATTER OF FACT**
>
> If all the water were squeezed out of the atmosphere, the resulting rainfall would be only 0.3 cm. Yet without water vapor in the atmosphere, there would be no life on land.

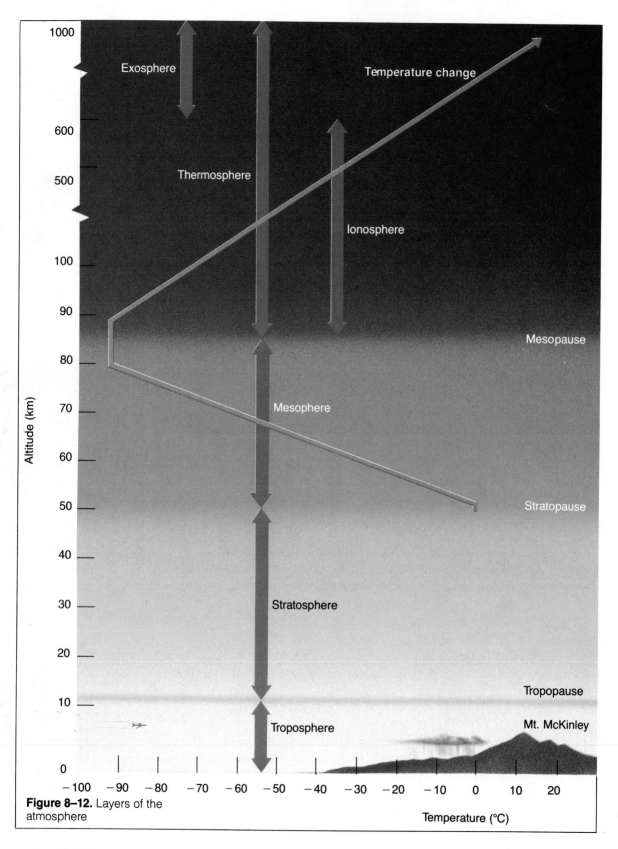

Figure 8–12. Layers of the atmosphere

Troposphere   The bottom layer of the atmosphere, the one nearest the earth, is called the *troposphere*. The troposphere is like the bottom layer of the wedding cake; it is here that much mixing of air occurs. Air near the ground is warmer and less dense than air above. Warm air rises and cold air sinks, forcing the troposphere to turn over. This overturning is a type of convection, which creates the weather systems of the world. Although it is not distinct, like the difference between the layers of a cake, the boundary between the troposphere and the next layer is called the *tropopause*.

Stratosphere   Above the tropopause is the second layer of the atmosphere, which is called the *stratosphere*. Like the second layer of the wedding cake, the stratosphere is cool on the bottom and warm at the top.

Most of the ozone of the atmosphere is located in the stratosphere. The concentration of ozone increases with altitude, reaching a maximum at 22 km. Ozone absorbs not only ultraviolet radiation but also some infrared and visible light. Most of the absorption of radiation occurs near the upper boundary of the stratosphere, called the *stratopause*.

Convection does not occur in the stratosphere because the less dense, warmer air is above the more dense, colder air. In fact, the stratosphere is named for its stable stratification, or layers.

Mesosphere   Above the stratopause lies the third layer of the atmosphere, which is called the *mesosphere*. Like the third layer of the cake, there is really nothing special about the mesosphere. The temperature drops constantly from the base of the mesosphere to the *mesopause*, at the top. This drop in temperature is due to a decrease in the concentration of ozone and an increase in distance from the warm surface of the earth.

Thermosphere   Above the mesopause lies the *thermosphere*, which goes all the way to the edge of the atmosphere. This is like the layer at the top of the cake.

Atmospheric gases are so thin in the thermosphere that the protons of the solar wind easily pass through. The atmospheric atoms and molecules hit by the protons move faster, causing a temperature rise. In addition, the solar wind removes electrons from atoms to form ions. This ionization occurs in the lower part of the thermosphere, called the *ionosphere*. Ionization is strongest at high latitudes because the ionizing particles are directed toward the poles by the earth's magnetic field. When the ions recapture their electrons, photons are released, and a colorful display known as an *aurora* is created. The upper

Figure 8–13. Although the layers of the atmosphere are not visible, as this photograph seems to indicate, they can be distinguished by temperature differences.

portion of the thermosphere is called the *exosphere*. Here the atmospheric gases are so thin that a gas molecule travels an average of 650 km before hitting another molecule. At sea level a molecule can travel only 1/100 000 mm before hitting another molecule.

Van Allen Belts   High-energy protons and electrons from beyond the solar system are trapped in a belt about 3200 km above the equator. A second belt, about 25 000 km above the equator, traps lower-energy protons and electrons from the sun. These belts are called the **Van Allen belts** after the American physicist, James Van Allen, who discovered them in 1958. These two belts act as holding areas for incoming protons and electrons and protect the earth from these high-energy particles.

○ *Name the layers of the atmosphere.*
○ *What characteristic is used to separate one atmospheric layer from another?*

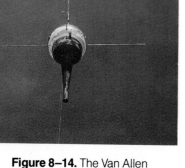

Figure 8–14. The Van Allen radiation belts (right) were discovered by a Vanguard satellite, launched in 1958 (above).

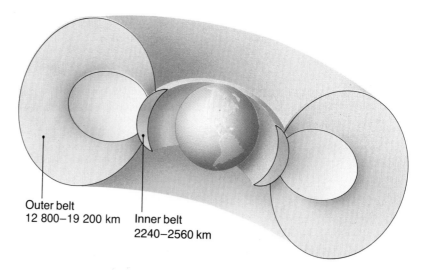

Outer belt
12 800–19 200 km

Inner belt
2240–2560 km

Section Review

READING CRITICALLY

1. What is a barometer and what does it measure?
2. Describe the temperature changes in each layer of the atmosphere.

THINKING CRITICALLY

3. If a barometer reads 760 mm at sea level and 660 mm at 1000 m above sea level, what might it read in Death Valley, which is nearly 100 m below sea level?
4. Why can auroras be seen only at high latitudes?

SECTION
4 Motions of the Atmosphere

SECTION OBJECTIVES

After completing this section, you should be able to:

- **Discuss** how the Coriolis effect determines the circulation of winds.
- **List** and **locate** the planetary wind belts.
- **Describe** the cause of land and sea breezes.

NEW SCIENCE TERMS

Coriolis effect
jet stream

8.7 The Coriolis Effect

In the nineteenth century, Gaspard Gustaf de Coriolis, a French physicist, studied the path of bodies in motion on the earth's surface. His explanation of the force that influences these motions is complicated. However, it can be described with a simple analogy.

Suppose that the earth did not rotate and was a smooth sphere with a very icy surface. Now imagine that you are on ice skates at the North Pole and wish to go to New Orleans. A friend gives you a mighty push, and you move south toward New Orleans. The distance between the North Pole and New Orleans is 6673 km. If you were traveling at 1000 km/h, you would reach New Orleans in just over 6.5 hours. However, the earth does rotate. In 24 hours the earth makes one full turn. During the time that you are moving south, New Orleans is moving east. By the time you get to New Orleans, it is gone and you will find the Pacific Ocean in its place.

In the Northern Hemisphere, moving objects tend to move to the right because of what is called the **Coriolis** (cawhr ee OH lihs) **effect.** If you want to get to New Orleans, you cannot just

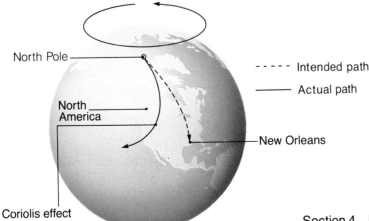

- - - - - Intended path

——— Actual path

Figure 8–15. In the Northern Hemisphere, moving objects are deflected toward the right by the Coriolis effect.

skate south. You have to go toward the east and chase after New Orleans. You have to continually change your direction. This change in direction is called *Coriolis acceleration.*

Now imagine you are at the South Pole and wish to skate to Rio de Janeiro. Your friend gives you another push, and you move north toward the equator. Traveling again at 1000 km/h, it takes you 7.5 hours to reach the latitude of Rio de Janeiro. By the time you get there, however, Rio de Janeiro is gone. In its place you find the South Pacific. Now, it seems you have veered to the left. In the Southern Hemisphere, moving objects tend to move to the left, again because of the Coriolis effect. If you wish to get to Rio de Janeiro, you will have to keep veering toward the east to follow the rotation of the earth.

The Coriolis effect occurs when Coriolis acceleration is insufficient or missing. A moving body is deflected to the right in the Northern Hemisphere and to the left in the Southern Hemisphere.

The Coriolis effect acts dramatically on the lower atmosphere. Since there is not enough friction between the air and the ground to force the air to follow the rotation of the earth, moving air masses are deflected to the right in the Northern Hemisphere and to the left in the Southern Hemisphere. These deflections cause the circular motions of the winds called *gyres* (JYRZ). Gyres are clockwise in the Northern Hemisphere and counterclockwise in the Southern Hemisphere.

○ **What is the Coriolis effect?**
○ **What causes the Coriolis effect?**

A MATTER OF FACT
Ocean currents are also affected by the Coriolis effect. In the Northern Hemisphere these currents are deflected to the right, the same as currents of air.

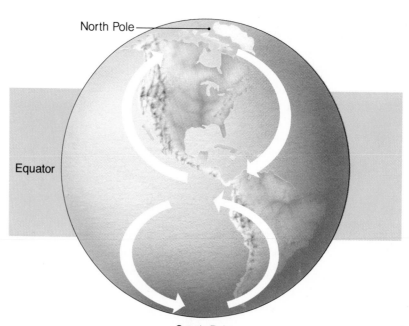

Figure 8–16. In the Northern Hemisphere, the gyres of the earth's winds are clockwise. These gyres are counterclockwise in the Southern Hemisphere.

North Pole

Equator

South Pole

8.8 Planetary Winds

Earth's atmosphere is constantly in motion. This motion is due to heat energy. The source of the heat that keeps the atmosphere in motion is the sun, which warms up the low latitudes much more than the high latitudes. Since hot air is less dense than cold air, you would expect the heated air to rise at the equator, travel to the poles, sink there, and return to the equator along the ground. However, this is not what happens.

The motion of the earth also adds motion to the atmosphere in the form of the Coriolis effect. Because of the Coriolis effect, hot air rising at the equator cannot travel straight to the poles. By the time it reaches latitude 30° North, the air has been turned eastward. The air cools and sinks. As it sinks, some of the air continues toward the poles. Most of it, however, spreads out to the north and south along the ground. Tne air spreading toward the equator forms a planetary wind belt called the *trade winds*. The air spreading northward forms another planetary wind belt called the *westerlies*.

The polar areas are very cold. Cold, dense air sinks over the poles and streams out at ground level. The streaming is clockwise as seen looking down over the Arctic; it is counterclockwise as seen looking down over the Antarctic. As a result, the polar winds blow from east to west in both hemispheres, forming the *polar easterlies*. The planetary wind systems of the world are shown in Figure 8–17.

DISCOVER

Creating Wind Belts

Place two fans about 1 m apart, blowing toward each other. Light a candle or a match and then carefully blow it out, so that there is a lot of smoke. Direct the smoke into the area in which the air from the two fans meets. What happens to the smoke in this area? What might happen in the atmosphere where two wind systems meet?

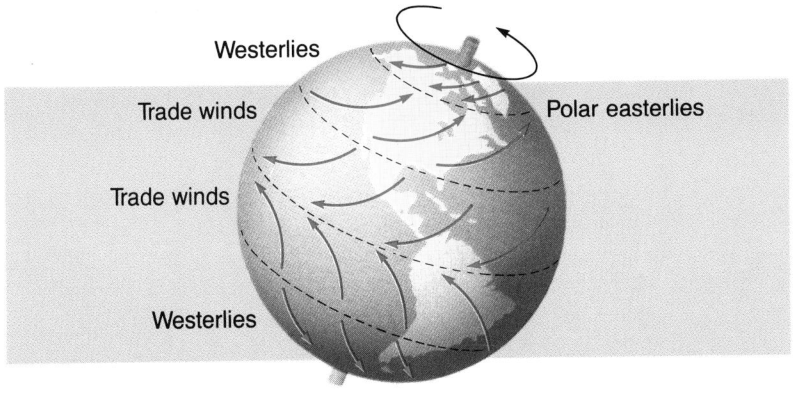

Westerlies

Trade winds

Polar easterlies

Trade winds

Westerlies

Figure 8–17. This diagram shows the planetary winds of the earth.

At latitudes below 40°, the tropopause is at an altitude of 16 km. Between latitudes 40° and 50°, the tropopause drops to an altitude of 10 km. There is a strong pressure difference across this drop. As air rushes toward the poles to balance the pressure, it is deflected eastward by the Coriolis effect. The result is a high-speed wind called the **jet stream,** circling each polar area at an altitude of 12 km.

The jet stream's speed averages 60 km/h in summer and 150 km/h in winter. The speed is greater in winter because the temperature contrast between the low latitudes and the high latitudes is much greater then. The jet stream does not circle the poles smoothly. Instead, it forms broad loops that may extend almost to the tropics. The jet stream strongly influences the weather at middle and high latitudes in both hemispheres.

Another planetary wind is the *monsoon*. A monsoon is characterized by a reversal of wind direction from summer to winter. Monsoons occur when air pressure over a large area reverses itself as the seasons change from summer to winter. The strongest monsoon is the Indian monsoon, which affects India and the northern Indian Ocean.

Monsoon circulation is not restricted to India, however. Weaker monsoons exist over the southwest United States, Australia, South America, and Africa.

The polar easterlies, the westerlies, the trade winds, and the monsoons form the planetary wind system. They are responsible for most of the weather in the middle latitudes and for distributing the heat of the tropics to the higher latitudes.

○ *What causes the planetary winds to curve?*
○ *Name the winds of the planetary wind system.*

Figure 8–18. The monsoon winds of Asia bring seasons of torrential rain to India.

8.9 Local Winds

In addition to the influences of planetary wind circulation, coastal areas often experience land breezes and sea breezes. During the day the land warms up more than the water. As the warmed air rises over the land, air from the ocean blows in to replace it. This flow of air is called a *sea breeze* and is shown in Figure 8–19. During the night the land cools more rapidly than the water, and a flow of air develops from land to sea. This flow is called a *land breeze* and is also shown in Figure 8–19.

Figure 8–19. These diagrams show the cause of a land breeze (right) and a sea breeze (left).

Sea and land breezes are common in the tropics during most of the year and at middle latitudes during the summer. Sea breezes usually produce a line of clouds just inland along a tropical coastline or above tropical islands.

When air blows down a mountain slope, pressure and temperature rise, and the air is very dry. This type of wind is a *katabatic* (kat uh BAT ihk) *wind.* Katabatic comes from a Greek word that means "moving down."

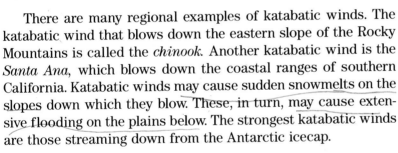

Figure 8–20. Sea breeze creates puffy clouds over tropical islands (above). Cold, dry winds stream down from the Antarctic ice cap (left). Both are examples of local winds.

There are many regional examples of katabatic winds. The katabatic wind that blows down the eastern slope of the Rocky Mountains is called the *chinook.* Another katabatic wind is the *Santa Ana,* which blows down the coastal ranges of southern California. Katabatic winds may cause sudden snowmelts on the slopes down which they blow. These, in turn, may cause extensive flooding on the plains below. The strongest katabatic winds are those streaming down from the Antarctic icecap.

○ *What causes land and sea breezes?*
○ *What are katabatic winds?*

Section Review

READING CRITICALLY

1. Why do winds turn to the right in the Northern Hemisphere and to the left in the Southern Hemisphere?
2. What causes the planetary winds to blow?

THINKING CRITICALLY

3. Why are monsoons part of the planetary wind system, while sea breezes are considered local winds?
4. If the earth did not rotate, what would the planetary wind system be like?

INVESTIGATION 8: Solar Radiation

PURPOSE

To examine the effect of solar radiation on light and dark surfaces

MATERIALS (per group of 3 or 4)

Scissors
Shoe box with lid
Black construction paper
White construction paper
Tape
Celsius thermometers (2)
Lamp
Stopwatch

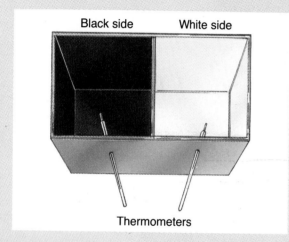

Black side   White side

Thermometers

PROCEDURE

1. Cut the lid of the shoe box as shown for use as a partition. Divide the shoe box in half with the partition.
2. Line one side of the box with black construction paper and the other side with white construction paper. Use tape to secure the linings.
3. Cut small holes in the box as shown and insert one thermometer in each side of the box.
4. Center the lamp directly above the box, about 3 cm above the rim.
5. Copy the chart shown.
6. Turn on the lamp and record the temperatures of both thermometers at time zero.

7. Record the temperatures again at one-minute intervals.
8. At the end of 10 minutes, turn off the lamp and continue recording until both thermometers show the same temperature as that recorded at time zero.

TABLE 1: RADIATION AND TEMPERATURE

Time	°C (dark)	°C (light)
0 min		
1 min		
2 min		
3 min		
4 min		
5 min		
6 min		
7 min		
8 min		
9 min		
10 min		

ANALYSES AND CONCLUSIONS

1. In which side of the shoe box did the temperature increase faster?
2. Which side of the shoe box was hotter at the end of ten minutes?
3. Compare the effects of white light on light and dark surfaces.

APPLICATION

Suppose you lived in a tropical area. What color car do you think would be the best to buy? Why?

SUMMARY

- Carbon dioxide, nitrogen, and sulfur dioxide formed most of the primitive atmosphere of the earth. (8.1)

- The removal of carbon dioxide from the primitive atmosphere left nitrogen as the most abundant atmospheric gas. (8.1)

- Nitrogen, oxygen, and argon form 99 percent of the modern atmosphere. (8.1)

- The gases of the earth's atmosphere that are heavier than helium do not escape into space because of gravity. (8.2)

- Ozone in the atmosphere absorbs dangerous ultraviolet radiation from the sun. (8.2)

- Fast protons from the sun heat up the upper part of the atmosphere. (8.2)

- Light consists of particles called photons, which travel from the sun to the earth as electromagnetic waves. (8.3)

- The sun emits light of all colors. The human eye sees this combination of colors as white. (8.3)

- The addition of carbon dioxide to the atmosphere may be enough to produce a strong greenhouse effect and melt the ice of Greenland and Antarctica. (8.4)

- Atmospheric pressure is the weight of the atmosphere. (8.5)

- The atmosphere is divided into layers, much like a multilayered cake. (8.6)

- From the earth outward, the layers of the atmosphere are troposphere, stratosphere, mesosphere, and thermosphere. (8.6)

- The Coriolis effect becomes evident when the Coriolis acceleration is inadequate. The Coriolis effect makes moving bodies veer to the right in the Northern Hemisphere and to the left in the Southern Hemisphere. (8.7)

- The planetary winds form broad belts that surround the earth at various latitudes. (8.8)

- Land and sea breezes are examples of local winds. (8.9)

Write all answers on a separate sheet of paper.

SCIENCE TERMS

Use each of the following terms correctly in a sentence.

atmosphere **(165)**
atmospheric pressure **(175)**
barometer **(175)**
Coriolis effect **(181)**
electromagnetic waves **(170)**
greenhouse effect **(174)**
jet stream **(183)**
ozone **(168)**
photons **(170)**
photosynthesis **(166)**
radiation balance **(172)**
solar wind **(165)**
sunspots **(168)**
Van Allen belts **(180)**

SCIENCE QUIZ

Modified True-False

Mark each statement *true* or *false*. If a statement is false, change the underlined term to make the statement true.

1. The atmosphere of the young earth consisted mainly of <u>oxygen</u>.

2. The distance between two crests of a wave is called the <u>wavelength</u>.

3. Infrared radiation has a <u>longer</u> wavelength than ultraviolet radiation.

4. Temperature <u>decreases</u> with altitude in the stratosphere.

5. The Coriolis effect causes bodies moving on the surface of the earth in the Northern Hemisphere to veer to the <u>left</u>.

6. The layer of the atmosphere in which all of the weather occurs is the <u>troposphere</u>.

continues

7. Land breezes and sea breezes are two kinds of <u>planetary</u> winds.

8. The aurora occurs in the <u>stratosphere</u>.

9. Oxygen in the early atmosphere was probably produced by <u>green plants</u>.

10. An increase in the temperature of the earth's atmosphere is called the <u>greenhouse effect</u>.

Multiple Choice

Write the letter of the choice that best answers the question or completes the statement.

11. The solar wind consists partially of
    a) dust.   b) protons.
    c) neutrons.   d) electrons.

12. The composition of the early atmosphere of the earth was changed into the modern one by the action of
    a) solar radiation.
    b) the earth's gravitational field.
    c) the earth's magnetic field.
    d) the activity of marine organisms.

13. The most abundant gases in the present atmosphere are
    a) hydrogen and helium.
    b) oxygen and nitrogen.
    c) argon and carbon dioxide.
    d) hydrogen and oxygen.

14. Light from the sun takes ____ minutes to reach the earth.
    a) 15.5   b) 30
    c) 8.3   d) 300

15. In the Northern Hemisphere, the trade winds blow from
    a) north to south.
    b) east to west.
    c) northeast to southwest.
    d) south to north.

Completion

Complete each statement by supplying the correct term or phrase.

16. The most abundant gases in the earth's atmosphere are ____ and ____.

17. The force that prevents the atmosphere from escaping into space is ____.

18. Most of the solar energy is emitted as ____ light.

19. The balance between incoming solar radiation and outgoing terrestrial radiation is called the ____ balance.

20. The lowest layer of the atmosphere is called the ____.

Short Answer

21. Why is the atmosphere of Earth now different from that of the primitive earth?

22. Why is it important not to disturb the earth's radiation balance?

23. Why does temperature decrease dramatically with altitude in the thermosphere?

Writing Critically

24. Describe how the Coriolis effect would affect the pattern of atmospheric and oceanic circulation if the earth were rotating faster than it does.

25. What would happen if the continued addition of industrial carbon dioxide to the atmosphere were to cause a large increase in the greenhouse effect? Explain.

EXTENSION

1. Research the atmospheric conditions of another planet. Find out what differences and similarities there are with Earth's atmosphere and report on these to the class.

2. Find out the direction of the predominant wind in your area and determine how this affects changes in your weather, both daily and long-range.

APPLICATION/CRITICAL THINKING

1. Why does the oxygen level in the earth's atmosphere remain nearly constant?

2. Describe what you think the Coriolis effect would be on Venus, which rotates very slowly, and in a direction opposite to that of Earth.

FOR FURTHER READING

Barth, M. and J. G. Titus, eds. *Greenhouse Effect* and *Sea Level Rise: A Challenge for This Generation.* New York: Van Nostrand Reinhold, 1984. This book discusses the possible global warming that could result from increased concentrations of carbon dioxide and other gases in the earth's atmosphere.

Heppenheimer, T. A. "Earth." *OMNI* (October 1985):18. This article presents an interesting comment on the marvels and inconveniences of wind turbines.

"Hole in the Sky." *Current Science* (December 1986):11. Scientists have detected a thinning of the ozone layer over the continent of Antarctica. The short article discusses the consequences of losing this layer.

Schneider, S. "Nuclear Winter: The Storm Builds." *Science Digest* 93(January 1985):48. The author examines the possibility that a nuclear war will plunge the earth into deadly cold by filling the atmosphere with smoke and dust.

Challenge Your Thinking

These athletes are extremely tired because of a lack of oxygen. The athletes at the 1968 Olympics suffered due to the thin atmosphere in 2240 m high Mexico City, yet many world records were set that year. Explain this apparent contradiction.

Water in the Atmosphere

Almost everyone knows what causes a rainbow, or at least that sun and rain are needed for one to form. However, do you know why it rains or how water gets into the atmosphere? Besides the obvious one, what role does the sun play in forming this beautiful picture?

Rainbows form as the sun shines through water droplets in the air.

1 Evaporation and Condensation

SECTION OBJECTIVES

After completing this section, you should be able to:

■ **Describe** how water evaporates from the ocean.

■ **Explain** how condensation produces clouds.

NEW SCIENCE TERMS

evaporation
saturated air
humidity
condensation
dew point
aerosols
cloud

9.1 Evaporation and Humidity

Look closely at Figure 9–1. What causes the dominant blue color of the earth's surface, or the white streaks in the atmosphere?

Both the dominant blue of the surface and the white streaks in the atmosphere are water. There is a great deal of water on the earth and in the atmosphere. Before clouds can form, however, some of the surface water must become water vapor, a gas.

Evaporation   The process of changing liquid water to water vapor is called **evaporation.** Most of the water in the atmosphere comes from the tropical oceans on both sides of the equator. The water temperature there is very warm, averaging 25°C. The warmer the water, the faster the water molecules move; the faster the molecules move, the more water evaporates. Why does the temperature affect the rate at which water molecules move?

Figure 9–1. One of Earth's dominant features, as viewed from space, is the cloud cover of the planet (right). Much of the evaporation, which eventually produces clouds, occurs over warm, tropical oceans (left).

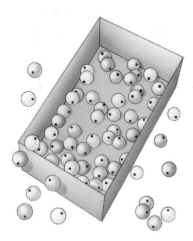

Figure 9–2. If you shake a box of plastic beads, some of the beads may have enough energy to escape the box. In a similar manner, water molecules, heated by the sun, may have enough energy to escape from the surface of the ocean.

Figure 9–3. Human bodies are cooled as sweat evaporates from the surface of the skin. A dog has no sweat glands, but water evaporating from its tongue as it pants has the same effect.

Imagine that you have a shoe box half-full of small plastic beads. If you supply a little energy to the box by shaking it gently, the beads move around a little, but nothing much happens. Similarly, molecules in a cold body of water do not receive much energy, and the water molecules move very little.

Now imagine supplying that box of plastic beads with more energy by shaking it vigorously. The beads move around a lot; some may even escape from the box. If a body of water receives a lot of heat energy from the sun, the water temperature increases. Some of the faster molecules escape from the water's surface, and the molecules that remain in the liquid also speed up. The energy used to speed up these molecules causes a decrease in the total amount of energy in the water. The decrease causes a loss of heat and lowers the water temperature.

You have probably experienced the cooling effects of evaporation, after splashing yourself with water on a warm day. In the same way, evaporation of sweat helps keep your body cool. Dogs, however, do not sweat; to cool their bodies, they pant. Water evaporating from the mouth and throat helps reduce a dog's body temperature.

Under certain conditions evaporation stops. For example, when the air above a pond is saturated with water, the number of molecules that leave the surface of the pond equals the number of molecules that return to the pond. **Saturated air** is air that is filled with as much water as it can hold. When air is saturated, no more evaporation takes place. For the same reason, water in a capped bottle does not evaporate; the air above the water surface is saturated. If the water molecules in the air above the surface are removed, however, evaporation continues.

Most ocean evaporation takes place in the tropics. There, high temperatures speed up water molecules and increase the rate of evaporation, and winds blow the moisture away before the air becomes saturated.

ACTIVITY: Measuring Evaporation

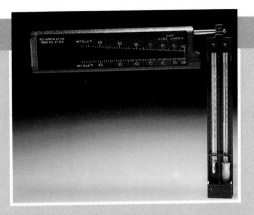

How can you show that evaporation is a cooling process?

MATERIALS (per group of 2 or 3)

thermometers in notched corks (2); ring stand; clamps (2); cotton or gauze; rubber bands; beakers, 100 mL (2); rubbing alcohol; water; wax pencil

PROCEDURE

1. Using rubber bands, carefully fasten a piece of cotton or gauze to the bulb of each thermometer.
2. Attach the clamps to the ring stand, 10–20 cm above the table. Insert the corks into the clamps and tighten.
3. Fill one beaker with rubbing alcohol and the other with water. Label each beaker with a wax pencil.
4. Read and record the temperature on each thermometer.
5. Immerse one thermometer in the alcohol and the other thermometer in the water. Be sure to saturate the cotton.
6. At 30-second intervals, record the temperature on each thermometer.

CONCLUSIONS/APPLICATIONS

1. After the cotton was saturated with water or alcohol, did either thermometer show a change in temperature?
2. If so, which thermometer showed a faster change in temperature? Explain.

Humidity   Very seldom is the air saturated with water vapor. The amount of water vapor in the air is called **humidity.** The actual amount of water vapor per cubic meter of air is called *absolute humidity.* Humidity, however, is usually expressed as *relative humidity,* or the percentage of saturation. For example, air at 30°C can hold a maximum of 26 g of water per cubic meter. If the air actually contains this amount of water vapor, the air is saturated, and the relative humidity is 100 percent. If the same air contains only 13 g of water per cubic meter, the relative humidity is only 50 percent.

Relative humidity affects the rate of evaporation from a body of water or from a human body. The higher the relative humidity, the slower evaporation occurs. The combination of heat and high humidity can be very uncomfortable, even dangerous. If there is no evaporation of sweat from the surface of your skin, your body temperature goes up. A rise of only a few degrees produces heat exhaustion. A rise of a few more degrees causes collapse and possibly death.

Figure 9–4. If sweat cannot evaporate from a person's skin, the body temperature continues to rise. If the temperature rises high enough, the person may collapse or even die.

The combined effect of heat and humidity, sometimes referred to as the "humiture," is called the *heat index*. The combined effect of cold and wind is called *wind chill*. These terms express what the temperature "feels like" to your body. High humidity can make a 35°C day feel warmer than a dry 40°C day. A 30 km/h wind can make 10°C feel like 0°C. A heat-index chart and a wind-chill chart are shown in the Reference Section on page 571.

○ **What is evaporation?**
○ **Why does evaporation stop when the air becomes saturated?**

9.2 Condensation and Dew Point

The warm air of day holds more water vapor than the cooler night air. As the temperature drops during the night, the relative humidity may reach 100 percent. If the temperature continues to fall, some of the water vapor in the air changes back to liquid. The change from water vapor to liquid water is called **condensation.**

When water condenses on the ground, it forms dew. The temperature at which dew forms is called the **dew point.** The dew point changes depending on the relative humidity of the air; humid air has a higher dew point than dry air. In the morning the temperature usually rises above the dew point and the dew evaporates. If the dew point is below 0°C, the dew freezes and becomes frost. As the temperature rises in the morning, the frost melts and then evaporates.

Have you ever seen a hot-air balloon floating gently in the cool morning breeze? The balloon rises because the warm air inside the balloon is less dense than the cool air outside the balloon.

The air in the balloon is warmed by a gas-fired burner. In nature, air is warmed by the ground, which is heated by the sun. As warm air rises, it cools, and its relative humidity increases. This causes the water vapor in the air to condense.

Figure 9–5. Water from the atmosphere often condenses on smooth surfaces, such as window panes. If the temperature is below 0°C, frost will form.

Figure 9–6. Hot-air balloons are usually flown early in the day because the air is cooler then and less heat is needed to make the balloons rise.

Condensation increases when microscopic solids, such as dust, smoke, salt crystals, and pollutants, are in the air. These solids in the air, called **aerosols,** act as nuclei for condensation, and water droplets grow around them. A **cloud** is a collection of water droplets in the atmosphere.

Fog is similar to a cloud except that it forms near the ground and stays there. If the temperature rises within a cloud or in an area where there is fog, the droplets evaporate and the cloud or the fog disappears.

Figure 9–7. As the morning sun warms the air, fog often lifts, forming clouds like those above. The fog lifts because the warmed air rises. Eventually the water droplets evaporate.

The droplets that form a cloud are so small that air currents keep them floating. Cloud droplets tend to combine with each other to form larger droplets and, eventually, raindrops. Raindrops are droplets that are too big to float. There are about one million cloud droplets in every raindrop.

○ *What is condensation?*
○ *Why does warm air rise?*

Section Review

READING CRITICALLY

1. Why do you feel chilly if, after a swim, you let your body dry in the wind?
2. What is the relationship between dew point and relative humidity?

THINKING CRITICALLY

3. How might an increase in pollution increase the amount of condensation in the atmosphere?
4. Why is frost more likely to occur on clear, calm nights than on cloudy, windy nights?

2 Clouds and Precipitation

NEW SCIENCE TERMS

cumulus clouds
stratus clouds
cirrus clouds
precipitation

SECTION OBJECTIVES

After completing this section, you should be able to:

- **List** the types of clouds.
- **Explain** the water cycle.

9.3 Types of Clouds

The different types of clouds form depending on conditions in the atmosphere. The atmospheric conditions that affect cloud formation are: air temperature; air pressure; humidity; and air currents, or wind.

TABLE 9–1: CLOUDS OF THE TROPOSPHERE

Cirrus	Elevation: 7 km to 13 km Thin, wispy clouds made of ice crystals. Associated with fair weather but may indicate approaching storms.
Cirrocumulus	Elevation: 7 km to 13 km Thin clouds, which look like ripples or waves. Associated with fair weather, but may indicate approaching storms.
Cirrostratus	Elevation: 7 km to 13 km Thin clouds that cause halos around the sun or moon. Associated with fair weather but may precede storms.
Altocumulus	Elevation: 2 km to 7 km Light gray clouds in rolls or patches. Associated with fair weather but may precede storms.
Altostratus	Elevation: 2 km to 7 km Gray clouds that usually produce light rain or snow. May precede warmer weather.

When hot, humid air rises, **cumulus** (KYOO myoo luhs) **clouds** form. Large cumulus clouds that rise to the top of the troposphere, where they spread out, are called *cumulonimbus clouds.* These are the familiar "thunderheads" of summer. **Stratus clouds** are extended cloud layers that may cover the entire sky. If the cloud is several kilometers thick, and produces rain or snow, the cloud is called *nimbostratus.* **Cirrus clouds** are the highest clouds in the troposphere. They are made of ice crystals, not water, and look like fine, white filaments. The clouds of the troposphere are summarized in Table 9–1.

Two types of clouds occur in the stratosphere. *Nacreous* (NAY kree uhs) *clouds* are translucent sheets of ice crystals. They are sometimes known as mother-of-pearl clouds because of their pearly appearance. *Noctilucent* (nahk tuh LOO sehnt) *clouds* are less dense than nacreous clouds, so they look very thin and wispy.

○ *Why do different types of clouds form?*
○ *Which clouds are formed in the stratosphere?*

Figure 9–8. Most clouds form from water droplets in the troposphere, but nacreous clouds, such as these, are formed of ice crystals in the stratosphere.

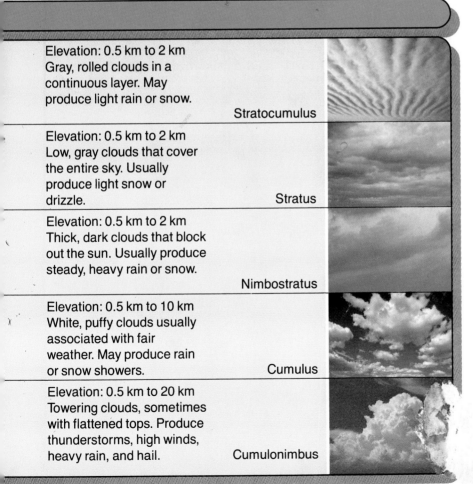

Elevation: 0.5 km to 2 km Gray, rolled clouds in a continuous layer. May produce light rain or snow. Stratocumulus	
Elevation: 0.5 km to 2 km Low, gray clouds that cover the entire sky. Usually produce light snow or drizzle. Stratus	
Elevation: 0.5 km to 2 km Thick, dark clouds that block out the sun. Usually produce steady, heavy rain or snow. Nimbostratus	
Elevation: 0.5 km to 10 km White, puffy clouds usually associated with fair weather. May produce rain or snow showers. Cumulus	
Elevation: 0.5 km to 20 km Towering clouds, sometimes with flattened tops. Produce thunderstorms, high winds, heavy rain, and hail. Cumulonimbus	

DISCOVER

Observing Clouds

Every day for the next week, write down the names of all the different kinds of clouds that you see. Along with the name of each kind of cloud, tell whether any precipitation is falling. At the end of the week, try to determine what kinds of clouds produce precipitation.

SKILL ACTIVITY: Reading a Wind-Chill Chart

BACKGROUND

Many times the easiest way to present a lot of information is by using a chart or table. The heat-index chart on page 571 of the Reference Section demonstrates this. Find the current temperature and the relative humidity and locate the point where these rows intersect. That point is the heat index.

There is a similar relationship between temperature and wind speed. Because moving air tends to speed up evaporation, and also to remove heat, a strong wind on a cool day makes it feel colder than it really is. This combination of temperature and wind is known as the wind chill.

PROCEDURE

Look at the part of a wind-chill chart presented here. The apparent wind-chill temperature is found by locating the column with the current temperature and then the row with the wind speed. Where the two intersect is the wind-chill temperature.

APPLICATION

Using the wind-chill chart, find the wind-chill temperature for each of the following conditions:

1.   15°F—10 mph
2.   10°F—10 mph
3.   5°F—10 mph
4.   0°F—10 mph
5. − 5°F—10 mph
6. − 10°F—10 mph
7. − 15°F—10 mph
8. − 20°F—10 mph
9. − 25°F—10 mph
10. − 30°F—10 mph

USING WHAT YOU HAVE LEARNED

Find a specific wind-chill temperature, such as − 25°F, and see how many combinations of wind speed and air temperature will produce this same apparent temperature. A complete wind-chill chart is shown on page 571 of the Reference Section.

WIND CHILL CHART (APPARENT TEMPERATURE)

Wind Speed		Cooling Power of Wind Expressed as "Equivalent Chill Temperature"															
Knots	MPH	Temperature (°F)															
Calm	Calm	40	35	30	25	20	15	10	5	0	−5	−10	−15	−20	−25	−30	−35	−4
		Equivalent Chill Temperature															
3	6	5	35	30	25	20	15	10	5	0	−5	−10	−15	−20	−25	−30	−35	−40	−
7	10	10	30	20	15	10	5	0	−10	−15	−20	−25	−35	−40	−45	−50	−60	−65	−7
11	15	15	25	15	10	0	−5	−10	−20	−25	−30	−40	−45	−50	−60	−65	−70	−80	−8
16	19	20	20	10	5	0	−10	−15	−25	−30	−35	−45	−50	−60	−65	−75	−80	−85	−9
20	23	25	15	10	0	−5	−15	−20	−30	−35	−45	−50	−60	−65	−75	−80	−90	−95	−10
24	28	30	10	5	0	−10	−20	−25	−30	−40	−50	−55	−65	−70	−80	−85	−95	−100	−110
			5									−75	−80	−90	−100		11

Figure 9–9. Some places, such as Mount Waialeale in Hawaii (left), receive precipitation nearly every day. Other places, such as parts of the Atacama Desert in Chile (right), seldom receive any precipitation.

9.4 Precipitation

Rain, snow, sleet, and hail are all precipitation. **Precipitation** is solid or liquid water that falls to the earth's surface. This completes the water cycle that began with evaporation from the surface of the ocean.

Precipitation varies widely around the world. For example, in some areas of the Atacama Desert of northern Chile, no precipitation has ever been recorded. At the other extreme, on Mount Waialeale, Hawaii, the average yearly precipitation is 11.45 m. That is more than enough water to completely cover a three-story building.

Rain Cloud droplets fuse together to form raindrops. In calm air the drops are very small and are usually referred to as drizzle. Some raindrops, carried by air currents, travel up and down inside a cloud several times, getting larger with each trip. The maximum size of a raindrop is, however, limited by evaporation and friction. Figure 9–10 shows the relative sizes of droplets and a raindrop.

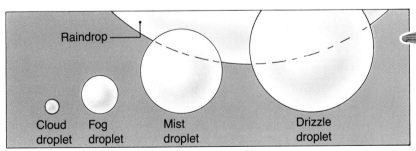

Raindrop

Cloud droplet | Fog droplet | Mist droplet | Drizzle droplet

Snow Ice crystals, which form at temperatures below 0°C, collect to form snowflakes. You may have heard that no two snowflakes are alike. That is probably not true, but there is a large variety of shapes possible among snowflakes.

Snow is much less dense than rain. One meter of snow equals only 10 to 15 cm of water. A heavy rainfall drops much more water than a heavy snowfall.

A MATTER OF FACT

The heaviest rainfall on record dropped almost 2 m of water in 24 hours on la Réunion, an island in the Indian Ocean, in 1952. The greatest 24-hour snowfall on record dropped almost 2 m of snow on Silver Lake, Colorado, in 1921.

Figure 9–10. Liquid precipitation, as the diagram shows, may range in size from microscopic cloud droplets to large rain drops that fall during thunderstorms. If condensation occurs at temperatures below 0°C, ice crystals or snowflakes form instead of rain.

Figure 9–11. Sleet (right) is rain that freezes as it falls. Hail (left) freezes in the clouds and then falls to the earth.

Sleet   Frozen rain falls when the air temperature near the ground is lower than the air temperature higher up. This frozen rain is called *sleet,* and it is different from snow. Sleet starts as rain; it then freezes before reaching the ground. Large accumulations of sleet are very rare. Usually either the temperature increases near the ground, changing the sleet to rain, or the temperature in the clouds decreases and the precipitation becomes snow.

Hail   In cumulonimbus clouds, ice crystals may grow into hailstones if the air in the lower part of a cloud is above 0°C. Ice crystals formed in the upper part of the cloud gather a layer of water as they fall. Strong updrafts carry the growing hailstones back up to where the temperature is low enough to freeze the accumulated water. Several round trips between the upper and lower parts of a cloud can produce hailstones the size of marbles. Baseball-sized hail is not uncommon. The largest hailstone on record fell in Kansas in 1970 and weighed 750 g. It was larger than a softball.

○ *What is precipitation?*
○ *What is the difference between sleet and hail?*

A MATTER OF FACT

Hail causes more economic damage by destroying crops in the United States than do tornadoes.

Section Review

READING CRITICALLY

1. What kinds of clouds usually produce precipitation?
2. Describe the process of hail formation.

THINKING CRITICALLY

3. What would happen if the water that evaporates from the ocean and falls on the ground as precipitation were not returned to the ocean? How would this affect the water cycle?
4. Evaporation and condensation are opposite processes. If evaporation is a cooling process, which absorbs energy, then describe the process of condensation.

VILHELM BJERKNES (1862–1951)

Vilhelm Bjerknes was one of the world's first meteorologists. He was born in Oslo, Norway, on March 14, 1862. He spent most of his youth working with his father, who was a mathematics professor. Together they studied the motions and actions of liquids, such as water, under many conditions of temperature and pressure.

When Bjerknes was 38, he began to work with Heinrich Hertz at the University of Bonn in Germany. They studied the behavior of electrical discharges, such as lightning, through gases. Some of their findings were later used in the development of the radio.

In 1897, Bjerknes became a physics professor at the University of Stockholm in Sweden. Over the next several years, he developed theories about the large-scale movements of the oceans and atmosphere. He began to predict the weather based on the known principles of physics. He found that some of the movements of weather systems are due to the Coriolis effect. An understanding of this effect allows for many weather predictions.

As a teacher, he was greatly respected and admired by his students. In 1904, many of his assistants joined him in founding a program for weather prediction in Leipzig and Bergen. In 1919, his son Jakob was among several researchers who discovered how cyclones begin. Cyclones begin as waves along fronts—the boundaries between different air masses.

WARREN M. WASHINGTON (1936–)

Warren M. Washington was born on August 28, 1936, in Portland, Oregon. Washington showed an early interest in science, especially in physics and in the work of Albert Einstein.

Washington attended Oregon State University, where he graduated with a bachelor's degree in physics in 1958. During the summer after his graduation, Washington worked as a mathematician at Stanford University. His job was to solve mathematical problems relating to the atmosphere. Afterward, Washington returned to Oregon State to continue his education in physics. He received his master's degree in 1960.

Washington's work at Stanford had sparked his interest in the atmosphere. He continued his education at Pennsylvania State University, receiving a doctorate in meteorology in 1964. He had combined physics, mathematics, and computer science in his study of meterology.

After graduating from Penn State, Dr. Washington began work at the National Center for Atmospheric Research (NCAR) in Boulder, Colorado. In 1972 he was appointed by President Richard Nixon to serve on the National Advisory Committee on Oceans and Atmosphere.

Dr. Washington is currently the director of the Climate and Global Dynamics Division at NCAR. He is best known for creating computer models of the atmosphere and the oceans, which are used to study changes that occur there.

Section 2   Clouds and Precipitation   **201**

INVESTIGATION 9: Determining Dew Point

PURPOSE

To determine the dew point and calculate the relative humidity

MATERIALS (per group of 3 or 4)

Small soup can
Ice cubes
Thermometer
Flashlight

PROCEDURE

1. Fill the can about half full of water and let it stand until the water reaches room temperature.
2. Record the air temperature.
3. Place the thermometer in the can and read and record the water temperature.
4. Add a few ice cubes to the water and stir carefully with a pencil.
5. The temperature of the ice-water mixture will drop and this, in turn, will lower the temperature of the can. Watch the outside of the can.
6. Have your lab partner observe the outside of the can so that he or she can be alert to when condensation first starts to appear. He or she should shine the flashlight on the can. This will assist in determining when the condensation first appears.
7. As soon as condensation is seen, your partner should say "Now!" At this instant the temperature of the water should be read and recorded. This temperature is the dew point.
8. If you have time, repeat the experiment outside the school building. Be sure that you are starting with water that is at air temperature.

ANALYSES AND CONCLUSIONS

1. Why did you have to let the water come to room temperature?
2. Why did the water and ice have to be constantly stirred?
3. Why was it important to notice the exact moment when condensation first appeared on the outside of the can?
4. Why did the temperature have to be read at the same moment as the first condensation appeared?
5. Why was the dew point different when you repeated the experiment outside?

APPLICATION

1. If the dew point were close to the air temperature, what could you say about the relative humidity?
2. If the dew point were below 0°C, what would form on the outside of the soup can?

SUMMARY

- The process of changing liquid water to water vapor is evaporation. (9.1)

- When water evaporates, cooling occurs. (9.1)

- Air is saturated when it can hold no more water vapor. (9.1)

- Water vapor in the air is humidity. (9.1)

- Relative humidity is the percentage of saturation of the air. (9.1)

- The change from water vapor to liquid water is condensation. (9.2)

- Clouds form when water vapor condenses around aerosol particles. (9.2)

- Fog is similar to a cloud except that it forms near the ground and stays there. (9.2)

- Different types of clouds form, depending on conditions in the atmosphere. (9.3)

- When hot, humid air rises, cumulus clouds form. Stratus clouds are extended cloud layers that cover the entire sky. Cirrus clouds are the highest clouds in the atmosphere. (9.3)

- Water that evaporates from the ocean precipitates onto the land and the ocean. (9.4)

- Rain, snow, sleet, and hail are all forms of precipitation. (9.4)

- Raindrops form when cloud droplets fuse together. Snow forms when cloud temperatures are below 0°C. Sleet is rain that freezes as it falls through a layer of cold air near the ground. Hail forms as rain freezes in the clouds. (9.4)

Write all answers on a separate sheet of paper.

SCIENCE TERMS

Correctly use each of the following terms in a sentence.

aerosols **(195)**
cirrus clouds **(197)**
cloud **(195)**
condensation **(194)**
cumulus clouds **(197)**
dew point **(194)**
evaporation **(191)**
humidity **(193)**
precipitation **(199)**
saturated air **(192)**
stratus clouds **(197)**

SCIENCE QUIZ

Modified True-False

Mark each statement *true* or *false*. If a statement is false, change the underlined word to make the statement true.

1. When water <u>condenses,</u> cooling occurs.

2. When the relative humidity of the cold night air reaches 100 percent, <u>dew</u> forms.

3. Water usually condenses around <u>aerosol</u> particles.

4. <u>Sleet</u> is frozen rain.

5. Snow forms when cloud temperatures are below <u>10°C.</u>

continues

Multiple Choice

Write the letter of the choice that best answers the question or completes the statement.

6. Which of the following are types of frozen rain?
   a) sleet and snow   b) sleet and hail
   c) snow and hail   d) drizzle and sleet

7. Which type of cloud usually produces rain?
   a) cumulus   b) stratus
   c) nimbostratus   d) cirrus

8. As warm air rises, water vapor in the air will
   a) evaporate.   b) condense.
   c) rain.   d) precipitate.

9. Aerosols form nuclei for
   a) evaporation.   b) condensation.
   c) precipitation. d) dew.

10. Water that falls to the earth is called
    a) snow.   b) dew.
    c) condensation. d) precipitation.

Completion

Complete each statement by supplying the correct term.

11. Evaporation is a _____ process.

12. _____ clouds are high, thin clouds made of ice crystals.

13. Precipitation that forms from ice crystals is called _____.

14. The amount of water that evaporates from the earth's surface is _____ as the amount of precipitation that falls on the earth's surface.

15. Particles of dirt in the atmosphere are called _____.

Short Answer

16. Explain the process of cloud formation.

17. How do snow and sleet differ?

18. What factors might limit the size of raindrops?

Writing Critically

19. Explain why precipitation is necessary to complete the cycle of water.

20. Since hail and sleet are both frozen rain, why is there no sleet in the summer and no hail in the winter?

EXTENSION

1. Make a diagram of the water cycle and show, with arrows and labels, why this is a complete cycle.

2. Take some close-up photographs of frost on a pane of glass. Bring these photographs to school and discuss with your classmates some possible reasons for the various patterns of frost on glass.

3. Photograph or draw all the different types of clouds you see in a week. Classify your examples according to height of the clouds.

APPLICATION/CRITICAL THINKING

1. What happens to the energy released during condensation?

2. Consider the following situation. Near the ground, the temperature is $-10°C$. At 100 m, the temperature is $10°C$, and at 500 m, where condensation is taking place, the temperature is $-25°C$. Describe the precipitation that would form and any changes to it that would occur as it fell.

3. The cloud shown here has a flattened top, sometimes called an *anvil* because of its shape. What do you think causes this anvil shape?

FOR FURTHER READING

Adler, D. *World of Weather.* Mahwah, New Jersey: Troll, 1984. This book provides basic weather information. The topic of evaporation and condensation is covered, as well as the topic of precipitation.

Cosgrove. M. *It's Snowing.* New York: Dodd, Mead, 1980. This well-illustrated book explains how ice crystals form in the atmosphere and grow into snowflakes. The book also describes some of the problems associated with heavy snowfalls in mountainous areas.

McFall, C. *Wonders of Dust.* New York: Dodd, Mead, 1980. This book provides many interesting descriptions of and facts about dust and its effect on the weather. Topics include dust storms, pollution, and a possible link between dust and hailstorms.

Schaefer, V. and J. Day. *A Field Guide to the Atmosphere.* Boston: Houghton Mifflin, 1981. This book describes the formation of clouds in the atmosphere and the types of precipitation they bring.

Vesilind, P. "Monsoons." *National Geographic* 166(December 1984): 712. This article is a brilliant photographic essay of the earth's most important storms.

Challenge Your Thinking

Recall from reading the chapter that hail grows in size by making several round trips between warm and cold layers inside a cloud. Most hailstorms reach an average size of 3 to 5 cm. Look at the number of hailstones in this picture. What must be happening inside the clouds to allow hail of this number to form?

Weather

The rain blows horizontally, and the wind screams. A hurricane unleashes its power on Miami. The real danger, however, is not the wind; it is the storm surge, the wall of water pushed ahead of the storm. The storm surge may be more than eight meters high, while most beach communities are less than three meters above sea level. In 1928, a hurricane in Florida killed nearly 2000 people.

A hurricane in south Florida

1 Frontal Weather

SECTION OBJECTIVES

After completing this section, you should be able to:

- **List** and **sketch** the different types of fronts.
- **Describe** air rotation around a cyclone in the Northern Hemisphere.
- **List** the major climatic zones of the earth.

NEW SCIENCE TERMS

air mass
front
cyclone
climate

10.1 Air Masses and Fronts

When you go home tonight, try to watch the local weather report on the evening news. See how many times the weather reporter mentions the terms *air mass* and *front*. These are terms that are used frequently in weather forecasting. What exactly are fronts? What influence do air masses have on weather?

An **air mass** is a large body of air with nearly uniform temperature and humidity. The characteristics of an air mass are determined by the nature of the surface beneath it. For example, a cool, moist air mass will form over a cold ocean, and a warm, dry air mass will form over a desert.

When different air masses meet, the surface between them is called a **front**. Perhaps, while studying history, you have heard the term *front* used in connection with war. The area where two opposing armies clash is referred to as a front. The term *front* has a similar meaning in meteorology, because there is often a clash between air masses along a front. For example, a front occurs where polar air meets tropical air.

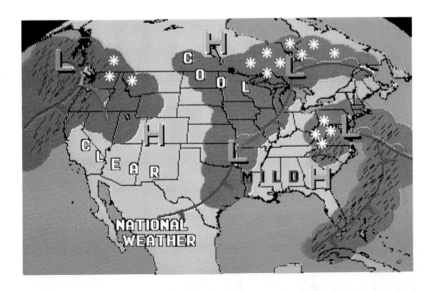

Figure 10–1. This weather map shows several fronts, which are the boundaries between air masses.

There are four types of fronts that commonly occur over North America: *cold fronts, warm fronts, stationary fronts,* and *occluded fronts.* A front is usually about 1 km wide, but it may be hundreds of kilometers long. Within this area the opposing air masses mix rapidly. The temperature difference between the air masses may be as much as 10°C.

Cold and Warm Fronts

Where a cold air mass moves in on a warm air mass, a *cold front* occurs. The cold air mass wedges itself under the warm air mass and raises it. The rising warm air expands and cools. Water vapor condenses and precipitation occurs. Figure 10–2 shows a typical cold front.

When a warm air mass moves over cold air, a *warm front* occurs. As the warm front continues to move, the warm air rises and cools. Water vapor condenses and precipitation occurs.

A cold front and a warm front move in different ways. A cold front moves toward a warm area, while a warm front moves toward a cold area.

Figure 10–2. These diagrams show four different fronts. A cold front (top left), a warm front (top right), a stationary front (bottom left), and an occluded front (bottom right). Notice the position of the different air masses in each type of front, and the direction that each air mass is moving.

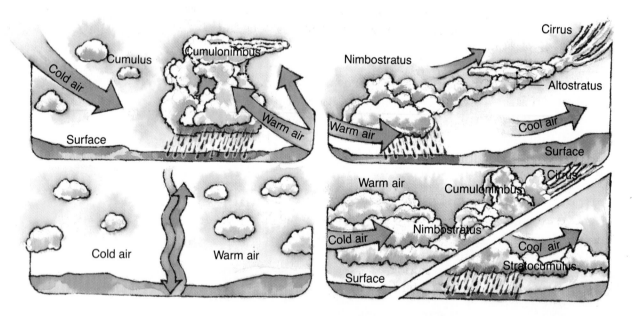

Stationary and Occluded Fronts

A front that does not move is called a *stationary front.* This type of front usually does not last more than a day. Stationary fronts separate the warm air of one mass from the cool air of another.

An *occluded front* develops when cold air moves faster than a warm front and overtakes it from behind. The warm air is forced upward and precipitation occurs. A stationary front and an occluded front are shown in Figure 10–2.

○ *How are air masses usually classified?*
○ *Where do fronts form?*

10.2 Cyclones

Have you ever seen the movie *The Wizard of Oz*? Dorothy and her dog, Toto, are carried away from Kansas by a "cyclone." The term *cyclone* is used by some people to describe especially severe storms, such as tornadoes. In meteorology the term **cyclone** is used to describe all weather systems that circulate around low-pressure centers.

If two air masses moving in opposite directions collide, they are deflected away from each other by the Coriolis effect. A low-pressure center forms between the two air masses. This low-pressure area attracts the air masses, and, in the Northern Hemisphere, a counterclockwise circulation develops. In the Southern Hemisphere, the circulation around a cyclone is clockwise.

In temperate regions of North America, frontal weather is usually associated with cyclones. Cyclones have a warm front ahead of them and a cold front behind them. Widespread precipitation occurs in advance of the warm front, and a narrow band of more intense precipitation occurs along the cold front. Eventually, the cold air overtakes the warm front, forming an occluded front.

Figure 10–3. This diagram shows the development of a cyclone along a cold front. In North America, cyclones move from west to east because of the prevailing westerlies.

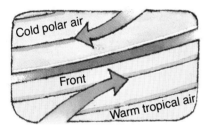

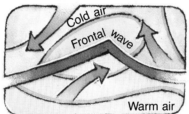

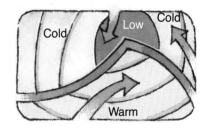

Cyclones and their frontal weather are frequent during the winter, moving eastward in both the Northern and Southern hemispheres. Cyclones follow the planetary wind patterns at a forward speed of about 40 km/h. A cyclone gets its energy from the heat released by condensation. Cyclones may last several days and drop several centimeters of water on areas they pass over.

Areas of low pressure are balanced in the atmosphere by areas of high pressure, called *anticyclones*. Circulation of air in anticyclones is opposite to that in cyclones—clockwise in the Northern Hemisphere and counterclockwise in the Southern Hemisphere. Also, instead of ascending, air in an anticyclone descends. As it descends the air warms and the relative humidity decreases. For this reason, the weather is usually fair within high-pressure areas.

Figure 10–4. The circulation around anticyclones, or high pressure areas, is opposite to that of cyclones.

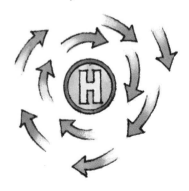

○ *What are cyclones?*
○ *What causes the circulation of cyclones and anticyclones?*

BACKGROUND

A meteorologist uses maps and symbols to describe weather patterns on the earth's surface. The symbols are used to represent cold fronts, warm fronts, stationary fronts, and occluded fronts. Air pressure and wind velocity as well as air temperature and local precipitation are also shown. The symbols are international symbols, and they can be interpreted by meteorologists all over the world.

PROCEDURE

1. Study the weather map and symbols in the Reference Section on page 584.
2. Try to determine the sky and weather conditions at each city.

APPLICATION

On a map of the United States, indicate the weather conditions for the nation as described in the accompanying paragraph. Use the correct symbols for each weather condition.

WEATHER REPORT

A cold front moving southeast is over California, Nevada, Idaho, and Montana. To the west of the front, there is a high-pressure area with a barometer reading of 760 mm. Associated with the front is a low-pressure area with a barometer reading of 749 mm. The winds over Colorado are from the south, accompanied by clouds and rain.

A warm front is moving northeast ahead of the cold front. The warm front reaches from the low over Montana through Wyoming, Colorado, and New Mexico. A high-pressure area, with a barometer reading of 765 mm, is over Kentucky and Illinois. Partly cloudy skies are found over Tennessee. The Midwest can expect clear skies today. There is fog along the New England coast. These are the conditions in the following cities:

Atlanta—clear, calm
Boise—snow, wind NW
Boston—fog, calm
Chicago—clear, wind SE
Denver—rain, wind S
El Paso—cloudy, wind SE
Knoxville—partly cloudy, wind E
Los Angeles—cloudy, calm
New York—fog, calm
Pittsburgh—clear, wind NE
St. Louis—clear, wind NE
Tampa—partly cloudy, wind SW

USING WHAT YOU HAVE LEARNED

Assuming that the fronts are moving east at 1000 km/day, what do you predict for tomorrow for Denver and Chicago? When will the cold front reach Philadelphia?

10.3 Climate

If you live along the Gulf Coast, in the Southwest, in Hawaii, or in Puerto Rico, you know the weather will be warm throughout most of the year. If you live in the Northeast or the Midwest, you know the summers will be warm, but the winters will be cold and snowy. These weather conditions describe the climate of your region.

Climate is the combination of all the weather conditions that characterize a region. Temperature and humidity are the two most important conditions that affect climate. Other conditions are the number of hours of sunshine, the amount of cloudiness, the amount of precipitation, the latitude, the altitude, and the wind speed and direction. Weather conditions may change from day to day, but climate is the average weather over a long period. Classification of the major climates of the world is shown in the Reference Section on page 578.

The areas near the equator have tropical, humid climates. Rain forests dominate these areas. North of the tropical belt are areas of high pressure and low relative humidity. The great deserts of the world, such as the Sahara, are found in these areas.

North of the desert belt lies the temperate area. This area includes the most fertile land in the world. In this area hot summers, cold winters, and moderate precipitation favor agricultural development. North of the temperate belt lies the polar region. Climate there is cold and dry.

South of the equator are the same climatic belts as are found in the Northern Hemisphere. However, in the Southern Hemisphere you move south from the equator toward the pole. Figure 10–5 shows examples of the major climatic zones of the world.

Figure 10–5. Anchorage, Alaska (left), has a cold climate, Savannah, Georgia (center), has a temperate climate, and Rio de Janeiro (right) has a warm climate.

Climate changes from tropical to polar with not only increased latitude but also increased altitude. Coastal New Guinea, for example, has a tropical humid climate. A little over 100 km inland lies Mount Wilhelm, which is 4700 m high. As you climb its slopes, the climate changes from tropical to polar.

Climate can also be affected by ocean currents. The Gulf Stream keeps the Gulf Coast, Florida, and the East Coast of the United States humid. Western Africa, on the opposite side of the Atlantic at the same latitude, is a desert. Further north the Gulf Stream passes near the northwestern coast of Europe. Ireland, England, and Norway have a temperate, humid climate due to the effects of this major current. Greenland and Labrador, directly across the Atlantic, are polar deserts.

Climate may also be affected by topography. All of central Asia is shielded from the ocean by high mountain ranges. Look at Figure 10–6. What do you think the effect is on the climate of central Asia as the warm, moist winds of the Indian Ocean are forced up over the Himalaya Mountains?

○ *List the different climatic zones of the world.*
○ *How might an ocean current affect a coastal climate?*

Figure 10–6. Because of the influence of a warm ocean current, Ireland (top) has a much warmer climate than Labrador (bottom), even though they are at nearly the same latitude. Much of central Asia (right) has a warm dry climate, because the warm, moist winds of the Indian Ocean and Arabian Sea are deflected by the Himalaya Mountains.

Section Review

READING CRITICALLY

1. How do cyclones develop? Why do they develop?
2. Describe a tropical climate.

THINKING CRITICALLY

3. Why would a cold frontal surface be steeper than a warm frontal surface?
4. Why is the climate of southern California moderate during most of the year while the climate of South Carolina shows great seasonal variations?

TECHNOLOGY: Using Computers to Model Climate

Computers producing climatic models for analysis

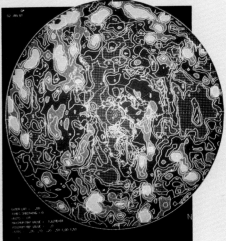

A computer-generated model of weather patterns

Scientists at the National Center for Atmospheric Research, in Boulder, Colorado, are using computers to model the world's climate. Drs. Stephan H. Schneider and Stanley L. Thompson have programmed the computer with vast amounts of weather data. Using mathematics and known principles of physics, they can simulate the present world climate. In addition, they are able to simulate the effects of volcanic eruptions, industrial pollution, and even nuclear war on worldwide climatic patterns.

Schneider and Thompson also simulate the climatic patterns that may have been present ages ago. They consider how the slow movement of the earth's crustal plates might have caused the climate to evolve to its present state. On their computer monitors, they watch ice ages come and go.

They examine how accumulating atmospheric carbon dioxide may change Earth's future climate. If industrial pollution slowly creates a greenhouse effect on the earth, will Earth's climate be like that of the planet Venus? Could nuclear war plunge the world into a nuclear winter, with a climate so cold that no living thing could survive? Computer models of the world's climate simulating these conditions are helping climatologists answer these and other questions.

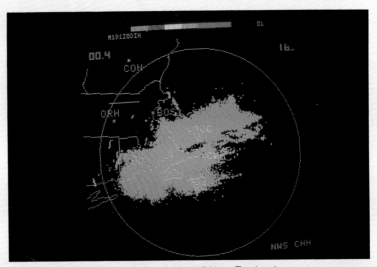

A computer-enhanced radar picture of New England

Violent Weather

NEW SCIENCE TERMS

thunderstorm
lightning
tornado
waterspout
hurricane

SECTION OBJECTIVES

After completing this section, you should be able to:

- **Explain** how thunderstorms get their energy.
- **Describe** the formation of tornadoes.
- **State** the reasons why hurricanes are dangerous.

10.4 Thunderstorms

Have you ever watched cumulonimbus clouds form on a hot, humid afternoon? They may reach several kilometers into the air in only a few minutes. You know that soon you can expect a thunderstorm. A **thunderstorm** is a violent storm that develops from cumulonimbus clouds. Thunderstorms get their energy from the condensation in the clouds. Figure 10–7 shows the development of a typical thunderstorm.

In fair weather, the ground is negatively charged to balance the positively charged ionosphere. During a thunderstorm, however, the negative charges that form in the cloud repel the negative charges in the ground. The negative charges are pushed deep into the ground, which becomes positively charged as a result.

If the difference in charge between the cloud and the ground is great enough, lightning occurs. **Lightning** is a stroke of electrons traveling at speeds of about 1000 km/s. The charge forms a

Figure 10–7. The illustration (right) shows the development of a thunderstorm. Thunderheads (above) show a characteristic flattened top.

tunnel between the cloud and the ground. Lightning can also occur within a cloud or between clouds. The temperature inside the tunnel through which the lightning travels rises to 30 000°C, and air pressure increases dramatically. This sudden increase in pressure causes a shock wave that travels through the atmosphere. You hear this shock wave as *thunder.*

Lightning emits not only visible light but also X rays, ultraviolet light, and radio waves. These radio waves can cause radios to crackle during a thunderstorm.

You see the lightning as soon as it shoots out of the cloud because light travels extremely fast: about 300 million m/s. Sound, however, travels much more slowly: only about 340 m/s. Count the number of seconds between the time you see the lightning and the time you hear the thunder. Divide that number by three and the result will be your distance from the lightning in kilometers.

Figure 10–8. Lightning can often be spectacular and dangerous. The Empire State Building in New York City (below) is protected by a huge lightning rod that safely channels the lightning into the ground.

You may have been told that lightning strikes the highest object in an area. This is the reason you should not stand under a tall tree during a thunderstorm. Lightning is likely to strike high structures, such as church steeples, factory chimneys, or ship masts. Because of this, it is unsafe to be in the open, to be on top of tall structures, or even to be near tall structures or trees during thunderstorms.

You may have seen structures that have metal rods on the top. These metal rods are called *lightning rods.* They are connected to the ground by means of a thick wire, a rod, or a pipe. If lightning strikes the structure, it will hit the rod first because that is the highest point. The rod will channel the electrical charge into the ground, where it will discharge harmlessly.

○ *Where do thunderstorms get their energy?*
○ *What causes lightning?*

10.5 Tornadoes

The "cyclone" in *The Wizard of Oz* was a tornado. A **tornado** is a small, extremely violent storm. Tornadoes are a kind of cyclone because the air is rapidly circulating around a small area of very low pressure.

Tornadoes occur where two very different air masses collide. One air mass must be warm and humid, the other cool and dry. The advancing cool air pushes the warm air up and away. Warm air rising rapidly along the front condenses into a line of thunderclouds called a *squall line.* The energy given off by condensation causes violent motions along the squall line, and many small, but very intense, low-pressure areas develop. Some of these areas may form tornadoes.

[plural, *vortices*]

A tornado starts as a center, or *vortex,* of low pressure. The tornado's vortex develops from the bottom of the thundercloud toward the ground. This downward extension is caused not by the air moving toward the ground, but by the air below the cloud being drawn into the vortex.

Air pressure at the center of the vortex is estimated to be 10 to 15 percent lower than the pressure outside the storm. This sudden pressure drop causes the air drawn into the vortex to become saturated with water vapor. When saturation occurs, water vapor in the cloud begins to condense, making the funnel visible. Figure 10–9 shows the development of a tornado.

In the Northern Hemisphere, the vortex spins counterclockwise as does that of a cyclone. The width of the vortex may be less than 20 m or more than 1 km. Tornadoes move parallel to the cold front at speeds of 40 to 60 km/h. The path of a tornado along the ground averages 5 km in length. However, tornado paths often are not continuous. The funnel may touch down and destroy one house on a street and not even disturb the house next door. A tornado lasts, on the average, only five to ten minutes.

Figure 10–9. This series of photographs shows the development of a tornado, from a cloud called a *hook* to a fully developed funnel.

Figure 10–10. A tornado, with winds as high as 500 km/h, can cause almost total destruction to anything in its path.

Tornadoes are so destructive because of their wind speeds. Horizontal wind speed within a tornado can be as high as 500 km/h. Upward wind speed may be as high as 300 km/h. There are few precise measurements of wind speed within a tornado because the measuring instruments are usually destroyed by the wind. Wind speeds are estimated from the damage caused by the tornado. Updraft speed is based on the fact that objects such as roofs, farm machinery, cattle, and even people have been lifted and carried for hundreds of meters.

Few structures can survive the fury of a tornado. When a tornado occurs, you should go to a room that has no outside openings, or stand in an inside doorway away from any windows. Why do you think you should stand in a doorway?

Although tornadoes occur in every state, they are most frequent in an area of the United States called *Tornado Alley.* As many as 300 tornadoes form there each year. Figure 10–11 shows the location of this area, which extends from northern Texas to southern Illinois.

Figure 10–11. The area of the United States from north Texas to southern Illinois is often referred to as *tornado alley* because of the large number of tornadoes that occur there each year.

Figure 10–12. Waterspouts are funnel clouds that develop over water. Waterspouts cause relatively little damage because the wind speed of a waterspout is usually not as great as that of a tornado.

A **waterspout** is a vortex that occurs over open water. Waterspouts are much less powerful than tornadoes because the temperature difference between air masses at sea is usually less than it is over land. Wind speed within waterspouts rarely exceeds 80 km/h. Waterspouts are common in the coastal waters of tropical and semitropical areas.

○ *Under what conditions do tornadoes develop?*
○ *Why are tornadoes dangerous?*

ACTIVITY: Making Weather Observations

How can you make weather observations with limited materials?

MATERIALS (per group of 3 or 4)
magnetic compass, thermometer, aneroid barometer, anemometer

PROCEDURE
1. Copy the chart below into your notebook. You will record simple weather observations for five days.

TABLE 1: WEATHER OBSERVATIONS

	Day 1	Day 2	Day 3	Day 4	Day 5
Temperature					
Pressure					
Sky					
Clouds					
Wind speed					
Precipitation					

2. Each day, take the outside temperature in a shady location and record it in the chart.

3. With the barometer, find the air pressure each day and record it in the chart.

4. Using the terms *clear, partly cloudy, partly sunny,* and *cloudy,* indicate on the chart the sky condition each day.

5. If there are clouds, determine the type from the photographs on pages 196 and 197 and record the cloud names on the chart.

6. Indicate on the chart any precipitation that is falling, or that has fallen since the previous observations.

7. Using the compass, and the anemometer, indicate on the chart the wind conditions each day.

CONCLUSIONS/APPLICATIONS

1. Is there any relationship between wind direction and air temperature?

2. Is there any relationship between air pressure and cloud cover or precipitation?

10.6 Hurricanes

What do you think is the most powerful storm in nature? Perhaps a tornado comes to mind. Although certainly destructive, tornadoes are confined to a relatively small area. The most powerful storm is a hurricane.

A **hurricane** is a large storm that develops over a tropical ocean. Hurricanes are also cyclones because of the way in which the air in the storms circulates—counterclockwise around areas of low pressure.

A hurricane develops as part of the planetary wind system. If you examine Figure 8–17 on page 183, you will see that there are several regions of the world where the planetary winds move in opposite directions. One of these regions is off the coast of western Africa. During summer, the southern trade winds cross the equator, where they collide with northern trade winds that blow in the opposite direction. The two are forced apart by the Coriolis effect. This creates low-pressure areas called *tropical depressions.* Although any of these depressions can become a hurricane, only about six, on the average, develop into hurricanes each year. Figure 10–13 shows areas of the world where tropical depressions form.

Figure 10–13. This map shows the areas of the world where tropical depressions are likely to develop.

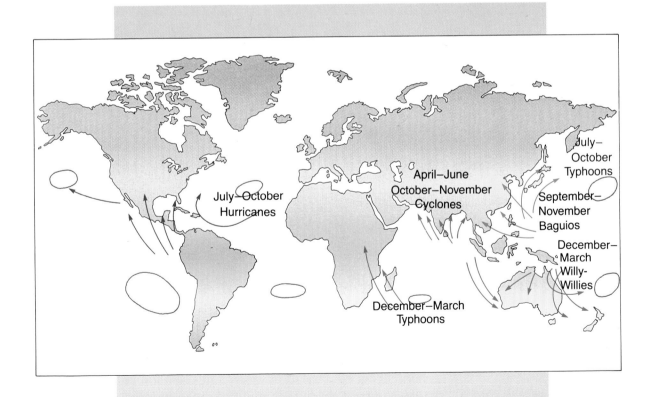

A fully developed hurricane is almost perfectly circular and is about 600 km in diameter. The pressure difference between the center and the outside of the vortex is about 10%. At the center of the storm is the *eye*—a core of warm, calm air about 10 to 20 km across. The highest winds circle the eye in a band 10 to 50 km wide that extends from sea level to the tropopause. Winds in this band may attain speeds of more than 350 km/h. Wind speed decreases rapidly about 50 km from the eye.

Figure 10–14. This diagram (right) shows a cross-section of a fully developed hurricane. Notice in the satellite photograph (above) that the center of the hurricane, called the *eye*, has no clouds.

DISCOVER

Reading About Storms

From a recent newspaper, find an article about a severe storm. Cut out the article, bring it to class, and discuss your article with your classmates. Determine the kind of damage each of the storms causes.

Outside the vortex are spiral bands of clouds. These clouds extend to the edge of the hurricane. The strength of the wind decreases from the center to the edge. The violent winds within a hurricane often cause tornadoes. These tornadoes increase the danger of hurricanes.

Like thunderstorms and tornadoes, hurricanes get their energy from condensation. Hurricanes develop only within the tropics, where absolute humidity is high and the sun provides heat for evaporation during most of the year.

Figure 10–15. These photographs show the damage caused by hurricane Camille, which struck the Gulf coasts of Mississippi, Alabama, and Florida in 1969. Camille had winds estimated to have been in excess of 400 km/h.

The destructive potential of a hurricane is tremendous. The sustained wind speed of most hurricanes is 100 to 120 km/h but may reach over 150 km/h. Hurricanes produce waves 3 to 5 m high in the open seas. These waves are piled on top of the *storm surge.* The storm surge is a dome of water 1 to 5 m high resulting from the hurricane's low pressure. When a hurricane crosses a coastline and moves onshore, the sea water pushes inland, causing extensive flooding. Strong winds, in addition to the flooding, can cause almost total destruction.

Movement of a hurricane, as it follows the general wind currents, is usually 5 to 25 km/h. Fortunately, the arrival of a hurricane can usually be predicted well in advance. People should follow the advice of authorities and promptly evacuate all low-lying coastal areas that are threatened.

○ *Where do hurricanes develop?*
○ *Why are hurricanes destructive?*

Section Review

READING CRITICALLY

1. What is the source of energy for thunderstorms, tornadoes, and hurricanes?
2. What causes lightning?

THINKING CRITICALLY

3. How do hurricanes create tornadoes? How are these tornadoes different from those created on land?
4. How do hurricanes distribute heat from the tropics to temperate regions?

INVESTIGATION 10: Air Masses and Temperature

PURPOSE

To demonstrate that cold air sinks and warm air rises

MATERIALS (per group of 3 or 4)

Scissors
Tape
Fish tank, small
Cardboard, larger than fish tank
Pencil
Thermometers (2)
Ring stands (2)
Test-tube clamps (2)
Paper towels
Beaker, 100 mL
Ice
Petri dish
Safety goggles
Laboratory apron
Candle
Crucible
Matches
Wire gauze
Beaker, 1000 mL

PROCEDURE

1. Using the scissors and tape, trim the piece of cardboard to make a lid for the top of the fish tank.
2. With a pencil, make two holes, one on each side of the lid, so that the thermometers can be inserted into the tank.
3. Secure the thermometers to the ring stands with clamps as shown in the diagram. Cut paper towels into small strips and pack them loosely into the 100 mL beaker.

4. Place the ice in the Petri dish and put it in one side of the tank.
5. **CAUTION: Be sure to wear safety goggles and laboratory apron during the rest of this investigation.**
6. Melt the bottom of the candle just enough so that it can be stuck inside the crucible.
7. Light the candle and put it in the other side of the tank.
8. Cut two holes, each about five to ten centimeters in diameter, in the cardboard. Position the lid so that one hole is over the ice and the other hole is over the candle.
9. Place the wire gauze on the hole over the ice.
10. Insert the thermometers, one on each side of the tank.
11. Light the paper towels in the small beaker and allow them to burn for several seconds. Place the small beaker on top of the wire gauze.
12. Invert the large beaker and use it to cover the small beaker.

ANALYSES AND CONCLUSIONS

1. What happened to the smoke that accumulated in the large beaker?
2. Describe how the smoke moved inside the tank.
3. What temperatures were recorded by the thermometers?
4. Explain why the clouds of smoke moved as they did.

APPLICATION

Explain how this investigation relates to the formation of the clouds that form along with a sea breeze.

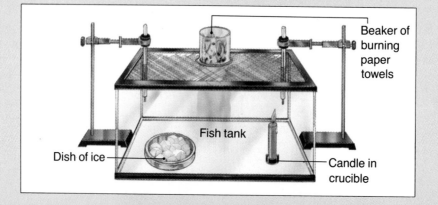

Beaker of burning paper towels

Fish tank

Dish of ice

Candle in crucible

222

SUMMARY

- An air mass is a large body of air with uniform properties. (10.1)

- Cold fronts wedge under warm fronts. As the warm air rises, it cools, and condensation occurs. (10.1)

- An occluded front occurs as a cold front moves faster than and overtakes a warm front. (10.1)

- Cyclones develop along differing air masses, mostly in winter. (10.2)

- Areas of low pressure are balanced by areas of high pressure called anticyclones. (10.2)

- The three major climatic zones are tropical, temperate, and polar. (10.3)

- The climate of northwestern Europe is temperate humid because of the Gulf Stream. Climate may also be affected by topography. (10.3)

- Thunderstorms are violent storms that develop on hot, humid days. (10.4)

- Lightning rods are used to conduct electric charges into the earth without damage to buildings or people. (10.4)

- Thunderstorms, tornadoes, and hurricanes derive their energy from the condensation of water vapor. (10.5)

- Tornadoes are small, violent cyclones. (10.5)

- Hurricanes are large cyclones that develop in the planetary wind system. (10.6)

- The highest wind speeds in a hurricane are around the eye of the hurricane. (10.6)

Write all answers on a separate sheet of paper.

SCIENCE TERMS

Correctly use each of the following terms in a sentence.

air mass **(207)**
climate **(211)**
cyclone **(209)**
front **(207)**
hurricane **(219)**
lightning **(214)**
thunderstorm **(214)**
tornado **(216)**
waterspout **(218)**

SCIENCE QUIZ

Modified True-False

Mark each statement *true* or *false.* If a statement is false, change the underlined word to make the statement true.

1. A <u>front</u> is the boundary between two air masses.

2. The most intense precipitation occurs along <u>warm fronts.</u>

3. The climate of an area is determined mainly by <u>temperature</u> and humidity.

4. An <u>air mass</u> is a large body of air with uniform conditions of temperature and humidity.

5. A <u>hurricane</u> is the most powerful of all storms.

Multiple Choice

Write the letter of the choice that best answers the question or completes the statement.

6. A cyclone gets its energy from the heat released as water in the atmosphere
   a) evaporates.   b) condenses.
   c) precipitates.   d) diverges.

continues

7. The climate of Scandinavia is
   a) tropical.   b) temperate.
   c) continental.   d) polar.

8. Lightning emits
   a) visible light.
   b) X rays.
   c) radio waves.
   d) all of the above.

9. As a hurricane approaches, people are usually asked to evacuate all
   a) low-lying coastal areas.
   b) inland areas.
   c) urban areas.
   d) rural areas.

10. The two most important factors in determining the climate of a region are
    a) temperature and air pressure.
    b) topography and humidity.
    c) temperature and humidity.
    d) prevailing winds and temperature.

Completion

Complete each statement by supplying the correct term.

11. Hurricanes usually form in the _____ zone.

12. Fronts that do not move are called _____ fronts.

13. Tropical forests of South America have a _____, _____ climate.

14. The most destructive natural force is the _____.

15. A thunderstorm gets its energy from _____.

Short Answer

16. Describe the formation of an occluded front.

17. Why are high places so dangerous during thunderstorms?

18. Describe the conditions of a tropical climate.

Writing Critically

19. Why do cyclones typically form along the polar front more in winter than in summer?

20. Why do thunderstorms occur more often in summer than in winter?

EXTENSION

1. Listen to a sequence of NOAA weather radio broadcasts to see what kinds of information are presented. Keep a record of forecasts for one week and check each the following day for accuracy. If you cannot receive the weather station, use a newspaper or television forecast.

2. Find out if there are any special procedures that must be followed during a weather emergency in your area. Some areas have special snow emergency routes, hurricane evacuation routes, or tornado drills. Make a list of the procedures for your area and discuss them with your family.

3. Go to your school library or public library and find several weather-related stories that were important enough to make the front page of your local newspaper. Try to determine the time of year when most of the storms occur.

APPLICATION/CRITICAL THINKING

1. Look at the weather map on page 207 and describe the current weather conditions in Georgia and Colorado. Predict the conditions for Denver, Colorado, and Atlanta, Georgia, over the next 24 hours.

2. If the overall climate in your area were to suddenly change, what immediate effects might it have on your life? What long-range effects might it have on your community?

3. What are the risks involved with building a house on a beach, next to a river, or on a steep mountain slope?

FOR FURTHER READING

Gibrilisco, S. *Violent Weather: Hurricanes, Tornadoes, and Storms.* Blue Ridge Summit, Pennsylvania: TAB Books, 1984. This book explains the formation of each of these weather conditions, their similarities and differences, and the destructive power of each.

Ludlum, D. *The Weather Factor.* Boston: Houghton Mifflin, 1984. A meteorologist gives a fascinating account of how the weather has influenced American history and life from colonial times to the Space Age.

Miller, P. "Tornado." *National Geographic* 171 (June 1987):690. This article presents a fascinating review of the team of scientists that chases tornadoes to study them.

"Tornado." *Current Science* (May 1987). This article gives a brief account of one family's encounter with a "twister."

Weisbird, S. "Stalking the Weather Bombs," *Science News,* Vol. 129, 5/17/86, pp. 314–317. This article explains how huge cyclones develop and discusses the extensive variety of instruments scientists use to track and study them.

Challenge Your Thinking

This photograph shows a frozen landscape near Barrow, Alaska. Some scientists have predicted that the climate of all of North America might be like this after a nuclear war. They call this cold climate that might result from a nuclear war *nuclear winter.* If you were to survive the war itself, what problems would you face because of the severe new climate?

UNIT 3 THE SCIENCE CONNECTION:

TOTO

People can do nothing to stop the awesome fury of a tornado. But with enough warning, they can escape the storm. A team of tornado chasers is helping to make tornado forecasts more reliable. This team from the National Severe Storms Laboratory (NSSL) travels the highways of "Tornado Alley" in western Oklahoma, looking for bad weather. With them they carry TOTO, or Total Tornado Observatory. TOTO is a metal barrel strengthened to withstand the intense winds of a tornado. Inside the barrel are instruments that measure temperature, wind speed, and atmospheric pressure.

The research team carries TOTO into the middle of violent Midwest thunderstorms. If they are lucky, they spot a tornado nearby. Then they drop TOTO in the path of the funnel cloud. As a tornado whips past, TOTO tells them more about the tremendous forces inside a tornado cloud.

The TOTO team begins by looking for weather conditions likely to produce tornadoes. Meteorologists from the National Severe Storms Center help the team locate unstable air masses and gathering thunderstorms. They look for storms with a *mesocyclone,* a rotating column of air inside the storm cloud. Out of this mesocyclone, a tornado funnel can drop to the ground without warning. The team leader yells out, "Tornado on the ground," and the team's vehicles race toward the funnel cloud. In two minutes the team can roll TOTO out of the truck and activate its instruments. However, a tornado can move quickly, too. The NSSL team must be ready to flee if a tornado threatens them. They want to get close to a tornado, but not too close.

Doppler radar screen

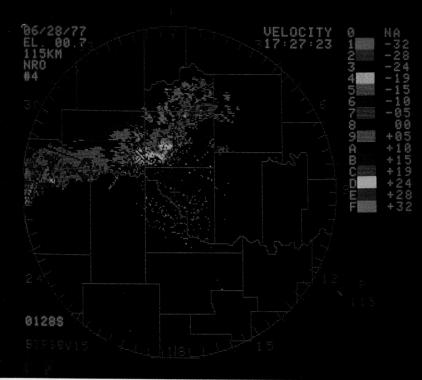

Computer-enhanced radar

Studying a tornado

The TOTO team is backed up by scientists using even more sophisticated weather instruments. Meteorologists at the National Severe Storms Center in Kansas City constantly watch thunderstorms across the U.S. Their aim is to give citizens a few minutes' warning before a tornado strikes. Those few minutes could mean the difference between life and death for the people in a tornado's path.

A special kind of radar, called *Doppler radar,* helps scientists locate tornado activity. Like all radar, a Doppler system sends out an electronic beep. The beep makes an "echo" when it bounces off raindrops and insects caught in a storm. By detecting differences in wave frequency, Doppler radar tells scientists in which direction the winds in the storm are moving and how fast. Rotating winds mean a tornado is brewing. On their radar screens, scientists look for bright red changing into bright green. These colors indicate twisting winds and a tornado on the ground.

Severe storm forecasters also rely on weather satellite photographs to spot bad weather. Infrared photos measure a storm's intensity by showing warm and cold air. Computers can flash several such photographs on a screen and combine them with other weather data. This shows scientists how rapidly a storm is developing. When they see a storm coming, forecasters issue a tornado watch, the first step in warning the public. By the 1990s, local weather stations will be linked up with a weather computer system. This system will give tornado warnings more quickly and reliably.

During the tornado season, this computer information is also passed to the TOTO weather chasers. The researchers use the information to decide in which direction their team should head to look for tornado weather. Often they find themselves driving through violent rain- or hailstorms. When everyone else is taking shelter, this well-trained and dedicated team is on the lookout — trying to learn more about one of nature's most powerful storms.

A Kansas "twister"

227

WEATHERING AND EROSION

This is the top of Triple Divide Peak in Glacier National Park. This peak marks the junction of three different drainage basins. Water falling on the northern slope of this mountain flows into the Arctic Ocean by way of Hudson Bay. Water falling on the western slope flows to the Pacific Ocean. Water falling on the eastern slope flows to the Gulf of Mexico and then to the Atlantic Ocean.

- Where do the glaciers come from that give the park its name?

- What are the divides that separate the major drainage basins of North America?

- How do wind, water, and ice help shape the land?

- Why are certain areas of North America set aside as national parks?

By reading the chapters in this unit, you will learn the answers to these questions. You will also develop an understanding of concepts that will allow you to answer many of your own questions about the study of the processes that shape the surface of the earth.

Triple Divide Peak

Weathering and Soil Formation

The rock formation in this picture looks as if it could have been created by some great sculptor; it was, in fact, created by the climate. As rain, wind, heat, and cold work together to make soil, they often produce beautiful works of art as well.

Weathered rocks

1 Weathering

SECTION OBJECTIVES

After completing this section, you should be able to:

- **Explain** the process of weathering.
- **List** the factors that affect the rate of weathering.
- **Compare** the processes of physical weathering and chemical weathering.

NEW SCIENCE TERMS

soil
weathering
exfoliation
frost action
abrasion
hydration
oxidation
carbonation

11.1 The Process of Weathering

If you have traveled much, you may have noticed that, except in mountainous areas, very little rock is visible on the earth's surface. In fact, in places like the Great Plains, you can travel for hundreds of kilometers and seldom see a rock, since most of the rock is covered by soil. **Soil** is a combination of small rock fragments and organic material in which plants can grow. Soil is formed when the bedrock of an area is broken into small fragments and organic matter is added to it.

Have you ever seen an unpainted house or barn that was "weathered" by the rain and sun, or read about an old sailor with a "weatherbeaten" face? The expression means that the house or the person has been changed by exposure to the weather.

Where rocks are exposed to the environment, such as on a mountain slope, they also weather. The processes that break rocks into smaller fragments, eventually producing soil, are called **weathering.** When rocks are weathered by rain and wind, the process is called *physical weathering.* If the rocks are chemically changed by acids in the air, rain, or soil, the process is called *chemical weathering.* Both types of weathering help to shape the surface of the earth.

Figure 11–1. This old barn (right) was weathered by exposure to sun, rain, and wind. These weathering agents are also responsible for changing rock into soil (left).

Figure 11–2. The cracks in these rocks may have been produced by cycles of freezing and thawing. If the seeds of some plant begin to grow in the cracks, the rocks may split even more.

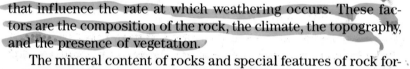

Although weathering is continuous, there are several factors that influence the rate at which weathering occurs. These factors are the composition of the rock, the climate, the topography, and the presence of vegetation.

The mineral content of rocks and special features of rock formations affect weathering. Soft minerals, such as calcite, are easily weathered; hard minerals, such as quartz, resist weathering. Certain features of rock formations, such as joints, faults, and layers, allow water to enter the formations. This water causes an increase in weathering.

The climate also affects the rate of weathering. In places that receive little rainfall, such as the southwestern United States, the weathering process is usually slow. The weathering is slow because water is the fastest weathering agent. Some areas have large temperature changes between seasons or even between day and night. These changes speed up the rate of weathering. Water entering the rock during the day may freeze at night, causing rocks to break apart.

Some features of topography may speed up weathering. For example, mountain slopes tend to have more rock exposed at the surface than do flat lands. Also, mountains are likely to have high winds and freezing temperatures. Together these conditions speed up the weathering process.

Even the vegetation of an area may be a factor in the rate of weathering. Plants protect rocks from wind and rain. However, decaying vegetation may speed up the rate of weathering, because the roots of many plants make acids that can dissolve rocks.

○ *What is weathering?*
○ *What factors determine the rate at which rock will weather?*

11.2 Physical Weathering

Physical weathering breaks rocks apart mechanically. Bedrock is broken into fragments in four ways: by pressure, cycles of freezing and thawing, plant growth, and abrasion.

The first type of physical weathering begins shortly after rocks form. Igneous rocks, forming at great depths, are under tremendous pressure because of overlying layers of rock. Once this pressure is removed, the rocks expand. This sudden expansion causes the surface of the rocks to crack and form joints. Expansion causes some rocks, especially granite, to flake off in sheets in a process called **exfoliation** (ehks foh lee AY shuhn). Temperature changes also cause exfoliation by loosening the mineral crystals in a rock. Figure 11–3 shows the typical rounded surface caused by exfoliation.

Have you ever placed a can or a bottle of liquid in the freezer to chill it and then forgotten about it for several hours? Besides the fact that the contents were frozen, you may have noticed that the can appeared swollen or that the bottle was broken. As water freezes, it expands and exerts a tremendous pressure on the container in which it is placed. Temperature changes also affect the water that collects in cracks in rocks. For instance, if a crack fills with water and the water freezes, the crack in the rock enlarges. This process of freezing and cracking is the second type of physical weathering, called **frost action**. Frost action mostly affects sedimentary rocks because they have many pore spaces, bedding planes, and fractures. Frost action also does severe damage to road surfaces during the winter. Why do you think frost action is so damaging to roads?

Figure 11–3. Exfoliation produces thin sheets of rock that flake off in layers.

Ice

DISCOVER

Demonstrating Frost Action

Fill a small plastic jar or bottle to the top with water. **CAUTION: Do not use a glass jar or bottle.** Replace the lid tightly, and place it in the freezer overnight. What happened to the container? What could have caused this to happen?

Figure 11–4. If you fill a bottle with water and place it in a freezer overnight, the bottle will break. In a similar manner, ice in a rock crack will break the rock.

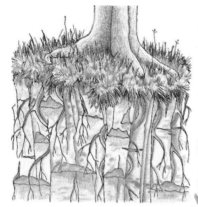

Figure 11–5. Roots growing in a rock crack exert tremendous force on the surrounding rocks. This force can expand the cracks or even cause pieces of the rocks to break off.

The third type of physical weathering occurs when the cracks started by frost action are expanded by plants. As seeds sprout and begin to develop in the cracks, the growing roots enlarge the cracks. This type of physical weathering is called *root-pry.* Another example of root-pry can be seen where the roots of a large tree close to a sidewalk or driveway have pushed up and cracked the concrete.

The final type of physical weathering occurs when rocks weather by physical contact with other rocks. As rock fragments bounce down a hill, roll in a stream, or are carried by the wind, they bump against each other. This weathering by physical contact is called **abrasion**. Abrasion knocks off sharp edges, producing rounded rock fragments. The longer that rock fragments undergo abrasion, the smaller and rounder they become.

○ *What are the agents of physical weathering?*
○ *Which two agents cause exfoliation?*

Figure 11–6. As the rocks in a flowing stream tumble and bump into each other, sharp edges are rounded off.

11.3 Chemical Weathering

During the process of chemical weathering, the minerals within a rock are chemically changed. Minerals may be added, removed, or changed into other minerals within a rock. Most chemical weathering involves water, because water has the ability to dissolve many minerals. Halite, for example, separates into sodium and chloride ions in water. Most minerals do not dissolve as rapidly as halite, and the process of chemical weathering is usually quite slow.

In addition to dissolving minerals, water can also combine with certain minerals to form new compounds in a process called **hydration**. Micas and feldspars, for instance, change to clay in the presence of water. This process is illustrated in Figure 11–7.

Not all chemical reactions require water, however. Another type of chemical weathering occurs when oxygen combines directly with an element in a process called **oxidation**. Rocks that have undergone oxidation can usually be identified because oxidation produces a color change. For example, oxygen combines with magnetite to form a reddish-brown rust, similar in appearance to the rust on an old nail. Even in this case, however, the presence of water speeds up the process. An unpainted bicycle will rust more rapidly in a humid climate than in the desert, because the oxygen dissolved in water combines more readily with iron than the oxygen in the air. In the same way, rocks weather more quickly in humid climates.

Figure 11–7. The red-clay soil of Georgia and certain other states results from hydration of the minerals in the soil.

Figure 11–8. Chemical weathering occurs to things other than rocks. Oxidation of the iron in this old car creates rust. Even a shiny new bicycle will weather if it is not protected.

Figure 11–9. Sulfur pollution in the atmosphere produces sulfuric acid when it combines with rain. The marble in the statue is being dissolved by acid rain.

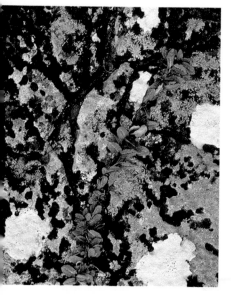

Figure 11–10. When lichens grow on rocks, they produce a type of acid that dissolves the minerals in the rocks.

Water is also involved in another chemical reaction. As rain falls, small amounts of carbon dioxide from the atmosphere are dissolved in the water. This forms a weak acid called *carbonic acid* (H_2CO_3). Carbonic acid reacts with the minerals in rocks such as limestone, dissolving them in a process called **carbonation.** The product of this reaction is washed away by the rain, leaving only the impurities from the original rock.

The decaying of plants also produces acids that chemically weather rock. Even bare rock surfaces may be weathered by plant acids. As more plants grow in an area, the effects of chemical weathering reach deeper, breaking down more rock and building up more soil. However, good soil takes thousands of years to form. Why do you think this is so?

○ *What happens to minerals in chemical weathering?*
○ *What are the agents of chemical weathering?*

Section Review

READING CRITICALLY

1. Describe the differences between chemical and physical weathering.
2. Why is water the most important agent of chemical weathering?

THINKING CRITICALLY

3. Why is there very little soil on mountaintops? Does that mean that little weathering occurs there? Explain your answer.
4. Explain the weathering process that would occur in a place with a lot of rain and large temperature changes.

② Mass Movement and Landforms

SECTION OBJECTIVES

After completing this section, you should be able to:

- **Explain** why gravity is necessary for mass movement.
- **Describe** the role of a cap rock in the formation of mesas and buttes.
- **List** the rock characteristics that would cause differential weathering.

NEW SCIENCE TERMS

mass movement
creep
flow
landslide
subsidence
differential weathering

11.4 Mass Movement

If you live in an area with steep hills and snow, you have probably rolled snowballs down a slope and watched them gain speed and snow. Rocks and soil are affected by gravity in much the same way in a process called *mass movement*. **Mass movement** is the downhill movement of rocks and soil due to the force of gravity. Mass movement is classified by the size of the rocks moved, by the amount of water involved, and by the steepness of the slope. The four types of mass movement are *creep, flow, landslide,* and *subsidence.*

Creep occurs when rock materials saturated with water move slowly downhill. The movement is often so slow that changes are hard to see. An example of creep is shown in Figure 11–11. This particular change took place over many years.

Figure 11–11. This fence was once straight and now it bends in the middle. The soil in which the fence posts sit is slowly moving by a process called *creep.*

237

Figure 11–12. The house on the left was caught in an earth flow. Soil that is saturated with water often flows down hill. The talus at the base of the mountain on the right resulted from a landslide—the movement of dry soil and rocks.

Figure 11–13. This sinkhole in Winter Park, Florida, is the result of subsidence. The limestone bedrock was dissolved by underground water, and the overlying soil caved in.

When water-saturated rock material moves at a noticeable rate, the process is called **flow.** There are two types of flows: earth flows and mud flows. Earth flows are the movement of water-saturated clay or silt down gentle hillsides. Mud flows are the movement of water-saturated rock debris down channels on steep hillsides. Mud flows, often called *mud slides,* are common along the California coast because of the clay soils.

A **landslide** occurs when dry soil or rock moves down a steep slope. The movement is usually rapid and unexpected, because, unlike mud flows, which are usually caused by heavy rain, most landslides have no direct cause. The force of gravity simply overcomes the friction between rocks, and the material moves. The fallen material collects at the base of the slope and forms a mass called *talus* (TAL uhs).

When rock material sinks to a lower level without sliding down a hillside, the process is called **subsidence** (suhb SY duhns). Subsidence can occur when minerals, such

as halite, are dissolved by water, and the overlying rock then collapses into the hollow. Sinkholes, such as the one shown in Figure 11–13, often result from subsidence. Subsidence can also occur as young rocks compact, especially if wells are pumping out water or oil. Many areas along the Gulf Coast are experiencing subsidence due to the compaction of sediments.

○ *What are the four types of mass movement?*
○ *What is the difference between creep and earth flow?*

11.5 Landforms and Weathering

Some interesting features of the land, or landforms, are created by weathering and mass movement. For example, landforms called *mesas* (MAY suhz) and *buttes* (BYOOTS) are the result of physical weathering. Mesas are lands with flat tops and nearly vertical sides. A butte is a lone, flat-top hill that rises abruptly above the surrounding flat land.

In mesas and buttes, the top rock, called the *cap rock*, is more resistant to weathering than the rocks below it. The less-resistant rock weathers away, leaving the formations shown in Figure 11–14. The cities of Mesa, Arizona, and Butte, Montana, are named after these prominent features. What places in your area are named after a particular landform?

Sometimes layers of rock that are resistant to weathering alternate with those that are less resistant. The easily weathered rock will form a gentle slope as it undercuts the resistant rock above it. The resistant rock will weather into vertical cliffs. The

Figure 11–14. Mesas, such as the one on the left, and buttes, shown on the right, result from the weathering of nonresistant rock layers. Rock that is resistant to weathering forms the caps of these formations.

Figure 11–15. Some of the unusual formations of the Grand Canyon are due to the differential weathering of rock layers. Different layers weather at different rates, even though they are exposed to the same environmental factors.

difference in weathering of different rocks in the same formation, exposed to the same environment, is called **differential weathering**. Differential weathering is responsible for creating many of the features seen in the Grand Canyon.

Differential weathering can sometimes be seen in a single formation. If weathering agents enter bedding plane cracks or fractures, differential weathering can form caves, ledges, or long columns of rocks. If the rock layers are not horizontal, the resistant rock will form ridges, as shown in Figure 11–16, while the more easily weathered rock will form valleys between the ridges.

Landforms resulting from mass movement are usually not as spectacular as those produced by differential weathering.

Figure 11–16. Resistant rocks sometimes form the tops of ridges, such as those shown in the photograph, while the less resistant rocks form the valleys between the ridges. Notice the location of mountains and ridges on the map.

Figure 11–17. Small, round lakes, such as these in northern Florida, are examples of karst topography. Karst topography results from a combination of chemical weathering and subsidence.

However, in Florida, a combination of chemical weathering of the limestone bedrock and subsidence has created a landscape dotted with small, round lakes. This type of landform is called *karst topography.*

○ *What are landforms?*
○ *What is differential weathering?*

Section Review

READING CRITICALLY

1. Explain why landslides and mud slides are not the same.
2. What is the difference between a mesa and a butte?

THINKING CRITICALLY

3. What would be the main agents of weathering in the formation of a mesa or butte? Explain your answer.
4. Imagine that you are considering buying a home in a subdivision where subsidence might be occurring. What should you look for in the area to decide if the house is safe to move into? Why?

ACTIVITY: Analyzing a Landform

How can you identify the effects of weathering on a landform?

PROCEDURE

Examine the topographic map on page 581. Study the contour interval and the map scale; then answer the following questions.

CONCLUSIONS/APPLICATIONS

1. What type of landform is shown in this map? Explain the features that led you to this conclusion.
2. How many resistant rock masses can you identify on this landform? Identify the resistant rocks by referring to the contour elevation marking the top of the layer.
3. Would any of these rocks be appropriately referred to as cap rocks? Explain your answer.
4. Use the term *differential weathering* to explain how the landform has weathered into this shape.
5. How does the type of landform shown help you determine the climate in this area?

GOING FURTHER

Use clay to build a model of the landform in this activity. Use one color of clay for the resistant rock layers and another color for the easily weathered layers.

SKILL ACTIVITY: Interpreting a Pie Graph

BACKGROUND

There are many methods available to scientists for presenting data. Sometimes scientists use a circle, or pie, graph to present information about the way something is divided or distributed. Usually the percentage of the whole is written on each division of the graph.

PROCEDURE

Examine the pie graphs for North and South America. Each one represents the uses of fertile soil on the continent. Determine how much of each continent is available for cropland and grazing land.

APPLICATION

1. Which continent has the greatest percentage of fertile soil in forested land?
2. Which continent has the greatest percentage of fertile soil in grazing land?
3. Which continent has the greatest percentage of fertile soil in cropland?
4. Which continent has the greatest percentage of fertile soil not used for cropland?

USING WHAT YOU HAVE LEARNED

Do some library research to find out the uses of land in your state; then make a pie graph for land use in your state. Label the percentage for each type of land use.

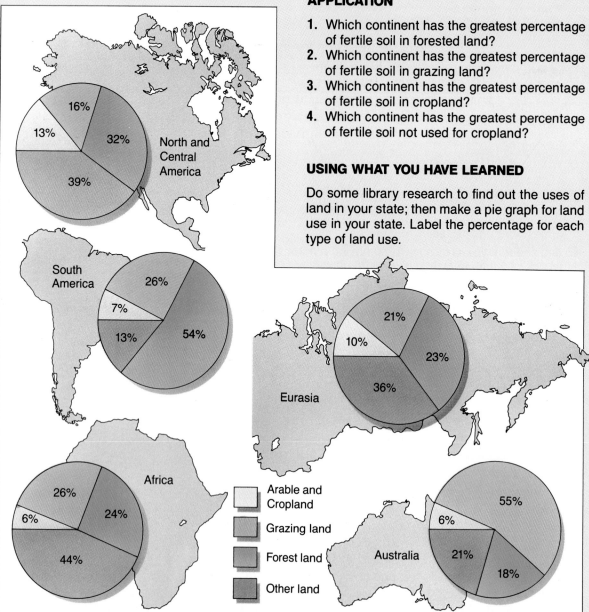

North and Central America: 16%, 32%, 39%, 13%

South America: 26%, 7%, 13%, 54%

Eurasia: 21%, 10%, 36%, 23%

Africa: 26%, 24%, 6%, 44%

Australia: 55%, 6%, 21%, 18%

Arable and Cropland
Grazing land
Forest land
Other land

3 Soils

SECTION OBJECTIVES

After completing this section, you should be able to:

■ **Identify** the factors that are involved in soil formation.

■ **Summarize** the characteristics of different types of soils.

NEW SCIENCE TERMS

humus

horizons

leaching

11.6 Soil Formation

The formation of soil is one of the most important aspects of the weathering process. Life as we know it would not be possible without soil for growing grains, fruits, vegetables, and pasture grasses. Care must be taken to preserve and protect the soil, because the process of soil formation is very slow. It takes from 100 to 1000 years for one centimeter of soil to form.

> **A MATTER OF FACT**
>
> A single rye plant can produce roots over 640 m long. When the plant dies, the root system decays and becomes part of the humus of the soil.

Figure 11–18. Deep, rich soil is needed to grow grains, fruits, and vegetables. The great variety of food available to North Americans is due to the abundance of good soil found on the continent.

Five factors are involved in the formation of soil: the parent material, the climate, the topography, the organisms in the soil, and time. The parent material is the original rock from which the soil is formed. Soils formed from shale, for instance, will be very different from soils formed from limestone.

The climatic factors that influence soil formation are temperature change and precipitation. Soils form more quickly in moist climates because weathering is faster in the presence of water. Soils also form more quickly where there are significant temperature changes. Related to climate is the topography, or shape of the land. Topography affects the drainage of an area. If you look at a steep slope, you will probably not find much soil, because a

Figure 11–19. The skimpy vegetation on steep slopes (right) is the direct result of the limited soil found in these areas. Gentle slopes (left) have more soil and, therefore, more vegetation.

steep slope allows water and mass movement to remove newly formed soil. A gradual slope allows soil to accumulate without being carried away.

The plant and animal life within the soil also influences soil formation. Tunnels made by ants and earthworms, for instance, create spaces for air and water. These spaces allow more weathering to occur; therefore, more soil will be formed. In addition, dead plants and animals decay, adding to the humus (HYOO muhs). **Humus** is the organic part of the soil.

The final factor in soil development is also the most obvious—time. The longer the time for soil formation, the more soil there will be. This will be true except in areas where environmental factors act to transport soil somewhere else. What environmental factors might act to transport soil?

○ *What are the five factors involved in soil formation?*
○ *What is humus?*

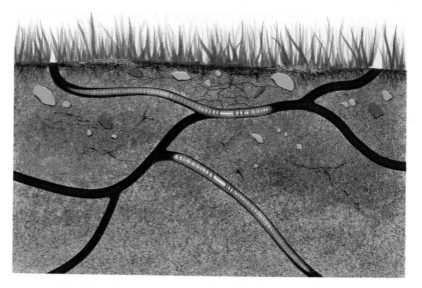

Figure 11–20. Worms, part of the soil biota, often make tunnels in the soil. These tunnels allow water and nutrients to reach the roots of plants.

TECHNOLOGY: Modern Farming

Most people in the United States live in urban areas. They get their food, wrapped in plastic and paper containers, from a supermarket. With each new day the demands on the land increase; more people must be fed, more trees must be cut for paper and building materials, and more land must be turned into areas where people can live.

Total land area is being carefully evaluated to make the best use of every available hectare.

Plants called *legumes* are being planted on marginally fertile soil because they produce nitrogen compounds that other plants can use. The contour of the land is studied, and crops are planted on terraces, or flat steps, rather than on slopes, to prevent loss of soil due to excess runoff.

Many farmers are now using a practice called *minimum till farming*. In this program, the soil is protected in the winter by the stubble of the previous crop. New crops are planted in narrow

Modern harvesting equipment

strips between the stubble to reduce the exposure of the soil to the drying wind.

New techniques in irrigation are turning semiarid parts of the world into productive cropland. In some places, water is dripped onto individual plants instead of being sprayed over entire fields, where much of the water evaporates before it reaches the roots of the plants.

Much work is being done with the practice of *hydroponics;* that is, growing food plants without soil. In hydroponics, water with essential plant nutrients is sprayed directly onto the roots of plants suspended in air. The roots absorb the nutrients, and the plants carry out photosynthesis as if they were growing in rich soil.

Efforts are also being made to "engineer" superior plants and animals in the laboratory. By cross-breeding organisms with many desirable characteristics, it is possible to develop plants and animals that produce more food in less time and at a lower cost than ever before.

If the world is to continue feeding its increasing population, further advances will be needed to produce even more food from the shrinking supply of land.

Contour plowing and drip irrigation (inset)

11.7 Soil Composition

Soil consists of inorganic (nonliving) and organic (once-living) material. However, at least 80 percent of all soil is inorganic—it is weathered rock.

Quartz and clay minerals are the most common minerals in soils, although other minerals may also be present. Soils also commonly contain compounds of potassium, phosphorus, and nitrogen, as well as air and water. Table 11–1 shows the composition of various soils.

TABLE 11–1: SOIL CHARACTERISTICS

Color	Drainage	Composition
Brown	Well drained	Silt, some humus
Red-brown	Well drained	Clay, iron minerals
Tan	Well drained	Sand, little humus
Gray	Poorly drained	Clay, few minerals
Black	Poorly drained	Muck, much humus

If you dug into the ground and then drew a picture of the soil layers you found, you would be drawing a *soil profile*. Soils develop in horizontal layers called **horizons.** Each horizon in a soil profile is labeled with a capital letter. Examine the soil profile in Figure 11–21.

The top layer is the A horizon. This layer contains humus and small amounts of clay and sand. The roots of many plants are restricted to this horizon. The next layer, the B horizon, contains coarse clay, sand, and a small amount of humus. The B horizon also contains dissolved elements that were washed from the

Figure 11–21. Good topsoil has a thick A horizon with plenty of humus and just the right proportion of sand and clay. Most of the roots of cover crops, such as grains and grasses, grow in this horizon.

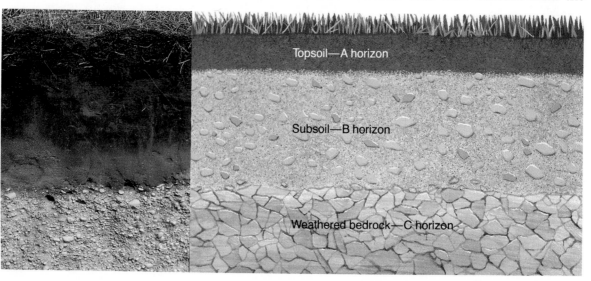

Topsoil—A horizon

Subsoil—B horizon

Weathered bedrock—C horizon

upper horizon by water. The movement of minerals from an upper horizon to a lower horizon is called **leaching.**

Below the B horizon is a thicker layer of partially weathered rock—the C horizon. Chemical weathering occurs in the C horizon, although here the process is very slow. Most plant roots do not reach this layer, and the exposure to air and water is also limited at this depth. Beneath the C horizon is bedrock.

Although most soils have the horizons just described, soils do vary from region to region. The climate of a region controls the rate of chemical weathering. Areas that are cold throughout most of the year have poorly developed soils because the frozen surface does not allow much weathering to occur. These areas also lack protective plant growth, so bare rock is found more often.

Humid, tropical areas develop thick soils because of the high temperature and abundant rain they receive. In such areas, however, many of the minerals are removed from the top layer of soil by water; thus the soil is not rich enough in nutrients to grow abundant crops.

In temperate climates, there is usually less rainfall and therefore less leaching of the soil than in the tropics. However, many temperate areas receive enough rainfall to develop thick soils. Ample plant growth in these regions produces soil that is rich and black with humus. Due to their richness, temperate soils are among the world's most fertile soils. Most of the world's food comes from soils of temperate areas, like those in the midwestern United States and Canada.

○ *What are soil horizons?*
○ *Why are tropical soils so poor in minerals?*

Figure 11–22. Poor crops, such as those shown in the photograph at left, often result if the topsoil has been leached of its nutrients. Notice the difference in color between the leached soil (top) and the rich soil (bottom).

Section Review

READING CRITICALLY

1. What are the differences among soil horizons?
2. What is mineral leaching?

THINKING CRITICALLY

3. If temperate climate is important in the development of good soils, where else in the world might you expect to find soils as good as those in North America?
4. Why are both inorganic and organic materials necessary for the development of good soils?

INVESTIGATION 11: Weathering by Carbonation

PURPOSE

To examine the effects of carbonic acid on the weathering of different rocks

MATERIALS (per group of 3 or 4)

Safety goggles
Laboratory apron
Laboratory balance
Small food jars with lids (4)
Tap water
Limestone
Carbonated water
Granite

PROCEDURE

1. Copy the chart shown.
2. **CAUTION: Wear safety goggles and a laboratory apron during this investigation.**
   Using the balance, determine the mass of the limestone specimen.
3. Fill a small jar half full of tap water, and add the limestone to the jar. Put the cap on the jar and gently shake the jar 200 times.
4. Remove the limestone from the jar, dry it, and determine its mass.
5. Repeat steps 3 and 4. After shaking the jar 400 times, record the mass.

6. Repeat steps 3 and 4 for a third time, shaking the jar 600 times. Record the results.
7. Compute the percentage mass change, using the following formula:
$$\frac{\text{Initial Mass} - \text{New Mass}}{\text{Initial Mass}}$$
8. Repeat steps 2 through 7, using limestone in carbonated water. Record your results.
9. Repeat steps 2 through 7, using granite in both tap water and carbonated water. Record your results.

ANALYSES AND CONCLUSIONS

1. Which combination of liquid and rock produced the greatest change in mass?
2. What kind(s) of weathering occurred in this investigation? List and describe the weathering in as much detail as possible.
3. In what way is the weathering process in this investigation similar to weathering in nature? In what way is it different?

APPLICATION

Using what you have learned in this investigation, predict two types of rock that would probably not weather very much. Predict one type of rock that would probably weather a lot. State the reasons for your predictions.

	Limestone		Granite	
	Water	**Carbonated Water**	**Water**	**Carbonated Water**
Initial mass				
Mass after 200 shakes				
Mass after 400 shakes				
Mass after 600 shakes				

TABLE 1: EFFECTS OF WEATHERING

SUMMARY

- Weathering is the process that breaks down rocks. (11.1)

- The two major types of weathering are physical and chemical. Factors that affect the rate of weathering are rock structure, climate, topography, and vegetation. (11.1)

- Physical weathering breaks rocks apart mechanically. (11.2)

- Exfoliation, frost action, root-pry, and abrasion are four types of physical weathering. (11.2)

- In chemical weathering, minerals within a rock are chemically changed. (11.3)

- Hydration, oxidation, and carbonation are types of chemical weathering. (11.3)

- Mass movement is the movement of rocks due to gravity. Mass movement is classified by the slope of the land, the size of materials, and the amount of water involved. (11.4)

- The four types of mass movement are creep, flow, landslide, and subsidence. (11.4)

- Differential weathering of rock layers results in the formation of distinctive landforms such as buttes and mesas. (11.5)

- Landforms resulting from mass movement are usually not as spectacular as those produced by differential weathering. (11.5)

- Soils are the product of the weathering process. (11.6)

- Soil formation is affected by the type of parent material, the climate, the topography, the soil organisms, and time. (11.6)

- The various layers, or horizons, of a soil are shown in a soil profile. (11.7)

- Most of the world's fertile soils are found in temperate regions, such as the United States. (11.7)

Write all answers on a separate sheet of paper.

SCIENCE TERMS

Correctly use each of the following terms in a sentence.

abrasion **(234)**
carbonation **(236)**
creep **(237)**
differential weathering **(240)**
exfoliation **(233)**
flow **(238)**
frost action **(233)**
horizons **(246)**
humus **(244)**
hydration **(235)**
landslide **(238)**
leaching **(247)**
mass movement **(237)**
oxidation **(235)**
soil **(231)**
subsidence **(238)**
weathering **(231)**

SCIENCE QUIZ

Modified True-False

Mark each statement *true* or *false.* If a statement is false, change the underlined term to make the statement true.

1. The most effective weathering process in a desert would be <u>chemical</u> weathering.

2. The soil layer richest in humus is the <u>A horizon</u>.

3. Mud slides are classified as <u>landslides</u>.

4. Subsidence is an example of <u>chemical weathering</u>.

5. Carbonic acid is formed when carbon dioxide combines with <u>calcite</u>.

continues

Multiple Choice

Write the letter of the choice that best answers the question or completes the statement.

6. Hydration is what type of weathering?
   a) physical   b) fast
   c) rock   d) chemical

7. The process of carbonation forms karst topography with what type of rock?
   a) limestone   b) quartzite
   c) granite   d) shale

8. Frost action is most effective when there
   a) are extended periods of subzero weather.
   b) is repeated freezing and thawing.
   c) is a rainy season.
   d) are large breaks in the rock.

9. Rounded quartz pebbles found along a stream bed would probably have been weathered by
   a) exfoliation.   b) hydration.
   c) abrasion.   d) mass movement.

10. A factor that is not important in the formation of soils is the
   a) topography.   b) amount of rainfall.
   c) type of bedrock.   d) age of bedrock.

Completion

Complete each statement by supplying the correct term or phrase.

11. Carbonation is a form of _____ weathering.

12. The main agent of chemical weathering is _____.

13. Frost action is an example of _____ weathering.

14. Decayed plant and animal matter within the soil is called _____.

15. Leaching of almost all soil minerals occurs in the _____.

Short Answer

16. Explain why mass movement is more closely related to physical weathering than to chemical weathering.

17. Why are soils made almost entirely of quartz and clay minerals?

18. Explain how vegetation can slow down some forms of weathering but speed up others.

Writing Critically

19. If you found a mineral that did not have a recognizable luster, why would you break it to look at a fresh surface?

20. Considering the length of time that it takes to form soils, do you think soils should be classified as renewable or nonrenewable resources?

EXTENSION

1. Write a report on the formation of potholes in the roads of your area. Take photographs of some potholes, analyze their locations, and identify the steps in their development.

2. Examine the exterior of several older stone buildings in your area. Determine if any chemical or physical weathering has occurred. Make an oral report to your class on your findings.

3. Describe several of the landforms in your area. Draw or photograph each. Identify the role that weathering has played in their formation.

4. Read John Steinbeck's book *The Grapes of Wrath,* and write a report on the effects of the "dust bowl" on farming in the 1930s.

5. If you live in a rural area, talk to one of the local farmers about the ways in which he or she prevents the loss of soil. Report your findings to the class.

APPLICATION/CRITICAL THINKING

1. In some parts of the country the soil has a definite reddish-brown color. What mineral might be the source of this color? Explain your answer.

2. Give an example of subsidence that might be caused by human activity. What steps might be taken to prevent subsidence in your example?

3. The Sphinx, a huge statue in the desert in Egypt, has withstood the processes of weathering for thousands of years. Yet in the last 50 years the decay of the statue's features has increased greatly. What might have caused this rapid increase in weathering? Explain your answer.

FOR FURTHER READING

Gerster, G. "Patterns of Plenty: The Art in Farming." *National Geographic* 166 (September 1984): 391. This photo essay on the subject of different crop patterns beautifully demonstrates the diversity of agricultural activities.

Gibbons, B. "Do We Treat Our Soil Like Dirt?" *National Geographic* 166 (September 1984): 350. This article analyzes the ways humans alter the soil with industrial and agricultural processes.

Harpstead, M., and F. Hole. *Soil Science Simplified.* Ames, Iowa: Iowa State University Press, 1980. This classic book presents a basic introduction to soil sciences without a strong emphasis on the technical aspects of the field.

Steinhart, P. "We Can't Grow When It's Gone." *National Wildlife* (March 1985): 74. This article impresses upon the reader the importance of good soil for growing abundant crops.

Challenge Your Thinking

This is a picture of the "dust bowl" of the 1930s. Poor farming practices and unusually dry weather created these conditions.

Although many steps have been taken to preserve our soil, scientists estimate that some areas of the country lost 10 cm of soil last year. What do you think could be done to prevent another dust bowl?

Running Water

Have you ever looked at a stream and tried to imagine the number of places the water had been? Some of the water may have spent thousands of years as ice in an Alaskan glacier. Some may have traveled the length of the Amazon River or fallen as sleet in the mountains of Montana. Some of the water may even have been part of the stream that carved this natural bridge.

A natural bridge, carved by water

1 Surface Water

SECTION OBJECTIVES

After completing this section, you should be able to:

- **List** the factors that affect runoff.
- **Describe** the various drainage patterns.

NEW SCIENCE TERMS

runoff

erosion

load

drainage basin

12.1 **Runoff**

You may recall from Chapter 9 that water molecules evaporate into the atmosphere from the surface of the ocean. Once in the atmosphere, water vapor may be carried over a landmass by the wind. After condensing into cloud droplets, the water molecules fall to the earth's surface as drops of rain. The movement of water is called the *water cycle.*

The moment a raindrop falls to the earth it begins its return to the sea. Some of the water soaks into the ground, but most moves across the surface toward a lower level. This movement of water on the surface of the earth is called **runoff.**

The shape of the land and the amount of rainfall affect runoff. The steeper the slope, the greater the speed of the runoff. The volume of runoff increases during heavy rain because there is not enough time for much water to soak into the ground. Runoff also increases if rain falls on ground that is already saturated with water, or if the water cannot get into the ground. Large cemented areas, such as those that are present in cities, provide little opportunity for water to get into the ground.

Figure 12–1. If a lot of rain falls in a short period, or if there is a sudden melting of a large accumulation of snow, a flood may result.

Figure 12–2. A muddy stream (left) is an indication of excess erosion. If soil is unprotected by a cover of plants, deep gullies can be eroded (right).

Runoff decreases in areas with extensive plant growth. Plants slow down the water's movement, so more water soaks into the ground. Some types of soil can also reduce the amount of runoff. If soil is loosely packed or has a large humus content, water can more easily soak into the ground.

As water rushes downhill, it carries sediment, which is then deposited at the bottom of the hill. If enough sediment washes away, a small channel, or *gully*, forms. The process that moves sediment and reshapes landscapes is called **erosion.** Water is not the only agent of erosion, but it is the most important.

Gullies usually form in places where the surface is not protected by a cover of plants, such as in a plowed field. With continued erosion, a gully develops into a *stream.* A stream carries water throughout most of the year, while a gully carries water only when it rains. As time passes and more land is eroded, a stream expands and develops into a *river.* The volume of water in most rivers is large because many streams empty into them.

Water has been called "the great leveler" because of its ability to erode highlands and fill in lowlands. Even a small stream can carry a tremendous quantity of eroded material.

The amount of eroded material carried by a stream or a river is called the **load.** The load a stream carries depends on the stream's water volume and the steepness of the slope down which it flows. A large stream or a fast-moving stream carries a larger load than a small or slow-moving stream carries. Scientists estimate that the rivers of the world carry 1000 metric tons of sediment to the sea each second.

○ *What is runoff?*
○ *What is erosion?*

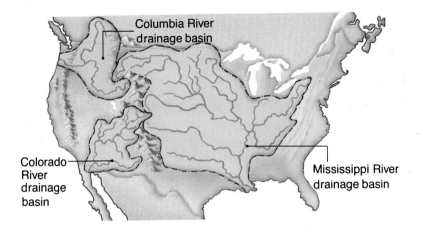

Figure 12–3. This map (left) shows the major drainage basins of the United States. A drainage basin includes all the main streams and tributaries that drain an area (right).

12.2 Drainage Basins

A large tree with its many leafy branches is part of a system for producing food. Similarly, a single river, called a *trunk*, and the streams that flow into it, called *tributaries*, act as a system for draining water from the land. The area drained by a trunk and its tributaries is called a **drainage basin.**

An imaginary line called a *divide* separates one drainage basin from another. For example, the continental divide, which runs the length of the Rocky Mountains, divides the basin that drains into the Pacific Ocean from the basin that drains into the Gulf of Mexico. Many regions of the country have divides that separate major drainage basins. Some of these drainage basins are shown in Figure 12–3.

The streams in a basin often develop unique drainage patterns due to the dominant landform of the area. For instance, streams flow in all directions away from a volcano, forming a radial drainage pattern. *Radial* means "coming from a central point," and as you can see in Figure 12–4, the volcano acts as the central point for this drainage pattern.

Drainage basins containing long, parallel ridges and valleys form *trellis* patterns. Unlike a flower trellis, which has branches originating from a single point, a trellis drainage pattern has tributary streams that flow parallel to each other and join the trunk at right angles.

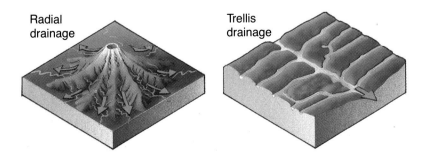

Figure 12–4. The streams in a radial drainage pattern all come from the same place, such as from a volcano (left). The tributaries of a trellis drainage pattern flow nearly parallel to one another, and join the trunk stream at right angles (right).

Section 1   Surface Water   **255**

Sometimes topography is the dominant factor in the development of a drainage pattern. Branching, or *dendritic*, patterns develop in basins where slopes are not very steep. Here the tributaries form fairly large angles where they join the trunk. The entire drainage system forms a pattern similar to the dendrites, or branched endings, of a nerve cell. Where slopes are steeper, *parallel* drainage may develop.

Finally, drainage patterns may be determined by special characteristics of the basin. In areas in which the bedrock is jointed or faulted, *rectangular* patterns develop. Some of the stream channels make sharp turns as they follow lines weakened by fractures in the rock. Patterns found in poorly drained areas tend to be irregular, or *deranged*. Deranged drainage basins have areas in which surface water collects, such as swamps, marshes, and lakes.

○ *What is a drainage basin?*
○ *Name four types of drainage patterns.*

Figure 12–5. Dendritic drainage patterns (bottom left) develop where slopes are gentle, whereas parallel patterns (top left) develop where slopes are steeper. Rectangular patterns (bottom right) occur where the bedrock is jointed or fractured. Deranged drainage patterns (top right) are found in areas where there is almost no slope at all to the land.

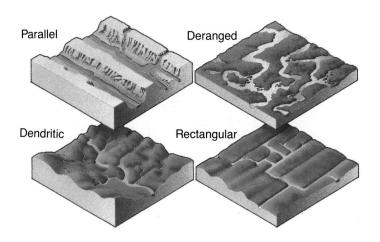

Parallel   Deranged

Dendritic   Rectangular

Section Review

READING CRITICALLY

1. What are three factors that affect runoff?
2. What types of drainage patterns develop around volcanoes and severely faulted areas?

THINKING CRITICALLY

3. From the map on page 255, determine how many major drainage basins there are in the United States.
4. Explain how the drainage pattern might change if a volcano were to develop in an area with trellis drainage.

EUGENE HILGARD
(1833–1916)

Eugene Hilgard was born in Bavaria in 1833. He immigrated to America with his family in 1836, when he was three years old. Hilgard attended schools and universities in both the United States and Europe, where he studied chemistry and geology. In 1853 he received a doctorate from the University of Heidelberg in Germany.

After returning to the United States, Hilgard spent much of his career studying the structure and composition of soils. Combining his knowledge of chemistry and geology, he was one of the first scientists to recognize the importance of soil analysis to agriculture. He understood that different types of plants require different soil conditions. For example, some plants grow only in well drained sandy soil, while others require wet soil with a lot of humus. At the time, most farmers did not understand the complex nature of soils. They only knew that a crop would produce smaller yields each year that it was planted. After several years of planting the same crop in the same location, they would abandon the field and begin planting in a new location.

Hilgard argued that soil should be considered a special geologic formation, deserving as much attention as the underlying rock formations. His writings brought him international recognition as a geologist and a soil scientist. His ideas lead to many changes in farming practices and to the recognition of soil as a complex material vital for human survival.

HELEN M. TURLEY
(1943–)

Helen Turley was born in Augusta, Georgia, in 1943. She received her undergraduate training from St. John's College in Annapolis, Maryland. She then attended the College of Agriculture and Life Sciences at Cornell University in Ithaca, New York. There she expanded her knowledge of soil science and biology.

In New York she was employed at the New York State Agriculture Experiment Station, where she studied the problems associated with growing grapes. Grapevines are commonly grown on hillsides, so the growers are especially concerned with soil erosion. In addition to concerns about soil erosion, the chemical and physical makeup of the soil is also of concern. If the soil chemistry is not right, the grapevines will not grow. Turley began concentrating on the needs of the grape-growing industry and soon became a specialist in the characteristics of soils and soil erosion.

With this expertise she relocated to serve the grape-growing industry in California. There she began a consulting business, applying her knowledge of soils and erosion to the problems found in the various grape-growing regions of California. She is presently conducting research on cover crops planted between the grapevines to protect the soil from erosion. Her research may lead to the selection of the best cover crops for preventing soil erosion in other agricultural areas as well.

② Erosion and Deposition

NEW SCIENCE TERMS

headward erosion
base level
flood plain
meanders
delta
alluvial fans
playas

SECTION OBJECTIVES

After completing this section, you should be able to:

- **Relate** a stream's slope, volume, base level, and velocity to its ability to erode the land.
- **Compare** the stages of valley development.
- **Explain** the processes involved in the formation of deltas.
- **Describe** the conditions necessary for the formation of alluvial fans and playas.

12.3 Headward Erosion

Think about some of the streams and rivers that you have seen in movies, on television, or while traveling. Some of these streams probably carry very little water and may flow only part of the year, while others carry large quantities of water year-round. You may have seen streams flowing down narrow, steep-sided valleys or across wide plains. Some streams are clear, with very little sediment, but others are always muddy because of the large amount of eroded material they transport.

In spite of these differences, all streams and rivers erode material from one place and deposit it somewhere else. As these processes of erosion and deposition continue, the shape of the stream itself changes.

Streams increase their length by a process called **headward erosion.** As more and more material is eroded from the origin, or *source*, of a stream, large gullies become part of the

Figure 12–6. Some streams (left) may be almost dry during part or even most of the year. Other streams run almost bank full (right), especially during seasons of heavy rain.

streambed. The other end of a stream, called the *mouth*, is where the stream flows into some other body of water. Deposits often occur at the mouth of a stream.

A stream or river cannot erode below the elevation of its mouth. This low point of the stream or river is called its **base level.** Sea level is the ultimate base level of all rivers and streams. A lake or a layer of resistant rock may act as a temporary base level.

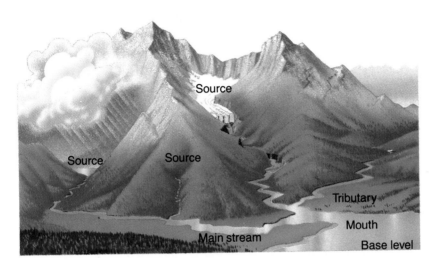

Figure 12–7. All streams have a source, where the runoff begins, and a base level which is the lowest point of erosion.

A MATTER OF FACT

The highest waterfall in the world is the Angel Falls in Venezuela, which is over 800 m high. The widest waterfall is the Khon Falls in Laos, Southeast Asia, which is over 10 km wide.

Figure 12–8. Resistant rock may form a temporary base level and a waterfall (left). The temporary base level of the Niagara River is beginning to erode away as this photograph of Niagara Falls shows (below).

Where a stream encounters a layer of rock that resists erosion, waterfalls or rapids often result. The resistant rock serves as a temporary base level for the water upstream. As the water slowly erodes the resistant rock, the waterfall or rapids move upstream. For instance, Niagara Falls, formed on resistant rock, has moved upstream more than 11 km in the last 12 000 years.

○ *What is headward erosion?*
○ *What is a stream's base level?*

12.4 Valley Formation

The development of a stream valley can be divided into three stages: youthful, mature, and old-age. Youthful streams have narrow, V-shaped valleys with steep walls, since most of the erosional energy is used to deepen the streambed. Areas between valleys are poorly drained because they are not really part of the drainage basin. Waterfalls or rapids may occur where the stream encounters layers of resistant rock.

As the youthful stream approaches its base level, it begins to erode the sides of its banks, producing a wider valley. Weathering of the valley walls and erosion by tributaries also help to widen the valley and to form a mature stream.

Figure 12–9. A youthful stream (above) is characterized by straight, steep-sided, V-shaped walls (bottom left). A mature stream (right) has many bends, and a wide, gently sloping valley (bottom right).

Youthful

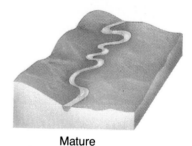

Mature

Mature stream valleys are characterized by a wide **flood plain,** the area over which a stream spreads in times of high-volume runoff. When a stream overflows, it deposits sediment on its flood plain, enriching the soil and making it a fine place to grow crops. Before the development of the Aswan Dam, the Nile River in Egypt flooded every spring, providing nutrients for the summer crops.

Flooding streams also deposit sediments next to the channel. Large rock fragments, such as boulders, can be carried only by fast moving water. As a stream begins to flood its valley, the water slows, and these sediments form *levees*, such as those shown in Figure 12–10, on either side of the stream.

The water in a mature stream erodes the banks, forming broad curves called **meanders.** A cross-sectional view of a meander shows that the outside of the curve is deeply eroded by fast-moving water, while the inside is shallow due to deposition in the slower-moving water.

An old-age valley is characterized by slow-moving water and a wide flood plain with many lakes, swamps, and meanders. The meanders often form such broad loops that one meander cuts into the next. Most of the water follows this shorter route. The abandoned meander forms an *oxbow,* a bow-shaped bend with a temporary lake. Some river valleys, such as the Mississippi, may be youthful at the source, while the section near the mouth may be in the old-age stage.

○ *What are the three stages of valley development?*

Figure 12–10. Natural levees (above) form as mature streams overflow their banks. To prevent flooding, people sometimes construct levees of sand and rocks (left).

Figure 12–11. Old-age streams have many meanders and cutoffs, called *oxbows* (below). The meanders form because old-age streams are near their base level, and most of the erosion occurs in the sides of the banks.

Old age

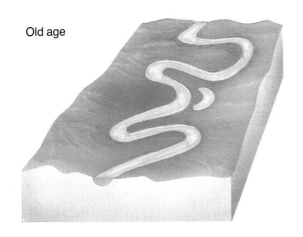

Studying Erosion

Find an area near your home or school where there seems to be some erosion due to runoff. Diagram the present characteristics of the area. After the next significant rain, return to the same area and diagram any additional erosion that has occurred because of the rain.

12.5 Water Deposits

Erosion is only part of the leveling process of water. Once eroded by running water, sediment is transported only as long as a stream maintains its volume or speed. If the water volume decreases, some of the load will be deposited in the streambed, forming *sandbars*. Before a channel was dug to improve river transportation, there were many of these deposits in the lower Mississippi River valley.

Deltas When a large, sediment-laden stream reaches its mouth, the flow spreads out and the water speed drops. As a stream slows down, it is unable to carry a large load of sediments, so the load is deposited near the mouth, forming a **delta.** Many streams form deltas, but the deltas located at the mouths of the Mississippi and Nile rivers are especially well known.

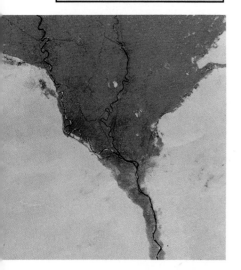

Figure 12–12. Deltas form at the mouth of the Mississippi River (right) and the Nile River (left) because of the large loads carried by these rivers.

As a delta builds around a river's mouth, deposits build up, and the mouth shifts slightly to one side. As this shifting back and forth of the mouth continues, the delta becomes fan-shaped. Sometimes the slowly moving stream splits into several channels as it moves through the delta. For example, the delta of the Mississippi River is 60 km across and has many channels flowing into the Gulf of Mexico. The many parts of the Mississippi River delta have been built during the past 5000 years—the present lobe in the past 500 years.

A small stream also deposits material at its mouth; however, it may not carry enough sediment to form a delta. Most small streams empty into rivers or lakes. When a stream flows into a lake, pebbles and sand carried in the water are deposited near the mouth. Silt and clay settle to the bottom farther out in the lake.

Alluvial Fans and Playas   Erosion by water occurs nearly everywhere. Even deserts receive some rain, and the runoff causes erosion and deposition. In dry, mountainous areas the runoff caused by a sudden storm carries large amounts of sediment down the slopes. At the base of the mountains, the water slows, leaving fan-shaped deposits called **alluvial fans.** Because there are no permanent streams to carry away these deposits, they form a thick accumulation of sediments, such as the one shown in Figure 12–13.

Although storms are rare in desert areas, they sometimes produce large amounts of runoff. The runoff collects temporarily on the desert floor in muddy pools on plains called **playas.** The water on these playas rapidly evaporates, leaving behind minerals such as gypsum and halite. The great salt flats east of the mountains in southern California are playas.

○ *What are deltas?*
○ *Describe two erosional features of deserts.*

Figure 12–13. Alluvial fans (left) and playas (right) form in dry, mountainous areas. The occasional heavy rains carry large amounts of sediment, which is deposited at the base of steep slopes.

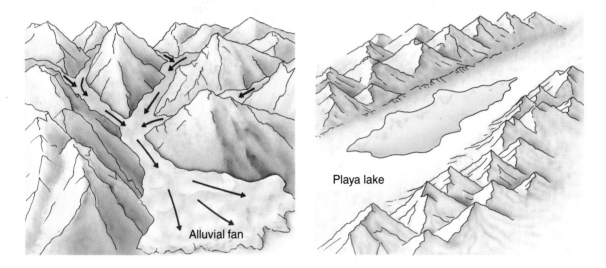

Alluvial fan

Playa lake

Section Review

READING CRITICALLY

1. Why are playas also called salt flats?
2. How is the formation of levees related to flood plains?

THINKING CRITICALLY

3. How could the amount of lake deposits show the volume of runoff of a stream?
4. What is the importance of slope and water volume to the amount of material that can be eroded by a stream?

SKILL ACTIVITY: Analyzing Diagrams

BACKGROUND

Diagrams are commonly used in textbooks to show relationships between objects, sequences of events, or other information best presented in a visual format. Many times the diagrams are simple and easily interpreted. Some diagrams, however, are more complex and require careful study in order to be fully understood.

PROCEDURE

Copy the stream pattern analysis table. Use the three steps described in the Procedure to analyze the diagram of stream patterns in Figures 12–4 and 12–5. As you follow each step, place your information on the Stream Pattern Analysis table. In analyzing diagrams about water-created landforms, the following steps should be used.

1. Study each pattern. List the types of landforms and water systems presented in each diagram. List any other features in the diagrams.
2. Identify relationships between the details. What is the relationship among the various tributaries in each diagram? In what direction do they flow?
3. Analyze similarities and differences. Examine the characteristics you have identified for each stream pattern. Select one or two characteristics that would allow you to distinguish each pattern from the others.

APPLICATION

1. If there is more than one trunk stream in a diagram, do they flow parallel to each other?
2. Do the trunk streams show any other pattern?
3. In each diagram, do the tributary streams enter the trunk streams at the same angle?
4. Which stream pattern has poor drainage, with considerable standing water?
5. Which stream pattern has tributaries flowing nearly parallel to the trunk stream?

USING WHAT YOU HAVE LEARNED

Analyze the drainage pattern shown, and decide which type of pattern it most closely represents.

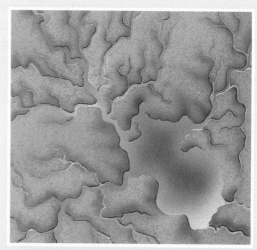

TABLE 1: STREAM PATTERN ANALYSIS

Stream Pattern	Relationship Between Trunk Streams	Relationship Between Tributaries	Relationship Between Tributaries and Trunk	Thoroughness of Drainage	Associated Landform
Dendritic					
Trellis					
Deranged					
Radial					
Rectangular					

SECTION
3 Ground Water

SECTION OBJECTIVES

After completing this section, you should be able to:

- **Define** the term *water table*.
- **Describe** the rock characteristics of porosity and permeability.
- **Explain** how karst deposits develop.

NEW SCIENCE TERMS

water table
porosity
permeability
aquifer

12.6 The Water Table

Although the cycle of water is usually completed by runoff, water is sometimes removed from the cycle temporarily. Not all of the water that falls on the earth runs off into streams. Some water soaks into the ground, where it enters tiny air spaces in the soil and rocks. As water moves through these air spaces in the soil, some of it sticks to the soil fragments. This creates an area in the soil with both air and water in the spaces, called the *zone of aeration*.

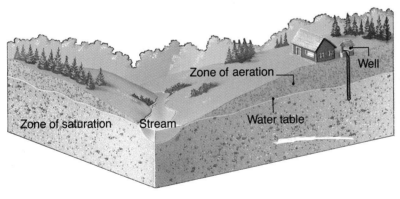

Figure 12–14. In swampy areas (above), the water table is very close to the surface. The water table is at the top of the zone of saturation (left).

Much of the water, however, continues moving down into the ground. There it reaches an area called the *zone of saturation*, where all the spaces are filled with water. This zone is the top of the **water table.**

The depth of the water table varies in different locations. Have you ever noticed that a hole dug on a beach quickly fills with water? In areas near the sea, the water table is at sea level, which may be just below the surface of the sand. In high desert regions, the water table may be hundreds of meters below the earth's surface. Every place where the earth's surface is below the water table, there is standing water such as a lake, a pond, or a swamp.

DISCOVER

Measuring Porosity

Using the following formula, determine the porosity of several different sponges.

$$porosity = \frac{\text{volume of pores}}{\text{total volume}}$$

To find the volume of the pores, soak each sponge in a bowl of water, and then wring as much water as possible out of each sponge into a graduate. To find the total volume of each sponge, measure and then multiply length × width × height.

Porosity   In most places the soil holds only a small fraction of the water that soaks into the ground; much of the water is held in the pores of the bedrock. It may seem strange that a solid object, such as a rock, can be porous. The structure of some rocks is similar to that of a sponge. If you try pouring a small amount of water into a sponge, you will find that most of the water will be trapped in the pores of the sponge. The amount of water that a rock can hold depends on its porosity. **Porosity** is the ratio of the volume of air space in the rock to the total volume of the rock.

The porosity of a rock depends both on the shape of the mineral grains and on the degree of cementing in the rock. For example, sand has more porosity than sandstone because some of the air space in sandstone is filled with cement. Porosity is always expressed as a percentage. An imaginary rock made of mineral spheres of uniform size that are packed as closely as possible will have 26 percent porosity. Porosities greater than 26 percent indicate poor packing of the mineral grains or irregular mineral shapes.

Permeability   A rock that is porous can hold a lot of water, but it will not necessarily allow the passage of water. If you look at a cork, you can see that it is very porous; however, if you put a cork into a bottle and turn the bottle upside down, none of the contents of the bottle will flow out through the cork.

Like cork, shale is porous, but water cannot move through shale because it is not very permeable. **Permeability** is the ability of rock to allow water to move through the rock. The permeability of a rock depends on how well the pores are connected to each other. Some rocks are permeable even though they are not porous at all. Limestone, for example, has almost no pores, but it is very permeable because of fractures that occur in the rock.

○ *What is the water table?*
○ *What is the difference between porosity and permeability?*

Figure 12–15. Porous rock, (left) such as sandstone, has many air spaces between mineral grains. Permeable rock (right), such as fractured limestone, has many interconnected pores or passages.

Porous permeable   Nonporous, permeable

12.7 Aquifers

When a permeable rock layer is between two nonpermeable layers, it forms an **aquifer** (AHK wuh fuhr). An aquifer transports water in much the same way as an underground pipeline does. Water enters an aquifer where the permeable rock comes to the surface. Some aquifers transport large amounts of water over great distances. For instance, much of the midwestern part of the United States lies on top of an aquifer that starts in the Rocky Mountains.

Water can be obtained from aquifers by drilling wells into them. Sometimes the water pressure in the aquifer is so great that the water emerges without being pumped. This type of well is called an *artesian well*. Sometimes a small aquifer comes to the surface naturally, where it forms a *spring*.

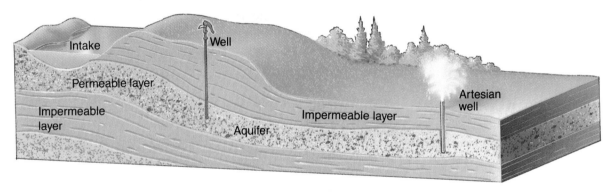

Ground water that has traveled through an aquifer is usually safe to drink because the rocks act as a filter. However, ground water may contain minerals that give it a distinctive taste, depending on the type of bedrock in the area.

Besides having an unusual taste, water with a lot of dissolved minerals may have undesirable characteristics. Water containing calcium, magnesium, or iron, called *hard water*, reacts chemically with soap, forming sticky lumps and film instead of suds. If the bedrock is limestone, dissolved calcium carbonate is redeposited as mineral layers in water pipes and drains.

○ *What is an aquifer?*
○ *What is hard water?*

Figure 12–16. An aquifer is a permeable layer of rock through which water passes. Some wells drilled into aquifers flow without being pumped. These wells are called *artesian wells*.

Figure 12–17. The minerals in hard water can accumulate on the inside of pipes and may even clog the pipe completely. Pipes carrying soft water do not show these obstructions.

12.8 Karst Topography

You may recall from Chapter 11 that karst topography occurs in areas with limestone bedrock. These areas have many ravines, sinkholes, underground streams, and caves. These features are the result of chemical weathering, mass movement, and water deposition.

Erosion As ground water travels down through fractures in the limestone, the rock is rapidly dissolved, leaving large underground passageways and caves. These passageways sometimes carry underground streams. These streams often come to the surface in the form of springs. After much water erosion, the ceilings of some caves collapse, forming sinkholes. The area around Mammoth Cave, Kentucky, shown in Figure 12–18, has karst topography.

Figure 12–18. This is a topographic map of the area around Mammoth Cave, Kentucky.

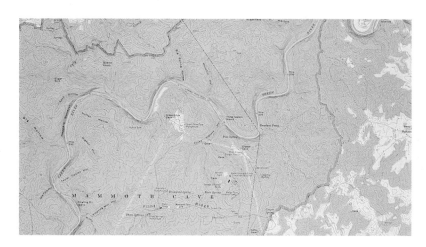

ACTIVITY: Analyzing Karst Landforms

What special features characterize karst landforms?

MATERIALS (per group of 3 or 4)

Topographic map of Mammoth, Kentucky

PROCEDURE

1. Review the features associated with karst deposits.
2. Examine the topographic map of the Mammoth Cave, Kentucky, area and answer the questions that follow.

CONCLUSIONS/APPLICATIONS

1. In general, is the topography in this area flat or hilly?
2. Locate the Green River and classify it as youthful, mature, or old-age. Explain your choice.
3. Locate two features on this map that support the hypothesis that this is a karst landform.
4. Assume that during heavy rains the various valleys probably contain temporary streams. What type of drainage pattern do these various valleys make?
5. Describe the sequence of events in the formation of karst landforms.

Deposits Underground deposits, called *karst deposits*, occur mainly on the inside of caves. Spectacular deposits form in some caves because ground water redeposits some of the calcium carbonate it had dissolved earlier from the limestone. As this water drips from cave ceilings, the water evaporates and limestone "icicles," called *stalactites*, form. Limestone also builds upward from the floor of a cave, forming *stalagmites*. There are many of these caverns, with their beautiful formations, in the limestone regions of the Appalachian Mountains and in New Mexico.

○ *How do karst deposits form?*

Figure 12–19. Karst deposits are common in limestone caverns, such as Mammoth Cave, Kentucky, (above) and Carlsbad Caverns, New Mexico (left).

Section Review

READING CRITICALLY

1. Explain how a rock can be permeable without being porous.
2. Why do karst deposits occur mainly in limestone formations?

THINKING CRITICALLY

3. Explain why water from a sandstone aquifer is not hard, while water from a limestone aquifer is hard.
4. Is sandstone porous or permeable or both? Explain your answer.

INVESTIGATION 12: Porosity and Permeability

PURPOSE

To compare the porosity and permeability of gravel, sand, and clay

MATERIALS (per group of 3 or 4)

Clear plastic tube
Hose and clamps, fitted to tube (2)
Beaker, 500 mL
Water
Gravel
Sand
Clay
Graduate
Stopwatch

PROCEDURE

1. Copy the table shown. Fill in the table as you complete this investigation.
2. With a clamp, attach the hose to the plastic tube.
3. Pour gravel into the beaker to the 250 mL level (slide the gravel down the side of the beaker so that you do not break the beaker); then pour the gravel into the plastic tube.
4. Secure a clamp on the hose so that water will not run out of the plastic tube; then slowly pour water from a graduate over the gravel in the plastic tube until the water reaches the top of the gravel.

5. Record the amount of water needed to fill all the pores in the gravel. Compute the porosity of the gravel using the following formula:

$$\text{porosity} = \frac{\text{amount of water} \times 100}{250 \text{ mL of gravel}}$$

6. Place the beaker under the plastic tube; then remove the clamp from the hose. Record the time it takes for the water to run out of the tube.
7. Repeat steps 2 through 6 for the sand and clay materials. .

ANALYSES AND CONCLUSIONS

1. Which of the materials was the most porous? Which was the least porous?
2. Using your permeability records, rank the materials in order of permeability. Use number 1 to indicate the most permeable material.
3. According to your data, what is the relationship between permeability and porosity? Explain why this relationship exists.
4. Would sandstone have a porosity different from the sand in this experiment? Explain your answer.

APPLICATION

How could this procedure be used to determine the porosity and permeability of a rock layer?

TABLE 1: POROSITY AND PERMEABILITY			
Characteristics	Gravel	Clay	Sand
Amount of Water in Pore Space			
Porosity			
Time for Water to Permeate			
Order of Permeability			

SUMMARY

- Surface runoff is responsible for considerable erosion of the land. (12.1)

- The amount of eroded material carried by a stream is called its load. (12.1)

- A drainage basin is an area drained by a single stream system. This system includes a trunk and tributaries. (12.2)

- Drainage systems form different shapes, depending upon the steepness of the slope and the underlying rock structure. (12.2)

- Streams increase their length by a process known as headward erosion. (12.3)

- The low point of a stream or river is its base level. A stream cannot erode below its base level. (12.3)

- Stream valley development is divided into three stages: youthful, mature, and old-age. (12.4)

- Youthful streams have V-shaped valleys. Mature streams have wide flood plains. Old-age streams have many meanders. (12.4)

- If a stream carries a large load of sediment, a delta forms at the mouth of the stream. (12.5)

- Playas and alluvial fans are features of dry, mountainous areas. (12.5)

- The amount of water that can enter the ground is determined by the porosity and permeability of the bedrock. (12.6)

- Porosity is the ratio of the volume of air space in a rock to the total volume of the rock. Permeability is the ability of a rock to allow the movement of water through the rock. (12.6)

- A permeable rock layer may act as an aquifer. (12.7)

- Stalactites and stalagmites are examples of karst deposits. (12.8)

Write all answers on a separate sheet of paper.

SCIENCE TERMS

Correctly use each of the following terms in a sentence.

alluvial fans **(263)**

aquifer **(267)**

base level **(259)**

delta **(262)**

drainage basin **(255)**

erosion **(254)**

flood plain **(260)**

headward erosion **(258)**

load **(254)**

meanders **(261)**

permeability **(266)**

playas **(263)**

porosity **(266)**

runoff **(253)**

water table **(265)**

SCIENCE QUIZ

Modified True-False

Mark each statement *true* or *false*. If a statement is false, change the underlined term to make the statement true.

1. Streams are able to erode both the <u>head</u> and the sides of their channels.

2. Mature streams mainly erode their <u>beds</u>.

3. Runoff is <u>increased</u> in areas of thick vegetation.

4. When a river overflows its banks, it forms a <u>delta</u>.

5. Aquifers are formed in layers of <u>permeable</u> rock.

continues

Multiple Choice

Write the letter of the choice that best answers the question or completes the statement.

6. Which of the following would not be porous?
a) sandstone c) obsidian
b) pumice d) scoria

7. What type of drainage pattern forms in an area of long parallel ridges?
a) dendritic c) deranged
b) parallel d) trellis

8. Deposits built up at the mouths of rivers are
a) meanders. c) natural levees.
b) distributaries. d) deltas.

9. The primary cause of hard water is
a) nearness to rocks.
b) depth of ground water.
c) relationship to deltas.
d) mineral content.

10. The Great Lakes could be considered to be
a) a temporary base level.
b) a playa.
c) an old-age valley.
d) formed by headward erosion.

Completion

Complete each statement by supplying the correct term.

11. A river is unable to erode below its _____.

12. A(n) _____ drainage pattern would form where slopes are not steep.

13. Water passes easily through most sandstones because they are very _____.

14. A(n) _____ is formed as water slows and drops sediments in fan-shaped deposits at the base of a steep desert slope.

15. Erosion by _____ is affected by both volume of water and steepness of slope.

Short Answer

16. How are deltas and alluvial fans similar? How are they different?

17. How is the base level of a stream affected by resistant rock formations?

18. Why do streams form natural levees as they overflow their banks?

Writing Critically

19. Why do flood plains occur only in the mature stage of stream development?

20. Identify each type of drainage pattern shown, and describe the situation that might produce each one.

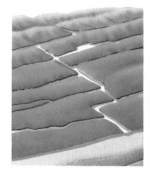

EXTENSION

1. Research and report on some impurities or pollutants that may enter ground water supplies and not be filtered out as the water passes through bedrock.

2. Do library research on major aquifers in the United States. Find out where their source is and how they flow. Report to your class on the aquifer in your region.

3. On a map of South America, locate the main drainage basins, and determine the divide that separates each basin.

1. Explain why meander cutoffs form a single river channel rather than several small channels.

2. A special type of drainage pattern develops over an area underlain by limestone. This type of pattern is called *centripetal,* or circular. Using your knowledge of karst topography, describe what you think the characteristics of centripetal drainage would be.

3. Many large farms are found in the hilly regions of the northeastern United States and in the foothills of the eastern mountains. Explain how modern contour plowing could decrease the amount of runoff into streams of these areas.

FOR FURTHER READING

Bain, I. *Water on the Land.* New York: The Bookwright Press, 1984. Major concepts about surface and ground water are clearly presented in this book.

Gunston, B. *Water.* Morristown, N.J.: Silver Burdett, 1980. This classic book presents a thorough treatment of water characteristics, water activity on land and in the atmosphere, and water uses by living things and industry.

Maltby, E. *Waterlogged Wealth: Why Waste the Earth's Wet Places?* Washington, D.C.: International Institute for Environment and Development, 1986. An excellent review of wetlands of the world.

Challenge Your Thinking

Flowing through the Grand Canyon, the Colorado River provides a puzzle as to its stage of development. The river shows the wide, looping meanders characteristic of an old-age river; however, the valley walls are steep and V-shaped, characteristics of a youthful river. How might this apparent conflict of characteristics be explained?

CHAPTER

13

Wind and Ice

Deserts, with their rolling hills of sand, have always been fascinating to writers and moviemakers searching for exotic settings. They are fascinating to geologists as well, because of the unusual landforms created by wind erosion. Since dry land covers so much of the earth, wind is second only to water as an agent of erosion and deposition.

The wind creates many desert landforms.

1 Wind

SECTION OBJECTIVES

After completing this section, you should be able to:

- **List** the factors that create arid environments.
- **Identify** the conditions that cause wind erosion.
- **Describe** the major deposit created by wind.

NEW SCIENCE TERMS

arid
rain shadow
saltation
deflation
barchan dune

13.1 Arid Environments

Viewed from space, our planet seems to be covered mostly by water. However, it may surprise you to learn that over one third of the land is desert. With so much water on Earth, special environmental conditions are needed to create dry, or **arid,** climates. Arid climates are usually caused by one of four factors: planetary winds, ocean currents, topography, or great distance from a water source.

You may recall from Chapter 9 that two dry planetary wind belts converge at the edge of the tropics. These are the areas where most of the world's major deserts can be found.

The circulation of ocean waters can also create arid climates. Along the western coast of some continents, the prevailing winds push the warm surface waters away from the land. This allows cold water to rise to the surface, chilling the air and condensing the water vapor. Rain and fog occur offshore, and the land remains dry. Ocean currents are discussed in detail in Chapter 19.

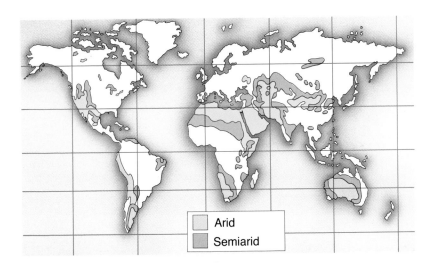

Arid
Semiarid

Figure 13–1. Many of the major deserts of the world are located just north and south of the tropics. Other deserts are created by topography, geography, or ocean currents.

Arid climates can also form because of special topography. For example, as moist air rises over a mountain range, it cools and the water vapor condenses. Rain falls on the side of the mountain facing into the wind, forming a rain shadow on the opposite side. A **rain shadow** is an area shielded from precipitation by a mountain or other topographic feature. Air descending on the other side is warm and dry, and the moisture in the land evaporates.

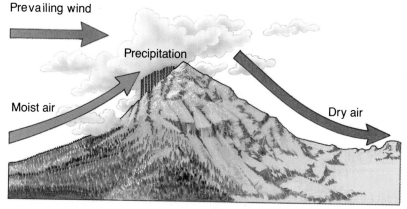

Figure 13–2. A rain shadow (right) often creates a desert such as the Sonoran Desert (left) of Arizona and northern Mexico.

Another factor in creating arid environments is the large size of some continents. As air masses travel great distances from the oceans, most of the water condenses and falls as precipitation, leaving little moisture in the air. For example, the prevailing westerlies drop most of their water on Europe, creating arid conditions farther to the east, in parts of Asia.

Figure 13–3. The Gobi Desert of central Asia (right) is hundreds of kilometers from the Pacific and Indian oceans. The dry conditions of the desert create a nomadic way of life for the people who live there.

276

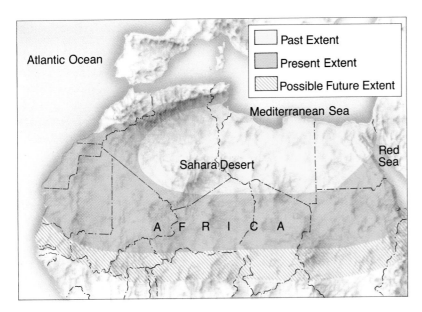

Figure 13-4. Desert areas sometimes increase in size. The Sahara desert (left) has been steadily increasing in size, creating food shortages and climatic problems for the people of Africa.

Deserts and other arid lands have not always been in their present locations. Deserts change location because the environmental factors that form them change, sometimes with human help. The deserts today are increasing in size. For example, the Sahara desert in northern Africa is constantly extending southward. Poor farming practices are one of the leading causes of this movement.

○ *What conditions create arid climates?*
○ *What is a rain shadow?*

13.2 **Wind Erosion**

Have you ever been to a beach on a very windy day and felt the sting of sand grains hitting your legs? Besides the pain, the blowing sand irritates your skin, and if you were to stay long enough, it could literally tear the skin from your legs. This same constantly blowing wind produces erosion of rocks and soil.

Wet soil tends to resist wind erosion; therefore, as the amount of water in the soil decreases, wind erosion increases. Part of this erosion occurs because a decrease in surface water means fewer plants grow. Loss of ground cover allows soil and rocks to be exposed to the wind, which then picks up and moves the unprotected soil.

Wind erosion can be temporary or seasonal. The creation of the "dust bowl" in the western United States in the 1930s was the result of several years of unusually dry weather. When the land dried out, the plants died, and the exposed soil was blown

A MATTER OF FACT
The sun is ultimately responsible for the erosion caused by wind. The sun heats the earth unevenly, causing pressure differences, which —in turn—cause the air to move.

Figure 13-5. A dry, plowed field loses much valuable topsoil to wind erosion.

away. However, as the area began receiving normal amounts of rain again, the plant cover returned, and the wind erosion stopped.

Areas influenced by monsoon winds may experience seasonal periods of wind erosion. When the monsoon wind blows from the dry continent, soil may be lost due to wind erosion. When the wind reverses and blows from the sea, torrential rains may cause erosion by runoff. Without water, wind becomes the dominant erosional force. As the wind blows over dry, barren land, it transports and deposits soil and rock fragments. Particles of clay and silt are carried to great heights, forming dust clouds. The dust clouds created in the dust bowl traveled all the way across the United States and deposited soil particles on the east coast. How could the western dust have been identified on the east coast?

Figure 13-6. The monsoon winds, which bring torrential rains to India, also bring a season of drought (right). Wind blown soil often obscures the sun, producing a red haze.

Sand-sized particles are usually blown along near the surface of the ground. If the wind speed is great enough, the sand may bounce as high as 2 m. This bouncing is called **saltation.** Saltation acts like a sandblasting machine, forming bizarre sculptures as different layers of rock are eroded at different rates. Saltation is the process that irritates your legs at the beach.

Tremendous amounts of soil, sometimes even all the soil in an area, can be eroded by the wind. The removal of all the soil of an area is called **deflation.** The resulting depressions in the earth's surface are called *deflation basins.*

○ *Why are the effects of saltation like sandblasting?*
○ *How are deflation basins created?*

Figure 13–7. The eroding power of wind blown particles produced this formation (left). This power is similar to that of commercial sandblasting (right).

Figure 13–8. A deflation basin, such as the one shown here, results when all of the loose particles of an area are removed by the wind.

Figure 13–9. When the supply of sand is plentiful and the wind blows constantly from the same direction, giant ridges of sand (right) can be produced. The area from which the sand is removed is often left barren, resulting in a hard surface called *desert pavement* (left).

13.3 Sand Dunes

Sand dunes are the best known form of wind deposit. Dunes are characterized by gently sloping sides, called the *windward side*, facing into the wind. The steep side, facing away from the wind, is called the *leeward side*. Dunes move, or migrate, as sand is rolled up the gentle slope and deposited over the crest onto the other side.

The most common type of dune is the **barchan** (BAHR kahn) **dune.** A barchan is a crescent-shaped dune with the bulging side facing into the wind. This type of dune forms in areas that have a limited supply of sand. As sand is separated from the larger pebbles and rocks in an area, the rocks that are left are called *desert pavement*. In areas with a large supply of sand or with variable winds, different kinds of dunes form. Some of these different dunes are shown in Table 13–1.

○ *What is a barchan dune?*

Section Review

READING CRITICALLY

1. List four factors that form arid lands.
2. Explain how a dune migrates.

THINKING CRITICALLY

3. How can a coastal environment be very much like a desert?
4. Explain why the term *shadow* is especially useful in picturing the concept of a rain shadow.

TABLE 13–1: TYPES OF DUNES

Shape	Name	Description
	Barchan	Forms where sand is limited and wind is strong and constant
	Star	Forms where sand is plentiful and wind is strong and shifting
	Linear	Forms primarily along seacoasts where the sea breeze and land breeze push the sand into long lines
	Transverse	Forms where sand is plentiful and wind blows from one direction
	Parabolic	Forms along seacoasts where vegetation holds the sand

ACTIVITY: Analyzing Dune Migration

How do changes in wind direction affect the shape of a barchan dune?

MATERIALS (per group of 3 or 4)

fine, dry sand; hair dryer; tray with low sides; goggles

PROCEDURE

1. Pour the sand into the tray and construct a barchan-shaped dune.
2. **CAUTION: Wear goggles to prevent sand from blowing into your eyes.** Turn the hair dryer on a low, cool setting and direct the air at a low angle toward the dune. Experiment with the dryer to determine how close you should be to the dune to make sand grains roll up the side of the dune.
3. Hold the hair dryer in a fixed position and study the way in which the dune slowly travels or migrates across the tray.
4. Shift the dryer about 10 cm away from your original position to change the wind direction. Study the way in which a wind shift causes the dune to change shape.

CONCLUSIONS/APPLICATIONS

1. Explain the differences in appearance between the two sides of a sand dune.
2. Why is the side away from the wind not vertical?
3. Describe the sequence of events that occurs as the dune migrates.
4. What happens to the dune when you shift the wind direction?

NEW SCIENCE TERMS

glacier
iceberg
firn
plucking
till
moraine
outwash plain
loess

SECTION OBJECTIVES

After completing this section, you should be able to:

- **Describe** the formation of glaciers.
- **Explain** the differences between direct and water-carried deposits.

13.4 Glacier Formation

Can you imagine a time when much of Canada and the northern United States was covered with ice 3 km thick? Can you imagine a warm, summer day with temperatures up to almost 0°C and only half a meter of new snow? For much of the recent history of the earth, glaciers covered large parts of North America and Europe. A **glacier** is a sheet of ice that covers a large continental area or a mountain region.

Today, 98 percent of all ice on Earth is located in the polar regions; the rest is found on high mountains. There are only two large *continental glaciers* remaining: one in Greenland and the other in Antarctica. However, many mountains around the world have smaller glaciers called *alpine glaciers.*

A continental glacier may be as much as 4 km thick and so heavy that it causes the rocks below it to sink into the mantle. If the ice sheet covering Greenland were suddenly to melt, the island would rise by nearly 1 km. The South Pole is under approximately 3 km of ice. However, there is no glacier at the North Pole. Glaciers form on land, but the North Pole is in the Arctic Ocean, where there is a thin layer of frozen sea water.

Figure 13–10. Continental glaciers, such as the one shown here, may be hundreds of meters thick and may cover areas as large as Antarctica.

Figure 13–11. The calving of an iceberg from a glacier is shown in this series of photographs.

There are pieces of glaciers, called *icebergs,* in the polar seas. An **iceberg** is a great mass of freshwater ice freed from a glacier by a process called *calving.* The process of calving is illustrated in Figure 13–11.

Why do glaciers form in some cold areas but not in others? Many areas of the world receive snow and ice during the winter months. However, in most areas the warm summer causes the snow and ice to melt. Glaciers form only in areas where the temperature remains too low for all of the snow to melt.

As the snow piles up, the pressure increases on the bottom of the snow pile, and the snow compacts. Snow that melts quickly refreezes, forming ice. Fresh snow is about 90 percent air. Once half of the air spaces in the snow pile have been filled with ice, the pile is called **firn.** Firn takes about a year to form.

Eventually firn becomes compressed and changes into a solid body of ice. This transition from snow to firn to ice may take three to five years in places where warm temperatures cause snow to melt, forming melt water. In colder parts of the world, this transition may take as long as 100 years. Why would this change take longer in places where it is colder?

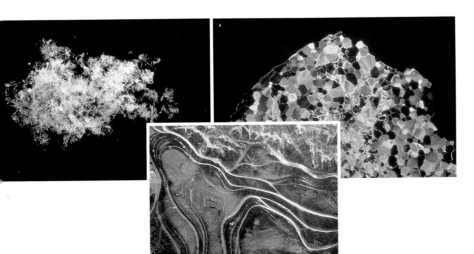

Figure 13–12. This series of photographs shows the transformation of snow (left) to firn (right), and then to glacial ice (bottom).

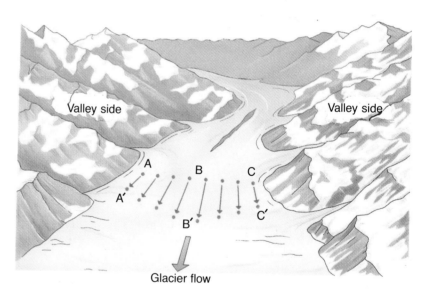

Figure 13–13. Crevasses, shown here, can be very dangerous to anyone exploring a glacier.

Figure 13–14. The movement of glaciers can be measured using this apparatus (right). Stakes placed along ABC will move to positions A'B'C' as the ice moves. Grooves, or striations, in rocks also show the movement of glaciers (above). Notice that the striations are all parallel. This shows the direction of movement of the glacier.

During the warmer months, the surface of a glacier may melt. Puddles of water or small channels often form on a glacier's surface. The water eventually flows through cracks, or crevasses, in the ice and into water channels within the glacier. In the summer, you can hear the sound of running water coming from within a glacier.

Even though glaciers are solid ice, they can still move and expand. As pressure builds up from the glacier's weight, some of the ice becomes deformed and flows like hot plastic. Individual ice grains respond to the pull of gravity by sliding past one another, allowing the ice to flow slowly downhill. Movement also occurs as melt water filters through the glacier and along the ground under the glacier. There it provides lubrication between the ice and the ground, allowing the glacier to slide smoothly on a film of water. Glacial movement increases during the warmer months because there is an increase in the amount of melt water.

○ *What are icebergs?*
○ *How do glaciers move?*

Valley side

Valley side

A

B

C

A'

C'

B'

Glacier flow

13.5 Glacial Deposits

Because of their mass, glaciers are especially effective at carrying rock materials. Some material simply falls from mountain slopes onto the surfaces of glaciers. However, glaciers also obtain rock material through a process called *plucking*. During **plucking,** glacial ice freezes around rocks and plucks them out of the ground as the glacier moves. Glaciers also move material much as a snowplow does. Rocks at the base of a glacier are picked up and pushed along in front of, and within, the glacier.

Figure 13–15. The deposits left by a melting glacier range in size from the large boulders in the middle of this photograph to the fine soil in the foreground.

Glaciers leave two different types of deposits. The first type is directly deposited by the ice. The other is eroded by ice and then reworked, or transported, by water as the ice melts. In this reworking, the materials are sorted by size, as are most water-carried materials. In contrast, the materials directly deposited by the ice are rough, irregular, and completely unsorted.

Direct Deposits   Because of their massiveness, glaciers are able to transport materials of all sizes. Glacial deposits are commonly found as heaps of boulders, sand, and fine silt mixed together. The shape of the piles of debris varies from ridges and hills to thick, lumpy blankets covering the land. This unsorted rock debris deposited directly by glaciers is called **till**.

Figure 13–16. Unsorted glacial deposits, such as those shown here, are called *till*.

Till sometimes contains large boulders called *erratics*. Some erratics are as big as houses. The finest material in till is called *rock flour*, or *glacial meal*. Although microscopic in size, rock flour is the result of physical weathering. Streams that flow from the front of some glaciers are milky-white because they carry so much rock flour.

As a glacier melts, it leaves a long mound of rock debris at its edge. A mound of unsorted rock materials that builds up along the edge of a glacier is called a **moraine**. Moraines appear as long ridges or hills on an otherwise flat landscape. In northwestern Pennsylvania and southeastern Wisconsin, state parks have been built on and between moraines.

The moraine marking the farthest advance of a glacier is called a *terminal moraine*. Those formed along the sides of a glacier are called *lateral moraines*. When two alpine glaciers join together, the lateral moraines combine to form a moraine down the middle of the glacial valley. This type of moraine is called a *medial moraine*.

Figure 13–17. The color of this water (left) is due to the presence of fine, white material called *rock flour*. The large boulders, seemingly placed on the landscape (right) are called *erratics*, because of their erratic occurrence.

Figure 13–18. Terminal moraines (right) are often found near streams and rivers. The streams may have carried the water of the melting glacier. Lateral and medial moraines can sometimes be seen near the end of an active valley glacier (left).

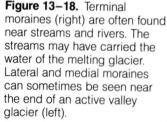

Water Deposits Large amounts of gravel, sand, and rock flour are carried away from glaciers by melt water. Much of this material is deposited on outwash plains. An **outwash plain** is a thick blanket of glacial sediments. The outwash plain is formed as water rushing from the melting glacier deposits material in front of a stationary glacier.

Figure 13-19. The flood of water from a melting glacier (right) often deposits a thick blanket of eroded material on an outwash plain.

Wind Deposits Rock flour in dry outwash is easily transported by wind. Thick layers of this fine glacial material form deposits called **loess** (LEHS). Much of the midwestern United States and Canada is covered by loess. When mixed with humus, loess forms a thick, rich soil, perfect for growing grains such as wheat and corn.

○ *What is glacial till?*
○ *What is an outwash plain?*

Figure 13-20. Much of the thick, rich soil of the Great Plains is loess. Loess is easily blown by the wind.

Section Review

READING CRITICALLY

1. Would the rocks in an outwash plain be sorted by size? Explain your answer.
2. How are moraines formed?

THINKING CRITICALLY

3. Explain why temperature is a controlling factor in the amount of time it takes firn to turn into ice.
4. Using the process that transforms snow into ice, explain why ice might be called a metamorphic rock.

CAREERS

SCIENCE TEACHER

If you enjoy studying science and working with people and ideas, you might consider a career as a *science teacher.* Usually, people studying to be science teachers take many different science courses, including biology, chemistry, physics, and geology, as well as general courses in mathematics and English. Eventually, science teachers specialize in one or more science areas to prepare themselves for teaching in their favorite field.

A bachelor's degree is required to begin teaching and, in most regions of the country, teachers are expected to continue taking college courses to remain current in their subject area.

For Additional Information
National Science Teachers
  Association
1742 Connecticut Avenue, NW
Washington, DC 20009

SNOW HYDROLOGIST

A *snow hydrologist* studies the movements and characteristics of ice and snow on all parts of the earth. This career blends a scientific background with the challenges of outdoor survival techniques. Snow hydrologists venture onto dangerous ice fields and spend long months enduring cold temperatures. However, they also get to enjoy the breathtaking beauty of glaciers.

This career requires a college degree in geology or hydrology. A strong background in foreign languages is helpful, too, since snow hydrologists have the opportunity to work with international teams of scientists. Training in mountain climbing and outdoor survival skills is also important.

For Additional Information
The American Geological
  Institute
5205 Leesburg Pike
Falls Church, VA 22041

COMPUTER OPERATOR

A *computer operator* provides an important link between a computer and the scientists who need data from the computers. Computer operators originate and experiment with new programs, make certain that programs are run in the proper order and on schedule, and monitor a program's progress.

Most computer operators receive on-the-job training after completing their high-school program. Many community colleges and trade schools offer courses in computer operation, programing, and repair.

There is much room for advancement, since many corporations hire computer programers to solve unique computer problems.

For Additional Information
Contact your local
  community college.

3 Glacial Landforms

SECTION OBJECTIVES

After completing this section, you should be able to:

- **Describe** the deposits formed by alpine glaciers.
- **Explain** how the deposits of continental glaciers are different from those of alpine glaciers.

NEW SCIENCE TERMS

There are no new science terms in this section.

13.6 Alpine Glaciers

Even after a glacier melts, it leaves permanent marks on the land. These marks create many landforms that can easily be identified as having been created by glaciers.

As an alpine glacier cuts into a narrow mountain valley, it carves out a wide, U-shaped trough. At the end of the valley, a terminal moraine is deposited, and lateral moraines line the sides of the valley. Also, the floor of the valley is layered with a type of till called *ground moraine*.

Sections along the valley floor that were deeply gouged by the glacier fill with melt water to form lakes. Such lakes have a characteristic bluish-white color due to the presence of rock flour.

An eroded mountain peak at the source of several glaciers forms a dramatic pointed spire called a *horn*. Horns are named for the famous Matterhorn on the border between Switzerland and Italy. In addition to horns, bowl-shaped depressions called *cirques* (SUHRKS) often form near the beginning of alpine glaciers.

Figure 13–21. The action of alpine glaciers produces many distinct features (left). The famous Matterhorn in the Swiss Alps (right) is one of those features.

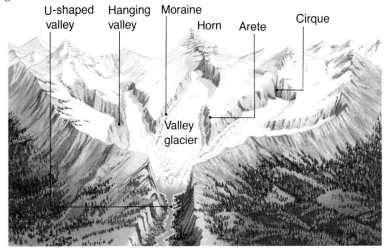

U-shaped valley · Hanging valley · Moraine · Horn · Arete · Cirque · Valley glacier

Figure 13–22. Many spectacular waterfalls, such as Yosemite Falls in California, are found in hanging valleys. Hanging valleys once contained tributary glaciers along the main valley glacier.

Figure 13–23. The Finger Lakes of New York (right) and the fiords of Alaska (top left), and Norway (bottom left) are valleys that once contained glaciers.

Along a glacial valley, small hanging valleys may be seen. These valleys form from tributary glaciers that joined the main valley glacier. They are much higher than the main valley and often contain small lakes drained by spectacular waterfalls. Yosemite Falls in California plunges 740 m from a hanging valley to the main valley below.

Some old glacial valleys are entirely filled with water, forming long, narrow "finger lakes." The Finger Lakes of upstate New York are old valleys, deepened and reshaped by glaciers. The fiords along the coasts of Norway and Greenland are also glacial valleys that were flooded by the rise in sea level after the glaciers melted.

○ *What is a horn?*
○ *What are hanging valleys?*

13.7 Continental Glaciers

Continental glaciers form long lines of terminal moraines at the glaciers' edges. These deposits are constantly moved and reworked by streams that meander back and forth across the outwash plain. As the glacier melts back, huge blocks of ice are sometimes isolated in the outwash plain. As deposits build around these blocks, the ice becomes buried. When the ice melts, a depression remains in the outwash. These gently sloping depressions, called *kettles*, fill with water, forming ponds and lakes. Thousands of kettle lakes can be found in Minnesota and Wisconsin.

The various piles of till left by continental glaciers are named for the shape of the deposit. Rounded mounds or hills of till are called *kames* (KAYMS). Kames are formed as melt water, traveling down through a crevasse in the glacier, piles up rock material at the bottom.

As melt water travels through crevasses and cracks in the glacier, it often develops a water channel within the glacier. Because melt water carries rocks, these channels become long, winding, rock-filled tubes. When the glacier melts, the rocks in these channels sink to the ground, leaving a winding, narrow ridge called an *esker*. Eskers can be found in the glaciated states of the Mississippi Valley, the north central states, New York,

DISCOVER

Finding Glacial Deposits

On a map of one of the states along the United States-Canada border, look for the names of places or landmarks that are related to glacier erosion and deposition. For example, in Wisconsin there is an area called Kettle Moraine, and in Pennsylvania there is Moraine State Park. Make a list of these names, and find out which state seems to have had the most glacial erosion and deposition.

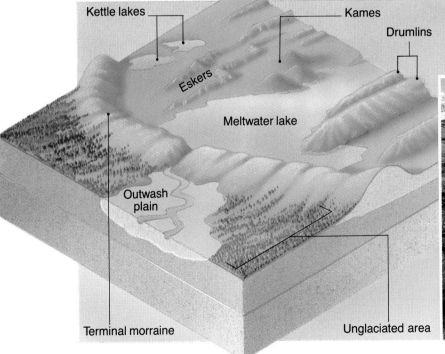

Kettle lakes — Kames — Drumlins — Eskers — Meltwater lake — Outwash plain — Terminal morraine — Unglaciated area

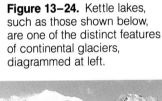

Figure 13–24. Kettle lakes, such as those shown below, are one of the distinct features of continental glaciers, diagrammed at left.

The famous Bunker Hill in Charlestown, Massachusetts, is a drumlin. Next to Bunker Hill is another drumlin, Breed's Hill. The Revolutionary War battle of Bunker Hill was actually fought on Breed's Hill.

and Maine. *Drumlins*, another type of continental glacier deposit, are long, narrow hills of unsorted material. One end of a drumlin, the side facing the glacier, is always steeper than the opposite end.

A continental glacier has a great effect on the drainage of the occupied region. Old river systems are blocked or buried, while depressions carved by the ice fill with water and form lakes. The most notable of these are the Great Lakes on the border between the United States and Canada.

○ **What are kettles?**
○ **What is the difference between a kame and an esker?**

Figure 13–25. Drumlins, such as Bunker Hill (left), are deposits left by continental glaciers. The Great Lakes of North America (right) were formed as the glaciers of the last Ice Age began to melt.

Section Review

READING CRITICALLY

1. How are the landforms produced by alpine and continental glaciers different from each other?
2. Why are hanging valleys high above the main glacier valley?

THINKING CRITICALLY

3. What causes continental glaciers to leave so many different kinds of deposits?
4. Explain why Minnesota has kettle lakes, while New York has elongated "finger lakes."

SKILL ACTIVITY: Constructing Models

BACKGROUND

Scientists often construct models in an attempt to interpret the relationships between known details and facts. Sometimes these models are actually built, while at other times they are computer models.

There are several steps to producing a model. The first step is to state the problem or identify the relationship that you wish to model. The second step is to identify all the known details that might be relevant to the problem. The third step is to let your imagination play with the details, experimenting with possible relationships. The fourth step is to test your idea to see if it will work in any situation. You may have noticed already that these steps are very much like the steps in the scientific method, because the building of a model is a part of scientific problem solving.

PROCEDURE

The following details are known about glaciers and special deposits called drumlins.

1. Drumlins are long, narrow hills with one steep end and one gently sloping end.
2. The steep end of the drumlin always faces toward the glacier.
3. Drumlins occur in groups, all pointing in the same direction.
4. The material in drumlins consists of unsorted rocks of varying sizes.
5. There is enough pressure at the bottom of a glacier to cause the ice to flow.
6. When ice in a glacier presses up against a solid mass, such as a large rock, the ice melts, flows around it, and then refreezes once the pressure is lessened.

Use this information, and the photograph at left, to build a model of a drumlin.

APPLICATION

Use your model to answer the following questions.

1. Are the materials in drumlins deposited in the streamlined shapes, or are they deposited in a different shape and then reworked by water or ice?
2. Are the drumlins shaped as the glacier extends itself or as it retreats?
3. Why are drumlins formed in groups?

USING WHAT YOU HAVE LEARNED

Look at the "ice pyramids" in the photo at right. Try making a model of these glacial features. Your model should be able to answer questions about ice pyramid formation.

INVESTIGATION 13: Examining Glacial Topography

PURPOSE

To compare the topography that develops in alpine and continental glaciated areas, using topographic maps

MATERIALS (per group of 3 or 4)

Topographic maps of Chief Mountain, Montana, and Ayer, Massachusetts, on pages 582–583 of the Reference Section.

PROCEDURE

1. Study the maps of the two glaciated areas. Chief Mountain was eroded by an alpine glacier. Ayer was eroded by a continental glacier.
2. Look for the typical landform features of the two types of glaciers.

ANALYSES AND CONCLUSIONS

CHIEF MOUNTAIN, MONTANA

1. How do the contour lines show that the valley containing Kennedy River is a U-shaped valley?

2. Explain the process that formed the string of lakes south of Glenns Lake.
3. Explain how the contour lines show that the lake located just north of Redgap Pass was formed in a hanging valley.
4. Helen Lake occupies what type of landform?
5. Locate and name a horn on the map.

AYER, MASSACHUSETTS

6. Pingy Hill is a kame. Identify three other kames either by name or by location on the map. Identify unnamed features by their highest elevation.
7. Explain why the drainage in this region is typical of a recently glaciated area.
8. Locate the esker running beside Beaver Brook. How long is this esker? Explain how the esker was formed.

APPLICATION

Look at a topographic map of your area and determine if any of the typical landforms of glaciation appear.

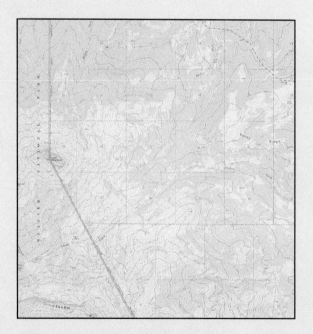

SUMMARY

- Over one third of the land area of the earth is covered by desert. (13.1)

- Arid conditions are the result of planetary winds, ocean currents, topographic features, and distance from a water source. (13.1)

- Wind erosion occurs where there is little vegetation to protect the soil. (13.2)

- Wind erosion may be temporary or seasonal, caused by factors such as dry monsoon winds. (13.2)

- The removal of all soil in an area is called *deflation.* (13.2)

- The most common dune is the barchan. (13.3)

- Presently, there are two large continental glaciers and many smaller alpine glaciers on Earth. (13.4)

- Glaciers are formed when layers of snow and ice accumulate over the years. (13.4)

- Alpine glaciers form in the mountains; continental glaciers form on flat topography. (13.4)

- Glaciers are powerful erosional agents. They erode by plucking and snowplowing large amounts of rock materials. (13.5)

- Glacial deposits may be left when a glacier melts, or they may be transported by water and wind. (13.5)

- Glacial deposits usually take the form of moraines and outwash plains. (13.5)

- Alpine glaciers form U-shaped valleys, horns, and hanging valleys. (13.6)

- Continental glaciers form kettle lakes, drumlins, kames, and eskers. (13.7)

Write all answers on a separate sheet of paper.

SCIENCE TERMS

Correctly use each of the following terms in a sentence.

arid **(275)**
barchan dune **(280)**
deflation **(279)**
firn **(283)**
glacier **(282)**
iceberg **(283)**
loess **(287)**
moraine **(286)**
outwash plain **(287)**
plucking **(284)**
rain shadow **(276)**
saltation **(279)**
till **(285)**

SCIENCE QUIZ

Modified True-False

Mark each statement *true* or *false*. If a statement is false, change the underlined term to make the statement true.

1. A <u>rain shadow</u> forms on the side of a mountain away from the wind.

2. An <u>esker</u> is a rounded hill formed of glacial rock deposits.

3. <u>Alpine</u> glaciers form when ice flows on a plain.

4. Firn consists of approximately <u>90</u> percent ice.

5. Wind erosion occurs in <u>arid</u> environments.

continues

Multiple Choice

Write the letter of the choice that best answers the question or completes the statement.

6. The formation of dunes along a coastline is due to
   a) the rain shadow effect.
   b) cold oceanic waters.
   c) poor farming practices.
   d) coastal winds.

7. Deep, rich soil deposits most likely are composed of
   a) loess.
   b) rounded sand grains.
   c) erratics.
   d) glacial till.

8. Glacial lakes are bluish-white in color due to the presence of
   a) outwash.
   b) glacial ice.
   c) rock flour.
   d) till.

9. Slow-melting blocks of glacial ice result in the formation of
   a) moraines.
   b) kames.
   c) eskers.
   d) kettles.

10. The bouncing of sand-sized particles is called
    a) firn.
    b) deflation.
    c) saltation.
    d) migration.

Completion

Complete each statement by supplying the correct term or phrase.

11. The process by which icebergs are formed is called ____.

12. The ____ is the type of dune that forms where there is a limited supply of sand.

13. A deposit formed from a stream flowing within a glacier is called a(n) ____.

14. The removal of large amounts of soil by wind erosion is called ____.

15. The two areas with large continental glaciers are ____ and ____.

Short Answer

16. Select four desert areas from the world map on pages 592–593. Explain why deserts probably developed there.

17. Compare and contrast the features associated with alpine and continental glaciers.

18. Describe the differences between till and the materials deposited on an outwash plain, and give a possible reason for the differences.

Writing Critically

19. If you found a landform that might have been created by ice, what details would you look for to be certain?

20. Why do the rocks that make up a desert pavement have very smooth, polished surfaces?

21. How would you explain the formation of several terminal moraines in the same general area?

22. Loess is a windblown deposit of glacial sediments. That is, it is first deposited by glaciers and then transported by the wind. Describe another case in which sediments deposited by one type of erosion are transported and redeposited by another type of erosion.

23. Use Table 13–1 on page 281 to explain how the amount of sand available and the characteristics of the wind act together to produce a variety of sand dune shapes.

24. Using Figure 13–14, explain how the direction of movement of continental and alpine glaciers can be determined.

EXTENSION

1. Read R. A. Bagnold's book *The Physics of Blowing Sand,* and report to the class about his poetic descriptions of arid environments.

2. Research the sinking of the *Titanic*. Pretend you are a survivor of the event and write a story about the tragedy.

3. Research and write a report about how the modern windmill could become an economical source of electricity in some areas of the world.

4. After reading about the earth's last Ice Age, make a poster showing the earth as it might have looked with much of the high latitudes covered with ice. Be sure to show how the vast expanses of ice would have affected the level of the oceans.

APPLICATION/CRITICAL THINKING

1. How has the erosion and deposition by glaciers influenced the drainage pattern of a glaciated region, such as upstate New York?

2. Explain how the lack of soil in the western United States shows the effect of erosion by wind.

3. Geologists claim that Long Island, which is off the coasts of New York and Connecticut, is actually a terminal moraine left from the last Ice Age. What evidence would you look for on Long Island if you were trying to prove this theory? What type of evidence might you look for on the mainland to help prove this theory?

FOR FURTHER READING

Bailey, R. *Glaciers.* Alexandria, Virginia: Time-Life Books, 1982. This book describes in detail the formation of glaciers and their effects on landforms.

Page, J. *Arid Lands.* Alexandria, Virginia: Time-Life Books, 1982. This book describes the processes that create and maintain arid environments.

Radlauer, R. and L. S. Gitkin. *The Power of Ice.* Chicago: Childrens Press, 1985. The book describes the experiences of a teen-age research assistant who spent two summers on the glaciers in Alaska.

Watson, L. *Heaven's Breath: A Natural History of the Wind.* New York: Morrow, 1985. This book is a comprehensive explanation of the impact of wind on the earth.

Challenge Your Thinking

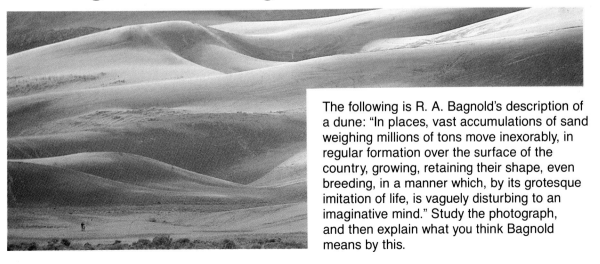

The following is R. A. Bagnold's description of a dune: "In places, vast accumulations of sand weighing millions of tons move inexorably, in regular formation over the surface of the country, growing, retaining their shape, even breeding, in a manner which, by its grotesque imitation of life, is vaguely disturbing to an imaginative mind." Study the photograph, and then explain what you think Bagnold means by this.

Geomorphology of North America

This shoreline, like all continental features, is the result of a struggle among geologic processes. Folding, faulting, uplifting, and orogeny build landforms, while the agents of weathering and erosion tear them down. The primary agents of erosion for this feature are the ocean waves.

The northern California coast

1 Geomorphic Regions

SECTION OBJECTIVES

After completing this section, you should be able to:

■ **Identify** the factors that create the landforms of a geomorphic province.
■ **Locate** the major geomorphic provinces in North America.

NEW SCIENCE TERMS

geomorphology
geomorphic province

14.1 Geomorphology

The word *geomorphology* sounds like a tongue twister, but it is made of three simple parts. *Geo* means "the earth." You have already been introduced to the word *geology*, which is "the study of the earth." The next part of *geomorphology* is *morph*, which means "form." Another word using this stem is *metamorphic*, which means "a change in form." You may recall that metamorphic rocks change their form. The final part of *geomorphology* is the ending *logy*. This ending is attached to many words and always means "the study of." For example, biology is "the study of living things." **Geomorphology** is the study of the landforms of the earth.

When you look at a photograph of the area in which you live, you may recognize a specific topographic pattern. This pattern is the result of a rock structure that has been eroded to a recognizable landform. Each region of the world has certain recognizable characteristics that set it apart from every other region.

○ *What is geomorphology?*

Figure 14–1. Gently rolling hills (left) and rugged mountains (right) are just two of the varied landforms that may be found in North America.

SKILL ACTIVITY: Drawing Profiles

BACKGROUND

When analyzing the geomorphology of an area, you need to be able to interpret both the rock structure underlying the area and the resulting shape of the land surface. Sometimes geologists draw a diagram called a *profile* to show accurately the shape of the land surface. A profile is a representation of an object as seen from the side. A profile of an area of land is like looking at hills outlined against the horizon. The contour lines on a topographic map are used to draw a profile.

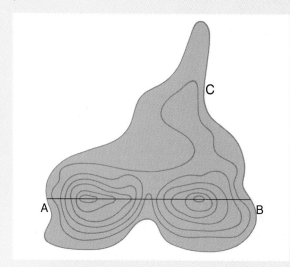

PROCEDURE

1. Identify the area to be profiled. The first step in drawing a profile is to draw a line across the map to show where the profile is to be taken. On the map, the line marked AB is drawn so that it crosses the peaks of two hills. Near the center of a piece of graph paper, draw a horizontal line. Fold your paper along this line, lay the fold along line AB on the map, and mark the ends of the line on your graph paper with dots labeled A and B.
2. Determine the vertical and horizontal scales. The horizontal scale is determined by the scale of the map. The scale of the map in this exercise is 1:20 000. The vertical scale is your choice; however, if you use a vertical scale that is the same as the horizontal scale, the slopes in your profile will be proportional to

the actual slopes of the land. On your graph paper, draw a vertical line upward from dot A. Make each line along the vertical a 20-foot contour interval starting with 0.
3. To mark the positions of the contour lines, place your line AB along line AB on the map. Mark on the horizontal where each contour line crosses line AB. Carefully label each mark with the elevation listed below the line.
4. Draw the profile. Find the mark that represents the position where the 20-foot contour line crossed your line AB. Directly above this mark, make a dot on the 20-foot line of your vertical scale. Mark the elevations of other contour lines in the same fashion. Connect the points on your graph paper with a smooth, curving line. You have now drawn a profile along line AB.

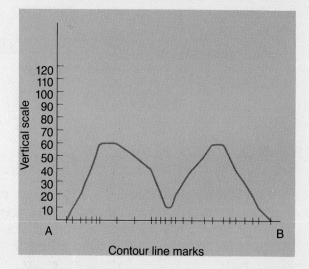

APPLICATION

Follow the same procedure and make a profile of the area between points C and A.

USING WHAT YOU HAVE LEARNED

Compare your profile to the contour-line pattern on the map.

1. Does it look as you expected? Explain.
2. Are the slopes steepest where the lines are closest together? Explain.

14.2 Geomorphic Provinces

North America is divided into regions called *geomorphic provinces*, which are given specific names. A **geomorphic province** is an area with a distinctive pattern of landforms. Sometimes a province covers an area that is hundreds of kilometers wide.

Geomorphic provinces can be divided into one of three categories: mountains, plateaus, or plains. Several examples of each of these provinces are found in North America. However, each example has some distinctive features that make it different from the others. The distinctive features that modify the provinces are rock type, folds, fractures, and faults. These features are then sculptured by climate, weathering agents, and erosion to produce the present landform.

Figure 14–2. This physical map of North America shows several large geomorphic regions. Each region is divided into geomorphic provinces, each with its own distinct features.

Figure 14–3. All rock structures can be divided into plateaus (left), plains (center), or mountains (right).

THEODORE ROOSEVELT (1858–1919)

Theodore Roosevelt was the twenty-sixth President of the United States. One of his many triumphs as president was in the area of conservation of the country's natural resources.

When Roosevelt was elected president, private interests were in control of most of the natural resources in the eastern part of the United States. Mineral rights, timber, waterways, and many other resources had been sold at prices below market value. There were few safeguards established to protect recreational sites and to guarantee the replenishment of renewable resources.

During Roosevelt's administration, a conservation program was developed to protect the vast western riches. Government-supervised programs for water and land usage were established. Public land was leased rather than sold for grazing and timber. Leasing contracts required the timberland to be replanted after cutting. Safeguards also were provided against overgrazing.

Roosevelt's focus on conservation and natural resources paved the way for the establishment of many national and state parks. The Theodore Roosevelt National Park was established in North Dakota in 1978. This section of sculptured badlands, petrified logs, and low-grade coal seams was described by Roosevelt as "barren, fantastic, and grimly picturesque."

MARJORIE STONEMAN DOUGLASS (1890–)

Marjorie Stoneman Douglass was born on April 7, 1890, in Minneapolis, Minnesota. She attended Wellesley College in Wellesley, Massachusetts, where she graduated in 1912 with a bachelor's degree. One of her first jobs was with the *Miami Herald* in Miami, Florida. She was a reporter and editor, and it was during this time that she learned first-hand about the beautiful and mysterious Everglades that lies just to the west of Miami.

After leaving the *Herald*, Ms. Douglass became an instructor at the University of Miami. While at the university she wrote and published many short stories for *The Saturday Evening Post*. Although she was never trained as a scientist, many of her stories were about the land she loved—the Everglades.

The Everglades is not a forbidding swamp as many people think; instead, it is a shallow river filled with tall grass. This river is over 100 km wide, but only about 10 to 15 cm deep, flowing southeast from Lake Okeechobee to the Atlantic Ocean.

With the publication of her novel, *The Everglades: River of Grass,* Ms. Douglass became a proponent of preserving the Everglades and was instrumental in the establishment and preservation of Everglades National Park. She has won many wildlife awards, and because of her work, the National Park Service established a special award in conservation and named it in her honor.

The age of the rocks is also important in shaping a province. In general, the older the rocks, the greater the effects of weathering and erosion. For instance, because of its recent eruption, Mount St. Helens has very few erosional features. On the other hand, all that is left of ancient mountain ranges in central and eastern Canada is their nearly level granite cores.

The overall pattern of rock structure in North America consists of relatively flat plains along the eastern and southern coasts, several mountain ranges running north-south, and a wide plain in the central part of the continent. The uniqueness of a region is quite often highlighted in areas set aside as national parks. As you study the different geomorphic provinces in this chapter, you will be introduced to some of these parks.

○ **What are geomorphic provinces?**
○ **What are the three basic categories of rock structure?**

DISCOVER

Using Words

See how many words you can write that have the prefix *geo*. Now see how many words you can write that have the stem *morph*. Finally, see how many words you can write that have the suffix *logy*. Compare your lists with those of your classmates.

Figure 14–4. The erosion of Mount St. Helens (left) is minimal, because the mountain is young. The erosion of the Canadian Shield (right) is extensive, because the mountains are millions of years old.

Section Review

READING CRITICALLY

1. What factors determine the shape of a region?
2. How does the age of a rock structure affect its shape?

THINKING CRITICALLY

3. Why are national parks often good places to study the landforms of a region?
4. Using your knowledge of plate tectonics, explain why the mountain ranges in North America run in a north-south direction.

2 Provinces of North America

SECTION OBJECTIVES

After completing this section, you should be able to:
- **Describe** the existing landforms of each province.
- **Identify** national parks that show unique landforms.

Figure 14–5. This map shows the region of the western geomorphic provinces.

14.3 The Western Provinces

The western provinces of the United States are primarily mountainous. In general, the mountain-building events started along what is now the eastern edge of the Rocky Mountain Province and moved westward. The western mountains have had a great deal of volcanic activity. On the eastern side of the mountains, there was much folding and faulting, and many intrusions formed.

The Rocky Mountain Province The Rocky Mountains are about 4500 km long and are part of a mountain chain extending from Alaska to the tip of South America. The summits of most of the mountains are over 1800 m above their bases. The rugged topography throughout the Rocky Mountains indicates that it is a young mountain range. Between the mountain peaks are U-shaped valleys carved by alpine glaciers.

The structure of the Rocky Mountains is both varied and complex. The southern part of the mountain chain has many

Figure 14–6. The Rocky Mountains contain landforms created by glacial erosion.

anticlines, batholiths, and volcanic formations. The middle part of the chain has lower mountain ranges that were formed by movement along faults. The northern part of the chain is made of thick sedimentary rocks and huge batholiths moved by thrust faults.

One of the parks in the southern part of the province is Rocky Mountain National Park in Colorado. The mountain peaks in this park are formed mostly of gneiss and granite.

The Colorado Plateau   The vast Colorado Plateau is an arid area characterized by deep canyons, buttes, and mesas. The upheavals that created the mountainous areas surrounding the Colorado Plateau had little effect on this region. The only change was that the region was raised just over one kilometer higher than it was before the mountains formed.

The Colorado Plateau has more national parks than any other province. Among the best known are Zion National Park and Bryce Canyon National Park. Most show interesting erosional features where wind and water have carved the brightly colored sedimentary rocks.

The Basin and Range Province   The Basin and Range Province includes a large area extending from southern Oregon southward into Mexico. The predominant structure in the Basin and Range Province is block faulting. The area broke into large blocks that tilted westward like a stack of fallen dominoes.

Several very well-known parks can be found in this province, including Death Valley National Monument of California and Nevada. Death Valley, located at the bottom of a fault block, is the lowest, hottest, and driest of all the national parks.

Figure 14–7. The western provinces include the Colorado Plateau (left), which was created by uplifting, and the Basin and Range Province. Death Valley National Monument (right) is located in a basin between the mountain ranges of this province.

A MATTER OF FACT

Eleven hundred species of insects have been identified in the fossils found in the volcanic ash formations in Florissant Fossil Beds National Monument in Colorado.

The Columbia-Snake River Plateau   This province, in Oregon and Washington, contains a variety of plains, hills, mountains, and plateaus. The predominant feature, however, is the extensive lava flow that blankets the province. The lava must have had a low viscosity, because it covered wide areas before solidifying. In some places, the lava deposits are more than 3000 m thick.

Although the region is mostly covered by lava flows, a few small volcanic cones do occur. The Craters of the Moon National Monument in Idaho is one such site. In this location, small cinder cones stand starkly on top of rugged blankets of lavas.

The Pacific Mountain System   The Pacific Mountain System forms a continuous line of young mountains paralleling the western coast of the United States. The volcanic origin of these mountains is evident, as some of the volcanoes are still active. This line of volcanic cones, known as the *Cascades*, is built on an eroded volcanic plateau. The Sierra Nevada, also in this province, formed from a single fault block.

The beautiful landscapes of the fault block and volcanic mountains may be admired in several national parks in this province, including Yosemite National Park in California, Crater Lake National Park in Oregon, and Mount Rainier National Park in Washington.

Figure 14–8. Craters of the Moon National Monument (below) is located on the Columbia-Snake River Plateau. Yosemite National Park (inset) is located in the Pacific Mountain System.

Figure 14–9. These giant redwood trees are located in Redwood National Park, part of the Pacific Border Province.

The Pacific Border Province The Pacific Border Province extends along the western edge of North America. The southern part of the province consists of the coast ranges and lowlands east of the mountains. The lowlands of this area, such as the Great Valley of California, are formed by faults. Earthquakes are frequent throughout this province as the western edge of southern California continues to move along the San Andreas fault.

In this province is Redwood National Park, which not only shows off the majestic redwoods but also provides beaches, tide pools, sea cliffs, and beach sand dunes.

○ *How did the Rocky Mountains form?*
○ *What evidence is there that the Pacific coastal mountains are volcanic?*

14.4 The Alaskan and Hawaiian Provinces

Alaska is so large that it actually contains many provinces within its borders. The southeastern sliver of Alaska is a mountainous coastal region with many islands and fiords. The Alaska Peninsula and the Aleutian Islands form a volcanic arc, and the Yukon Basin is a plateau and lowland area drained by the Yukon River.

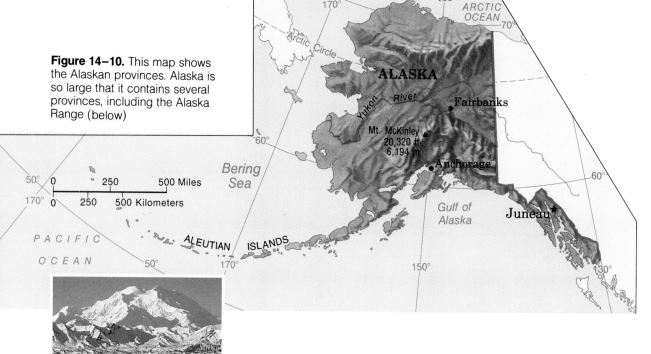

Figure 14–10. This map shows the Alaskan provinces. Alaska is so large that it contains several provinces, including the Alaska Range (below)

South central Alaska has a coastal strip leading to a band of mountains known as the Alaska Range. The Brooks Range in northern Alaska is roughly parallel to the Alaska Range. The most northern area is a gently sloping plain called the *Arctic Slope*, which borders the Arctic Ocean. Mount McKinley, in the Alaska Range, is the highest mountain in North America. The peak is nearly 6200 m high and has many glaciers.

Figure 14–11. The Arctic Slope (left) and the Alaskan coast near Sitka National Park (right) are very different from the mountainous provinces in the center of the state.

The Hawaiian Islands have developed over a hot spot below the Pacific plate. The most northwestern volcanoes are the oldest and have been weathered down to *atolls*, or round coral islands, and guyots.

The southernmost volcanoes are actively building up the present islands, and southeast of the island of Hawaii are submerged volcanoes that will eventually reach the ocean surface. These islands are all shield volcanoes with characteristic gentle slopes and relatively quiet eruptions. A great deal of the scientific knowledge about volcanoes has come from research done in Hawaii Volcanoes National Park, which is on the island of Hawaii.

○ *Describe the provinces of Alaska.*
○ *What are atolls?*

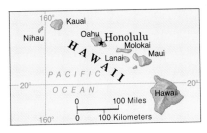

Figure 14–12. The Hawaiian Islands (above) comprise a province of their own. The structure of mid-plate volcanoes can be studied at Hawaii Volcanoes National Park (left).

14.5 **The Plains and Lowland Provinces**

In the heart of the continent are the great plains and prairies that stretch from the Arctic Circle nearly to the Tropic of Cancer. The lowland provinces of North America are primarily located along the Atlantic and Gulf coasts. Although these regions all seem the same, there are many subtle differences that make each one unique.

Figure 14–13. The famous Wisconsin Dells are located in a glaciated area of the Great Plains.

The Great Plains and Central Lowlands   The Great Plains and the Central Lowlands extend from the Rocky Mountains to the eastern mountains, and from the arctic to the Gulf of Mexico. They form a vast plain composed mostly of sedimentary rocks that lie relatively undisturbed by structural events.

Three notable interruptions to the nearly horizontal orientation of the rocks are the Black Hills of South Dakota, the Central Mineral Region of Texas, and the Baraboo Dome of Wisconsin. Each of these locations is a dome formed as granite rocks were forced upward by intrusions into the crust.

Many of the parks in the plains states are small state-owned areas showing off lovely streams or glacial features, such as the Kettle Moraine area in Wisconsin.

Figure 14–14. Many karst features, such as these caverns, occur on the Coastal Plains.

The Coastal Plains   The Coastal Plains, formed of gently sloping layers of sedimentary rock, are located along the Atlantic Ocean and the Gulf of Mexico. These young layers of rock blanket the older rock structures of the interior.

In this region there are many excellent examples of karst topography, with sink holes, underground rivers, and many cave systems. For instance, Florida has karst topography because of its limestone bedrock, but much of the state was underwater about 100 000 years ago. For this reason, Florida is often referred to as "the land from the sea."

○ *Name one of the dome structures of the Central Lowlands.*
○ *What type of rock is found on the Coastal Plains?*

Figure 14–15. This map of the eastern half of North America shows the location of the plains and central lowlands and the Appalachian provinces.

14.6 The Appalachian Provinces

Scientists believe that about 225 million years ago compressional forces folded and raised the area west of the Atlantic coast, forming the Appalachian Mountains. Since that time, these mountains have been deeply eroded, supplying vast amounts of sediments to the Coastal Plains and regions between the mountains.

Figure 14–16. Many small farms are located in the valleys among the ridges of the Blue Ridge Mountains.

The Blue Ridge Mountains   The Blue Ridge Mountains extend from Virginia to Tennessee. Although this area must once have been of very high elevation, it has been extensively eroded. All that remain of the original rocks in this province are very old granites, with a few slightly metamorphosed sandstones and conglomerates.

There are two beautiful national parks in the Blue Ridge Mountains. Shenandoah National Park is located in northern Virginia, and Great Smoky Mountains National Park is in North Carolina and Tennessee. Both provide beautiful landscapes as well as important historical details about the westward growth of the United States.

The Piedmont Plateau   The rocks of the Piedmont Plateau, east of the Blue Ridge Mountains, are a complex mixture of metamorphic rocks, such as slates and marbles. The structure shows both faulting and folding, but the area has been eroded to such an extent that the underlying rock structure is not reflected in the plateau.

The Valley and Ridge Province   The Valley and Ridge Province of northern Virginia, Maryland, and eastern Pennsylvania provides a classic example of folded mountains that have been extensively eroded. The accordion-like ridges and valleys run in long, narrow lines along a northeast-to-southwest axis. The ridge lines are composed of the most resistant rocks; the valleys the most easily eroded rocks.

Figure 14–17. Hardwood forests (right) are dominant in the Valley and Ridge Province, which stretches from northern Virginia to eastern Pennsylvania (left).

Figure 14-18. Many rivers erode the landscape of the Appalachian Plateau (right). Hot Springs, Arkansas (below), is located in the Ouachita Mountains.

The Appalachian Plateau

The Appalachian Plateau is a large province to the west of the Valley and Ridge Province. The Appalachian Plateau includes an area equal to the Piedmont, Blue Ridge Mountains, and Valley and Ridge provinces combined. Its rock layers are nearly horizontal or only slightly folded. The plateau's surface dips westward, forming a cliff along its eastern edge that overlooks the neighboring Valley and Ridge Province.

The Ouachita and Ozark Mountains

An extension of the Appalachian Provinces occurs in Oklahoma and Arkansas. The rock structures of the Ouachita (WAHSH uh taw) and Ozark mountains of these states are predominantly tightly folded sediments. The entire area has been thrust northward for a distance of at least 30 km, which probably occurred at the same time as the folding and faulting of the Appalachian Mountains.

Continental glaciers did not reach as far south as the Ouachitas or the Ozarks. Instead, their topography is due to differential weathering, mass wasting, and stream erosion.

Hot Springs National Park is located in the Ouachita Mountains. The spring water in this area is unique since it does not have an offensive odor or taste like most other hot springs.

The New England Province

The rock structure of the New England Province is similar to that of the Piedmont Plateau; however, its topography is very different. There are high mountains, such as the Green Mountains of Vermont, made of very old gneiss. Other mountains in this province are large granite intrusions. The glaciers of the Ice Ages carved and recarved the New England Province, rounding the highest ridges and gouging out river valleys. Glacial erratics and till can be found everywhere.

Figure 14-19. The erosion of glaciers and ocean waves helped shape the coast of the New England Provinces (right). The coast of Maine is composed of granite, eroded smooth by Ice Age glaciers (left).

Acadia National Park of Maine is located in this province. The park offers an excellent opportunity to view landforms shaped by both the sea and glaciers.

The Canadian Shield   Extending from the Canadian Great Plains all the way to the east coast of Canada lies the old, stable, granite base on which much of the continent is built. The great ice sheets that originated on the shield have eroded it to a flat plain. Due to its low elevation and poor rock porosity, standing water forms many lakes and marshes.

○ *Why are the Appalachian Mountains no longer very high?*
○ *What type of erosion created the Valley and Ridge Province?*
○ *What is unusual about the erosion in Acadia National Park?*

Section Review

Figure 14–20. The granite base of the Canadian Shield is all that remains of the once-great mountain range upon which much of North America rests.

READING CRITICALLY

1. Compare and contrast the Rocky Mountain Province and the Pacific Mountain System.
2. Compare the Columbia Plateau to the Appalachian Plateau.

THINKING CRITICALLY

3. If you were to draw a cross section of the continent from San Francisco, California, to Norfolk, Virginia, how would the elevation of the different provinces change?
4. The valleys in the Blue Ridge Mountains are not accumulating many sediments. Explain what happens to the sediments.

ACTIVITY: Planning an Informational Brochure

How can you design a brochure to provide tourists with useful details about the geology of a park?

MATERIALS (per group of 3 or 4)

paper, colored pencils, rulers, local topographic and geologic maps

PROCEDURE

1. Pretend that you have been hired to develop a brochure for a park in your region. Review a topographic-geologic map of the area you wish to write about.

2. Design a booklet or brochure to inform the public about the rock types, rock structures, topography, and erosional history of the park. Design your booklet to have at least six pages, including pictures to assist you in your presentation.

CONCLUSIONS/APPLICATIONS

1. How would this type of brochure be useful to tourists visiting the park?
2. In addition to the required information, what else might you include in the brochure to make it more informative?

INVESTIGATION 14: Correlating Features on Maps

PURPOSE

To use geologic and topographic maps to interpret the geomorphology of an area

PROCEDURE

1. Locate streams, rivers, mountains, and other features represented on the topographic map.
2. Study the rock structure for the area, and compare it with the position of the landforms on the topographic map.
3. Determine which rock structures make valleys and which rock units make ridges and mountains by comparing the geologic map to the topographic map.

ANALYSES AND CONCLUSIONS

1. Is the rock structure between Black Creek Mountain and Jack Mountain an anticline or a syncline? Explain your conclusion.
2. Which two units seem to be the most resistant to weathering?
3. What type of drainage pattern is found in the Jackson River Valley (left of Jack Mountain)? Describe its features.
4. Relate the drainage pattern to the structural pattern in the area.

APPLICATION

How could an engineer use this map and rock column to design a new bridge across the Jackson River?

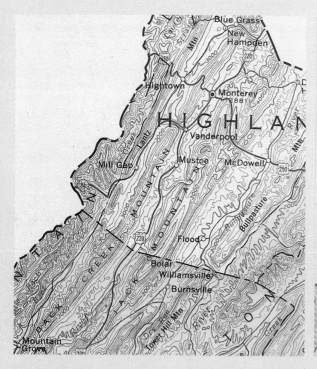

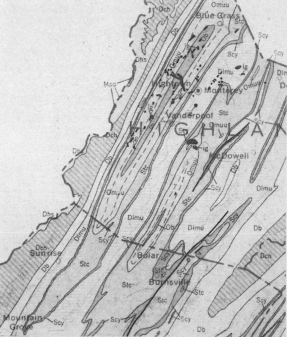

SUMMARY

- Rock structure, erosional agents, and time influence the geomorphology of an area. (14.1, 14.2)

- North America is divided into regions called geomorphic provinces. (14.2)

- The western provinces of the United States are characterized by relatively young mountains with many folds, and faults, intrusive magma bodies, and extrusive volcanic formation. (14.3)

- The western provinces include the Rocky Mountain Province, the Colorado Plateau, the Basin and Range Province, the Columbia-Snake River Plateau, the Pacific Mountain System, and the Pacific Border Province. (14.3)

- Alaska is composed of many different provinces. (14.4)

- The Hawaiian Islands are part of an arc of volcanic islands, many of which remain submerged. (14.4)

- The central provinces are composed primarily of thick layers of undisturbed sedimentary rocks. (14.5)

- The Plains and Lowland provinces include the Great Plains and Central Lowlands and the Coastal Plains. (14.5)

- The eastern mountains of the United States are very old and highly eroded. (14.6)

- The Appalachian Provinces include the Blue Ridge Mountains, the Piedmont Plateau, the Valley and Ridge Province, the Appalachian Plateau, the Ouachita and Ozark mountains, the New England Province, and the Canadian Shield. (14.6)

Write all answers on a separate sheet of paper.

SCIENCE TERMS

Correctly use each of the following terms in a sentence.

geomorphic province **(301)**
geomorphology **(299)**

SCIENCE QUIZ

Modified True-False

Mark each statement *true* or *false.* If a statement is false, change the underlined term to make the statement true.

1. <u>Geology</u> is the study of the landforms of the earth.

2. Geomorphic provinces can be divided into <u>ten</u> main categories.

3. Rocky Mountain National Park is located in the <u>western provinces</u>.

4. <u>Plateaus</u> can be formed by folding and faulting.

5. Florida has cave systems because of its <u>karst topography</u>.

Multiple Choice

Write the letter of the term that best answers the question or completes the statement.

6. Which one of the following has little in common with the others?
   a) Hawaiian Islands
   b) Rocky Mountain National Park
   c) Cascade Mountains
   d) Colorado Plateau

continues

7. What do the Black Hills and the Central Mineral Region have in common?
   a) dome formation
   b) old granite cores
   c) formed at the same time as the Rocky Mountains
   d) All of the choices are correct.

8. How are the Rocky Mountains and the Appalachian Mountains different?
   a) One is folded; the other is faulted.
   b) One contains ridges: the other does not.
   c) One is older than the other.
   d) One has been eroded; the other has not.

9. Redwood National Park is located in the
   a) Pacific Border Province.
   b) Colorado Plateau.
   c) Basin and Range Province.
   d) Pacific Mountain System.

10. Which one of the following has little in common with the others?
    a) Rocky Mountains
    b) Canadian Shield
    c) Colorado Plateau
    d) Valley and Ridge Province

Completion

Complete each statement by supplying the correct term or phrase.

11. The province in the Appalachian Mountains characterized by repeated folds is the _____ Province.

12. The Cascades are formed from _____.

13. One type of structural interruption in the Great Plains and the Central Lowlands is a _____.

14. The Alaskan Peninsula and Aleutian Islands are formed from _____.

15. Maine's Acadia National Park is located in the _____ Province.

Short Answer

16. Why might it be argued that the source of lava for the Cascades is the same as the source for Mount St. Helens?

17. Why are the Black Hills not a separate province?

18. What do the Great Valley and Death Valley have in common?

Writing Critically

19. Explain why the eastern mountains are more eroded than the western mountains.

20. Why is there karst topography in the state of Florida?

EXTENSION

1. Visit a local, state, or national park in your area and ask for an informational brochure. Read the brochure and write a report on the geological characteristics of the park.

2. Go to the library and research the geomorphology of your region. Determine the dominant rock type and structure and note any unusual landforms in your area. Make careful notes and report your findings to the class.

3. This photograph shows a Landsat image of south Florida, including Everglades National Park. Read about the Everglades and report to the class on its special features and environmental problems.

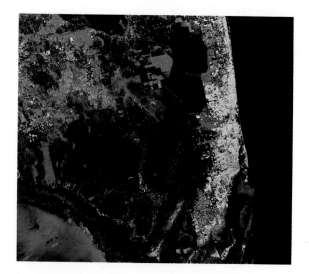

APPLICATION/CRITICAL THINKING

1. Look at the Landsat photograph of south Florida on page 316. The large lake in the picture is Lake Okeechobee, which supplies water to most of the people of southeast Florida, as well as to part of Everglades National Park. The survival of wildlife in the park depends on a constant supply of water from the lake. The cities of southeast Florida are increasing in population faster than any other section of the country, which means greater demands for water in the future. How can the water needs of the people and the park best be satisfied in the immediate and distant future?

2. What might happen to the Everglades wildlife and the cities of south Florida if the water that flows into Lake Okeechobee is cut off?

FOR FURTHER READING

Chronic, H. *Pages of Stone: Geology of Western National Parks and Monuments.* Seattle, Washington: Mountaineers, 1984. This well-illustrated book presents the geology of the western United States through a study of the national parks and monuments.

Harris, B. *Landscapes of America,* 2d ed. New York: Crescent Books, 1986. This book contains a spectacular collection of photographs that show a wide range of American landscapes.

Maltby, E. *Waterlogged Wealth: Why Waste the Earth's Wet Places?* Washington, D.C.: International Institute for Environment and Development, 1986. An excellent review of wetlands of the world.

Challenge Your Thinking

In addition to preserving special geological features, why were some national parks and monuments created? What is the importance of preserving interesting and unique landforms when there is a critical shortage of land for human needs in many parts of the continent?

The majestic beauty of Glacier National Park provides many contrasts. Glacier-carved mountain peaks rise to heights of over 3000 m, forming horns and long, pointed ridges. Many hanging valleys, where tributary glaciers once joined the main valley glaciers, may be seen along upper valley walls. Beautiful waterfalls plunge hundreds of meters to the bottoms of U-shaped glacier valleys.

The valley floors are sprinkled with cold, blue lakes fed by the many waterfalls spilling from higher lakes in the hanging valleys. These sapphire-blue lakes, clouded with rock flour, stand in sharp contrast to the glistening white of the 50 glaciers found throughout the park. Although these glaciers still erode the valleys, their effects do not compare to those of the massive bodies of ice that sculptured the original valleys during the last Ice Age.

The largest glacier in Glacier National Park is Grinnell Glacier. The ice of this glacier moves down slope about 10 m each winter and then melts back about the same distance in the summer. The massive rock walls that form towering peaks and ridges above the glacier contrast with the piles of rubble and loose rock at its base. Long, rolling mounds of this moraine line many of its valleys.

Glacier National Park straddles the Continental Divide, which separates the drainage basins of the Pacific Ocean and Gulf of Mexico. The high mountains of the divide also create distinctly contrasting climates on the east and west sides of the park. The western side receives warm, moist winds. Vegetation thrives and the temperatures are moderate. As the air rises to cross the mountains, it cools and loses much of its water content. Thus, the eastern side of the park, in a rain shadow, has a drier, colder climate. Vegetation struggles to survive the cold winds of winter and the hot, drying winds of summer.

A hanging valley waterfall

Purcell Sill

• Glacier
National
Park

Chief Mountain

The rocks in Glacier National Park show a history of calm periods of sedimentation alternating with times of violent upheaval. The rocks show evidence of extremely old deposits as well as relatively recent changes.

Most of the rocks found in Glacier National Park are about a billion years old. These ancient red and gray-green mudstones and siltstones reveal amazing details about the environment in which they formed. Some have ripple marks made by wave motion, while others show mudcracks and impressions made by raindrops. Still other rocks contain stromatolites. These details indicate that during the time the sediments were deposited, this part of the continent was a shallow mudflat at the edge of a sea. Occasionally, the sea must have dried out, causing the mud to crack. Thick limestones were also deposited in this area, indicating that during some periods it was a deep-water marine environment.

The imprint of flowing lava can also be found in the park. Iron-rich magma forced its way into cracks in the limestone, leaving dark vertical dikes and horizontal layers of igneous rocks. One of the most visible of the lava formations is the Purcell sill, a thin, black layer of igneous rock sandwiched between layers of white limestone. This sill is a dramatic feature in the valley wall high above Grinnell Glacier.

About 70 million years ago, as the Rocky Mountains were forming, a large block of land slid eastward as a low-angle thrust fault. This block moved eastward about 55 km and came to rest on top of layers of much younger rocks. This overlying block confused scientists for years. They could not figure out why the mass of billion-year-old rock that forms Chief Mountain rests on rocks only 100 000 000 years old. The roots of Chief Mountain lie over 50 km to the west. Chief Mountain is actually an isolated mountain; the rest of the thrust block has eroded away.

The contrasting features of Glacier National Park make it a beautiful and interesting place to visit. As conservationist John Muir said, "Give a month at least to this precious reserve. The time will not be taken from the sum of your life. Instead of shortening it, it will indefinitely lengthen it, and make you truly immortal."

Grinnell Glacier

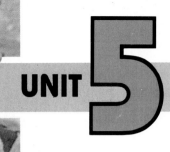

EARTH HISTORY

In 1977 geologist Luis Alvarez set out to study a thin layer of clay rock about 65 million years old. He concluded that the clay had been deposited when a giant meteor struck the earth. Alvarez knew that at about that same time nearly all the dinosaurs had vanished in a mysterious mass extinction. Alvarez and his scientist father put these facts together and stated a new theory to explain the extinction of the dinosaurs. This theory started a scientific controversy. The Alvarezes claimed that the dinosaurs died when a cloud of meteor dust covered the earth.

- How can fossils be dated to show the evolution of animal and plant life?

- What may have caused the evolution and extinction of animal species?

- How will studying Earth's past enable humans to prepare for the future?

By reading the chapters in this unit, you will begin to learn the answers to these questions. You will also begin to develop an understanding of concepts that will help you to answer many of your own questions about the history of the earth.

Fossil bones

321

Historical Geology

You may not think of the Grand Canyon as a history book, but to a geologist it is. The pages of the book are the layers of rock piled one on top of the other. Fossils from the early history of the earth are found in layers near the bottom of the canyon; more recent history can be read in the layers nearer the top. The Grand Canyon contains one of the most complete fossil records on Earth.

The Grand Canyon, in Arizona

1 The Record in Rocks

SECTION OBJECTIVES

After completing this section, you should be able to:

- **State** the principle of uniformitarianism.
- **Use** the principle of superposition to find the relative ages of layers of sedimentary rocks.
- **Relate** some characteristics of sedimentary rocks to the environment in which they formed.

NEW SCIENCE TERMS

principle of uniformitarianism
principle of superposition
crossbedding

15.1 Principles of Geology

Have you ever noticed how uniform or consistent some things are? For example, if you take a slice of bread from anywhere in a loaf, you notice its uniform texture. From top to bottom and from side to side, the texture in that loaf of bread is the same.

A slice of bread from a different loaf of the same brand still shows the same uniformity of texture. The bread-making process insures the uniformity of the product. The same processes, even at different times, produce the same results.

Uniformity is an underlying principle of many geological processes. In the eighteenth century, James Hutton, a Scottish physician and geologist, proposed that the geological processes that form rocks today formed similar rocks in the past. He called this process *uniformitarianism*. The **principle of uniformitarianism** states that similar processes, in different times, produce similar results.

DISCOVER

Studying Clues

Play a game of "twenty questions" with a classmate. You need to think of a common rock or mineral. Be especially careful that you know for sure where this rock or mineral is found. Your partner will ask questions about your rock or mineral to try to guess where it is found. The questions asked must be able to be answered with a "yes" or "no" and must not include the name of the rock or mineral. The object is to find out where the rock or mineral is found, using the fewest number of questions possible.

Figure 15–1. In formulating his principle of uniformitarianism, Hutton may have studied the White Cliffs of Dover, in England.

Figure 15–2. These alternating layers of sandstone and limestone illustrate the principle of superposition. The younger layers are on top of the older layers.

Years after Hutton developed the principle of uniformitarianism, scientists continue studying various geological processes. Observations confirm Hutton's theory. For example, sediments may be carried by water and wind and deposited in new places, but they are always deposited in similar ways. Over long periods the sediments are compressed and cemented, forming sedimentary rocks.

Sediments are deposited on top of each other. Look at the formation in Figure 15–2. The layers of sandstone and limestone alternate, but the younger rocks are always on top; the rocks on the bottom must have been deposited first. The fact that younger rocks are deposited on top of older rocks is called the **principle of superposition.** Geologists use this principle to help them determine the ages of the rock formations.

○ *What is uniformitarianism?*
○ *How can the principle of superposition be used to date the relative ages of formations?*

15.2 **Records of Environmental Change**

Fossil seashells are sometimes found on mountains, and ice scratches are often found on desert rocks. For this to be true, either the rocks have moved or the environment in which the rocks formed has changed.

Geologists can "read" environmental changes as easily as you read a mystery story. In some ways, rocks are like a mystery story; they contain clues about the environment in which they were formed. These clues help scientists reconstruct the past. For example, fossil seashells in mountain rocks indicate that the rocks formed under water and were later uplifted.

Fossils give information about past conditions. For instance, limestone formed in shallow water often contains shells of animals such as clams or snails. Deep-water limestone contains fossils of microscopic organisms that lived in the open ocean.

Figure 15–3. Fossils of ancient sea life (left) and mud cracks (right) provide evidence that these areas were once covered by water.

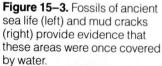

Figure 15–4. The Dakota sandstone (left) formed on a beach, while the Cody shale (right) probably formed in deep, still water.

Mineral texture may also help geologists determine what sort of environment existed when a rock formed. For instance, in sandstone, the rounded quartz grains probably formed on a beach. Geologists know this because the action of waves rounds the sand grains, and sand-sized sediments are usually deposited near a shore.

The size of the fragments in sedimentary rocks is another clue to the conditions under which the rocks formed. Consider certain rocks that are found only where large bodies of water once existed. You may recall from Chapter 4 that large fragments are deposited first as a stream slows. In a similar way, the movement of waves also separates fragments by size. Large fragments are deposited close to shore, while smaller ones settle farther offshore, in deeper water. Ocean currents may carry clay-sized fragments far from land. In quiet waters, away from the beach, the clay settles to the bottom, eventually forming shale.

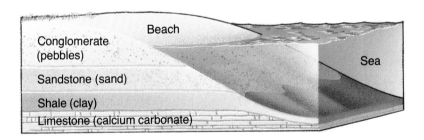

Figure 15–5. The shoreline profile (left) shows the relationship between sediment size and location of deposition.

Figure 15–6. The pattern of crossbedding (below) shows that the sand in this formation was blown by winds that constantly changed direction.

Still other environmental clues are found in rocks formed on land. Sandstone formed on land might show crossbedding patterns. **Crossbedding** is a design in rocks that shows directional changes in wind patterns. Notice the crossbedding in the sandstone shown in Figure 15–6. Rocks formed on land contain fossils of land organisms. For instance, the bones of dinosaurs, horses, or even humans would be found only in rocks formed on land. Fossils of seashells would not be found in rock formed on land.

Figure 15–7. This diagram shows the types of rocks that form in areas where the sea level is dropping.

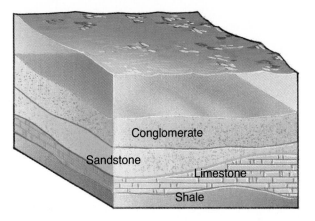

Conglomerate

Sandstone

Limestone

Shale

A series of several different formations can show environmental changes that occurred over a long period. In Figure 15–7 the bottom deposit is shale. On top of the shale is coral limestone. This suggests that between the formation of the shale and the limestone, the sea level probably dropped. Geologists would know this because shale often forms in deep water and coral limestone in shallow water. If sandstone were found on top of limestone, geologists might conclude that the sea level fell even further.

If there were a column of sediments deposited continuously since the formation of the earth, the entire history of the planet could be reconstructed. Unfortunately no such column exists. Where sediments are missing, a break in the sedimentary record occurs. Breaks result in gaps in the record that may range from a few years to hundreds of millions of years. Breaks in the sedimentary record are called *unconformities.*

○ *How do geologists read environmental changes in rock formations?*

○ *What is an unconformity?*

Figure 15–8. An unconformity (above) is a break that occurs where layers of sedimentary rocks are missing.

Section Review

READING CRITICALLY

1. What is uniformitarianism?
2. Suppose you find a sedimentary rock that has fossil corals. Was this rock formed in shallow water or deep water? Explain.

THINKING CRITICALLY

3. Imagine that you are doing field work and are studying the Cody shale and the Dakota sandstone shown in Figure 15–4. What was the probable environment in which these two rocks formed?
4. Describe a rock formation that would result from the gradual rising of sea level. Explain why this formation develops.

BACKGROUND

Classification systems are used by scientists to organize information. Being organized makes it easier to find information when it is wanted. Also, organization makes it easier to understand the objects being classified.

Imagine what it would be like to list all of the plants and animals in the world. You could list them in alphabetical order, or by color, size, or shape. There are many ways to classify things. In this activity you will create a classification system for some objects in your classroom.

PROCEDURE

1. List six objects in your classroom.
2. Divide the objects into two groups, so that each object in a group has the same characteristics as every other object in that group.
3. Divide each large group into smaller groups. Once again, all objects within a group should have some of the same distinctive characteristics.
4. Continue this procedure until each object is separated from all of the others.
5. Name each object according to the groups to which it belongs, for example, a pencil is also named a hard, small, skinny, pointed, yellow tool.

APPLICATION

Look at the pictures of the seven fossils. You are to classify these in the same manner. If you do not know the names of the fossils, name them as you might have named a pencil in procedure 5.

USING WHAT YOU HAVE LEARNED

1. What are some of the difficulties scientists encounter in trying to classify objects?
2. What characteristics did you choose to classify the fossils?
3. Which characteristics were common to the most fossils?

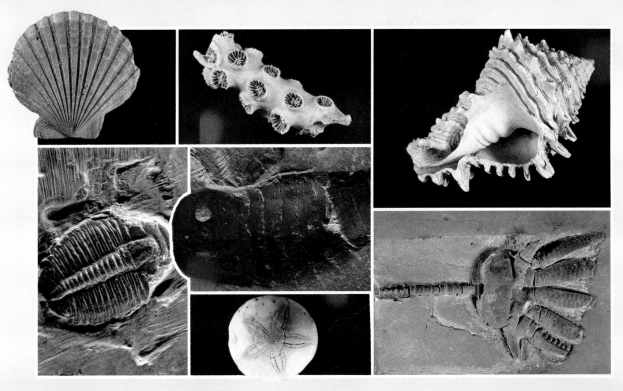

② The Geologic Clock

NEW SCIENCE TERMS

radioactive decay

half-life

absolute dating

relative dating

SECTION OBJECTIVES

After completing this section, you should be able to:

■ **Discuss** how some elements decay.

■ **Explain** various methods of dating rocks.

15.3 **Radioactivity and Absolute Dating**

You may recall from Chapter 3 that atoms are made of protons, neutrons, and electrons. The lighter atoms tend to have similar numbers of protons and neutrons in their nuclei. For instance, calcium has 20 protons and 20 neutrons. The heavier atoms tend to have many more neutrons than protons. The heaviest naturally occurring atom is an isotope of uranium. Uranium-238 has 92 protons and 146 neutrons in its nucleus.

If a nucleus has a large number of neutrons and protons, it may be unstable. Unstable nuclei may give off particles or energy, producing stable atoms. When this happens the atoms are said to undergo **radioactive decay.** For example, lead, a stable element, forms from the decay of unstable Uranium-238.

Figure 15–9. Uranium is mined from large deposits of uranium ore.

Some radioactive nuclei are very unstable and decay rapidly; others decay more slowly. The time it takes for half of the atoms of a radioactive isotope to decay is called **half-life.** Each successive half-life reduces the remaining number of atoms by one-half. Some isotopes have a half-life shorter than one second; others have a half-life of billions of years.

The fact that radioactive isotopes decay at known rates can be used to find the age of geologic material. **Absolute dating** is the process of determining the age of a sample using the radioactive isotopes in the sample. Rocks, minerals, fossils, water, and ice are dated using this method. The isotopes most commonly used in absolute dating are listed in Table 15–1.

TABLE 15–1: ISOTOPES COMMONLY USED IN ABSOLUTE DATING			
Isotope	Symbol	Half-life (years)	Material to which applied
Tritium	H-3	12.3	Ground water, sea water, ice
Carbon-14	C-14	5730	Wood, bones, shells
Potassium-40	K-40	1.3×10^9	Rocks, minerals
Rubidium-87	Rb-87	48×10^9	Rocks, minerals
Thorium-232	Th-232	14×10^9	Rocks, minerals
Uranium-235	U-235	704×10^6	Rocks, minerals
Uranium-238	U-238	4.5×10^9	Rocks, minerals

Geologists can determine the age of water by using an isotope of hydrogen called *tritium* (TRIHT ee uhm). Tritium forms in the upper atmosphere and combines with hydrogen and oxygen atoms to form water. This radioactive water falls as rain. Scientists can calculate the age of surface or ground water by measuring the amount of tritium it contains. For example, if a sample of the ground water of the Mississippi River Valley has only half as much tritium as rain does, this water must be 12 years old. Explain why this is true.

Carbon-14, or radiocarbon, also forms in the upper atmosphere, where it combines with oxygen to form carbon dioxide. Carbon dioxide becomes part of plant tissue during photosynthesis. Animals eat plants, so radiocarbon becomes part of animal tissue as well. This process is shown in Figure 15–10.

During the life of an organism, the level of radiocarbon stays about the same. When an organism dies, however, the amount of radiocarbon decreases by one half every 5730 years. A piece of wood with half the radiocarbon of a tree would be 5730 years old.

○ *What is radioactive decay?*
○ *What is meant by the half-life of a radioactive isotope?*

Figure 15–10. This diagram shows the decay of carbon-14, which has a half-life of 5730 years.

Carbon-14 remaining	Time
1/2	5730 years
1/4	11 460 years
1/8	17 190 years
1/16	22 920 years

How can radioactive decay be used to determine the age of a sample?

MATERIALS (per group of 3 or 4)

100 pennies, shoe box, pencil and paper, clock with second hand

PROCEDURE

1. Copy the data table shown below.

TABLE 1: HALF-LIFE

Shake	Time	Pennies in box
0	_____	100
1	_____	_____
2	_____	_____
3	_____	_____
4	_____	_____
5	_____	_____

2. Place 100 pennies in a shoe box so that all pennies are heads up.

3. Record the time.
4. Cover the shoe box and shake it vigorously so that the pennies are well mixed.
5. Open the shoe box and remove all pennies that are tails up.
6. Record on your data table the number of pennies remaining in the box.
7. Record the time, cover the box, and shake it again.
8. Repeat Steps 5–7 until only one penny remains. Record the time for each trial.

CONCLUSIONS/APPLICATIONS

1. Approximately what fraction of the remaining pennies was removed from the box after each shaking?
2. How many times did you shake the box before only one penny remained?
3. If someone stopped you and counted only 12 pennies remaining, could he or she have calculated the time that you started the experiment? Explain.
4. Imagine that the shoe box is a fossilized bone that contains 24 pμg (picomicrogram) of radiocarbon. When it was buried, it contained 100 pμg of radiocarbon. If the half-life of radiocarbon is 5730 years, how old is the bone?

15.4 Dating Minerals and Rocks

Tritium has a short half-life; therefore, it cannot be used to date material much older than 100 years. After 100 years, there would not be enough tritium left in the material to measure. Radiocarbon has a longer half-life, so it can be used to date materials as old as 50 000 years.

There are radioactive isotopes, such as uranium-238, that have a much longer half-life—long enough to date rocks and minerals as old as the earth itself. Most igneous rocks contain traces of these isotopes. One problem with dating rocks is that scientists do not know the amounts of the radioactive isotopes that were in the rocks when they formed. Therefore, the age of the rocks cannot be found by determining how much of the isotopes remain.

What scientists can do, however, is measure the amounts of both the parent isotope (such as uranium) and the decay product (lead). The ratio of decay product to parent isotope

A MATTER OF FACT

Scientists believe that about 65 million years ago a worldwide disaster destroyed more than two-thirds of all plant and animal life on Earth, including dinosaurs. Geologists are examining rocks to find out what might have caused this extinction.

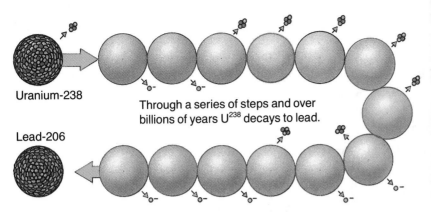

Uranium-238

Through a series of steps and over billions of years U^{238} decays to lead.

Lead-206

gives the age of the sample. The older the mineral or the rock, the more decay product, in this case lead, is present. Figure 15–11 shows how this process can be used to date a rock sample.

When uranium-238 decays, for example, particles shoot out in different directions, producing tracks in the surrounding material. Since this type of decay is similar to fission, these tracks are called *fission tracks*. Each fission track represents a nucleus that has decayed. The tracks can be counted. Then the remaining amount of parent isotope is determined in the laboratory. The ratio of these two quantities is used to determine the age of the sample. This method of dating is called *fission-track dating*.

Figure 15–11. The diagram (left) shows the decay of uranium into lead. The scientist (right) is using a Geiger counter to locate radioactive rocks. Some rocks can be dated by determining the ratio of uranium to lead in a rock sample.

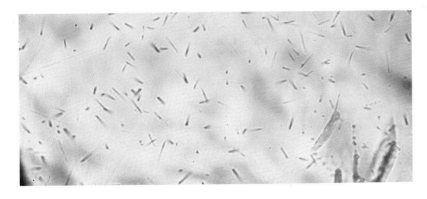

Figure 15–12. Fission tracks in rocks can be used to date some types of rocks.

Absolute dating cannot be used to date sedimentary rocks. Radioactive isotopes in a sedimentary rock can be used to date the minerals in the rock, but not the rock itself. You may recall from Chapter 4 that sediments can be transported by water and wind. Therefore, sedimentary rocks may be composed of pieces of many different parent rocks.

Metamorphic rocks cannot be dated by radioactive processes either. Temperature and pressure changes affect the amount of both parent isotope and decay product, and fission tracks are sealed over by the heat of metamorphism.

A MATTER OF FACT

Over a thousand metric tons of meteors enter the earth's atmosphere every year, but most of them never hit the ground because they burn up in the atmosphere. Those that do reach the earth are used to determine the age of the earth and the solar system.

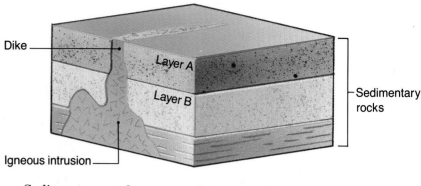

Figure 15–13. This igneous rock intrusion may be helpful in determining the relative ages of the surrounding sedimentary rocks.

Sedimentary and metamorphic rocks can be dated by relative dating. In **relative dating,** a nearby igneous formation is dated absolutely, and then the age of the neighboring formation is estimated by its position relative to the igneous rock.

○ *How can relative dating be used to date a sedimentary formation?*
○ *What is fission-track dating?*

15.5 Fossils

Fossils are the remains or traces of organisms that lived in the past. Hard parts of organisms, such as bones and shells, may be preserved intact for millions of years.

Buried in sediments, fossils often undergo chemical changes. Underground water may deposit silica in the pores of bone or wood, transforming it into stone. Often the silica replaces soft tissue. Replacement takes place slowly so that much of the original tissue structure is preserved. Replacement may be so complete as to allow scientists to study the structure of ancient cells.

Sometimes whole organisms are preserved nearly intact. Figure 15–15 shows an insect fossilized in hardened tree resin, called *amber.* The animal is preserved by the chemicals in the resin. Figure 15–15 also shows one of many Ice Age mammoths preserved for thousands of years in the frozen soil of Siberia. They were so well preserved that when thawed, their tusks were sold for the ivory and their meat was used to feed sled dogs.

Figure 15–14. Fossils, such as brachiopods, enable scientists to study organisms that lived millions of years ago.

Figure 15–15. The mammoth (right) was preserved in the frozen soil of Siberia for thousands of years, whereas the insect (left) was preserved in amber, or hardened tree resin.

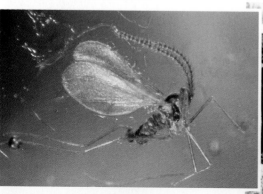

Plant and animal tissue buried under thick sediments undergoes a process called *carbonization*. In the absence of oxygen, part of the tissue becomes methane and other gases, and the rest remains as pure carbon. Most of the fossils found in coal formed by carbonization.

Shells buried in sediments may be dissolved by water and replaced with debris or mineral crystals that preserve the shape of the original shell. This type of fossil, shown in Figure 15–17, is called an *external cast*. There are also internal casts. An *internal cast* is made when the internal cavity of a shell becomes filled with material, and then the shell dissolves. Internal casts often show the internal structure of the organism in great detail.

Figure 15–17. This series of drawings shows the formation of two types of fossils—internal casts and external casts.

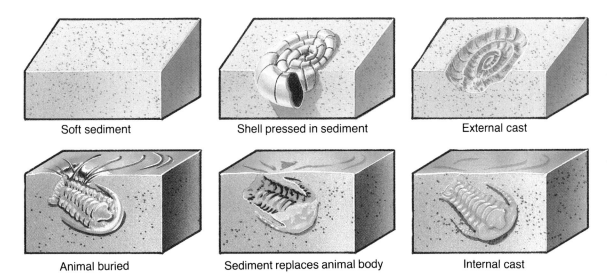

Soft sediment   Shell pressed in sediment   External cast

Animal buried   Sediment replaces animal body   Internal cast

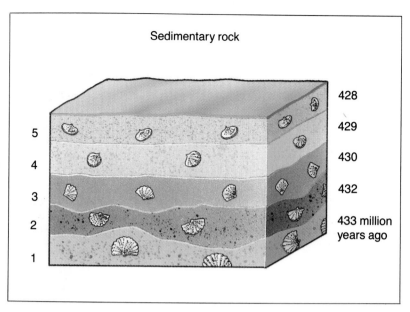

Figure 15–18. The diagram (right) shows a method for dating rocks using fossils. The scientist (left) is looking for fossils in layers of sedimentary rocks.

A MATTER OF FACT
Fossilized bones of prehistoric hominids show that early human ancestors began to walk upright about 4 million years ago.

Plant and animal species exist for a limited time. Because life evolves, fossils can be used as time indicators. Different fossils appear in sedimentary rock of different ages. By dating the fossils absolutely, or by dating similar fossils that occur elsewhere, the age of sedimentary rock can be determined.

The major geologic events in Earth's history can be dated and placed on a time line. The time line begins with the formation of the earth and continues to the present. The line is divided into sections representing specific events in geologic time. In the next chapter, you will study the history of Earth and geologic time.

○ *What are fossils?*
○ *How can fossils be used to date sedimentary rock formations?*

Section Review

READING CRITICALLY

1. What makes certain atoms radioactive?
2. State two reasons fossils are important to the study of the geologic record.

THINKING CRITICALLY

3. If a piece of wood contains 1/32 of its original radiocarbon, how old is that piece of wood?
4. How would you explain the occurrence of a formation where the younger fossils were underneath the older ones?

TECHNOLOGY: New Techniques in Radiocarbon Dating

Radiocarbon dating is the most common way to date fossils. Scientists use this method to find the age of fossils less than 50 000 years old. The use of this method depends on two things: the age of the fossil and how well it is preserved. In some cases, the fossil is too old for this dating method—there is not enough radiocarbon for instruments to detect. Background radiation, which occurs naturally, often interferes with the detection of very small amounts of radiocarbon.

To avoid this problem, a larger sample must be used, but this usually destroys the fossil. Some fossils are simply not preserved well enough for radiocarbon dating. Still other fossils are contaminated with recent carbon sources. This makes the radiocarbon method inaccurate.

Recently scientists have discovered a dating method that requires a very small sample of fossil. The new method is called *accelerator mass spectrometry*. In this process, the atoms in the sample are stripped of one electron each, and then the electrically charged ions are accelerated to high speeds. The ions are forced to move in a curved path. The heavier ions in the sample, such as radiocarbon, have greater energy than the lighter ions.

Searching for fossils

Because of their mass, the radiocarbon ions cannot be made to change direction as easily as the lighter ions. Therefore, they will separate from the lighter ions. After separation, the radiocarbon ions can be counted directly, and the age of the sample determined in the usual manner.

Some scientists think that this new method will at least double the range of radiocarbon dating, because it will allow detection of very small amounts of radiocarbon.

Examining fossils in the laboratory

INVESTIGATION 15: Fossilizing Bones

PURPOSE

To prepare internal and external casts of bones and shells

MATERIALS (per group of 3 or 4)

Plaster of Paris
Tray
Water
Petroleum jelly
Vise
Small saw
Needle-nosed pliers
Small hammer
Chicken or lamb bones
Snail shells or seashells

PROCEDURE

1. Secure one of the bone or shell samples in the vise. Use the saw to expose the interior of the sample. Be careful not to splinter the sample.
2. Remove the sample from the vise and wash and dry it thoroughly to remove any soft organic tissue.
3. Coat the surface of each sample with a thin layer of petroleum jelly.
4. Prepare a mixture of plaster of Paris according to the directions on the package and pour the mixture into the tray.

5. Submerge all of the samples completely in the plaster. Be sure that all cavities in the sample are filled.
6. Allow the plaster to dry for at least 24 hours.
7. Use the small hammer to crack the dried plaster of Paris and expose the sample. Do not break up the plaster.
8. Use the pliers to gently remove the sample from the plaster.
9. Repeat steps 1 through 8 for as many samples as you have.

ANALYSES AND CONCLUSIONS

1. What are some of the problems geologists might have in removing fossils from rocks?
2. Distinguish between the internal and external casts made with your samples.
3. Ask other students to try to identify what samples you used to make your plaster casts.

APPLICATION

How could these procedures be modified to make casts of fossil imprints, such as dinosaur footprints?

SUMMARY

- The principle of uniformitarianism states that today's geological processes are similar to those of the past. (15.1)

- The principle of superposition states that younger rocks are deposited on top of older rocks. (15.1)

- Fossils and sedimentary features provide evidence about past environmental conditions. (15.2)

- A series of formations can provide a continuous record of environmental changes over a long period of time. (15.2)

- The time it takes for half of the atoms of a radioactive isotope to decay is called half-life. (15.3)

- Some unstable atomic nuclei undergo radioactive decay. (15.3)

- Radioactive isotopes can be used to absolutely date geologic materials. (15.3)

- Igneous rocks can be dated absolutely, but sedimentary and metamorphic rocks must be dated relative to nearby igneous formations. (15.4)

- As a radioactive isotope decays, it may leave fission tracks in the surrounding material. (15.4)

- Fossils are the remains, or traces, of organisms that lived in the past. (15.5)

- Fossils can be used to date sedimentary rock. (15.5)

Write all answers on a separate sheet of paper.

SCIENCE TERMS

Correctly use each of the following terms in a sentence.

absolute dating **(329)**
crossbedding **(325)**
half-life **(328)**
principle of superposition **(324)**
principle of uniformitarianism **(323)**
radioactive decay **(328)**
relative dating **(332)**

SCIENCE QUIZ

Modified True-False

Mark each statement *true* or *false*. If a statement is false, change the underlined term to make the statement true.

1. The nucleus of an atom contains protons and <u>electrons</u>.

2. Isotopes of an element have different numbers of <u>protons</u> in the nucleus.

3. The oldest rocks can be dated using <u>uranium-238</u>.

4. The age of <u>fossils</u> can be found by absolute dating.

5. Radiocarbon is commonly used to date <u>rock</u> material less than 50 000 years old.

6. The age of <u>sedimentary</u> rocks can be estimated from the age of fossils found in the rocks.

Multiple Choice

Write the letter of the choice that best answers the question or completes the statement.

7. The isotope of hydrogen known as *tritium* has an atomic number of three, so it probably has
   a) 2 protons and 1 neutron.
   b) 1 proton and 1 neutron.
   c) 1 proton and 2 neutrons.
   d) 2 protons and 2 neutrons.

continues

8. What type of particles make radioactive isotopes unstable?
   - a) protons and electrons
   - b) neutrons and electrons
   - c) protons and neutrons
   - d) electrons

9. As the number of particles inside a nucleus increases, the stability of the nucleus often
   - a) increases.
   - b) decreases.
   - c) remains the same.
   - d) cancels the forces of gravity.

10. The best isotope to use in dating wood, bone, and shell is
    - a) tritium.
    - b) radiocarbon.
    - c) rubidium-87.
    - d) uranium-235.

11. Absolute dating can be used to find the age of
    - a) igneous rocks.
    - b) sedimentary rocks.
    - c) metamorphic rocks.
    - d) all rocks.

12. Unconformities in the rock record are due to
    - a) fission.
    - b) radiocarbon.
    - c) missing sediments.
    - d) extra neutrons

Completion

Complete each statement by supplying the correct term.

13. An element whose nucleus loses a particle containing 2 protons and 2 neutrons and energy has undergone _____.

14. The process of decay of the nucleus of an atom is called _____.

15. Another name for carbon-14 is _____.

16. Water containing tritium will lose half of its radioactivity in about _____ years.

17. Sedimentary rocks can be dated by _____ dating.

18. The geologic principle that states that younger sediments are deposited on top of older sediments is the principle of _____.

Short Answer

19. Explain the meaning of the term *half-life*.

20. What is an unconformity?

21. Explain why radiocarbon would not be useful in dating very old samples.

22. Explain the principle of uniformitarianism.

Writing Critically

23. How might excess neutrons and protons make an atomic nucleus unstable?

24. Explain why fossils are not found in igneous rocks.

25. According to the principle of superposition, sediments are deposited in horizontal layers. However, in some places rock layers are nearly vertical. Explain this apparent contradiction between the principle and facts.

EXTENSION

1. At your local or school library, research the types of land fossils that have been found on the east coast of South America and the west coast of Africa. Compare and contrast these fossils, and then use tracings and drawings to support the theory that these two continents were once joined.

2. Use the library to find information about the Apollo moon missions of the 1970s. Explain why these space missions were helpful in determining the age of the earth and the solar system.

3. Trace the evolution of the horse over the past 50 million years. Display this information in class in the form of a time line.

4. From the list For Further Reading, read the *National Geographic* article "Fossils: Annals of Life Written in Rock" and report to the class on the article.

APPLICATION/CRITICAL THINKING

1. A geologist finds a fossilized bone in a canyon wall. Carefully digging out the fossil, the geologist takes a small sample of the bone and discovers that the sample contains about 2 pμg of radiocarbon. If the geologist assumes that a similar amount of living tissue would contain about 16 pμg of radiocarbon, what would most likely be the age of the bone?

2. Scientists do not have a complete record of the living forms that have inhabited our planet. There are many gaps in the fossil record. Discuss the processes that may have caused gaps in the record.

3. The sedimentary rock walls of the Grand Canyon have often been referred to by geologists as a "history book of the earth's geological past." Explain what is meant by this phrase, and tell why the walls of the Grand Canyon, unlike any other place on Earth, provide so much fossil evidence of past geological processes.

FOR FURTHER READING

Cairne-Smith, A. G. *Seven Clues to the Origin of Life: A Scientific Detective Story.* New York: Cambridge University Press, 1985. The author uses a Sherlock Holmes approach to develop a possible scenario for the origin of life during the early history of Earth.

Jeffery, D. "Fossils: Annals of Life Written in Rock." *National Geographic* 168 (August 1985): 182. This article gives a colorful explanation of how Earth's ancient life has been preserved in rocks.

Weaver, K. "The Search for Our Ancestors." *National Geographic* 168 (November 1985): 560. This article contains brilliant photographs and a review of the most recent evidence of human evolution.

Challenge Your Thinking

This photograph shows a break in a sedimentary rock formation. At many points in the sedimentary record, there are places where fossils just seem to disappear. What do you think could cause these gaps in the rock record?

16

Earth's Past

These scientists are studying history—Earth history. The study of Earth history is a bit different from the history you may be familiar with. Studying Earth history is a physically demanding job, but it is one that offers great rewards. Imagine discovering, as these men have done, the skeleton of an animal that no human has ever seen before.

Examining a new fossil find

1 The Cryptozoic Eon

SECTION OBJECTIVES

After completing this section, you should be able to:

- **Describe** how the earth's crust formed and grew through time.
- **Discuss** the role of bacteria in the evolution of life on Earth.
- **Explain** when and how multicellular animals evolved.

NEW SCIENCE TERMS

planetesimals
fermentation
cellular respiration

16.1 Divisions of Earth History

The history of the earth covers such a long period that most people have trouble imagining when some important event occurred. If you try to imagine the whole of Earth's past compressed into one calendar year, it becomes easier to understand. Like a calendar year, which is divided into months of different lengths, the calendar of Earth's history can be divided into unequal segments.

Earth's history has two main divisions, or *eons* (EE ahnz): the Cryptozoic (krihp toh ZOH ihk) Eon and the Phanerozoic (fan IR uhz OH ihk) Eon. Just as a calendar year is divided into months, eons are divided into *eras* (EHR ahz). Like months, eras are not all the same length.

The calendar of Earth history begins with the *Cryptozoic Eon*. This eon includes the time between the formation of the solar system, probably about 4.6 billion years ago, and the appearance on Earth of abundant fossils, about 590 million years ago. The Cryptozoic Eon is followed by the *Phanerozoic Eon*, which covers the period from about 590 million years ago to the present.

Figure 16–1. In the distant past, the surface of the earth was probably very bleak (left). As plants and animals evolved, lush forests covered much of the land (right).

The Cryptozoic Eon covers 87 percent of the earth's history. On a scale of a one-year period, this would approximate the time from New Year's Day to Thanksgiving on the calendar.

The Cryptozoic Eon is divided into three eras: the Hadean (HAY dee uhn), the Archean (AHR kee uhn), and the Proterozoic (proht uhr oh ZOH ihk). The *Hadean Era* is the time between the formation of the solar system and the age of the oldest known rocks on Earth. In a calendar year, this era would last about two months. The *Archean Era* ranges from about 3.8 billion years ago to about 2.7 billion years ago, or approximately three months. The *Proterozoic Era* was the period from about 2.7 billion years ago to about 590 million years ago; this would equal about 5.5 months on a year's calendar.

○ *What are the two main divisions of Earth history?*

16.2 The Hadean Era

During the first 100 million years of the Hadean Era, the earth and the other planets probably formed. Scientists believe that the planets formed from small bodies of gases and solids called **planetesimals** (plahn uh TEHS uh muhlz). The planetesimals circled the sun in rings, much like the rings that now circle Saturn and other planets. The largest planetesimal within each ring probably captured the smaller ones, and an early planet formed. Why would the largest planetesimal capture the smaller ones?

As the earth formed, its surface probably resembled that of today's moon. Craters and large plains of basalt marked the surface. The earth was hot, and there were large pools of molten

Figure 16–2. Some of the oldest rocks on Earth have been found in Greenland. These rocks are nearly 3 billion years old.

A MATTER OF FACT

The idea that the solar system formed from a rotating disk of dust and gas is not new. This was first proposed by the great French philosopher René Descartes in the year 1644.

Figure 16–3. In the Hadean Era, the surface of the earth was probably much like the present surface of the moon.

lava. There was a thick atmosphere of water vapor, carbon dioxide, and sulfur dioxide. As the earth cooled, the water vapor probably condensed, and torrential rains began. Water collected in craters and other surface depressions. The carbon dioxide and sulfur dioxide remained in the atmosphere.

The earth's original surface, or crust, consisted of silicate rock. Basalt, a dense rock, formed the crust of low-lying areas, such as the ocean basins. Less-dense granite formed the elevated areas—the continents. Scientists believe the crust greatly increased in size during the Hadean Era and is, in fact, still growing. Almost all of the continental crust, however, probably formed during the Cryptozoic Eon.

The interior of the earth remained hot due to heat trapped during its formation. In fact, the interior was so hot that convection currents developed. Hot rocks from the interior rose to the surface, and cooler surface rocks sank. The convection currents brought together blocks of granite to form crustal plates. These crustal plates formed a lid and trapped the heat below.

By the end of the Hadean Era, scientists believe, the surface of the earth had large oceans. There were barren continents with mountain ranges, plains, rivers, and volcanoes. There were many clouds, strong winds, and torrential rains, which were acidic due to the sulfur dioxide in the atmosphere.

○ *Name three events that occurred during the Hadean Era.*

Figure 16–4. The earth's primitive atmosphere probably contained carbon dioxide, sulfur dioxide, and water vapor. These gases may have escaped from the interior of the earth through volcanic vents (left), creating torrential rains that filled the earth's oceans.

16.3 The Archean Era

In early March on the one-year calendar of Earth history, the Archean Era began. Many important events occurred during the Archean Era, the most important of which, according to scientists, was the evolution of life.

Among the many different molecules on the surface of the primitive earth, complex organic molecules containing carbon were probably common. However, to progress from complex molecules to even the simplest living organism was a very long process.

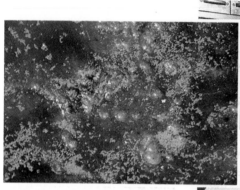 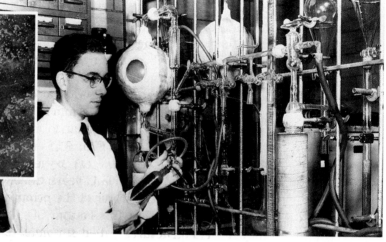

Figure 16–5. During the Archean Era, the waters of the earth probably contained a fermenting organic broth (left). Dr. Stanley Miller (right) tested the hypothesis that organic chemicals could be produced from simple chemicals in the earth's atmosphere. By simulating conditions of the primitive atmosphere, Miller proved that organic molecules could be produced from simple chemicals.

Among the first, and most important, chemicals of life were the *carbohydrates*. Carbohydrate molecules are made of carbon, hydrogen, and oxygen atoms. Carbohydrates are easily broken down by simple organisms. Primitive bacteria, probably some of the first living organisms to evolve, used carbohydrates in a process called *fermentation*. **Fermentation** is the partial breakdown of sugar, a simple carbohydrate, to release energy. This process takes place without oxygen, and it is not very efficient. However, it is likely that it was the basis for most early forms of life. Today simple organisms such as bacteria and yeasts get their energy through fermentation.

Another important group of compounds probably present on the early earth were the *amino acids*. Amino acids are molecules made of carbon, hydrogen, oxygen, and nitrogen atoms. These amino acids probably joined together by chance to make long complex molecules known as *proteins*. Much of the tissue of living organisms is made of proteins. In addition, many chemical processes require the presence of special proteins.

Perhaps the most important organic molecule produced was *chlorophyll*. Chlorophyll allowed early organisms to make their

A MATTER OF FACT

The sequence of amino acids in similar proteins can be used to trace the evolution of organisms from one species to another.

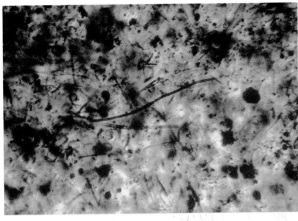
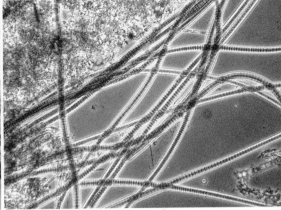

own food. Molecules of chlorophyll consist of a magnesium atom surrounded by nitrogen, carbon, hydrogen, and oxygen atoms. Solar energy was used by chlorophyll-containing organisms to produce carbohydrates from carbon dioxide and water.

Photosynthesis first appeared in organisms known as *cyanobacteria*. An example of an early fossil cyanobacterium is shown in Figure 16–6. These ancient bacteria were similar to modern cyanobacteria, such as *Nostoc*, also shown in Figure 16–6.

Scientists believe these cyanobacteria gave off oxygen as a byproduct of photosynthesis. For about 2 billion years the oxygen combined with iron dissolved in the water of the primitive earth. Iron oxide layers accumulated on the bottom of lakes and oceans, where they formed structures called *stromatolites*. As most of the available iron was oxidized, what do you think happened to the oxygen produced by photosynthesis? Why?

○ **What organisms appeared during the Archean Era?**
○ **Why was chlorophyll important in the evolution of new organisms?**

Figure 16–6. Ancient cyanobacteria (left) were probably the first organisms to use chlorophyll to make their own food. *Nostoc* (right), a modern cyanobacterium, produces its own food in much the same way.

A MATTER OF FACT

Oxygen present in the atmosphere when cyanobacteria began to photosynthesize was only one one-thousandth of its present level.

Figure 16–7. Stromatolites, such as these found in the Bahamas, were formed as oxygen from photosynthesis combined with iron in the rocks on the bottom of the ocean. These layers alternate with layers of sand, algae, and calcium carbonate.

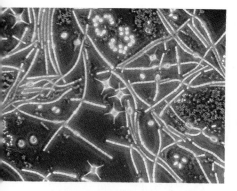

Figure 16–8. Diatoms are single-celled organisms that produce much of the oxygen needed for cellular respiration.

Figure 16–9. The life style of *Euglena* (bottom left) may provide a clue to the evolution of animals. In sunlight, *Euglena* produces food by photosynthesis like the green alga (top left). Without sunlight, Euglena can gather food like *Paramecium* (bottom right).

16.4 The Proterozoic Era

The longest "month" on the earth-history calendar is the Proterozoic Era. Many important events occurred during this era.

Scientists believe that during the Proterozoic Era life forms continued to evolve and occupy new habitats. Bacteria began to form simple groups. Some of these groups became enclosed in protein membranes, forming single complex cells. These cells were much more complex than the early bacteria. Most of these organisms continued the process of photosynthesis. In appearance, they were probably similar to modern green algae.

During this era some algae lost their ability to photosynthesize; they evolved a process for releasing energy from carbohydrates. This process, known as **cellular respiration,** is the breakdown of carbohydrates in the presence of oxygen. Respiration releases almost ten times more energy than fermentation does.

Having lost the ability to photosynthesize, these organisms, which were probably similar to present-day protozoa, had to depend on food made by cyanobacteria and green algae for their survival. Figure 16–9 shows a possible evolutionary link between algae and protozoa.

In a period of about 150 million years—or less than two weeks on the earth-history calendar—protozoa evolved into multicellular organisms. Multicellular organisms were more than just collections of simple cells. There was much specialization of function and process among the cells of these organisms.

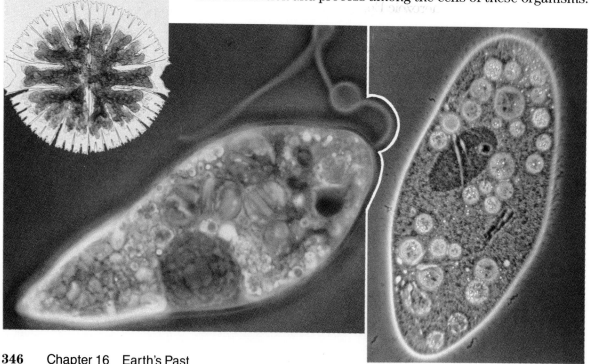

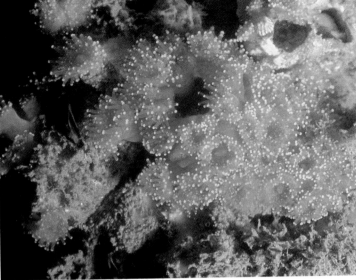

Most of these organisms were similar to the group of animals to which modern corals belong. Others were like present-day marine worms. There was nothing simple about these early animals; they were as complex as their modern counterparts.

These animals probably fed by eating algae or each other. As this food-hunting practice, known as *predation*, developed, a number of organisms evolved protective body coverings. These coverings were hard and left many fossils. The appearance of abundant fossils in the geologic record marks the beginning of the next eon, the Phanerozoic Eon.

○ *What is cellular respiration?*
○ *What type of complex organisms evolved during the Proterozoic Era?*

Figure 16–10. The earliest animals may have been similar to simple sponges (left) and corals (right).

Section Review

READING CRITICALLY

1. List several kinds of organic compounds and their functions in living organisms.
2. What is photosynthesis? What effect did it have on the development of life?

THINKING CRITICALLY

3. Why were the rains during the Hadean Era acid rains?
4. Even though multicellular organisms preyed on each other, what provided the original energy for all life on Earth? Explain your answer.

BACKGROUND

Scientists often have to analyze data collected from different sources before they draw any conclusions about a problem. Sometimes the data must be organized before it becomes meaningful. In this activity you will be gathering data about plant and animal characteristics to establish evolutionary relationships.

Multicellular organisms have existed for only about the last fifth of the history of life on Earth, having evolved from single-celled organisms prior to the Phanerozoic Eon. Yet they have adapted to a variety of environments in a relatively short time. Plants and animals are the principal organisms considered in this activity.

PROCEDURE

1. Use tracing paper to make a copy of the evolutionary tree shown below.
2. Using a life-science or biology textbook, examine the physical features of the plants and animals labeled in this activity. Note their similarities and differences.

A) Cyanobacteria or blue-green bacteria

B) Red algae

C) Brown algae

D) Seed plants

E) Green algae

F) Fungi

G) Slime molds

H) Annelids

I) Arthropods

J) Vertebrates

K) Primitive chordates

L) Echinoderms

M) Mollusks

N) Coelenterates

O) Sponges

P) Ciliates

Q) Amoeboids

R) Bacteria

3. Check the position for each group of organisms on the evolutionary tree. Decide which are most likely to have had a common ancestor.
4. Write the name of each plant or animal group above the lettered circle on the tree.

APPLICATION

1. Which group of organisms has evolved most independently of the others?
2. List the organisms that could be considered plants. List the animals. Are there any organisms that could be on both lists?
3. Compare your list with your teacher's list. Do you think that physical characteristics alone are sufficient to accurately classify all of the organisms pictured here? Explain.

USING WHAT YOU HAVE LEARNED

Organize the fossil data from Investigation 16, on page 358, in a similar fashion, then try to answer the Application questions again for this new data.

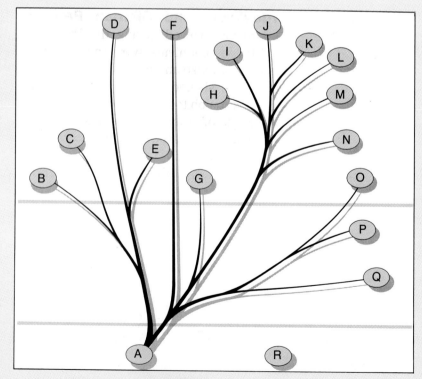

② The Phanerozoic Eon

SECTION OBJECTIVES

After completing this section, you should be able to:
- **List** the characteristics of trilobites.
- **Compare** the mass extinctions that occurred at the end of the Ordovician and Devonian Periods.
- **Describe** the ice ages of the Cenozoic Era.

NEW SCIENCE TERM

glaciation

16.5 The Paleozoic Era

Many of the familiar events of Earth history occurred during the *Phanerozoic Eon.* This eon begins the last month or so of the year-long geologic calendar. As with the Cryptozoic Eon, the Phanerozoic Eon has been divided into three eras: the *Paleozoic* (pay lee oh ZOH ihk), the *Mesozoic* (mehs uh ZOH ihk), and the *Cenozoic* (see nuh ZOH ihk). These eras are the final weeks of the geologic calendar year. Each era is further divided into periods, making it easier to study the many events that occurred. To see the relationship between eras and periods, time boundaries, and the major events of each period, look at the geologic time scale on page 574 of the Reference Section.

The Cambrian, Ordovician, and Silurian Periods

The earliest shelled organisms evolved during the Paleozoic Era. Most impressive of these organisms were the *trilobites* (TRY luh byts). Trilobites were crablike animals with highly complex, segmented bodies. These shelled animals were like the horseshoe crabs of today. Most trilobites were a few centimeters long, but some reached a length of several decimeters.

Trilobites were an important part of the shallow-water ocean communities of the first Paleozoic period, the *Cambrian* (KAM bree uhn) *Period.* Trilobites lived in shallow ocean water from about 570 million years ago to about 250 million years ago.

DISCOVER

Comparing Relatives

Compare the photograph of a horseshoe crab shell with the illustration of the trilobites. Note the similarities and differences between these organisms. The horseshoe crab has remained nearly unchanged for millions of years. What differences between the two organisms might have allowed horseshoe crabs to survive, and caused the trilobites to become extinct?

Figure 16-11. A modern horseshoe crab (right) is similar in appearance to the trilobites of the Paleozoic Era (left).

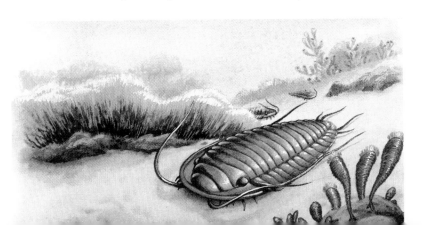

Figure 16–12. Ordovician fishes (right) may have been similar to modern fishes (left).

Only a few species of trilobites survived into the next period, the *Ordovician* (ohr duh VIHSH uhn). During the Ordovician Period, an ice age began, and sea level was greatly lowered. The shallow marine environment was reduced in area, and many habitats were destroyed. The extinction of many organisms probably occurred due to these changed conditions. How would an ice age lower the sea level?

The earliest fishes appeared late in the Ordovician Period. Because their bodies were covered with protective bony plates, they were probably poor swimmers. As they evolved, the bony plates disappeared. During the next period, the *Silurian* (sih LUHR ee uhn), all major groups of fishes—sharks, rays, and bony fishes—appeared.

During the Cambrian and Ordovician periods, there were no land plants. Land plants first appeared in the Silurian Period. These earliest plants had no true roots, stems, or leaves. Instead, they had flexible structures that served some of the same purposes. These plants reproduced by spores.

Figure 16–13. Modern liverworts (left) have many of the same characteristics as the first land plants (right).

The Devonian, Carboniferous, and Permian Periods

During the *Devonian* (deh VOH nee uhn) *Period* amphibians and seed-bearing trees first appeared. Amphibians were the first vertebrates that could live on dry land. A temperature increase occurred toward the end of the Devonian Period that may have caused the extinction of many animal species.

The last two periods of the Paleozoic Era were the *Carboniferous* (kar bah NIHF ur uhs) and the *Permian* (PUR mee uhn). Many species of land plants evolved during these periods, and vast forests of seed-bearing trees covered the land.

Reptiles, which evolved during the Carboniferous Period, differ significantly from amphibians and fish in their method of reproduction. Amphibians and fish deposit eggs that can survive only in water. Reptile eggs can survive on land. Due to their ability to reproduce away from water, the reptile populations spread rapidly during the Permian Period.

○ *What may have caused the extinction of the trilobites?*
○ *What important groups of plants and animals evolved during the last half of the Paleozoic Era?*

Figure 16–14. During the Carboniferous Period, great forests covered much of the land (left). The dominant plants of these forests were probably the tree ferns, similar to tree ferns growing today in Australia (right).

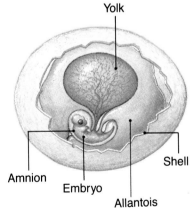

Figure 16–15. Reptile eggs (above) are protected by several internal membranes. These eggs, such as those of the sea turtle (left), can incubate and hatch on dry land.

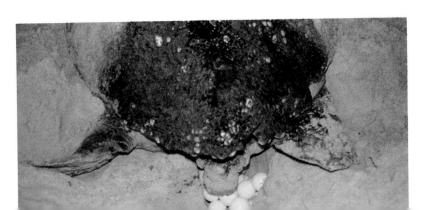

351

16.6 The Mesozoic Era

The geologic calendar is now into the middle of December. This is the *Mesozoic Era,* the age of reptiles.

By far, the most impressive group of animals of the Mesozoic Era were the reptiles. They evolved into a large number of species that occupied all major habitats. Most reptiles lived on land, but there were also marine and flying reptiles. Some reptiles evolved into dinosaurs, birds, and mammals in the *Triassic* (try AS ihk) *Period,* the first period of the era.

Mammals and dinosaurs evolved at about the same time. Both were small animals at first, but dinosaurs spread much faster and forced the mammals out of many habitats. The major advantage dinosaurs had over the mammals of the time was their agility.

Dinosaurs faced the problem of overheating during the warm *Jurassic* (joo RAS ihk) *Period.* To combat this problem, dinosaurs evolved elaborate structures to cool their blood. These structures included back plates, collar plates, and ducts in the skull. These structures made some of the dinosaurs look quite strange, as shown in Figure 16–16.

Figure 16–16. The Mesozoic Era was the age of reptiles (top). Some of the dinosaurs of this era developed elaborate structures to keep themselves cool in the warm climate of the time.

The dinosaurs became extinct about 65 million years ago, at the end of the *Cretaceous* (krih TAY shuhs) *Period,* the final period of the era. Many other organisms also became extinct at this time. Many shallow-water marine animals, reef-building corals, protozoa, and algae died. High-latitude animals and animals living on the deep sea floor suffered less. Some scientists believe that a period of high temperature was the cause of this mass extinction. Other scientists believe just the opposite: that a short period of unusually cold weather killed most of the tropical plants of the world, and the animals soon died from starvation.

○ *Why is the Mesozoic Era called the age of reptiles?*
○ *What might have caused the mass extinction at the end of the Mesozoic Era?*

Figure 16–17. There are many locations where fossilized bones may be found.

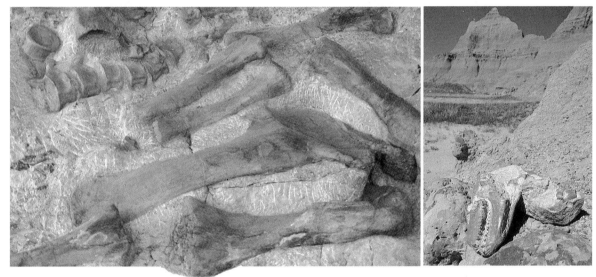

16.7 The Cenozoic Era

We are now at the end of the geologic calendar year. The present era is the *Cenozoic Era.* Although the Cenozoic Era started about 65 million years ago, it would be represented on the geologic calendar by the last few days of the year. During the Cenozoic Era, the mammal populations expanded, filling the habitats of the extinct dinosaurs.

The Cenozoic Era is divided into two periods: the Tertiary Period, which lasted from about 65 million years ago to about 3 million years ago; and the Quaternary Period, which is the present period. Each period is subdivided into epochs.

About 3 million years ago, ice began to push southward across the northern continents. Since then great sheets of ice have advanced and retreated every 100 000 years. Each time the ice formed, it would take about 75 000 years to reach its southernmost point. The ice would remain fewer than 10 000 years

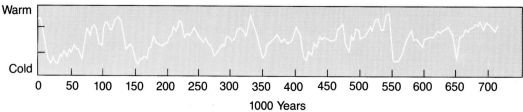

Figure 16–18. The graph (above) is a record of Earth's temperature changes over the last 700 000 years. During this time, periods of glaciation (top left) alternated with periods of moderate climate (top right).

Figure 16–19. This time line shows the divisions of Earth's history.

and would retreat over another 10 000 years. The ice would then disappear from all northern continental areas except Greenland. The continents would remain free of ice for about 5000 years, and then the ice cycle would repeat.

Figure 16–18 shows the global changes in temperature that have taken place during the past 700 000 years. Each peak represents conditions similar to those of today. Each valley represents a glaciation. **Glaciation** is a period in which great continental ice sheets advance and retreat repeatedly.

During glaciations, summer temperatures were probably similar to the winter temperatures of today. Summers were stormy; winters were even stormier. Periods between glaciations were short. Warm temperatures, like those of today, lasted about 5000 years or less. If the past is an indication of the future, the world should be entering a new glaciation within the next thousand years.

Cryptozoic Eon						Phanerozoic	
Precambrian or Hadean, Archean, and Proterozoic eras	Paleozoic Era						
	Cambrian Period			Ordovician Period		Silurian Period	Devonian Period	Carboniferous Period
	Early	Middle	Late	Early	Late		Early	Middle	Late	Mississippian	Pennsylvanian

Figure 16–20. Modern tarsiers (left) and lemurs (right) may resemble the earliest primates.

The evolution of present-day plants and animals, including humans, occurred during these extreme weather cycles. Humans evolved with a group of mammals known as *primates*. Primates appeared soon after the dinosaurs became extinct, about 65 million years ago, or about December 24 of the geologic year. Modern lemurs and tarsiers, shown in Figure 16–20, are very similar to primitive primates.

Monkeys and apes appeared later, about 30 million years ago. The first *hominids*, or humanlike primates, appeared about 3.5 million years ago, or December 29. *Homo sapiens*, the present human species, evolved about 0.125 million years ago, at approximately 11:30 P.M., the last half-hour of the geologic calendar year. The geologic year shows how short a time humans have existed, compared to the age of the earth.

At the end of the last glaciation, humans were hunters, mainly of large mammals. More than 200 species became extinct

Eon																
		Mesozoic Era								Cenozoic Era						
Permian Period		Triassic Period			Jurassic Period			Cretaceous Period		Tertiary Period					Quaternary Period	
Early	Late	Early	Middle	Late	Early	Middle	Late	Early	Late	Paleocene	Eocene	Oligocene	Miocene	Pliocene	Pleistocene	Holocene

Figure 16–21. All people, regardless of individual characteristics, evolved from the same ancestral hominids over 3.5 million years ago.

as a result of this hunting. About 12 000 years ago, humans discovered that there was little left to hunt. Therefore, for about 4000 years life was probably hard, since food was scarce. Only when agriculture was developed, about 8000 years ago, did food again become plentiful.

Humans are an important part of the earth environment. Together with every other living organism, humans probably evolved from bacteria that lived more than 4 billion years ago. What kinds of organisms do you think will evolve over the next 4 billion years?

○ *What is glaciation?*
○ *What is a hominid?*

Section Review

READING CRITICALLY

1. How did the climate of the Cenozoic glaciations differ from today's climate?
2. When did dinosaurs, mammals, and birds come into existence?

THINKING CRITICALLY

3. Do you think humans are still evolving? Explain.
4. How might the dinosaurs' body-heat problems have led to their extinction?

ACTIVITY : Making a Geologic Time Line

How can you compare the relative length of eras and periods of the Phanerozoic Eon?

MATERIALS (per group of 3 or 4)

construction paper, tape, markers, metric ruler, scissors

PROCEDURE

1. Cut the construction paper along its length into strips measuring 5 cm in width.
2. Tape the strips end to end so that you have a ribbon approximately 4 m in length.
3. Roll up the ribbon so that it is easy to handle, and unroll it as required.
4. Using Figure 16–19 on pages 354–355 and using a scale that 5 mm equals 1 million years, measure the length of each era and

period of the Phanerozoic Eon. Draw a single line to mark the beginning and end of each period. Use a double line to separate eras.
5. Label the eras and periods.
6. Add important events in Earth history or make drawings of fossils in the appropriate places.

CONCLUSIONS AND APPLICATIONS

1. How long was the Paleozoic Era? The Mesozoic? The Cenozoic?
2. How many times longer than the Mesozoic was the Paleozoic?
3. On your measurement scale, what problems arose when you were trying to measure the lengths of the periods of the Cenozoic Era?
4. According to Figure 16–19, for what percent of the Phanerozoic Eon has *Homo sapiens* inhabited this planet?

BIOGRAPHIES: Then and Now

HAROLD UREY (1893–1981)

Most scientists become interested in only one or two fields of science, such as biology or chemistry. Other scientists, however, have undying curiosity about all sciences. They try to learn as much as they can about many scientific fields. Harold Clayton Urey was the second kind of scientist.

He received his bachelor's degree in 1917 from Montana State University. At that time his major interests were zoology and chemistry. After he was awarded a doctorate from the University of California at Berkeley, he traveled to Denmark. There he worked from 1923 until 1924 with Niels Bohr, the famous atomic physicist.

Urey's work in atomic research over the next decade led to his discovery of deuterium (doo TEER ee uhm), a heavy isotope of hydrogen. In 1934 he received the Nobel prize for his work.

In the early 1950s, Urey turned his attention to the studies of geochemistry, astrophysics, and the origin of life. He wanted to know how the earth and solar system had come to be. His vast knowledge of biology, physics, and chemistry helped him conduct research and write many articles on geochemistry. He reviewed many theories on how the sun and planets were formed. He studied the chemical reactions of gases that existed in Earth's primitive atmosphere, and he was the first to show that amino acids could have formed in the atmosphere. Although he never proved how life originated, he did add evidence to the theory that life could have started by itself on the primitive earth.

LYNN MARGULIS (1938–)

Imagine that you are standing knee-deep in a pond of pink, green, and purple ooze. You turn to the person next to you, a scientist from Boston University. She tells you that you are soaking in some of the most interesting bacteria in the world. You might scream and run for dry land. Professor Lynn Margulis would stay right where she is, studying the microscopic creatures of the murky pool.

Dr. Margulis considers herself a "microbiological evolutionist," and her work has shed a great deal of light on the origin of life. Because the subject is so complex and difficult, it requires interest and knowledge in many scientific fields.

She has tried to explain how primitive organisms might have evolved into more complex life forms. She has declared that the evolution of complex animals might have relied more on biological cooperation, or *symbiosis,* than was previously thought. Symbiosis occurs when organisms live together for their mutual benefit.

Margulis received a doctorate in genetics from the University of California at Berkeley in 1965. She is currently teaching at Boston University. In 1985 she was named "The Wizard of Ooze" by author Richard Wolkomir in an article in *Omni* magazine.

Dr. Margulis has suggested that the millions of kinds of microorganisms may be adding gases other than oxygen to our atmosphere. "Kill off animals and plants and the planet will recover," she has said, "but kill off the microbes and in weeks the earth will be just as sterile as the moon."

INVESTIGATION 16: Classifying Fossils

PURPOSE

To classify fossils based on similar characteristics

MATERIALS

Set of ten numbered fossils
Hand lens or stereomicroscope

PROCEDURE

1. Copy the table for recording your observations.
2. Make four or five detailed observations of each fossil specimen and record your description in the table. You may use a hand lens or a stereomicroscope to help see details.
3. Separate the fossils into groups with similar characteristics.
4. List the fossils by number on the table. Give several reasons for your grouping.

5. Compare the fossils in your groups to those in your classmates' groups.

ANALYSES AND CONCLUSIONS

1. Is there any one correct way to group these fossils? Explain.
2. On what similarities were your groupings based?
3. Are there different ways to group these fossils? What are some other ways?
4. What information about the fossils helped you in grouping them?
5. How do you think scientists might group them?

APPLICATIONS

1. How could you use your data table to help you classify any fossils you might find in the field?
2. How would you use your data table to develop a classification key of fossil types?

TABLE 1: FOSSIL GROUPS			
Number	**Description**	**Group**	**Reasons**
1.			
2.			
3.			
4.			
5.			
6.			
7.			
8.			
9.			
10.			

SUMMARY

- Earth's history has two main divisions, or eons: the Cryptozoic Eon and the Phanerozoic Eon. (16.1)

- The Cryptozoic Eon is the time interval from the formation of Earth to about 590 million years ago. (16.1)

- The Cryptozoic Eon is divided into three eras: the Hadean, the Archean, and the Proterozoic. (16.1)

- Most of the earth's crust was formed during the Hadean Era. Dense basalt formed the oceanic crust while the lighter granite formed the continental crust. (16.2)

- The fundamental chemical reactions necessary to living cells were developed by bacteria during the Archean Era. (16.3)

- Multicellular animals evolved during a glaciation that took place in the Proterozoic Era. (16.4)

- The Phanerozoic Eon is divided into the Paleozoic, Mesozoic, and Cenozoic eras. (16.5)

- Trilobites were marine animals that lived during the Paleozoic Era. (16.5)

- Great forests developed during the Carboniferous Period. (16.5)

- Reptiles were the first animals to deposit their eggs on dry land. (16.5)

- Dinosaurs became the dominant animals during the Mesozoic Era. (16.6)

- A major extinction of plants and animals occurred at the end of the Mesozoic Era. (16.6)

- *Homo sapiens* evolved about 0.125 million years ago. (16.7)

Write all answers on a separate sheet of paper.

SCIENCE TERMS

Correctly use each of the following terms in a sentence.

cellular respiration **(346)**

fermentation **(344)**

glaciation **(354)**

planetesimals **(342)**

SCIENCE QUIZ

Modified True-False

Mark each statement *true* or *false*. If a statement is false, change the underlined term to make the statement true.

1. The Cryptozoic Eon <u>ended</u> 4.6 billion years ago.

2. The earth's original crust was made of <u>silicate</u> rock.

3. Trilobites looked a lot like <u>mammals</u>.

4. The last <u>era</u> of the Paleozoic is the Permian.

5. <u>Stromatolites</u> were formed from iron oxides and silica.

Multiple Choice

Write the letter of the choice that best answers the question or completes the statement.

6. The earth's primitive atmosphere contained
   a) water vapor.
   b) carbon dioxide.
   c) sulfur dioxide.
   d) All of the above are correct.

7. The time between the formation of the solar system and the age of Earth's oldest known rocks is called the
   a) Hadean Era. b) Proterozoic Era.
   c) Archean Era. d) Cryptozoic Eon.

continues

8. Photosynthesis produces
   a) carbon dioxide. b) water vapor.
   c) oxygen. d) lipids.

9. When did the earliest protozoa appear on Earth?
   a) about 3.8 billion years ago
   b) about 900 million years ago
   c) about 1.5 billion years ago
   d) about 590 million years ago

10. A major extinction occurred at the end of the
    a) Paleozoic Era. b) Cenozoic Era.
    c) Hadean Era. d) Archean Era.

Completion

Complete each statement by supplying the correct term.

11. Multicellular animals appeared during the _____ Era.

12. Photosynthesis requires a magnesium-containing molecule called _____.

13. Amino acids link together to form _____.

14. Trilobites lived during the _____.

15. The Mesozoic Era is sometimes called the age of _____.

Short Answer

16. Give one characteristic feature of each era of the Cryptozoic Eon.

17. Describe the formation of stromatolites.

18. Name one dominant form of animal life for each of the eras of the Phanerozoic Eon.

Writing Critically

19. Compare and contrast two theories of geologic history that try to explain the extinctions at the end of the Mesozoic Era.

20. Explain why modern animals could not have evolved before green algae and plants.

EXTENSION

1. Research the process of fermentation and the types of single-celled organisms that use it to make energy. What are the chemical products of fermentation? Do these products have any uses?

2. Obtain information on the work conducted in the early 1950s by Harold C. Urey at the University of Chicago. In what ways did his research help explain how life might have started on our planet?

APPLICATION/CRITICAL THINKING

1. Radioactive dating methods reveal that the oldest known rocks on Earth are about 3.8 billion years old. The same methods have shown that some rocks brought back from the moon are 4.6 billion years old. What might account for this difference?

2. Describe how glaciation might cause living organisms to evolve.

3. The following statements summarize the geologic history of the earth during the Phanerozoic Eon. However, some of the events are not in the correct sequence. Copy the statements on another sheet of paper, and put them in the correct sequence.

   a. Shallow seas cover the land.
   b. Fish appear, and simple animals inhabit the seas.
   c. Dinosaurs are dominant, and mammals begin to evolve.
   d. Ice covers large parts of the continents, and hominids evolve.
   e. Great forests cover the land.
   f. Trilobites inhabit the seas.
   g. Reptiles evolve protected eggs.
   h. Reptiles and amphibians dominate.
   i. Many sea creatures, including trilobites, become extinct.
   j. Dinosaurs become extinct, and mammals dominate.
   k. Corals and sponges dominate the seas, and simple land plants evolve.
   l. Birds, mammals, and dinosaurs evolve.

FOR FURTHER READING

Gore, R. "The Planets: Between Fire and Ice." *National Geographic* 167 (January 1985): 4. This article contains colorful illustrations of how the solar system might have looked during its creation.

Halliday, T. R., and K. Adler. *The Encyclopedia of Reptiles and Amphibians.* New York: Facts on File, 1986. This book tells almost everything you could ever want to know about snakes, lizards, frogs, turtles, and related animals.

Leakey, R., and A. Walker. "Homo Erectus Unearthed." *National Geographic* 168 (November 1985): 624. Famous archaeologist Richard Leakey gives evidence that humans' most recent ancestor was around over 1 600 000 years ago.

Sattler, H. R. *Pterosaurs, The Flying Reptiles.* New York: Lothrop, Lee & Shepard, 1985. The author describes pterosaurs, which were flying reptiles that ruled the skies during the age of the dinosaurs.

Wolkomir, R. "The Wizard of Ooze." *Omni* 7 (January 1985): 48. The work of micro-biological evolutionist Lynn Margulis is explored in this picturesque display of creatures whose ancestors may have lived in Earth's primordial ooze.

Challenge Your Thinking

This is a picture of a manatee, sometimes called a *sea cow.* Manatees and elephants are believed to have had a common ancestor. What kinds of environmental changes might have occurred that resulted in the development of two such different-looking animals?

Earth's Future

What will the future really be like? No one can be certain, but it is possible to imagine because geological processes, such as mountain building and erosion, remain constant. What is less certain, however, is the future of human life-styles. If humans use all the mineral resources of the earth, the life-styles of the future will be very different from those of today.

Spaceship Earth, EPCOT Center, Florida

1 Managing Natural Resources

SECTION OBJECTIVES

After completing this section, you should be able to:

- **Distinguish between** renewable and nonrenewable resources.
- **Explain** the importance of water, air, and land, and **describe** how pollution can endanger their usefulness.
- **Discuss** the recycling of natural resources.

NEW SCIENCE TERMS

pollution
smog
pesticides
recycle
conservation

17.1 Natural Resources

When you think of natural resources, you probably think of coal, oil, iron, and various minerals. While it is true that these are important resources, the most important natural resources are usually taken for granted. These resources include forests, water, air, land, and soil. These resources are vast, and at least partly *renewable*. With careful management they may be used over and over. However, waste and misuse of these resources seriously threaten their futures. Forests can be replanted, but water, air, and land abuse must be controlled to ensure that these resources will serve the future needs of humans.

Other natural resources, including metals, minerals, and fossil fuels, are *nonrenewable*. Nonrenewable resources are limited in quantity. When they are used up, they cannot be replaced in the foreseeable future. Even though these resources cannot be replaced, there are ways to extend their use. However, as with the renewable resources, waste and misuse will shorten the time of their usefulness.

Figure 17–1. This rusting scrap metal (left) represents a waste of precious resources as does air, water, and land pollution, which wastes the natural beauty of the earth (right).

○ *What are renewable and nonrenewable resources?*

Figure 17–2. Many places on Earth have too little water. In Holland, however, there is too little land. The solution is to reclaim land from the North Sea. Dikes are built (left) and the seawater is pumped out using the energy of the wind (right).

17.2 **Water**

Nearly three-fourths of the earth's surface is covered by water. Through the water cycle, this resource is purified and returned to the land. Where precipitation is limited, humans have brought in water through elaborate irrigation systems. In some places, such as Holland, there is not enough land, so a series of dikes and levees is used to reclaim land from the sea.

Plants and animals, including humans, require fresh water for survival. Unfortunately, 97 percent of the world's water is salty. Much of the fresh water is unavailable for use because it is frozen in glaciers or trapped deep underground. Surface water supplies most of the water needed for industry, homes, and irrigation.

In the United States, about 300 L of water are used every day by every person for drinking, cooking, and sanitation. If industrial uses are included, about 5000 L of water per day per person are used.

With so many people using the relatively limited supply of surface water, much of it must be reused. For example, it is estimated that by the time the water in the Mahoning River flows from its origin in central Ohio to the city of Youngstown—a distance of less than 200 km—it has been used 10 times.

Unfortunately, in many areas, surface water is polluted. **Pollution** is the unwanted dirt and waste that fouls water, air, and land. Surface water is polluted by untreated organic wastes, sediments, chemicals, and heat.

Organic waste, or sewage, comes from untreated human and animal wastes. The wastes of food-processing plants, including blood, hair, feathers, and bones, also pollute. These wastes require oxygen as they decay, and the loss of oxygen may kill

fish and other organisms. This type of pollution can be decreased if organic wastes are removed by sewage treatment plants. These plants break down the waste before it is released into the water.

Sediments, although natural in surface water, may be increased by the construction of roads and buildings, or by runoff from bare soil or hillsides stripped of their protective cover of plants. Sediments cover water plants, preventing photosynthesis, and clog the gills of fish and other animals. The amount of sediments can be reduced by allowing the water to stand in settling ponds. As the sediments fall to the bottom, the water becomes clear again.

Chemical pollutants are usually of two different types. Plant nutrients come from fertilizers and detergents. Toxic, or poisonous, chemicals come from industrial processes. Excess nutrients cause an increase in algal growth. This sudden population explosion, called a *bloom*, creates problems for fish living in the water. When algae die, much of the oxygen in the water is used in the process of decay. The resulting lack of oxygen causes large fish-kills. This is especially true in summer because warm water cannot hold much dissolved oxygen.

Toxic chemicals include acids, lye, metals, cyanide, and even radioactive elements. These are dangerous to plants and animals that live in or use the water. Chemicals can be removed by settling or by special types of treatment plants. Some water may be reused within a factory, thus retaining the chemical pollution and reducing the need for water resources.

Figure 17–3. A modern sewage treatment plant (left) cleans waste water by aeration (center) and by filtration (right).

Figure 17–4. Even seemingly clean water (bottom) can be deadly to fish and other organisms if it contains dangerous chemicals. Industrial wastes can be removed by filtration and sedimentation (below).

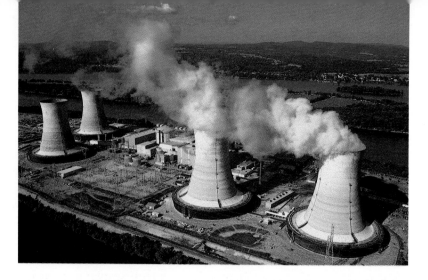

Figure 17–5. Thermal pollution can be eliminated through the use of cooling towers (right). In a cooling tower, excess heat is removed by the evaporation of water (left).

Excess heat, called *thermal pollution,* is created by industrial processes such as the manufacture of steel. Power plants, which use large quantities of water to produce steam to drive electrical generators, also produce thermal pollution. Heated water may kill some organisms directly. Heated water also decreases the amount of oxygen in the water, causing more fish to die. Water can be cooled in evaporation towers or retained in ponds before being discharged into streams or lakes.

The quality of water has been gradually improving since the mid-1970s. However, more must be done to ensure a clean, sufficient supply for the future.

○ *What is pollution?*
○ *Name four sources of water pollution.*

ACTIVITY: Chlorinating Water

How can laundry bleach be used to purify water?

MATERIALS (per group of 3 or 4)

binocular microscope, Petri dish, pond water (containing microscopic organisms), laundry bleach (containing 5 percent sodium hypochlorite), small beaker, medicine dropper

PROCEDURE

1. Place the Petri dish on the microscope stage; then pour the pond water into the dish and observe the microorganisms. Draw what you observe.
2. Your teacher will put a few milliliters of laundry bleach, containing a 5 percent solution of sodium hypochlorite, into the beaker.

3. Use the medicine dropper to add the bleach to the Petri dish, one drop at a time, while observing the solution's effect on the movement of the organisms.

CONCLUSIONS/APPLICATIONS

1. What happened to the microorganisms as the bleach was added?
2. What other steps might be taken to make sure a water sample is safe to drink?

17.3 Air

Unlike water, air is evenly distributed over the earth. Except for high elevations where the atmosphere is thin, there is plenty of air for all organisms. However, there are no barriers to hold pollution in one place. Local and planetary winds constantly mix the air. Moreover, any pollution in the atmosphere tends to remain there until it is washed out by precipitation.

There have always been natural sources of air pollution, such as volcanic eruptions, forest and grass fires, and wind-blown soil or dust. In fact, some natural pollution is necessary for the formation of condensation nuclei for rain and snow. However, as you may recall from Chapter 8, too much smoke and carbon dioxide can lead to a greenhouse effect. If the worldwide temperatures were to rise only a few degrees, the polar icecaps would melt, and many coastal areas would be flooded by rising seas.

Pollution caused by human activity has a much more immediate effect on the atmosphere. The burning of fossil fuels has increased the amount of carbon dioxide in the atmosphere. The burning of fossil fuels can also produce a type of air pollution known as smog. **Smog** is a chemical fog produced by the reaction between sunlight and pollutants. Smog is particularly serious in places like Los Angeles and Denver, where it is sunny nearly every day and where thousands of automobiles produce tons of pollution every day. Smog is also potentially dangerous to human health. Since the 1940s, smog is thought to have caused the deaths of thousands of people, both in big cities and small towns.

Figure 17–6. Natural air pollution includes sea salt, dust, volcanic ash, and smoke from forest fires (above).

Figure 17–7. In Los Angeles (left) pollution from automobiles is converted to smog by a chemical reaction started by the sun (right).

Figure 17–8. Acid rain can cause damage to statues (left) and to trees (right).

There is also evidence that the burning of fossil fuels is responsible for producing acid rain. The sulfur dioxide released by the burning of some coal and oil becomes sulfuric acid (H_2SO_4) when it combines with moisture in the air. The acid precipitation that results damages buildings, plants, and even fish. Even the ancient Sphinx in Egypt, which has withstood natural weathering for over 5000 years, is now in danger of being destroyed by acid rain.

As has happened with water, the quality of air has been slowly improving since the mid-1970s. Enforcement of laws by state, federal, and international environmental agencies have reduced some sources of pollution. Modern equipment, such as air scrubbers for industry and catalytic converters for automobiles, has also helped reduce pollution. However, continual improvement is needed to keep up with the pollution resulting from an increase in numbers of cars and factories.

○ *What is smog?*
○ *What causes acid rain?*

Figure 17–9. Some large, industrial cities still suffer from pollution caused by automobiles and factories.

368

17.4 Land and Soil

Dry land covers less than one-third of the earth's surface, or about 15 billion hectares. Of that, only about 2 billion hectares can be farmed. The rest is either too rocky, or the climate is too wet, too dry, or too cold for farming. In the United States and Canada, the same land is used for crops each year; the nutrients used by the plants are replaced by chemical fertilizers. So efficient are farmers in the United States and Canada that only about one person in ten is involved in farming. These farmers produce enough food to more than meet the needs of all their countries' people.

Figure 17–10. In many industrialized countries, modern equipment is important to food production (left). In underdeveloped countries, farming depends more heavily on the labor of humans and animals (right).

The United States, Canada, Australia, and Argentina are the great food exporting nations of the world. In most other countries, and especially in Asia, Africa, and Central America, not enough food can be produced to meet the people's needs. Many underdeveloped countries cannot afford the technology to produce large crops on the same land each year. When good crops can no longer be produced, people abandon the land, so new land must be found each year to produce food. The abandoned land is left without a protective layer of plants, and more soil is lost to wind and water erosion.

Figure 17–11. In countries such as the United States, Canada, and Australia, surplus food is sometimes given away (left), or stored until it is needed (right).

369

Figure 17–12. The survival of species such as the bald eagle (above) was once threatened by the use of a pesticide called *DDT*.

In countries with advanced technologies, improper use of chemicals can poison the soil and make it unfit for growing crops. **Pesticides,** chemicals used to kill crop-destroying insects, can accumulate in the soil. There, the pesticides can become a hazard to animals other than the insects for which they were intended. In the late 1960s, a pesticide called *DDT* had to be banned from general use because it threatened some species of birds, such as the brown pelican and the bald eagle, with extinction.

Another important use of land and soil is for the grazing of animals. Land used for range or pasture may not be fertile enough for crops, or it may be too hilly for farming. Range land may either be planted with special grasses for feeding cattle, sheep, or goats, or the animals may graze on native plants. Sometimes, especially during dry periods, the land may be overgrazed; then the plants die, and the soil may be eroded by wind and water.

Figure 17–13. In some areas, overgrazing by cattle has destroyed the grasses covering the soil (right). Much erosion can be prevented if trees are planted to hold the soil. Millions of tree seedlings, ready for transplanting on barren ground, are grown at tree farms (below).

Forests also occupy large areas of marginally fertile land. This renewable resource provides lumber for buildings, fuel for stoves or for making charcoal, and pulpwood for paper products. However, many forests are threatened as more land is used for urban development and the expansion of industry.

○ *State three uses for land and soil.*
○ *Name three of the world's major food-producing countries.*

370

17.5 Mineral Resources

How many nonrenewable resources have you used this week? Have you had a drink from a metal can and thrown the can away? Have you thrown away a plastic cup? If you have, you have wasted mineral resources. You may recall from earlier in this chapter that part of the definition of nonrenewable resources is that once they are used up, they are gone—forever. Metal deposits and fossil fuels take millions of years to develop but only a few seconds to be destroyed.

Although nothing can be done to replace the resources that have been used up, it is possible to extend the time of usefulness for those that remain through recycling. To **recycle** means to use something again. There are many examples of recycling in nature. You may recall the discussion of the water cycle in Chapter 9. All the water on Earth goes through a series of processes that eventually return it to the sea. The water cycle is an example of natural recycling.

Figure 17–14. The carbon-oxygen cycle (left) is an example of natural recycling. Many metals and other mineral resources can be recycled as well (below).

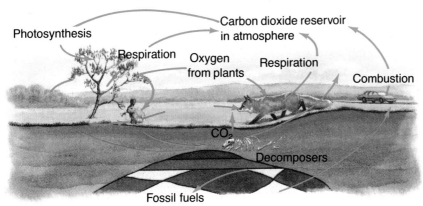

Many elements and compounds are also recycled in nature. For example, water and carbon dioxide are taken in by green plants and, through the process of photosynthesis, become carbohydrates. At the same time, oxygen is released into the atmosphere. Animals use the oxygen and the carbohydrates in cellular respiration. As these are used to provide energy for animals, water and carbon dioxide are released into the environment. This completes the cycle—the elements have been recycled and are ready to be used again.

Some nonrenewable resources may also be recycled. For example, the aluminum in beverage cans can easily be recycled to produce new beverage cans. Many aluminum companies will pay a few cents per kilogram for recycled aluminum. Making cans from recycled aluminum is much cheaper than producing cans from aluminum ore. In addition to saving money, recycling cans saves aluminum ore and energy resources. Melting aluminum cans uses far less energy than refining ore.

Figure 17–15. Abandoned cars (left) and old newspapers (right) can be recycled.

You would probably never think of just throwing away gold jewelry or silver coins. Like gold and silver, most metals should be recycled. Old cars can be melted and their iron used to produce new steel, thus saving much iron ore and energy.

Many nonmetals can also be recycled. Glass can be ground up, melted, and formed into new products. Although trees are renewable resources, many can be saved for other uses by recycling newspapers and cardboard.

Recycling is not the only way to save nonrenewable resources. With some resources, such as fossil fuels, recycling is impractical if not impossible. Sometimes, however, conservation of resources can extend their use. **Conservation** means using materials wisely. Energy resources can be conserved by careful use. Energy resources can also be extended by using alternatives to fossil fuels. Some of these alternative energy resources will be discussed in the next section.

○ *What is recycling?*
○ *Name three advantages of recycling aluminum cans.*

Figure 17–16. Riding a bicycle to work or to school is a good way to conserve energy resources.

A MATTER OF FACT

Three metric tons of copper ore are required to provide enough copper (2.5 kg) for the wiring of a single automobile.

Section Review

READING CRITICALLY

1. Explain why wood is a renewable resource.
2. How can conservation increase the availability of a nonrenewable resource?

THINKING CRITICALLY

3. How does pollution threaten the quality of air, water, and soil?
4. If certain pesticides are dangerous to birds and other animals, why are they not completely banned from use?

SKILL ACTIVITY: Analyzing Bar Graphs

BACKGROUND

Bar graphs are very useful for presenting information that must be quickly and accurately compared. Seeing information presented in a bar graph often makes the information easier to remember than reading the information in print. When using a bar graph, the first task is to determine what information is being presented, and then to determine what each bar on the graph represents.

PROCEDURE

1. Look for the title of the graph. The title tells you what information is being presented in the graph.
2. Look at the bottom of the graph. There you will find the names of the items that are being compared.

3. Look at the side of the graph. There you will find the quantities that are being compared.
4. The colored bars on the graph help to visually present the comparisons.

APPLICATION

1. What is being compared in this bar graph?
2. What units are used for this comparison?
3. Which area consumes the most energy? How much energy is used there?
4. Which area consumes the least energy? What is the difference between the most energy consumed and the least energy consumed?

USING WHAT YOU HAVE LEARNED

Make a bar graph using Table 17–1 on page 374. Be sure to include a key showing the items that are being compared.

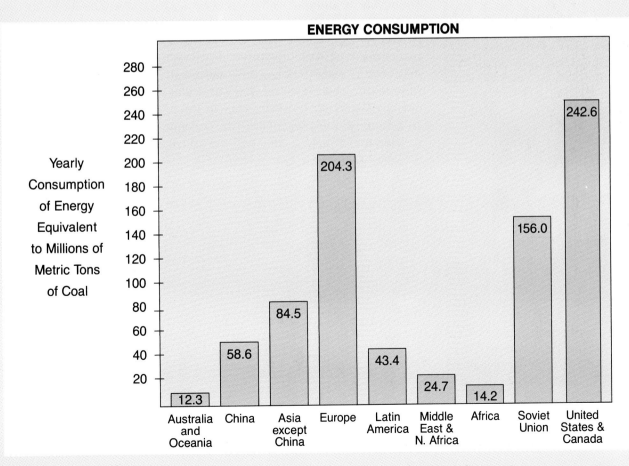

ENERGY CONSUMPTION

Yearly Consumption of Energy Equivalent to Millions of Metric Tons of Coal

Australia and Oceania: 12.3
China: 58.6
Asia except China: 84.5
Europe: 204.3
Latin America: 43.4
Middle East & N. Africa: 24.7
Africa: 14.2
Soviet Union: 156.0
United States & Canada: 242.6

② Managing Energy Resources

NEW SCIENCE TERMS

hydroelectric power
tidal power
geothermal power
solar power
solar cell
biomass
fission
fusion energy

SECTION OBJECTIVES

After completing this section, you should be able to:
- **Explain** the difference between renewable and nonrenewable energy sources.
- **List** five energy sources that may be used in the future.
- **Discuss** the advantages of fusion energy over fission energy.

17.6 Nonrenewable Energy Sources

Most scientists agree that in the future, industry, transportation, and residential needs will be supplied almost exclusively by electrical power. At present, however, most electricity is produced from nonrenewable energy sources, such as fossil fuels and uranium. These nonrenewable energy resources are quickly being exhausted. The resources that took 500 million years to accumulate are being exhausted by humans in just a few decades.

The amounts of known world reserves of fuels are compared with the present yearly consumption in Table 17–1. The table shows that the reserves of natural gas and coal are far from being exhausted; however, this is not true for petroleum and uranium. At the present rate of consumption, there is enough petroleum for only 28 years of use and enough uranium for only 40 years.

Figure 17–17. Oil shale (below), the rock that burns, could provide an alternative source of petroleum (inset).

TABLE 17–1: WORLD RESERVES OF FUELS VERSUS YEARLY CONSUMPTION			
Fuel	**Known World Reserves**	**Yearly Consumption**	**Yearly Consumption as a Percentage of Reserves**
Petroleum	72.0×10^9 tons	2.5×10^9 tons	3.5 percent
Natural gas	5.0×10^{13} m³	1.5×10^9 m³	0.003 percent
Coal	7.5×10^{12} tons	3.7×10^6 tons	0.00005 percent
Uranium	1.5×10^6 tons	38 000 tons	2.5 percent

The problem is even worse for the United States. At the present rate of consumption, the United States' reserves of petroleum probably will be exhausted in 15 years and those of uranium in 18 years.

When these sources are used up, they will be gone forever. In the past, geologists have been able to discover new fuel reserves to balance those consumed. If humans are to continue present life styles, new, inexhaustible sources of energy must become available before these reserves run out.

○ *Which fuel has the largest reserves?*
○ *How soon will the world run out of petroleum?*

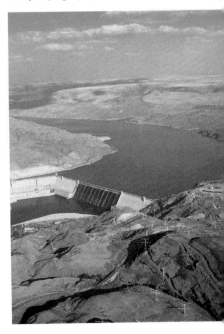

Figure 17–18. Coal is the main fuel used to produce electricity (above). Many coal mines are abandoned (left) as they are mined out.

17.7 **Renewable Energy Sources**

Renewable energy sources are those that are produced from inexhaustible supplies, such as the sun. Solar, wind, hydroelectric, and ocean tidal energy are all examples of renewable resources. At the present time, hydroelectric energy is the only renewable resource contributing significantly to the world's power production.

Figure 17–19. Hydroelectric power is clean and inexpensive, but its use is limited to places with large water supplies and steep topography.

Hydroelectric Power   Power produced from the energy of falling water is called **hydroelectric power.** Before falling water was used to produce electricity, water power was used for such functions as turning millstones to grind grain. In the production of electricity, the falling water turns turbines. Hydroelectric power is clean and nonpolluting. However, it requires a natural waterfall or a large river where a dam can be built. Unfortunately, the places where dams can be easily and inexpensively built are few. One of the largest hydroelectric power plants in North America is the Grand Coulee Dam in Washington state.

Wind Power   In some areas of the world, it is possible to use wind as an energy source to produce power. Wind power has been used for thousands of years to lift water and to grind grains. In areas where the wind is strong and constant, wind-driven electrical generators are practical. A windmill is connected to the generator's shaft, and the electricity produced is

Figure 17-20. Wind turbines (left) can provide power for rural areas, while large wind-turbine "farms" (right) can provide power for entire cities.

used directly or is used to charge storage batteries. The electricity produced is limited and is most practical in rural areas where it is not profitable to distribute commercially generated electricity. However, in California there are many wind farms, where wind turbines produce significant amounts of electricity for commercial use.

Tidal Power   Along some coastlines, in France for instance, the ocean tides rush up bays that extend many kilometers inland. At the mouth of the bay, the tide can be directed through barriers with openings leading to turbines, producing electric energy. At high tide, the water returns to the ocean through the same turbines. This electric energy is produced by tidal power. **Tidal power** is produced by the energy of changing tides.

A French power station at the mouth of the Rance River produces 150 megawatts (MW) of electricity. A huge tidal power project under construction in northeastern Canada is discussed in the Science Connection on pages 386–387.

A huge tidal power project under construction in northeastern Canada is discussed in the Science Connection on pages 386–387.

A MATTER OF FACT

Volcanic eruptions and lightning flashes are not useful as sources of power because they are too uncertain.

Figure 17-21. Tidal power may produce much electricity for coastal cities in the future.

Geothermal Power

In some regions, such as in Italy, New Zealand, and Iceland, hot igneous rocks produce usable energy within the earth. Rainwater penetrates porous rocks near the heat source and is converted to steam. The steam may surface through natural vents, or it can be extracted by drilling. The most famous steam field is at Larderello in Tuscany, Italy. Steam from Larderello is used to drive turbines that produce electricity. This steam power from the earth is called **geothermal power.** Geothermal power accounts for only 0.4 percent of the power used in Italy, but in New Zealand geothermal power accounts for 5 percent of the power produced.

Hot water and steam from geothermal fields are also used as a source of direct heat for buildings. The entire city of Reykjavik, Iceland, is heated in this way.

Solar Power

On a cold day, you have probably tried to stay in the sun to keep warm. The power produced by energy from the sun, called **solar power,** can be used directly as a source of heat or to produce electricity. The most common use of direct solar energy is for heating water. An array of dark-colored pipes placed on the roof of a building will provide hot water for personal use, and it may provide some heat for the building as well. Why would the pipes need to be dark in color?

Figure 17–22. Geothermal energy is widely used in Iceland (right). In California some electricity is produced from steam fields such as the one shown here (left).

Figure 17–23. Solar energy can be collected passively to heat water (right) or entire homes (left).

Figure 17–24. Solar energy can provide electricity directly for a variety of uses.

A **solar cell** can produce electricity from sunlight. A solar cell, such as those shown in Figure 17–24, includes a junction between two semiconductors. Photons falling on the junction produce electricity.

One disadvantage of solar power is the size of the solar cells needed to produce large amounts of electricity. A solar power plant with a capacity of 100 MW would cover a surface of 4 km^2. A solar power plant might better be constructed in space. There, the panels of solar cells would be free from clouds and air pollution, with no limit to the number of panels that could be displayed. In addition, the assembly could be made to face the sun at all times. The power generated could be transmitted to Earth by means of microwave beams.

Solar energy already has been used extensively to power spacecrafts. Panels of solar cells keep batteries charged for radio and data transmission from space.

Biomass   Another renewable energy source is **biomass,** or once-living matter. Wood and plant waste can be burned as fuel. In the United States much of the plant material is first converted to methane or alcohol, but in many countries the waste is burned directly.

About 10 percent of the energy consumed in the world comes from biomass. The use of biomass as a source of energy may not be apparent in industrialized nations, but in underdeveloped countries biomass is a major source of energy.

○ *Name five sources of renewable energy.*
○ *What are solar cells?*
○ *What is biomass?*

Figure 17–25. In some countries peat is cut, stacked, and dried so it can be used as fuel.

17.8 Nuclear Energy

You may recall from Chapter 15 that some atoms, such as those of uranium, tend to break apart due to natural radioactive decay. Some radioactive isotopes, however, can be forced to decay. When the nuclei of these atoms are bombarded with neutrons, the atoms undergo fission. **Fission** is the splitting of the nucleus of one atom into smaller nuclei. Fission releases a tremendous amount of energy, which can be used to produce steam to drive turbines. There are, however, some serious problems with fission power. Besides the possibility of accidental releases of radioactivity, such as occurred at Chernobyl, USSR, in 1986, the world supply of uranium for fission is very limited.

There is a type of nuclear energy that is potentially so abundant as to be considered truly inexhaustible. This energy, called **fusion energy,** is produced when small atomic nuclei fuse to form new atoms. In a fusion reaction, two isotopes of hydrogen, deuterium (doo TIHR ee uhm) and tritium (TRIHT ee uhm), fuse to form helium. This is the same process that produces the energy of the sun.

The main advantage of using fusion energy as compared to fission energy is that no dangerous radioactive isotopes are produced. The reaction produces only harmless helium that diffuses into the atmosphere and escapes into space. Also, the deuterium and tritium for the reaction can be obtained easily from sea water.

Figure 17–26. Although inexpensive and plentiful, there are some dangers to using nuclear energy to generate electricity, as the accident at Chernobyl (left) showed.

Figure 17–27. When construction of this fusion reactor is completed, it may help make sea water a source of abundant electric power.

The main disadvantage of fusion energy is the very high temperature needed for the fusion reaction to take place. While fissioning of uranium can be started at room temperature, a temperature of 100 million °C is needed to initiate a fusion reaction. No known matter can withstand temperatures this high, so an energy field must be developed to contain the reaction. Scientists in the United States are working hard on these problems because fusion holds the hope of providing an infinite amount of energy for the future.

○ *What is the difference between fission and fusion?*
○ *What is the main problem in producing fusion energy?*

Section Review

READING CRITICALLY

1. What are the major sources of energy used to produce electrical power?
2. What limits the expansion of hydroelectric power?
3. How is fusion energy produced?

THINKING CRITICALLY

4. Why is it important to develop energy alternatives to fossil fuels and uranium?
5. How could coal be used to extend the reserves of petroleum and natural gas?

TECHNOLOGY: Fusion Power

In the past few decades humans have begun to tap the energy of the atom. Vast amounts of energy are released when the nuclei of atoms are split (fission) or fused (fusion).

The reality of this energy source was demonstrated with the explosion of the first atomic bomb in 1945. The uncontrolled chain reaction of splitting atoms produced a violent explosion. Since then nuclear scientists have learned how to control the rate at which atoms split. Today, fission reactors are used in many parts of the world to generate electricity.

Scientists are now trying to control nuclear fusion reactions in the same way they can control fission reactions. Controlled fusion can produce much more energy with less pollution than fission reactions. The problem with fusion is that a great amount of heat is necessary to start the reactions. For example, a tem-

Reactor control room

perature in excess of 100 million°C is needed to start a fusion reaction.

However, some scientists are currently working on a method of producing fusion reactions at very low temperatures. Experiments have shown that hydrogen fusion can take place at temperatures as low as 13 K (−260°C) or at room temperature.

The key to the success of these low-temperature reactions lies in the development of a process called *muon-catalyzed fusion,* or cold fusion. In order for the positively charged nuclei of deuterium to fuse, they must be brought close together. To overcome their natural electrical repulsion, most experimental techniques use high temperatures. However, muon particles make that unnecessary.

A *muon particle* is a subatomic particle that is about one-fifth the size of a proton, but with a negative electrical charge. Be-

Producing a fusion reaction

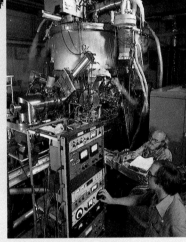

Testing the equipment

cause it has the same charge as an electron, it can replace an electron in a molecule of deuterium. The added mass of the muon causes the nuclei of the deuterium atoms to be drawn close together. When the nuclei of the atoms are close to one another, they can be more easily fused. If this process of cold fusion can be perfected, a relatively inexpensive alternative to nuclear fission could be producing electricity in the future.

Section 2   Managing Energy Resources   **381**

INVESTIGATION 17: Generating Electricity

PURPOSE

To construct a working model of an electric generator

MATERIALS (per group of 3 or 4)

Bar magnet
Insulated bell wire (10 m)
Electrical tape
Plastic straw
Compass
Wooden block (5 cm × 10 cm × 10 cm)
Thumbtacks (4)
Pencil
String
Hole punch

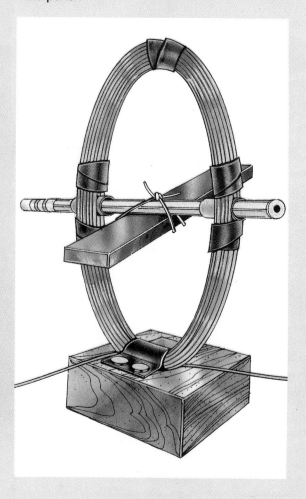

PROCEDURE

1. Place the bar magnet on the table, and prepare an oval coil of wire (about 30 turns) so that the bar magnet will fit into it lengthwise.

2. Use electrical tape wrapped at several places as shown to prevent the coil from unwinding.

3. Cut two small pieces (5 cm) from a plastic straw. Insert a piece of straw into each long side of the coil and secure it with electrical tape as shown.

4. Cut a piece of electrical tape and use it to mount the coil, with thumbtacks, to the wooden block as shown.

5. Insert a pencil through the holes in the coil, and use electrical tape or string to secure the bar magnet as shown.

6. Twist together the exposed ends of the wire coil.

7. Twirl the pencil to rotate the magnet while observing the behavior of a compass placed at varying positions around the closed loop.

ANALYSES AND CONCLUSIONS

1. Twirling the pencil is an example of mechanical energy. What other forms of mechanical energy could be used to rotate the magnet?

2. How did the compass behave when placed near the coil? Near the attached ends of the loop?

3. Draw a diagram of a generator in which the coil, and not the magnet, is rotated.

APPLICATION

Connect the ends of the wire coil to a galvanometer or a small flashlight bulb. Explain what happens.

SUMMARY

- Partly renewable natural resources include forests, water, air, land, and soil. (17.1)

- Metals and fossil fuels are nonrenewable resources. (17.1)

- In the United States, each person uses about 300 L of water daily. (17.2)

- Much water is polluted by organic waste, sediments, chemicals, and heat. (17.2)

- Polluted water can be treated, and the quality of water has improved since the mid-1970s. (17.2)

- Smog is a chemical fog produced by the reaction of sunlight and air pollutants. (17.3)

- Acid rain affects buildings, plants, and even fish. (17.3)

- Only a small percentage of the earth's land surface is suitable for farming. The rest is either too rocky or has a climate too severe for growing crops. (17.4)

- Chemical fertilizers improve the soil in countries with advanced technologies, but some chemical pesticides may be harmful to more than pests. (17.4)

- The use of nonrenewable resources may be extended by recycling. (17.5)

- Where it is not practical to recycle, conservation of resources may extend their usefulness. (17.5)

- Nonrenewable energy sources, such as fossil fuels and uranium, are quickly being exhausted. (17.6)

- Inexhaustible energy alternatives for the future may include hydroelectric power, wind power, tidal power, geothermal power, solar power, and biomass. (17.7)

- Nuclear energy from fission may be limited by the small reserve of uranium and by the danger of radioactivity. (17.8)

Write all answers on a separate sheet of paper.

SCIENCE TERMS

Correctly use each of the following terms in a sentence.

biomass **(379)**
conservation **(372)**
fission **(379)**
fusion energy **(379)**
geothermal power **(377)**
hydroelectric power **(375)**
pesticides **(370)**
pollution **(364)**
recycle **(371)**
smog **(367)**
solar cell **(378)**
solar power **(377)**
tidal power **(376)**

SCIENCE QUIZ

Modified True-False

Mark each statement *true* or *false*. If a statement is false, change the underlined word to make the statement true.

1. Metals and fossil fuels are <u>nonrenewable</u> resources.

2. Water is recycled for repeated use through the <u>purification</u> cycle.

3. In the United States, a single person uses about <u>20 liters</u> of water every day.

4. Another name for chemical fog is <u>acid rain</u>.

5. Without conservation, petroleum reserves in the United States will be exhausted in <u>15 years</u>.

6. Hydroelectric power comes from <u>falling water</u>.

continues

Multiple Choice

Write the letter of the choice that best answers the question or completes the sentence.

7. Which of the following is not a renewable resource?
   a) air
   b) water
   c) land
   d) coal

8. Which country exports the least food to other nations of the world?
   a) Soviet Union
   b) United States
   c) Canada
   d) Australia

9. The resource consumed most per year by percentage of its remaining reserves is
   a) uranium.
   b) natural gas.
   c) petroleum.
   d) coal.

10. Using the tides to generate electricity would be an example of _____ power.
    a) geothermal
    b) tidal
    c) biomass
    d) lunar

11. The resource that could be used to provide fuel to generate electricity using nuclear fusion is
    a) uranium.
    b) coal.
    c) water.
    d) All of the choices are correct.

12. Which of the following energy sources could be considered truly renewable?
    a) nuclear fusion
    b) hydroelectric power
    c) geothermal power
    d) biomass

Completion

Complete each statement by supplying the correct term or phrase.

13. Nuclear _____ produces less radiation than nuclear fission.

14. In underdeveloped nations, _____ is a major source of energy.

15. Much of the city of Reykjavik, Iceland, is heated by _____.

16. Fossil fuels required more than _____ years to accumulate.

17. Unwanted and unnatural chemicals found in the environment are called _____.

18. Natural waterfalls can sometimes be used as sources of _____ power.

Short Answer

19. Explain the difference between renewable and nonrenewable resources.

20. Explain why forests are considered less renewable than air or water.

21. In addition to fossil fuels, list four alternative sources of power.

22. Describe the two ways that solar energy may be used by humans.

Writing Critically

23. How could air pollution affect the renewability of forests?

24. Discuss the multiple uses of land and soil.

25. When electricity is generated by burning fossil fuels, much waste heat is produced. This waste heat could be used to heat buildings in a process called *cogeneration*. Describe some benefits and some possible drawbacks to using cogeneration as an alternative energy source.

EXTENSION

1. Go to the library and read about alternatives to chemical pesticides. Write a brief report describing the effort of scientists to use biological forms of pest control.

2. From the electric meter for your home, find out how much electricity is used each day by your family. Your electric power company can send you a booklet on how to read the meter. Then look around your home for ways by which electricity could be saved, such as turning off lights. After locating several areas of waste and correcting the problems, check again to see how much electricity has been saved. Report to the class on your success.

APPLICATION/CRITICAL THINKING

1. Compare and contrast the processes of nuclear fission and fusion. Because of the nuclear reactions involved, fusion would be a considerably cleaner form of energy production. Referring to the different reactions, explain why this is so.

2. Coal can be converted to a gaseous fuel by a chemical process, and the coal gas can be used instead of natural gas. What environmental problems might develop if coal gas is used as a replacement for natural gas?

3. What types of common items could be recycled that are not presently being recycled? Why do you think they are not now being recycled?

FOR FURTHER READING

Caulfield, C. *In the Rainforest.* Chicago: University of Chicago Press, 1986. This book describes the rain forests of Africa, South and Central America, and Asia. The reasons for the destruction of these fragile ecosystems are discussed.

League of Women Voters Education Fund. *The Nuclear Waste Primer.* New York: Nick Lyons Books, 1985. This book offers the nonexpert a brief introduction to nuclear wastes.

McKay, A. *The Making of the Atomic Age.* London: University of Oxford Press, 1984. This comprehensive text follows the development of the nuclear age from the discovery of the electron to the industrial development of nuclear power.

Challenge Your Thinking

This is a photo of a nuclear breeder reactor in France. A breeder reactor makes electricity and at the same time makes plutonium fuel, which can also be used to produce more electricity. This eliminates somewhat the dependence on nonrenewable uranium as a source of fuel. The problem is that the plutonium is very dangerous to living organisms and produces wastes that take hundreds of thousands of years to decay. Discuss the advantages and disadvantages of using breeder reactors as an alternative way of generating electricity.

No one knows for certain why the dinosaurs suddenly disappeared 65 million years ago. Fossils provide few clues about the cause of the extinction.

Scientists are currently debating a startling new theory. Some scientists now believe that a giant meteor or comet struck the earth near the end of the Mesozoic Era. The meteorite would probably have exploded as it struck the earth, causing global climatic changes. The sky would have darkened as ash clogged the air, blocking the sun's rays and lowering the temperature. Millions of animals would have died from the cold or lack of food, and many species would have become extinct. The animals that survived became the evolutionary ancestors of present life forms on Earth. Some scientists believe that catastrophes like this occur every 26 million years. They say that the catastrophes may be caused by an unknown planet or star passing through the solar system and carrying along many comets. Each time Earth is visited by this alien body, there is a rain of comets, and millions of creatures die.

Luis Alvarez developed a new theory after observing some sedimentary rock formations in Italy. Sediments in central Italy contain one of the best records of the boundary between the Cretaceous Period of the Mesozoic Era and the Tertiary Period of the Cenozoic Era. Alvarez discovered a curious layer of clay between the rocks of these two periods. This clay had a high concentration of the element iridium, which is very rare in the earth's crust. "Why," asked Alvarez, "would the iridium level in this layer of clay be 25 times higher than normal?" Alvarez hypothesized that a meteorite slammed into the earth about 65 million years ago. Alvarez based his hypothesis on the fact that extra-terrestrial bodies sometimes contain 1000 times more iridium than does the earth's crust.

Alvarez reasoned that iridium-rich dust from the collision spread through the atmosphere. The dust was deposited as the layer of sediment that separates the rocks of the Cretaceous Period from those of the Tertiary Period.

Volcanic ash darkens the sky

Iridium-rich layer (marked by coin)

Looking for fossil clues

Jurassic Period dinosaur

Alvarez believed that this gigantic collision and the extinction of the dinosaurs happened at about the same time. First, he calculated how much dust a meteorite collision would have thrown into the air, and then he determined how much sunlight would have been blocked by the dust. Under these conditions, he said, plants and marine organisms would have died almost immediately. Soon large animals such as the dinosaurs would have had nothing to eat, and would have died too. Eventually the meteorite dust would have fallen to Earth like black snow, burying more than half of Earth's animal species.

The scientific community debates these theories as it searches for new evidence. At the same time, scientists in other fields have been inspired by the Alvarez research. Some scientists have theorized that a nuclear war might cause the same kind of global winter as the Cretaceous meteorite. Evolutionary biologists, who once thought that dinosaurs had died out because of their small brains, now realize that there may have been other reasons. Astronomers are searching for the cause of the catastrophe. Was the meteor jolted into Earth's path by some unknown star or mysterious planet? Does a catastrophe like this happen every 26 million years, as some fossil records suggest?

The answers to these questions may affect the way humans view themselves. Evolutionary biology encourages humans to believe that they survive because they are the most intelligent species on Earth. Could it be that without the intervention of a meteorite from space, dinosaurs might still be ruling the earth? The answers to this and other questions lie some-where in Earth's history, waiting perhaps for you to find them.

Alvarez in his lab

OCEANOGRAPHY

The sun, the moon, and the ocean combine to create one of Earth's most reliable forces—tides. Pulled by the gravity of the sun and the moon, tides flow in and out of bays and inlets. Engineers dream of harnessing tidal power to generate inexpensive electricity. On the Rance River in France the dream is a reality. In Canada the dream is rapidly becoming a reality. Canadian engineers are planning a giant tidal power project on Canada's Bay of Fundy, which has the highest tides in the world.

- How can tides be used to produce electricity?
- How are tides, waves, and currents produced?
- What causes the ocean to be salty?
- What important resources does the ocean hold?

By reading the chapters in this unit, you will learn the answers to these questions. You will also develop an understanding of concepts that will allow you to answer many of your own questions about the oceans of the earth.

Moonrise over a bay

18

The Water Planet

Water covers nearly three fourths of the earth's surface. Yet humans have explored more of the surface of the moon than they have of the floor of the ocean. Unique in the solar system, Earth is truly the water planet. The ocean links the continents and provides many of the raw materials needed for human survival. The final frontier may not be space; it may be the oceans.

Sunset over the Pacific Ocean

1 Bodies of Water

SECTION OBJECTIVES

After completing this section, you should be able to:

- **Locate** the oceans of the earth, and **identify** the differences between oceans and seas.
- **Describe** some characteristics of the ocean floor, and **compare** the floors of the Pacific and the Atlantic.
- **Discuss** the importance of large bodies of fresh water on the earth.

NEW SCIENCE TERMS

oceans
seas
sonar
continental shelf
continental slope
lakes

18.1 Oceans and Seas

The surface of the Earth is actually one huge ocean with large islands, or continents, scattered here and there. The ocean is so large that you could sail on it for weeks and not see any land. The ocean is also very deep, reaching depths of more than 10 km in some places. From space the ocean looks calm, but if you walk along a beach, you can see that the ocean is constantly in motion.

If the earth were completely smooth, its surface would be covered evenly with water to a depth of over 2.5 km. Only about 60 percent of the Northern Hemisphere is covered with water since most of the continents are found there. About 81 percent of the Southern Hemisphere is covered with water.

Oceans are major expanses of salt water with surface areas larger than 10 million km². Even though the earth has only one large ocean, most geographers usually divide it into four oceans.

Figure 18–1. Before the twentieth century, the oceans influenced nearly all aspects of human society. Cities were built near the shore and the oceans provided a transportation link between countries. Humans also depended on the oceans for salt and for much of their food.

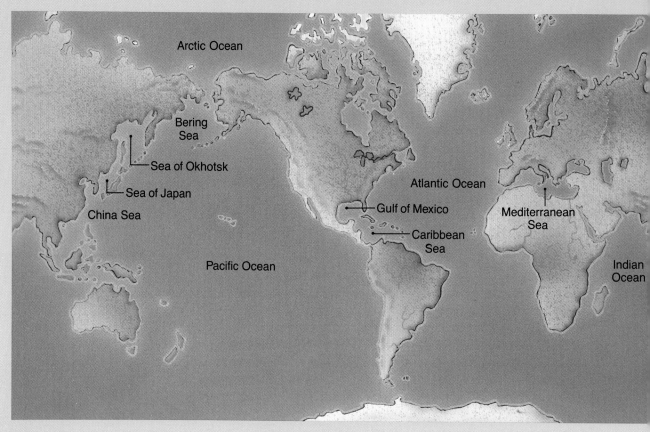

Figure 18–2. This map shows the location of the earth's oceans and major seas.

The area around the continent of Antarctica is sometimes called the *Antarctic* or *Southern Ocean,* but as the map in Figure 18–2 shows, the Southern Ocean is really just the southernmost portions of the Atlantic, Pacific, and Indian oceans. Figure 18–2 also shows many of the earth's seas. **Seas** are expanses of salt water less than 3 million km² in area. Seas are partially enclosed by land and are therefore somewhat separated from the oceans.

TABLE 18–1: THE WORLD'S OCEANS		
Ocean	**Area ($\times 10^6$ km²)**	**Average depth (m)**
Pacific	166.2	4188
Atlantic	86.6	3736
Indian	73.4	3872
Arctic	12.2	1117

TABLE 18–2: THE LARGEST SEAS		
Sea	Area ($\times 10^6$ km^2)	Average depth (m)
China Sea	2.8	1437
Caribbean Sea	2.5	2575
Mediterranean Sea	2.5	1502
Bering Sea	2.3	1492
Gulf of Mexico	1.5	1614
Sea of Okhotsk	1.4	973
Sea of Japan	1.0	1667

The Caribbean Sea and the seas off eastern Asia are separated from the open ocean by island arcs. Sea water passes freely between the islands, making the saltiness of these seas close to that of the oceans. Other seas, such as the Black Sea and the Baltic Sea, receive a great deal of fresh water from rivers that flow into them. Therefore, their saltiness is lower than that of the oceans. The Mediterranean Sea, on the other hand, joins the ocean only through the narrow, shallow Strait of Gibraltar. The saltiness of the Mediterranean is higher than that of the open ocean because of a high evaporation rate.

Six million years ago, the passage between the Atlantic and the Mediterranean Sea closed due to the movement of Africa and the lowering of the water level in the oceans. Over the next thousand years, the Mediterranean nearly dried up; only in the eastern Mediterranean did some small salt lakes remain. These lakes were fed by water spilling down from the Black Sea and from the Nile River. The rest of the Mediterranean was a dry sea bed, covered with a layer of salt 50 m thick. The Mediterranean remained dry for several thousand years, and then the ocean level rose again. This cycle was repeated many times over the next million years. Each time the ocean level rose, water poured over the Strait of Gibraltar, forming a great system of rapids and waterfalls that quickly refilled the Mediterranean.

○ *Name the four oceans.*
○ *How is a sea different from an ocean?*

Figure 18–3. The famous Strait of Gibraltar (below) separates the Atlantic Ocean from the Mediterranean Sea (left).

18.2 The Ocean Floor

While the surface of continents is easy to study, studying the ocean floor presents unique problems. Scientists began to map the ocean floor in the 1940s using sonar, which stands for *SO*und *NA*vigation *R*anging. **Sonar** is the process of bouncing sound waves off a solid surface. A sonar instrument sends a series of sound waves through the water. These sound waves strike the ocean floor and bounce back to the device. This is similar to what happens when you hear your voice echo in an empty room or in a mountain canyon.

Just as you can guess the size of a dark room or unlit cave by how fast your echo returns, the depth of the ocean floor can be determined by the time it takes for the sonar echo to return. From this information, a picture of the ocean floor, such as the one in Figure 18–5, can be produced.

Figure 18–4. Sonar operators receive echoes from the ocean floor. Oceanographers use sonar to chart the depth of the ocean.

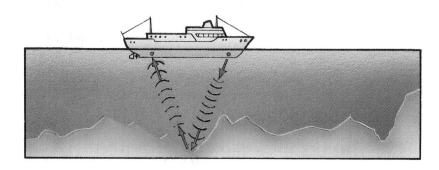

Sonar readings show that the North American continent extends for several kilometers into the Atlantic Ocean. This extension forms a gradually sloping bottom called the **continental shelf.** Farther offshore, the continental shelf is replaced by a steeper slope called the **continental slope.** The North American continental slope forms one side of the Atlantic Ocean basin. What do you think forms the other side of the Atlantic Ocean basin?

Figure 18–5. This sonar screen (below) shows a profile of a small section of the ocean floor diagrammed at right.

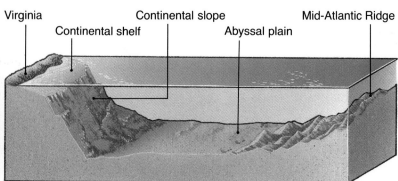

Virginia   Continental slope   Mid-Atlantic Ridge

Continental shelf   Abyssal plain

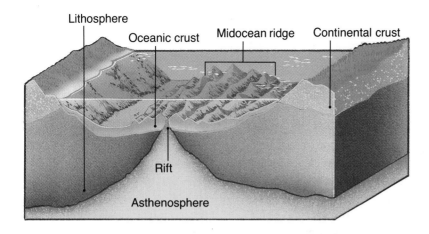

Lithosphere
Oceanic crust
Midocean ridge
Continental crust
Rift
Asthenosphere

Figure 18–6. Along the Mid-Atlantic Ridge (left) the crust is spreading with the addition of magma from below the crust. Venting gases, rich in chemicals, provide nutrients for the growth of many special organisms (right).

Nearly in the middle of the Atlantic Ocean is a long underwater mountain range called the Mid-Atlantic Ridge. The mountains of this range are parallel to the edges of the continents on either side of the ocean. This mountain range is 14 400 km long and is part of the midocean ridge system. Parts of this midocean ridge are found in the other oceans as well.

Recent studies of the Mid-Atlantic Ridge show that a deep valley occurs along the ridge's center. A cross section of this ridge and valley is shown in Figure 18–6. These mountains do not form a continuous line; they are separated by cracks along the chain, where many small earthquakes occur. Lava escapes from the mantle, forming new crust on the ocean floor. Dating has shown that the rocks nearest the center of the ridge are the youngest; older rocks are found farther away from the center, on either side of the ridge.

The rocks on either side of the ridge show another interesting pattern. The basalt of the ocean crust contains magnetic minerals that act like compass needles. These minerals usually line up with the earth's magnetic field before the rock solidifies. Geologists discovered, however, that not all the minerals point toward the magnetic north pole. In some places, the minerals are magnetized so that they point toward the magnetic south pole. Earth's magnetic field reverses periodically. When these minerals formed, the earth's magnetic field must have been reversed.

Magnetic mapping of the ocean floor along the ridges shows a pattern like the one shown in Figure 18–7. One strip of rock is magnetized toward the present North Pole, and the next strip is magnetized toward the South Pole. There are matching patterns on opposite sides of the ridge. The ocean floor is spreading from the center of the ridge. As the ocean floor spreads, it forces the continents on either side of the Atlantic farther apart. This evidence was used to support the theory of continental drift discussed in Chapter 6.

A MATTER OF FACT

The cooling and shrinking of the upper layer of the earth can pull down an island by more than a kilometer in less than a million years.

Figure 18–7. A record of the earth's magnetic reversals can be found in the rocks on both sides of the Mid-Atlantic Ridge.

North

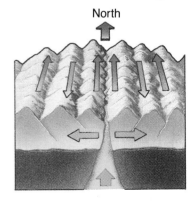

Figure 18–8. The ocean floor has recently been explored with the use of small submersible vessles (left). These vessels are transported to the research site by a support ship (right).

Figure 18–9. The floor of the Pacific Ocean is different from the floor of the Atlantic Ocean. The continental edges are much steeper, there is no midocean ridge, and there are many deep trenches.

The Pacific Ocean is the largest and deepest of all the oceans, covering nearly one third of the earth's surface. The floor of the Pacific is quite different from that of the Atlantic. Instead of gently sloping continental shelves, deep trenches occur all along the edges of the Pacific Ocean where the oceanic lithosphere is subducted under the continents. Trenches such as these are rare in the Atlantic and other oceans. These trenches trap much of the sediment that flows in rivers from the continents. Perhaps the Pacific Ocean is so deep because it is not flooded with these sediments.

Although there is no mountain ridge in the middle of the Pacific, an extension of the midocean ridge forms the Eastern Pacific Rise. Other chains are made up mostly of volcanoes, and in some places they rise above the surface, forming islands such as the Hawaiian and Aleutian islands.

○ *What is the midocean ridge?*
○ *Where are most ocean trenches located?*

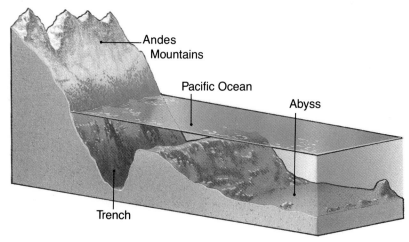

SKILL ACTIVITY: Charting the Ocean Floor

BACKGROUND

In the past few decades, oceanographers have learned more about the floor of the ocean than at any other time in history. Long ago, seafarers measured the depth of the shallow seas by throwing overboard a mass of lead tied to a rope. The rope had knots tied in it at regular intervals. They lowered the lead until it touched bottom. Then, they pulled it back on board and counted the knots of the rope that had gone into the water. If they ran out of rope, the water's depth in that area remained unknown.

PROCEDURE

Today, scientists use *echo sounding*, a form of sonar, to measure the depth of the ocean. Since scientists know that sound travels through water at about 1500 m/s, depth can be calculated by using the following formula.

$$\frac{s \times t}{2} = d$$

Here s is the speed of sound, 1500 m/s; t is the time it takes for the sound produced by the scientist to leave the ship and echo from the bottom; and d is depth of the water. Calculate the depth of the water if the time is 2 seconds.

APPLICATION

1. Prepare a table like the one shown.

2. Calculate and record the depth to the ocean bottom for each recorded sounding, using the given formula.

3. Use a piece of graph paper to draw the ocean bottom. Plot ocean depth on the vertical axis and distance from origin on the horizontal axis. Connect the plotted dots with smooth lines to diagram the features (i.e., mountains, valleys, and plains) of the ocean bottom.

TABLE 1: CHARTING THE DEPTH OF THE OCEAN

Sounding	Distance from Origin (km)	Echo-Sounding Time (seconds)	Calculated Ocean Depth (m)
1	2	1.2	
2	4	2.0	
3	6	3.6	
4	8	4.8	
5	10	5.3	
6	12	2.3	
7	14	3.1	
8	16	4.5	
9	18	5.0	
10	20	5.1	
11	22	4.9	
12	24	4.8	
13	26	6.7	
14	28	3.6	
15	30	3.0	

USING WHAT YOU HAVE LEARNED

How could you use soundings to help locate a deep channel where you could safely sail your boat?

18.3 Lakes and Rivers

Large bodies of fresh or salt water completely surrounded by land are called **lakes.** Lakes receive water from rivers and streams. Some bodies of water that are called seas are actually large lakes. For example, the Caspian and Aral seas in Asia are really lakes, even though they have salty water.

The Caspian Sea is the world's largest lake. The Caspian has an area of over 370 000 km^2 and an average depth of more than 1000 m. The largest freshwater lake is Lake Superior, on the border between the United States and Canada. Lake Superior, with an area of nearly 85 000 km^2 and an average depth of about 400 m, is much smaller than the Caspian Sea.

Figure 18–10. The Great Lakes (right) form a northern coast for the United States that stretches nearly halfway across the continent. The shores of the lakes are similar in many ways to an ocean shore (above).

The North American chain of the five Great Lakes, which includes Lake Superior, contains 20 percent of the world's available fresh water. These lakes, shown in Figure 18–10, were formed about 12 000 years ago at the end of the last glaciation. All the lakes are connected by natural and artificial channels, and they empty into the North Atlantic through the Saint Lawrence River. These channels form a 3058-km-long waterway called the *Saint Lawrence Seaway* which reaches nearly halfway across North America. The seaway allows for many ocean ports in inland cities, such as Toronto, Cleveland, Detroit, and Chicago.

Figure 18–11. From Lake Erie to the Atlantic Ocean, the elevation of the St. Lawrence Seaway drops by about 300 m.

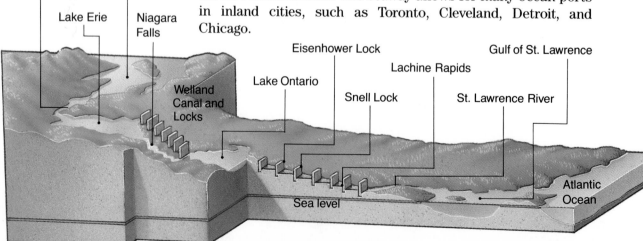

Rivers and lakes are very important to civilization. Besides being a source of fresh water, they provide a means of transportation within and between countries. Many early human settlements developed along the banks of rivers. The names, locations, and lengths of some of the world's major rivers are shown in Table 18–3.

Figure 18–12. Large ocean-going vessels (left) use the locks (right) of the St. Lawrence Seaway to reach inland ports.

TABLE 18–3: SOME MAJOR RIVERS OF THE WORLD		
River	**Location**	**Length (km)**
Nile	Africa	6695
Amazon	South America	6276
Chang Jiang	Asia	4989
Congo	Africa	4394
Mississippi	North America	3779
Missouri	North America	3726
Yukon	North America	3185

○ *What is a lake?*
○ *What is the largest saltwater lake and the largest freshwater lake on Earth?*

Section Review

READING CRITICALLY

1. Explain how sonar is used to map the ocean floor.
2. What happened when the water level in the ocean rose and the passage between the Atlantic and the Mediterranean reopened?

THINKING CRITICALLY

3. What process created the trenches in the Pacific Ocean?
4. Why could the oceans and the ocean floors be called the last frontier?

CAREERS

OCEANOGRAPHER

Oceanography is the study of the oceans. An *oceanographer* is a scientist who studies the oceans. Oceanographers are concerned with the interaction between the atmosphere and the waters of the world, the formation and features of the ocean floor and its sediments, and the influence of tidal action and weather on coasts and estuaries.

An oceanographer must study many branches of science, including geology, meteorology, biology, chemistry, mathematics, and phys-ics. Usually oceanographers have a master's degree or a doctorate.

Oceanographers conduct many experiments at surface research stations and at submerged ocean-ographic stations. They gather data that help them to better understand the oceans and the seas.

For Additional Information
American Oceanographic
　Organization
P.O. Box 2249
Springfield, VA 22152

MARINE CONSERVATIONIST

Bird watchers, marine biologists, recreational boaters, waterfront property owners, and many other citizens are concerned about the health and vitality of the oceans.

A *marine conservationist* helps all of us stay aware of the condition of our marine environment. Because the oceans and water-ways of the world are so dynamic, marine conservation is a challenging and exciting pursuit. A bachelor's degree or higher is usually necessary, because a marine conservationist must keep abreast of the current research reported by oceanographers in order to understand how the oceans are affected by both natural and human pressures.

For Additional Information
National Coalition of Marine
　Conservationists
P.O. Box 23298
Savannah, GA 31403

UNDERWATER PHOTOGRAPHER

Underwater photography is both an enjoyable and difficult task. A professional *underwater photographer* must be well trained in two very important skills: scuba diving and photography.

Underwater photographers not only have to contend with the sometimes hazardous conditions of the water but also must know how the water affects the quality of their work. Sophisticated equipment is used to account for the dif-ferent wavelengths of light that are absorbed at different ocean depths. Many types of filters and film must be used to produce quality images. Only a high-school diploma is needed to be an underwater photographer.

For Additional Information
Underwater Photographic
　Society
P.O. Box 2401
Culver City, CA 90231

SECTION OBJECTIVES

After completing this section, you should be able to:

- **Describe** the two main processes that shape coastlines.
- **Identify** some of the common coastline features and **explain** how they are formed.
- **Analyze** the problem of coastline erosion and **evaluate** efforts being made to preserve coastlines.

NEW SCIENCE TERMS

primary coast
fiord
bay
secondary coast

18.4 Coast Development

Have you ever placed a seashell or a cup against one ear and heard what sounded like the ocean? If you have been to the seashore, you know that the sound you hear in the shell is similar to the sound of the ocean meeting the land. Areas where the ocean meets the land are very special places called *coasts*. Coastal areas are in a state of constant change as they are shaped by erosional agents such as wind, glaciers, running water, and waves. Some coasts have wide, flat, sandy beaches, while others have rocky shores or jagged cliffs.

Coastal areas can be classified into two types. The first type, called a *primary coast*, is formed by erosional processes of the land. The second type, called a *secondary coast*, is formed by erosional and depositional processes of the sea.

Figure 18–13. Erosional agents of the land cause the features found along primary coasts (top left and top right). Secondary coasts (bottom left and bottom right) are eroded by the sea.

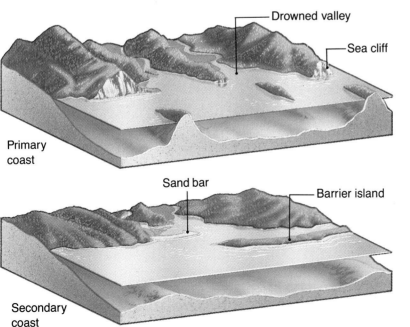

Drowned valley

Sea cliff

Primary coast

Sand bar

Barrier island

Secondary coast

Figure 18–14. Chesapeake Bay (left) formed as a drowned river valley. Where glacial valleys are flooded by the ocean, deep channels called *fiords* form (right).

A **primary coast** often results from erosional processes of the land and changes in sea level. For instance, the sea level was lower during the last glaciation because so much water was trapped in glaciers. When the glaciers melted, the sea level rose, covering, or drowning, many old river valleys and glacier valleys. The coasts of these drowned valleys show characteristics of stream and glacial erosion, rather than erosion by waves. The coasts of Norway, Greenland, New England, and Alaska formed this way. When the sea level rose, water entered many glacier valleys, forming fiords (fee OHRDZ). A **fiord** is a long, deep, narrow, U-shaped sea channel that was originally formed by ice.

The sea has also traveled up the mouths of many old river valleys, forming bays and estuaries. A **bay** is an indentation of the sea, partially enclosed by land. If the bay is a submerged river valley, it is called an *estuary*. The Chesapeake and Delaware bays and New York Harbor are all examples of estuaries.

The drowned valley of the Hudson River slopes into the sea near New York City, forming part of the harbor. This is a very deep valley, so the waters several kilometers upstream are a mixture of fresh and salt water. This mixture of waters, called *brackish waters*, are important nursery areas for many kinds of ocean fish and invertebrates. The living resources of the sea are discussed in Chapter 20.

Figure 18–15. The estuaries that occur along much of the east coast (right) are probably the most productive waters in North America. Many estuaries, such as New York Harbor (below), are less productive because of pollution.

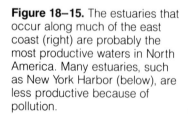

Figure 18–16. A strong sea breeze and plentiful sand have created many large dunes along the Atlantic coast, such as Jockey Ridge on North Carolina's Outer Banks.

Winds also help shape primary coasts. A dune line forms as wind-blown sediments accumulate. This dune line may migrate inland if winds blow constantly offshore. For example, the dunes along the Outer Banks of North Carolina are moving westward because of the prevailing sea breezes from the Atlantic Ocean.

Forces from within the earth may also shape some coasts. Faults may raise or lower parts of the land or create fractures, which are easily eroded. Much of the California coast is on the San Andreas fault or branch faults. The sea erodes inland along the faults, forming bays such as Tomales Bay near San Francisco.

A coast that is shaped by erosional or depositional processes of the sea is a **secondary coast.** Secondary coasts are usually so modified by the sea that the primary-coast features, such as river and glacier valleys, are completely gone. Even the hardest rock is undercut and broken into fragments by the pounding of waves. If the rocks are uniformly resistant to wave erosion, a straight coast will form. If the rock formations along a coast differ in resistance, the waves form an irregular coast. In these cases, headlands of hard rock jut into the sea, and areas of softer rock erode rapidly, forming sandy coves and inlets.

Erosion by the sea creates some very distinctive coastal features. As waves continue to erode headlands, some of the rocks become isolated from the shore. These form islands called *sea stacks.*

DISCOVER

Settling Sediments

To a jar half filled with water, add a handful of sediments of various sizes (soil, sand, and gravel). Stir the mixture and slowly pour it into one end of a shallow pan of water. Observe the settling of the sediments in the pan. Which sediments settled closest to the point where they entered the pan? Which sediments settled farthest away? How is this similar to the settling of sediments at the point where a stream enters a lake or the ocean?

Figure 18–17. Along the west coast of the United States, there are many spectacular arches and seastacks.

Have you ever been to a beach and seen people many meters offshore, standing in shallow water? Often sand washed away from a beach is deposited a short distance offshore. These offshore deposits, which parallel the coast, are called *sand bars.* Sometimes sand bars curve away from a beach, forming attached deposits called *sandspits.* If the sand bars build up enough to break the surface of the water, they form *barrier islands.* There are many large barrier islands along the Atlantic and Gulf coasts from Fire Island, New York, to Miami Beach, Florida, to Padre Island, Texas. Why do you think that there are no barrier islands on the Pacific coast?

○ *How are primary coasts formed?*
○ *How are secondary coasts formed?*

Figure 18–18. The Atlantic and Gulf coasts of the United States, from New York (above), around Florida, and to the Texas border with Mexico (right), are protected by barrier islands.

Figure 18–19. Groins, jetties (top), seawalls, and breakwaters (bottom) are constructed to provide protection against the eroding action of waves.

18.5 Preserving Coasts

Coastal areas are considered prime recreational and development areas. As a result, their erosion is of major concern. Various structures have been designed to protect beach and coastal areas. *Groins,* vertical walls of rock placed perpendicular to the shore, are designed to trap sand carried by currents. Groins do succeed in trapping sand on one side, but the erosion on the opposite side is usually increased.

Seawalls are another type of structure designed to provide protection from erosion. Seawalls are usually made of rocks or concrete and are placed parallel to the shore. Seawalls are placed between the water and the shore to protect the shore from erosion. However, waves sometimes wash over and destroy seawalls by eroding away their support.

Rocky projections built to extend out into the sea are called *breakwaters* and *jetties.* They are usually built to protect harbors and beaches from the effects of storms. The water behind breakwaters and jetties usually remains calm. However, jetties collect soil and sand deposition. As a result, the next beach downshore from a jetty receives no sand and soon erodes.

Figure 18–20. Natural vegetation is the best protection against erosion of coasts. The addition of fences, however, sometimes helps hold the sand until the plants become established.

Designing coastal structures that correct one problem without creating other problems is very difficult. Each coast is different and presents its own unique set of problems. The best protection is to plant vegetation along the shore and to restrict building close to the sea.

○ *What is the major concern to coasts?*
○ *What are groins, seawalls, and jetties?*

Section Review

READING CRITICALLY

1. What is the difference between a primary coast and a secondary coast?
2. What new problems do the building of groins and jetties produce?

THINKING CRITICALLY

3. How would planting vegetation along a coast prevent erosion?
4. When do coastlines stop changing? Explain your answer.

ACTIVITY: Taking Core Samples

How can you take core samples of near-shore sediments?

MATERIALS (per group of 3 or 4)

sediments, soda straw, marking pen, razor blade

PROCEDURE

1. Use the sediments prepared for the Discover on page 396, or follow the directions there for making sediments. Gently push a soda straw into the sediments until it reaches the bottom.
2. Mark the top of the layer of sediments on the straw with the marking pen.

3. **CAUTION: Razor blades are very sharp; use extreme caution.** Carefully cut the straw and core lengthwise with the razor blade.
4. Make a drawing of your core sample. Be sure to label the top and bottom of the sample, and indicate each layer of sediments.

CONCLUSIONS/APPLICATIONS

1. How many layers are there in your core sample?
2. Which sediments were deposited first? Which sediments were deposited last? Explain your answers.
3. If your core sample was from the deep ocean instead of near-shore, how would the layering of the sediments be different?

INVESTIGATION 18: Wave Action

PURPOSE

To show how the action of waves erodes coasts

MATERIALS (per group of 3 or 4)

Coarse soil or sand
Large aquarium
Bucket
Water
Large beaker, 1500 mL
Metric ruler

PROCEDURE

1. Pour coarse soil or sand into the aquarium to a depth of 1 or 2 cm. This will serve as the floor of the "ocean."
2. Add more sand to one side of the tank. This pile will represent a coast, and should be 6 or 7 cm high. Make the coast slope gently to the "ocean floor."
3. Fill a bucket with water; then use a large beaker to fill the ocean side of the tank with about 4 or 5 cm of water.
4. Place a ruler on edge, at the bottom of the tank, on the side away from the coast. Gently move the ruler back and forth to create waves. Try to keep the wavelengths constant, and note the effect of the waves on the coast.
5. Repeat the motions several times, creating waves of varying wavelength. Note the effect

of the different wavelengths on the erosion of the coast. Compare your results with those of other groups in the class.
6. Use the beaker to empty the aquarium, returning the water to the bucket. Reshape the coast, duplicating as much as possible the original features. Then refill the ocean.
7. Place the ruler flat on the bottom this time and move it up and down, creating waves of different heights. Note the effect of waves of different heights on the coast, and compare your results with the results of other groups.

ANALYSES AND CONCLUSIONS

1. How did waves of varying wavelengths affect the coast?
2. How did waves of different heights affect the coast?
3. What had the most effect on the coast, waves of varying wavelengths or of different heights? Explain your answer.

APPLICATION

How could engineers use a coastal model, such as the one you constructed, to show the effects of waves on various coastal construction projects?

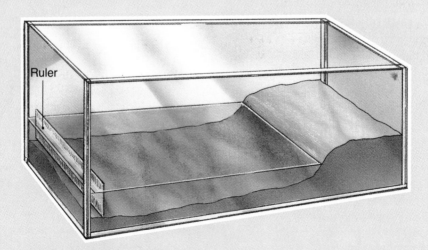

Ruler

SUMMARY

- The earth has one large ocean that is divided into four oceans and numerous seas. (18.1)

- Seas may be less salty than the oceans if they receive a large amount of fresh water, or they may be saltier than the oceans if there is a high rate of evaporation. (18.1)

- The detailed study of the ocean floor began in the 1940s after the invention of sonar. (18.2)

- Between the continents and the Atlantic Ocean floor are the gently sloping continental shelf and the steeper continental slope. (18.2)

- In the middle of the Atlantic Ocean is an underwater mountain range called the Mid-Atlantic Ridge. (18.2)

- The floor of the Pacific Ocean has many deep trenches and volcanic mountains. (18.2)

- Lakes are large bodies of water completely surrounded by land. (18.3)

- Rivers and lakes are important as sources of fresh water and as means of transportation. (18.3)

- Coasts are areas where the ocean meets the land. Coasts may be classified as primary or secondary. (18.4)

- Primary coasts show erosion by land agents, such as glaciers and running water. Secondary coasts show much erosion by waves and wind. (18.4)

- Drowned valleys are common along primary coasts, while secondary coasts have sand bars and barrier islands. (18.4)

- Groins, seawalls, jetties, and breakwaters are all used to prevent erosion of coasts. (18.5)

Write all answers on a separate sheet of paper.

SCIENCE TERMS

Correctly use each of the following terms in a sentence.

bay **(402)**
continental shelf **(394)**
continental slope **(394)**
fiord **(402)**
lakes **(398)**
oceans **(391)**
primary coast **(402)**
seas **(392)**
secondary coast **(403)**
sonar **(394)**

SCIENCE QUIZ

Modified True-False

Mark each statement *true* or *false*. If a statement is false, change the underlined word to make the statement true.

1. The saltiness of the Mediterranean is the result of a high rate of <u>evaporation</u>.

2. The Pacific Ocean has nearly <u>twice</u> the area of the Atlantic Ocean.

3. The steep incline off the continental shelf is called the <u>ocean basin</u>.

4. The largest lake in the world is <u>Lake Superior</u>.

5. Barrier islands are built up from <u>sand bars</u>.

6. The system of rivers, canals, and lakes that allow ocean ships to reach cities such as Chicago is the <u>St. Lawrence Seaway</u>.

continues

Multiple Choice

Write the letter of the choice that best answers the question or completes the sentence.

7. The percentage of the earth's surface that is covered by water is about
   a) 25 percent.   b) 50 percent.
   b) 10 percent.   d) 75 percent.

8. Which ocean is the deepest?
   a) Pacific   b) Atlantic
   c) Indian   d) Arctic

9. Which is the largest sea?
   a) Sea of Japan
   b) Bering Sea
   c) Mediterranean Sea
   d) China Sea

10. What is the longest river in the world?
    a) Nile   b) Amazon
    c) Yangtze   d) Congo

11. Coasts can be reshaped by
    a) wind.
    b) glaciers.
    c) running water.
    d) All of the choices are correct.

12. Which type of feature is not found along secondary coasts?
    a) sand bars   b) barrier islands
    c) estuaries   d) sea stacks

Completion

Complete each statement by supplying the correct term or phrase.

13. The deepest portion of the Pacific Ocean is the _____.

14. The idea that the earth's magnetic poles reverse can be supported by the mapping of _____.

15. Coasts eroded by the action of the ocean are called _____.

16. A mixture of fresh and salt water is _____.

17. A body of water partially enclosed by land is called a _____.

18. The water around the continent of Antarctica is sometimes referred to as the Antarctic or _____ Ocean, but this water is really just part of the Atlantic, Pacific, and Indian oceans.

Short Answer

19. Explain the difference between a bay and a lake.

20. List four features of the ocean floor.

21. Describe four factors that affect the shape of coasts.

22. Explain how groins, jetties, and breakwaters, which are meant to protect coasts, may actually do more harm than good.

Writing Critically

23. Explain how the industrial use of rivers can pollute the oceans.

24. Compare and contrast some of the methods used to preserve coasts.

25. Why is it important to protect estuaries and other areas of brackish water?

EXTENSION

1. Use a map to determine the elevations of several major coastal cities. If the level of the sea were to rise 6 m, which cities would be flooded?

2. Read about the differences in the coasts of Maine and Florida, and prepare a poster that shows the different features and the processes that created them.

3. The level of Lake Superior is over 300 m above sea level, yet ocean vessels dock at ports along the lake. Find out about the locks along the Saint Lawrence Seaway, and report your findings to the class.

4. Early fur-trading routes followed nearly the same path as the St. Lawrence Seaway. Read about these routes, and report on their importance in the development of the interior of North America.

APPLICATION/CRITICAL THINKING

1. If one sound from an oceanographic vessel takes 1.6 seconds to echo back to the sounding instrument, a second sound takes 2.4 seconds, and a third sound takes 4.8 seconds, what might that section of the ocean bottom look like? If sound travels 1.5 km/s in ocean water, what are the differences in depth of the three locations?

2. Explain why the melting of glaciers might be responsible for changes in sea level in the future.

FOR FURTHER READING

Gilbert, S. "America Washing Away." *Science Digest* (August 1986). This article cites many examples of how coastal erosion is affecting homes and recreation areas. It clearly explains some of the factors involved in coastal erosion.

Gilbreath, A. *The Continental Shelf: An Underwater Frontier.* Minneapolis: Dillon, 1986. This book provides an overview of the geology and ecology of the continental shelf. All the species living on the shelf are discussed.

Glaser, M. *The Nature of the Seashore.* Fiskdale, Mass.: Knickerbocker, 1986. This field guide covers sand, wind, waves, and animal and plant life of the seashore.

Jeffery, D. "Maine's Working Coast." *National Geographic* 167 (February 1985): 209. This article tells a lively story of the pleasures and perils of living and working along the coast of Maine.

Challenge Your Thinking

From your knowledge of the processes that shape coasts, what process or processes were probably involved in creating these features? Is this a primary or a secondary coast? How can you tell? Where in North America might you find a coast such as this?

Ocean Waters

Over 70 percent of the earth is covered with salt water. If the water were to suddenly evaporate, the remaining deposits would be 13 meters thick. These deposits, however, would not be just salt. They would contain many of the minerals that have eroded from the land since the first torrential rains began filling the oceans billions of years ago.

Pacific Ocean "tube"

1 Characteristics of Ocean Waters

SECTION OBJECTIVES

After completing this section, you should be able to:

- **Name** the major chemicals dissolved in ocean water.
- **Explain** how heat is stored in sea water.
- **Describe** the factors that affect the density of ocean water.

NEW SCIENCE TERMS

salinity
thermohaline
thermocline

19.1 Composition of the Oceans

If you have ever been swimming in the ocean, you know that the water is salty. When the oceans first began filling, rivers carried countless tons of dissolved compounds from the land to the oceans. As a result, the oceans became salty. Almost all of the elements of the earth's crust can be found in today's ocean water. For example, 1 km^3 of ocean water contains over 10 kg of gold. Could this gold be extracted from the water? The answer is yes; however, the concentration of gold is only about 10 parts of gold per trillion parts of water. Extracting gold from the water would cost more than the gold is worth. Although it does not make good economic sense to "mine" gold from the ocean, other elements, such as bromine, magnesium, and salt, can be extracted economically from the ocean waters.

If 1 kg (about 1 L) of normal ocean water is allowed to evaporate, about 34.5 g of salts are left behind. The most abundant of these salts is sodium chloride (NaCl)—common table salt. Sodium chloride makes up 77.4 percent by weight of the total salts in ocean water. Table 19–1 lists some of the salts found in ocean water.

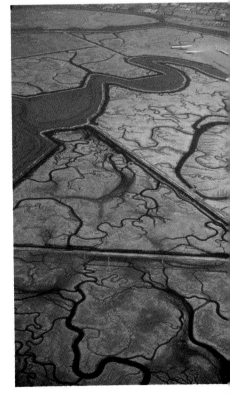

Figure 19–1. Probably the most important mineral resource humans get from the ocean is salt.

TABLE 19–1: SALTS DISSOLVED IN OCEAN WATER			
Name	**Chemical Formula**	**Concentration (g/kg)**	**Percent**
Sodium chloride	NaCl	26.70	77.4
Magnesium sulfate	$MgSO_4$	3.25	9.4
Magnesium chloride	$MgCl_2$	2.20	6.4
Calcium chloride	$CaCl_2$	1.10	3.2
Potassium chloride	KCl	0.70	2.0
Sodium bicarbonate	$NaHCO_3$	0.20	0.6
All others		0.35	1.0

Figure 19–2. The salinity of the North Atlantic Ocean is below average partly because of the addition of fresh water from melting icebergs.

Salinity is the measure of the amount of salts dissolved in ocean water. Salinity is usually expressed in grams per kilogram of sea water. The average salinity of the ocean is 34.5 g/kg. Areas where evaporation is high have higher-than-average salinity. Areas of high salinity include tropical oceans, the Mediterranean Sea, and the Red Sea. Areas where a lot of rain, melting ice, or river water enters the oceans have lower-than-average salinity. Examples of low-salinity areas are the Arctic Ocean, the Baltic Sea, and the Black Sea.

The water that flows from the land is called *fresh water.* Fresh water is not the same thing as pure water. Pure water contains water molecules and nothing else. Fresh water contains elements and compounds eroded from the rocks and soil by rainwater. Rainwater flowing across the land carries these chemicals to the oceans. Some of the chemicals stay dissolved in ocean water. Others precipitate out of the ocean water and accumulate on the ocean floor.

Of all the chemicals dissolved in Earth's waters, only silica and nitrate are more concentrated in fresh water than in ocean water. Silica and nitrate are not as concentrated in ocean water because they are used by ocean organisms. Silica, or silicon dioxide (SiO_2), is used by some organisms to make their shells. Nitrate (NO_3^-) is removed from sea water by ocean plants to make proteins.

There are two elements that occur in much greater quantities in ocean water than in fresh water—chlorine and sodium. These two elements stay dissolved in ocean water, making it "salty."

Although sodium and chlorine are continually added to ocean water, ocean salinity does not increase. Ocean water reached its present salinity more than a billion years ago. Excess sodium and chlorine precipitate out of ocean water and accumulate in ocean sediments. From there they are carried into the mantle along subduction zones. Both reappear at the surface as products of volcanoes or in minerals. There they are eroded by rain and returned to the sea, completing the cycle.

Figure 19–3. These organisms live near a volcanic vent along a midocean ridge. Their food chain is based on chemosynthesis rather than on photosynthesis, because no sunlight reaches this depth.

412

Ocean water also contains dissolved atmospheric gases. You may recall from Chapter 8 that most of the carbon dioxide on Earth is dissolved in the oceans; only a small fraction of Earth's carbon dioxide is in the atmosphere. The opposite is true of nitrogen and oxygen. Very little nitrogen and oxygen are dissolved in the ocean; most is in the atmosphere.

○ *What is the most abundant salt in ocean water?*
○ *What areas of the oceans have higher-than-average salinity?*
○ *What is fresh water?*

19.2 Temperature of the Oceans

Have you ever been to a desert where it is hot during the day and cold at night? Why do you think the temperature changes so much more in the desert than in other places? The reason is the lack of water in the desert. Large bodies of water help to moderate temperature changes.

If the earth had no oceans, the air temperature would rise above 100°C during the day and plunge below −100°C at night. In the polar regions, the temperature would drop to −200°C during the winter. Fortunately, the earth does have oceans that help to regulate the temperature. The oceans absorb a lot of solar energy when the sun is high and release that stored energy very slowly when the air is cold. Aided by the mixing effect of the planetary winds, this tends to keep the overall climate of the earth moderate.

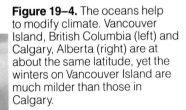

Figure 19–4. The oceans help to modify climate. Vancouver Island, British Columbia (left) and Calgary, Alberta (right) are at about the same latitude, yet the winters on Vancouver Island are much milder than those in Calgary.

The ability of sea water to absorb and release energy results from the unique structure of the water molecule and the dissolved salts in the water. The water molecule has a somewhat triangular shape. Because of covalent bonds, water molecules have a positive side and a negative side. This causes water molecules to bond together. A lot of heat is needed to break these bonds. In the oceans, weak bonds form between the positive ions in the water and the negative ends of the water molecules. Bonds also form between negative ions in the water and the positive ends of the water molecules. This bonding tends to keep the water molecules in ocean water from bonding to each other. As a result, ocean water freezes at a lower temperature than fresh water, that is, at $-1.872°C$ instead of $0°C$. Water that is saltier than ocean water freezes at an even lower temperature. The Great Salt Lake in Utah, for instance, freezes at $-20°C$. Salt is put on icy streets to lower the freezing point and melt the ice and snow. Explain another way that you could use this principle.

Figure 19–5. Spreading salt on an icy street causes the ice to melt. Melting occurs because salt water freezes at a lower temperature than does fresh water.

The freezing of ocean water forces the salt ions out of the water, so the ice that forms is almost freshwater ice. Ocean ice is generally no thicker than 3 m. The ice insulates the water under it from the colder air temperature above; therefore, the water below the ice never gets cold enough to freeze.

Figure 19–6. The Arctic Ocean does not freeze solid. Each summer, patches of open water can be found near the coasts.

The water immediately below the ice is not only cold but it also contains a high concentration of salts. This high concentration of salts makes the water denser than surface water. The denser water sinks to the bottom of the ocean, causing a vertical circulation of the ocean water. This type of circulation is called **thermohaline** (thur moh HAY lyn), referring to both temperature (thermo) and salinity (haline). Because of this circulation, the deep water of the oceans has about the same temperature as the coldest surface water.

In tropical regions the ocean is warmed by the sun. Ocean waters near the equator may have a surface temperature as high as 30°C. Near the poles the surface water temperature is about 0°C. Swimmers know that the water temperature is warmer off the coast of Florida than off the coast of Maine.

The surface layer of water, which is penetrated by solar radiation, is about 100 m deep. This water is warmer and less dense than the colder water below. Because of this temperature difference, there is little mixing between the two. The zone separating the warm surface water and the cold water below is called the **thermocline** (THUR muh klyn). As you can see in Figure 19–8, temperature changes rapidly through the thermocline.

○ *How does salt affect the freezing point of water?*
○ *What is a thermocline?*

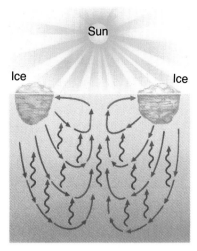

Figure 19–7. Cold, salty water is very dense, so it sinks to the ocean bottom, causing warmer, less-salty water to rise. This is a thermohaline circulation.

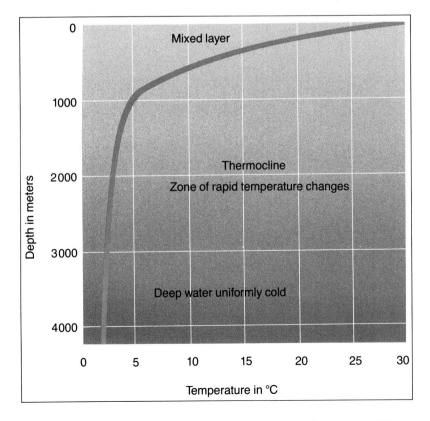

Figure 19–8. This graph shows the change in ocean temperature from the warm, sunlit surface to the cold, dark depths.

ACTIVITY: Relating Water Temperature and Density

How can you show that cold water sinks while hot water remains near the surface?

MATERIALS (per group of 3 or 4)

safety goggles, laboratory apron, 500-mL beaker, red and blue food coloring, crushed ice, test tubes (2), test-tube holder, Bunsen burner

PROCEDURE

1. **CAUTION: Put on safety goggles and a laboratory apron and leave them on throughout this activity.**
2. Fill the 500-mL beaker nearly to the rim with tap water at room temperature.
3. Place four or five drops of red food coloring in one test tube. Place four or five drops of blue food coloring in the other test tube. Add about 2 cm³ of crushed ice to this tube.
4. Fill both tubes about two-thirds full with tap water.
5. Using the test-tube holder, heat the red water to a near boil over the Bunsen burner. Turn off the Bunsen burner when you are done.
6. Slowly pour the contents of both test tubes over the edge of the beaker at opposite sides.

CONCLUSIONS/APPLICATIONS

1. What happened to the cold water?
2. What happened to the hot water?
3. Did the colors mix or tend to remain separate? Explain.
4. Write a general statement about the movement of ocean water.

19.3 **Pressure and Density of the Oceans**

You may recall from Chapter 8 that air pressure is due to the weight of the atmosphere and that as you go up in the atmosphere, the air pressure decreases. Water also exerts pressure, but water pressure is much greater than air pressure, because water is denser than air.

Have you ever seen a movie about deep-sea diving? The heavy suit and metal helmet that deep-sea divers wear protect them from the tremendous pressure of the water. The pressure of the air pumped into the diving suit from the surface must equal the water pressure on the outside of the suit. If the air pressure in the suit were to suddenly drop, the diver could be crushed from the water pressure.

Figure 19–9. Deep-sea divers (left) were the first to explore the ocean bottom. Today scientists protect themselves in thick-walled submersible vessels (right).

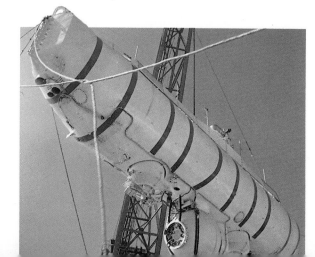

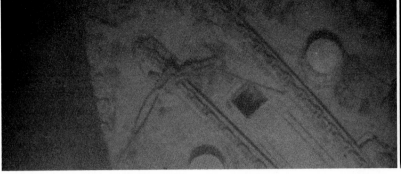

Figure 19–10. The foredeck of the Titanic, on the bottom of the North Atlantic Ocean, is shown on the left. Parts of the ship are well preserved due to the lack of oxidation of the ship's steel (right).

Water pressure increases as you go down from the surface. The greatest water pressure is at the bottom of the deepest part of the oceans. The area of greatest pressure is in the Challenger Deep in the western Pacific, which reaches a depth of over 10 900 m. Despite the tremendous pressure, the bottom of the Challenger Deep was visited by humans on January 23, 1960. Swiss scientist Jacques-Ernst Piccard and United States Navy Lieutenant Donald Walsh reached the bottom of the trench inside the pressurized cabin of the submersible vessel *Trieste*. The steel used to make the *Trieste* cabin is 15 cm thick. The one small window for viewing is also 15 cm thick. The descent to the bottom took nearly five hours.

How does the great pressure at the bottom of the ocean affect the density of the water there? Water, like all fluids, is not easily compressed because the molecules are in contact with each other. Density, therefore, does not increase much with pressure. Density at the bottom of the Challenger Deep is only 7 percent greater than the density of the surface water.

For many years scientists thought that pressure would have a greater effect on the density of ocean water than it actually does. In fact, it was once thought that sinking ships would descend only halfway to the bottom and float forever on a layer of denser water. Scientists now know that this is not the case.

○ *How does pressure affect density?*

Section Review

READING CRITICALLY

1. How does water hold heat?
2. Why is sea water more dense than fresh water?
3. What causes the differences in salinity at different depths in the oceans?

THINKING CRITICALLY

4. What is the relationship among temperature, pressure, salinity, and density?
5. If there were no oceans in the Northern Hemisphere, how might the climate be different?

MATTHEW FONTAINE MAURY (1806–1873)

Few people have contributed as much to the science of oceanography as Matthew Fontaine Maury. Born in Virginia on January 14, 1806, he was the fourth son in a family of four boys and four girls. At the age of 12, he entered Harpeth Academy in the frontier town of Franklin, Tennessee. In 1825 he followed in the footsteps of one of his elder brothers and became a United States naval officer.

Maury began his career as an oceanographer by making three long ocean voyages. The first voyage took him across the Atlantic to France. Along the way he made many measurements about the ocean waters. On board was the famous French general, the Marquis de Lafayette. Lafayette had helped the colonists win their independence from Great Britain. On his second voyage, Maury sailed around the world and began to draw charts of the winds and ocean currents of the world.

In 1836 Maury published his first major work on oceanography, *A New Theoretical and Practical Treatise on Navigation*. The book contained charts and maps of winds and ocean currents. It contained information about all of the oceans of the world. Maury claimed that by following his charts, navigators could save many wasted days of sailing. He claimed that a trip from New York to Rio de Janeiro could be decreased by 10 to 15 days. His challenge was accepted by several shipping companies who were pleased to discover that he was right.

ROBERT DUANE BALLARD (1942–)

Although he was born in Wichita, Kansas, Robert Ballard grew up in Montana, where he worked on a ranch. Some have called him a "high-tech cowboy." Ballard claims he's a cowboy who feels just as comfortable training porpoises as chasing wild horses.

He graduated from the University of California at Santa Barbara in 1965. With a degree in chemistry and geology, he joined the United States Army as a second lieutenant. On duty in Hawaii, he continued his studies at the Institute of Geophysics. He became interested in the use of small submarines called *submersibles* to explore the depths of the ocean.

In 1966 the United States Navy asked him to use a submersible to retrieve an unexploded hydrogen bomb from the floor of the Mediterranean Sea. Ballard has logged more hours in the deep than any other marine scientist. He has made over 200 dives, and he compares his journeys to the dark depths to those of astronauts who have visited the surface of the moon.

In September of 1985 Ballard led a team of American and French scientists from the Woods Hole Oceanographic Institution to the lost wreckage of the *Titanic*. The great luxury liner, which sank after striking an iceberg on April 15, 1912, was resting more than 4 km below the surface of the North Atlantic. For weeks Ballard's team made videotapes and photographs of the wreck. Since then, French explorers have recovered material from the *Titanic*.

SECTION
2 Motions of the Ocean

SECTION OBJECTIVES

After completing this section, you should be able to:

- **Diagram** the location of major ocean currents.
- **Describe** the characteristics of a sea wave.
- **Explain** how ocean tides are produced.

NEW SCIENCE TERMS

currents
geostrophic currents
waves
rip current
tides

19.4 Ocean Currents

The next time you are in a swimming pool, slowly push the water with your hand. You will notice that you can produce a stream, or current, of water within the pool. **Currents** are streams of water that move like rivers through the oceans. Most currents are created by planetary winds blowing along the surface of the ocean. On land, the wind may move loose sediments, forming dunes. At sea, the wind drags along the surface of the ocean. The friction between water molecules causes whole columns of water to move at the same time, forming currents. As these currents move, they are influenced by the Coriolis effect described in Chapter 8.

Large ocean currents transport millions of cubic meters of sea water per second. By comparison, the Amazon River, the largest river by volume in the world, moves only 120 000 m³ of water per second.

A MATTER OF FACT

The major ocean currents are often hundreds of kilometers wide and hundreds of meters deep.

Figure 19–11. Notice the flow of red-colored water that identifies the Gulf Stream current in this infrared photograph.

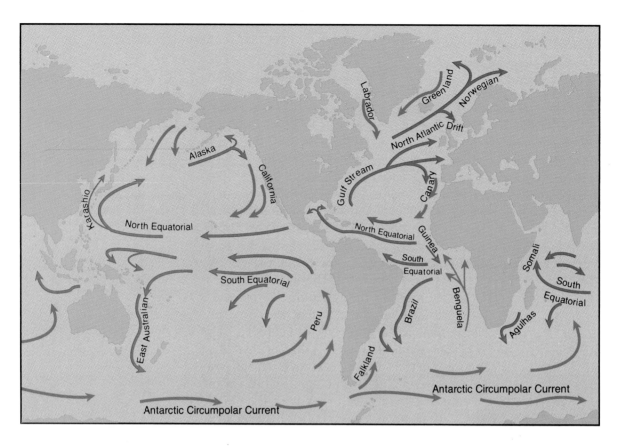

Figure 19–12. This map shows the major surface currents of the world's oceans.

The largest ocean current, the Antarctic Circumpolar Current, flows eastward around Antarctica and transports 125 million cubic meters of water per second. This great current is caused by the westerly winds that blow steadily from west to east around Antarctica. In the northern oceans, there is no current that circles the earth because the continents are in the way. However, the winds over the North Atlantic and North Pacific drive ocean currents in a similar manner, forming great circular systems called *gyres*. Water in each of the gyres tends to move toward the center of the gyre because of the Coriolis effect.

The currents shown in Figure 19–12 are surface currents. In the deep water below many surface currents are countercurrents. Countercurrents flow in the opposite direction to currents, keeping water from piling up in one place.

There are areas of the ocean where surface currents come together, or converge. There are also areas where a current may separate, or diverge. A major area of convergence occurs around Antarctica. There the Antarctic Circumpolar Current and parts of the great gyres of the south Pacific, Atlantic, and Indian oceans come together. Surface water at this convergence sinks to the bottom, carrying oxygen to the depths.

One major area where currents diverge is along the equator. There the Coriolis effect forces surface currents on either side of the equator to turn in opposite directions. A current may also diverge from a coast, again because of the Coriolis effect. The California Current, for instance, turns to the right and moves away from the California coast. Deep water rises to the surface to replace the water of this diverging current. Deep water is rich in nutrients, so areas of divergence have many marine organisms.

Often where currents converge, the water surface is 1 m higher than the average sea level. Where currents diverge, the water level is about 1 m lower. Water moving out of these high areas in the surface of the ocean and into low areas forms **geostrophic currents.** Geostrophic currents also flow under the influence of the Coriolis effect.

Ocean currents help move heat from low to high latitudes. Although winds account for 80 percent of the heat transport, ocean currents account for most of the remaining 20 percent. Heat transport across latitudes has the effect of warming the higher latitudes, making them suitable places for people to live. For instance, much of Europe is at higher latitudes than the northeastern United States, yet the climate is similar. This is due to the effects of the Gulf Stream. The Gulf Stream carries warm Atlantic water from near the equator northward along the American coast. The current then turns eastward, bringing warm water to the coast of Europe.

○ *What causes most surface currents?*
○ *What are convergences and divergences?*

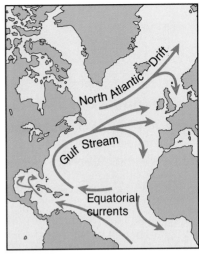

Figure 19–13. The Gulf Stream diverges in the Atlantic Ocean between North America and Eurasia, and the equatorial currents converge between Africa and South America.

Figure 19–14. Because of one gyre of the Gulf Stream, in winter the beaches of the Mediterranean are crowded (left), while the New Jersey shore, at about the same latitude, is nearly deserted (below).

SKILL ACTIVITY: Mapping Ocean Currents

BACKGROUND

The wind, the earth's rotation, and the water temperature are the main factors that determine the path of the oceans' surface currents. Subsurface, or counter currents, however, flow in the opposite direction, and their paths are determined by other factors. These deep ocean currents circulate mainly because of density differences and the friction between moving layers of water.

PROCEDURE

Use page 420 of your textbook to locate the main surface currents of the world. On a piece of tracing paper, sketch a rough outline of the oceans. Then trace the warm currents with a red pencil and the cold currents with a blue pencil.

APPLICATION

On another sheet of tracing paper, again sketch a rough outline of the oceans, and trace the coun-

ter currents from below with a purple pencil. Hold the two drawings together so the outlines of the oceans match.

1. How do the locations of the surface and subsurface currents compare?
2. How does the direction of flow between the surface and subsurface currents compare?
3. Does it seem to matter if the surface current is warm or cold? Explain.
4. How can you account for any differences in location or direction of flow between the two sets of currents?

USING WHAT YOU HAVE LEARNED

If you were to make a third drawing, this one of the planetary wind patterns, how do you think it would compare to the drawings of the surface and subsurface currents? Explain.

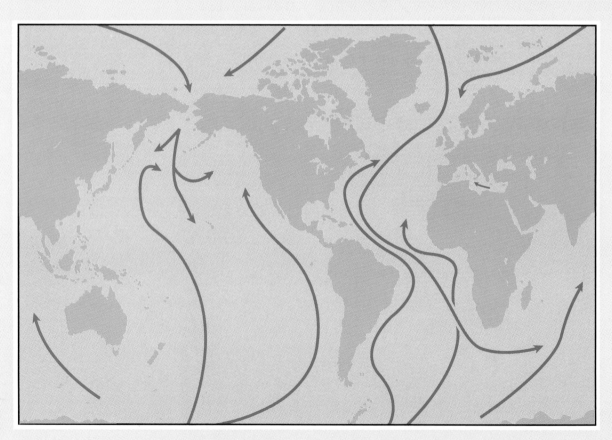

19.5 **Ocean Waves**

While surface currents are produced by planetary winds, winds also produce the surface motions called **waves.** Waves are produced by friction between the wind and the ocean surface. This friction causes ripples that are heightened by the impact of wind on one side of the wave and air suction on the other side. How is the energy of the wind transferred to the waves?

Figure 19–15. On a calm day, the surface of the ocean can be nearly as smooth as glass (right). A strong wind, however, can whip up large waves on the same beach (left).

Regardless of size, all waves are characterized by *wavelength, height,* and *period,* illustrated in Figure 19–16. *Wavelength* is the distance from the crest of one wave to the crest of the next wave. The average wavelength of ocean waves is 60 m to 120 m. The *height* of a wave is the vertical distance between the bottom, or trough, of a wave and its crest. The average height of ocean waves is 1 m to 2 m. The *period,* or frequency, of a wave is the time it takes two successive waves to pass the same point. The standard period for ocean waves is 6 seconds to 9 seconds.

Figure 19–16. Deep-water waves do not break (left). Instead, successive crests and troughs roll past any given point (right).

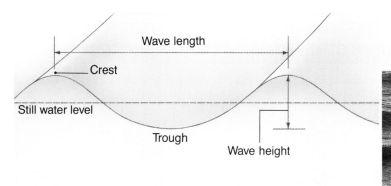

Figure 19–17. Swells are open-water waves with long wavelengths.

Figure 19–18. In deep water, the water molecules in a wave travel in a circle. In shallow water, the water molecules at the surface move faster than those at the bottom (right). The top of the wave falls over the bottom, forming a breaker that spills onto the beach (left).

Waves develop best in the open ocean where there is no interference with the ocean bottom. Large storms are capable of producing waves with long wavelengths. These waves, called *swells*, have wavelengths of several hundred meters. Swells sometimes cross an entire ocean at speeds of tens of kilometers per hour without losing much energy. Waves generated by storms around Antarctica, for instance, often cross the entire Pacific and reach Alaska.

If you watch waves in the open ocean, they seem to move. However, the only thing that really moves is the wave form. The water that makes up the wave only moves up and down, as diagrammed in Figure 19–18. Water moves in the direction of the wave motion only if the bottom of the ocean interferes with the traveling wave. Where the bottom becomes shallow enough to affect the wave, friction slows the lower part of the wave. The wave then becomes unbalanced, breakers form, and the water crashes onto the beach. Beaches that face into the prevailing winds of the open ocean have the high, curling breakers that are perfect for surfing. Beaches in Hawaii, Australia, and the west coast of the United States and Mexico often have these kinds of waves.

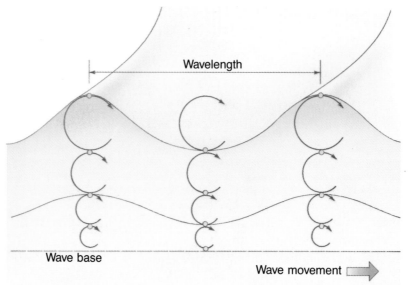

Wavelength

Wave base

Wave movement ▷

Water crashing onto the beach quickly returns to the sea. Depending on the shape of the coastline, the returning water may form a **rip current** flowing out along a channel at right angles to the beach. The speed of a rip current, or undertow, may be as fast as 1 m/s—about the same as a fast-paced walk. Some rip currents are strong enough to drag even good swimmers out to sea. Sometimes people caught in a rip current panic and drown. If you are ever in this situation, all you have to do is swim parallel to the beach. Rip currents are usually very narrow, so you can easily swim out of them and return to shore. Never try to swim against a rip current.

Occasionally waves over 10 m high come ashore, causing tremendous damage and killing many people. These waves, called *tsunamis* (soo NAH meez), are caused by earthquakes in coastal areas or beneath the ocean floor. Tsunamis have very long wavelengths, often more than 100 km. In the open ocean, tsunamis are only about 0.5 m high, but they travel at the speed of several hundred kilometers per hour. A tsunami generated by an earthquake in Alaska can reach Hawaii in less than eight hours. As the ocean bottom becomes shallow, the tsunami becomes higher. Even though the height of a tsunami in the open ocean is small, the wave contains a lot of water because of its great wavelength. A tsunami hits land with tremendous force. Few structures can survive such a force.

○ *What causes waves?*
○ *What are the three characteristics of a wave?*
○ *What are tsunamis?*

Figure 19–19. These arrows show the location of rip currents on this beach.

Figure 19–20. In 1946, a tsunami caused almost total destruction of this waterfront area of Hilo, Hawaii.

19.6 Ocean Tides

If you are a sailor or a fan of high-seas adventure films, you have probably heard someone say, "We sail with the tide." As the earth turns on its axis, the gravity of the moon pulls up a crest of ocean water, forming **tides.** Tides occur because the gravitational pull of the moon is strong on the side of the earth facing the moon, and weaker elsewhere on the earth. The height of the tide varies with the position of the moon. Tides are very long period waves, usually 24 hours long.

As the earth-moon system rotates about its center of mass, the force of this moving system produces two tidal crests, one on the side of the earth facing the moon and one on the opposite side. These tidal crests circle the earth in 24 hours and 50 minutes. Why does it take the tide more than a day to circle the earth? The extra 50 minutes are needed because the moon is traveling around the earth in the same direction in which the earth rotates. This means that the earth has to turn for 24 hours and 50 minutes in order for the same place to be directly under the moon. For example, if you look out your bedroom window and see the moon straight ahead, it will not return to that exact position until 50 minutes later the next night.

Figure 19–21. Because it is the closest body to the earth, the moon has the greatest influence on producing tides (right). The revolution of the moon around the earth causes the tides to be about 50 minutes later each night (above).

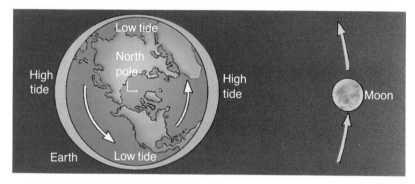

The extra 50 minutes in the lunar day is an ideal case. That is, the earth would have two tides every 24 hours and 50 minutes if there were no continents. The continents, however, slow down the progress of the tides. Special charts show the times of the day when high tides or low tides occur at different locations on Earth. These charts are important to navigation, especially in shallow water and in the entrance to harbors.

The average tide is 0.55 m high in the open ocean. However, the height may reach 2 m to 3 m as the tide runs into a coast. If the shape of the coast funnels the water into a small area, the tide will be even higher. The world-record tide occurs at the head of the Bay of Fundy, in Nova Scotia, Canada. There the tide may reach a height of 14 m.

Figure 19–22. The greatest tidal differences occur in the Bay of Fundy, in Nova Scotia, Canada.

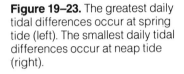

The sun also produces tides. The sun, of course, is much larger than the moon, but it is also much farther away. The gravitational attraction of the sun is much stronger than the moon's, but it does not change much from place to place. As a result, solar tides are only about half as high as lunar tides.

When the moon and sun are in a straight line with the earth, the heights of the lunar and solar tides add up. The tides that result from this alignment are called *spring tides*. The name "spring tide" does not refer to the spring season, but to the "springing up" of the water. Spring tides occur twice during a lunar month: at full moon and at new moon. When the moon and sun are at right angles to each other, as seen from the earth, very low tides, called *neap tides*, occur. Neap tides also occur twice a month—midway between the spring tides. Figure 19–23 shows the Spring and Neap tides.

Under special circumstances, strong, dangerous tidal currents can develop. These strong currents develop most often in narrow passages connecting bodies of water that have tides at different times of the day. The most famous of these is Charybdis, a tidal vortex that develops in the Strait of Messina, between Sicily and mainland Italy.

○ *What causes tides?*
○ *What are spring tides and neap tides?*

Figure 19–23. The greatest daily tidal differences occur at spring tide (left). The smallest daily tidal differences occur at neap tide (right).

Figure 19–24. Shown here is an artists view of the tidal vortex in the Strait of Messina, between mainland Italy and Sicily.

Section Review

READING CRITICALLY

1. How does the depth of the ocean floor affect waves?
2. What are geostrophic currents?
3. What is a rip current?
4. Why are lunar tides more than twice as high as solar tides?

THINKING CRITICALLY

5. Clams live in the sand under shallow water. When would be the best time to collect clams? Explain your answer.
6. Why are storm waves more dangerous during spring tides?

INVESTIGATION 19: The Effect of Temperature on Water

PURPOSE

To test the hypothesis that fresh water freezes at a higher temperature than salt water

MATERIALS (per group of 3 or 4)

Beaker, 250 mL
Crushed ice
Salt
Wax pencil
Test tubes (2)
Tap water
Beakers, 100 mL (2)
Balance
Stirring rod
Graduate, 10 mL
Thermometers (2)
Stopwatch

TABLE 1: EFFECTS OF TEMPERATURE ON WATER

Time	Temperature (°C)	
	Fresh Water	Salt Water

PROCEDURE

1. Prepare a data table similar to the one shown. Make the table large enough for 30 entries.
2. Fill the 250-mL beaker with crushed ice and salt.
3. Use the wax pencil to label the test tubes "F" (for fresh water) and "S" (for salt water).
4. Gently insert the test tubes into the ice about 3 cm apart. The test tubes should be touching the inside of the beaker, 3 cm from the bottom.
5. Put about 100 mL of tap water in one of the 100-mL beakers.
6. Measure 4 g of salt and add it to the other 100-mL beaker. Fill the beaker with water. Stir for 30 seconds.
7. Use the graduate to measure 2 mL of fresh water. Pour it into test tube "F."
8. Measure an equal amount of salt water and pour it into test tube "S."
9. Carefully insert a thermometer into each test tube. **CAUTION: Do not drop or hit the thermometers against the walls of the test tubes. They may crack the test tubes or break them.**
10. Record in your data table the original temperature of each solution.
11. Take temperature readings from each test tube once a minute for 30 minutes.
12. Graph your results.

ANALYSES AND CONCLUSIONS

1. Which thermometer records a faster drop in temperature? Explain why.
2. Do your results show a pause in the temperature drop at any time during the experiment? If so, what do you think was happening to the water during this pause?
3. At what temperature did the fresh water finally begin turning to ice?
4. At what temperature did the salt water finally begin freezing, if at all?

APPLICATION

How might this investigation explain why salt is added to the ice in a home ice cream freezer?

SUMMARY

- Sodium chloride is the most common salt in sea water. (19.1)

- The average salinity of the ocean is about 34.5 g/kg. (19.1)

- Salt water freezes at a lower temperature than fresh water. (19.2)

- Cold, salty water sets up a circulation in ocean water called thermohaline. (19.2)

- The thermocline is a layer below the surface of the ocean through which temperature decreases rapidly. (19.2)

- The pressure of the ocean water increases with depth. (19.3)

- Many surface ocean currents are caused by planetary winds blowing on the surface of the ocean. (19.4)

- The largest ocean current in the world is the Antarctic Circumpolar Current. (19.4)

- Where currents converge, the water level is about 1 m above the average sea level; where currents diverge, the water level is about 1 m below the average sea level. (19.4)

- Geostrophic currents are currents that move out of highs in the surface of the ocean and into lows. Like all currents, they are influenced by the Coriolis effect. (19.4)

- Ocean currents help to distribute heat from low latitudes to high latitudes. (19.4)

- Waves are produced by friction between local winds and the ocean surface. (19.5)

- The water in a wave does not move forward; only the wave form moves. (19.5)

- The tides are produced by the attraction of the moon and the sun on different parts of the earth's surface. (19.6)

- Spring tides occur when the sun and moon align. Neap tides occur when the sun and moon are at right angles. (19.6)

Write all answers on a separate sheet of paper.

SCIENCE TERMS

Correctly use each of the following terms in a sentence.

currents **(419)**
geostrophic currents **(421)**
rip current **(425)**
salinity **(412)**
thermocline **(415)**
thermohaline **(415)**
tides **(425)**
waves **(423)**

SCIENCE QUIZ

Modified True-False

Mark each statement *true* or *false*. If a statement is false, change the underlined term to make the statement true.

1. Sodium chloride accounts for <u>less than</u> three-quarters of the salt in the world's oceans.

2. Ocean temperature increases rapidly <u>above</u> a thermocline.

3. Ocean water freezes at a <u>higher</u> temperature than fresh water.

4. Most ocean currents are driven by the <u>wind</u>.

5. The high salinity of the Mediterranean Sea is a result of high <u>evaporation</u>.

Multiple Choice

Write the letter of the choice that best answers the question or completes the statement.

6. The sides of the earth away from the moon will be in
   a) low tide.   b) high tide.
   c) spring tide.   d) neap tide.

continues

7. The percentage of the earth's surface that is covered by salt water is about
a) 40 percent.
b) 50 percent.
c) 60 percent.
d) 70 percent.

8. Surface currents are caused by
a) the wind.
b) waves.
c) the tides.
d) the ocean.

9. As salinity increases, the freezing point of sea water
a) increases.
b) decreases.
c) varies.
d) remains the same.

10. Which salt is not common in ocean water?
a) NaCl   b) KCl
c) $MgCl_2$ d) $FeCl_2$

11. If there were no oceans, the air temperature range over much of the earth would be
a) $-50°C$ to $50°C$.
b) $-100°C$ to $100°C$.
c) $0°C$ to $50°C$.
d) $-20°C$ to $20°C$.

12. The deepest part of the ocean is the
a) Pacific Ocean.
b) Challenger Deep.
c) thermocline.
d) thermohaline.

13. The ocean currents caused by winds are
a) surface currents.
b) geostrophic currents.
c) divergences.
d) convergences.

14. The time between the passage of two successive waves is the
a) period.
b) wavelength.
c) height.
d) crest.

15. High tides that occur monthly are called
a) spring tides.   b) neap tides.
c) tidal crests.   d) seasonal tides.

Completion

Complete each statement by supplying the correct term or phrase.

16. Silica and nitrate are taken out of sea water by _____.

17. During a first-quarter moon, high tides are at their _____.

18. Swimmers should always be aware of _____ currents near the shoreline.

19. The _____ of ocean water increases with a decrease in temperature.

20. The deepest portion of the oceans is in the _____.

Short Answer

21. Draw diagrams to show how the earth, moon, and sun would be aligned during spring tides and neap tides.

22. Tell how salinity in ocean water affects its freezing temperature.

23. Describe the movement of water in an ocean wave.

Writing Critically

24. Describe the problems of docking a boat in the Bay of Fundy.

25. If the water in a wave does not move, how do surfers ride the waves for a distance?

EXTENSION

1. Read about the Sargasso Sea and report to your class about where it is located, why it exists, and what effect it has had on sailing during the past 500 years.

2. In a shallow tub of water, try to create long- and short-period waves, and then diagram the differences on a sheet of paper.

3. If you live near the ocean, study the tide tables for a week. See if you can figure out the pattern for the daily difference in time for high and low tide. After you know the pattern, determine the times for high and low tide for the following week. Then check the newspaper to see how accurate you are.

APPLICATION/CRITICAL THINKING

1. Benjamin Franklin was one of the first persons to describe the flow of the Gulf Stream. Without the modern equipment that is used today, how do you think he was able to accurately diagram this ocean current?

2. What do you think the pattern of surface currents would be like if there were no continents to block the flow of ocean waters?

3. In some fast-moving streams, waves develop that do not move but remain stationary in the middle of the stream. Explain how these waves, called *standing waves,* might develop.

FOR FURTHER READING

Ballard, R. "How We Found *Titanic.*" *National Geographic* 168 (December 1985): 696. This article gives an exciting, minute-by-minute account of the discovery of the lost ocean liner.

Ballard, R. "A Long Last Look at *Titanic.*" *National Geographic* 170 (December 1986): 697. This article describes the exploration of the *Titanic,* using deep-sea submersibles such as *Alvin.*

Severin, T. "Jason's Voyage: In Search of the Golden Fleece." *National Geographic* 168 (September 1985): 406. This article tells how the author and his shipmates sailed the Black and Aegean seas, mapping the course taken over 3000 years ago by Jason and the Argonauts.

Challenge Your Thinking

Look at these photographs of the Arctic Ocean in winter and summer. Much of the winter ice melts in the summer because it averages only 3 m in thickness. Even at the North Pole, only a thin layer of the Arctic Ocean is frozen. This is because ice is less dense than water, so the ice floats on the surface, insulating the water below and preventing it from freezing. If, like most substances, solid water were denser than liquid water, what would happen to the Arctic Ocean? What effect might this have on the climate of the world and on the organisms that live on Earth?

Ocean Resources

Earth's human population increases every minute. Each new person requires a share of Earth's raw materials. More fresh water is needed and more food is needed. How can the ocean help meet these needs? Humans only now are beginning to tap the vast resources of the sea. In the future these resources will be required to maintain human life on Earth.

A commercial shrimper at work

1 Ocean Sediments

SECTION OBJECTIVES

After completing this section, you should be able to:

- **Discuss** how the world's iron, coal, and oil resources formed.
- **List** the minerals obtained from sea water and the processes used to obtain them.
- **Describe** how salt domes form.
- **Discuss** how deep-sea sediments reveal the climatic history of the earth.

NEW SCIENCE TERMS

brine

salt domes

20.1 Iron, Coal, and Petroleum

Have you ever been swimming where the bottom sediments were soft and slimy? This is not very pleasant for swimming, but bottom sediments are important because of what they may contain.

Ocean sediments, in particular, are important because they contain many of the mineral resources of the world. For example, about 3 billion years ago the ancient sea floor accumulated layers of iron oxide that alternate with layers of silicates and carbonates. The most important deposits of this type are found around Lake Superior, in eastern Canada, in the Soviet Union, in South America, and in Australia. These deposits form the major iron resources of the world today.

Fossil records show that large deposits of coal, a major energy resource, formed during a period in which the sea repeatedly invaded low, swampy lands and then retreated. Each time the sea washed in, a layer of ocean sediments was deposited on top of decaying plant matter. As these sediments accumulated, pressure and heat changed the plant matter into coal. More than a hundred coal layers, alternating with ocean sediments, are found between the Appalachian Mountains and the Mississippi River. Similar coal layers are found in England, the Soviet Union, China, South Africa, and Australia.

Figure 20–1. These deposits of iron ore (left) and coal (right) were formed millions of years ago.

Many ocean sediments are rich in the remains of once-living organisms. Sediments washed into the sea from the land contain many nutrients. Marine algae use these nutrients to produce food for themselves and other organisms. When these marine organisms die, their remains are added to the debris that falls to the ocean floor. Where the ocean bottom is *anaerobic*—that is, without oxygen—organic matter does not decay. Instead, it accumulates with the sediments. As more sediments accumulate, the buried organic matter becomes petroleum.

The largest known petroleum deposits on Earth formed when the African continent drifted northward and pushed organic-rich sediments against Asia. These sediments form the large petroleum reserves of the Middle East. Other deposits of petroleum are found on the north shore of Alaska, beneath the North Sea, and along the coast of the Gulf of Mexico.

○ *Where are Earth's major deposits of iron ore, coal, and petroleum?*

○ *How are the formation of iron, coal, and oil similar?*

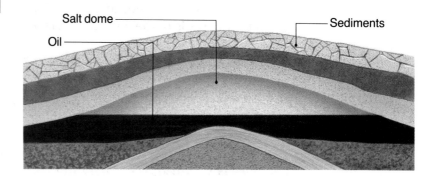

Figure 20–2. Oil is often found under a dome of rock salt.

20.2 Fresh Water and Salt

Two of the ocean's most valuable resources are also the most common: water and salt. The problem is that they must be separated before they can be used. The mineral most valuable to ancient civilizations was salt. Salt, which was used primarily to preserve food, was traded like money. Salt is the major mineral resource of the oceans. Sea salt is obtained by evaporating sea water or by mining salt deposits. About one-third of the salt used in the United States is obtained by evaporation; the rest is obtained by mining.

Figure 20–3. In the United States, some salt is obtained by the evaporation of salt water, but most is mined from huge deposits such as this.

The evaporation process uses the energy of the sun to remove the water from brine. **Brine** is the extremely salty water that accumulates in areas that are totally or partially isolated from the open ocean. As the water evaporates, salt beds form. These beds may be as thick as 1000 m.

If salt beds are buried, the deposits may be squeezed by the weight of other sediments, forming salt domes. **Salt domes** are mounds of salt, 2 km to 3 km across and 5 km to 10 km high. Most salt domes remain buried under other deposits. However, in some places, such as the floor of the Gulf of Mexico, salt domes rise above the surrounding sediments.

Fresh water is also obtained from sea water by evaporation. Although evaporation is an inexpensive method of *desalination*, or removing the salt from sea water, it is practical only in areas that receive plenty of solar energy.

The most direct way to get fresh water from sea water is to squeeze out the fresh water. In this process, called *reverse osmosis*, sea water is placed in a special container that allows the water to pass through but not the salt. Unfortunately, this method of making fresh water is expensive because of the high pressure needed and because the walls of the container must be cleaned frequently.

Desalination is an important area of continuing research and development. As world population increases, water resources for agriculture and other human uses dwindle. Desalination could provide enough water for humans all over the world if a practical, inexpensive method could be developed.

○ *How is salt obtained?*
○ *Other than salt, what else is obtained from evaporated sea water?*
○ *Why could fresh water from sea water be important to the world?*

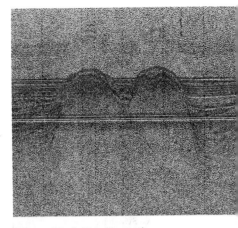

Figure 20–4. This illustration shows a sonar tracing of the Sigsbee Knowles—large salt domes—in the Gulf of Mexico.

Figure 20–5. Fresh water can be obtained by removing the salt from ocean water in a desalination plant.

20.3 Deep-Sea Sediments

Most ocean resources are obtained from shallow water since the difficulty of recovering deep-sea resources adds to their cost. The only major mineral resources that can be gathered from the deep ocean floor are manganese nodules, which are lumps such as the ones shown in Figure 20–6.

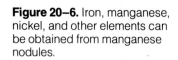

Figure 20–6. Iron, manganese, nickel, and other elements can be obtained from manganese nodules.

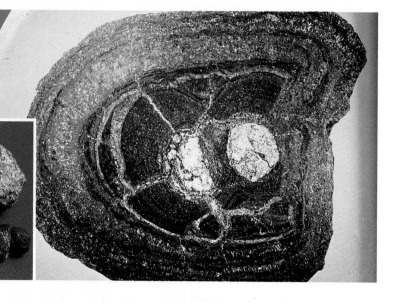

Manganese nodules, which average 5 cm to 10 cm in diameter, form slowly, adding only about 1 mm every million years. In these nodules, layers of manganese and iron oxides are mixed with sedimentary particles. Scientists estimate that in addition to manganese and iron, the manganese nodules in the Pacific contain 15 billion tons of nickel and 5 billion tons of cobalt. Plans are being made to mine these resources; however, the manganese nodules are in international waters, and mining can be done only by international agreement. The average composition of manganese nodules is shown in Table 20–1.

TABLE 20–1: AVERAGE COMPOSITION OF MANGANESE NODULES	
Material	**Percentage**
Iron	16.0
Manganese	14.0
Nickel	4.0
Cobalt	3.0
Phosphorus	1.5
Clay, carbonate, quartz	61.5

Figure 20–7. These scientists (left) are taking a core sample of the sediments on the ocean floor. The core sample is then taken to a laboratory for study (right).

In addition to being a source of valuable resources, deep-sea sediments contain a precious record of the geological and climatic changes that have occurred on Earth. Scientists collect many samples from the deep ocean floor and study them in great detail. Information gathered about shifts in populations of marine organisms and changes in seashell chemistry make it possible to reconstruct the climatic history of the earth. For example, scientists have discovered that 60 million years ago the temperature in the deep sea was 20°C. Since then the temperature has dropped to about 0°C.

○ *What minerals can be mined from manganese nodules?*
○ *What information do deep-sea sediments contain?*

A MATTER OF FACT

The oceans supply humans with over 65 million metric tons of fish, plants, and shellfish every year.

Section Review

READING CRITICALLY

1. How did iron, coal, and petroleum deposits form?
2. How are salt deposits used?
3. Why are deep-sea sediments important?

THINKING CRITICALLY

4. Much of the fresh water on Earth comes from the oceans without humans having to use desalination. Explain.
5. How might information about past climates be useful in predicting future climatic changes?

TECHNOLOGY: Submersibles

The ocean depths not only present many hazards but also provide many treasures. In order to claim the resources in the ocean, humans must explore these deep regions. This exploration has been made possible by the invention of the submersible. Submersibles are small vessels capable of diving to great depths. There are several types of submersibles, including tethered vehicles, roving submersibles, observation bells, and diving gear.

Tethered vehicles usually carry a single passenger, or they may be unpiloted and operated by remote control. Tethered vehicles receive their power and directions from surface craft. They are mostly used for photographing, manipulating, and repairing undersea structures or installations. The first tethered, remotely operated vehicles were used to bury underwater communication and power cables in the 1950s.

Roving vehicles make it possible for an explorer, or group of explorers, to venture far from the surface. They are self-contained vessels that provide internal air pressure equal to that at sea level, despite the enormous pressure being exerted on the craft by the water. These vehicles are battery powered and built to make long, deep journeys to the sea floor. This type of submersible has been used to explore the deepest trenches of the oceans. The Mariana Trench, however, was explored by a tethered submersible. A submersible can also be used to locate sunken vessels, as was done by oceanographer Robert Ballard in finding the lost wreck of the luxury liner *Titanic.* Sometimes, these vehicles can also be used to retrieve sunken treasure.

Another type of submersible is the observation bell. Although some observation bells are capable of independent movement, most are designed to stay at a particular station. There, the bell's mechanical arms and hands are used for the delicate manipulation of underwater objects. The passengers in an

A personal submersible vessel

observation bell are able to view a wide panorama of the sea-floor landscape through wide, pressure-resistant windows.

A different type of submersible consists of a wide range of diving gear. Simple scuba gear allows divers to explore shallow waters. The term *scuba* is actually an acronym that stands for *S*elf-*C*ontained *U*nderwater *B*reathing *A*pparatus.

More elaborate diving suits, called *atmospheric diving suits* (ADS), provide a diver with a pressurized environment for deep-water exploration. An ADS may contain a power unit, or it may be tethered to a support vessel. The diver maneuvers by walking or by using thrusters attached to the suit. These pressurized suits are operational to a maximum depth of about 600 m.

Without submersibles, many of the mysteries and treasures of the deep would forever remain unknown.

A scuba diver and submersible (inset)

SECTION

② Ocean Life

SECTION OBJECTIVES

After completing this section, you should be able to:

- **Identify** the categories of sea life and **give** some **examples** of each.
- **Discuss** some uses of algae.
- **List** the major sources of ocean pollution and **describe** the effects of ocean pollution on human and sea life.

NEW SCIENCE TERMS

plankton
nekton
benthos

20.4 Living Resources

If you have ever gone snorkeling or diving, or if you have visited a marine-life park, you know that the oceans and seas of the world are full of living organisms. These organisms can be grouped into three categories: *plankton, nekton,* and *benthos.*

Plankton   Microscopic plantlike and animal-like organisms that drift with the tides and currents are **plankton.** The tiny, plantlike organisms are called *phytoplankton;* the animal-like organisms are called *zooplankton.*

Phytoplankton are the beginning of most food chains in the ocean. Phytoplankton also produce much of the earth's free oxygen. Zooplankton feed on phytoplankton. Small fish live on plankton and in turn are food for larger fish. A typical ocean food chain is shown as a pyramid of mass in Figure 20–8.

Phytoplankton can only grow where sunlight is strong enough to support photosynthesis. This area, known as the *euphotic zone,* reaches a depth of about 100 m. If the water is not clear, this zone may be even shallower.

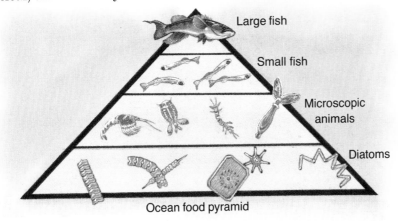

Ocean food pyramid

Large fish

Small fish

Microscopic animals

Diatoms

Figure 20–8. These diatoms (above) are at the bottom of most of the food chains in the ocean (left).

SKILL ACTIVITY: Organizing Information

BACKGROUND

Many food products are available that contain seaweed as an ingredient. One gelatinous mixture, called *alginic acid* or *algin,* is extracted from brown kelp and used in the making of jellies and medicines. *Agar* is another extraction of kelp used in making jellies and for growing bacterial cultures for medical and other scientific purposes. A red alga, *carrageen,* is also used in foods, as well as in hand and face lotions.

PROCEDURE

In this activity you will compile a list of products, such as those shown, that contain these ocean resources: algin (or alginic acid), agar (or agar-agar), and carrageen. Make a copy of the table and fill it in as you check each product.

TABLE 1: ALGAE PRODUCTS

Item	Algin	Agar	Carrageen

INGREDIENTS: SOYBEAN OIL; WATER; HIGH FRUCTOSE CORN SYRUP; VINEGAR; TOMATO PASTE; SALT; GARLIC*; ONION*; SPICE; XANTHAN GUM AND ALGIN DERIVATIVE FOR CONSISTENCY; LEMON JUICE*; OLEORESIN PAPRIKA FOR COLOR; SUGAR; CALCIUM DISODIUM EDTA TO PRESERVE FRESHNESS.
*DEHYDRATED

APPLICATION

Visit your local supermarket and see which popular products contain these ingredients. Use a table like the one shown to record your observations.

USING WHAT YOU HAVE LEARNED

What do all of the products containing algae have in common? What role do you think the algae have in producing this characteristic?

INGREDIENTS: PARTIALLY SKIMMED MILK, MALTODEXTRIN, SUGAR, NONFAT DRY MILK, SODIUM CASEINATE, CORN OIL, ARTIFICIAL FLAVOR, MONO AND DIGLYCERIDES, ASCORBIC ACID, MAGNESIUM OXIDE, CARRAGEENAN, SOY LECITHIN, ALPHA TOCOPHERYL ACETATE, FERRIC ORTHOPHOSPHATE, ZINC SULFATE, NIACINAMIDE, D-CALCIUM PANTOTHENATE, VITAMIN A PALMITATE, COPPER GLUCONATE, MANGANESE SULFATE, ARTIFICIAL COLOR, VITAMIN B_{12}, VITAMIN D, PYRIDOXINE HYDROCHLORIDE, RIBOFLAVIN, THIAMINE HYDROCHLORIDE, FOLIC ACID, D-BIOTIN.

Nekton While plankton can only drift, **nekton** are sea animals that swim about freely. These include all the fish, such as cod, tuna, and sharks, as well as whales, seals, octopuses, and others.

Fish and shellfish have long been food for humans, especially for people living near the coasts. Today, it is not necessary to live near the ocean to enjoy seafood. Fish can be shipped fresh, dried, canned, or frozen. Seafood is an important source of protein, unsaturated fats, and many other nutrients.

Fish can live almost anywhere there is water, from the Arctic to the tropics. In the North Atlantic, fish such as cod, herring, and haddock are harvested. Grouper and pompano are taken off the coasts of the southeastern and Gulf states, while tuna and salmon are caught in the eastern Pacific.

A MATTER OF FACT

Three of the most desirable ocean resources are crabs, shrimps, and lobsters. These creatures are scavengers of the deep. They feed on the waste and decaying materials of other organisms.

As the population of the world continues to grow, fish may become more and more important as a protein source for many people. With such modern technology as sonar, radar, and powerful winches for pulling in large nets, more fish can be caught. The increase in technology raises the danger of overfishing in some areas. Restrictions have been placed on the size of the catches of some species so that the populations can rebuild. Rare species must be protected so they do not become extinct.

Figure 20–9. Redfish were almost eliminated from the Gulf of Mexico because of the popularity of these fish in cooking (above). Now, more plentiful species of schooling fish are used (left).

Figure 20–10. Oysters (left) and shrimp (right) are shellfish harvested from the Atlantic Ocean.

Benthos   Organisms that live in or on the bottom of the ocean are called **benthos.** These include the invertebrates, such as worms, sponges, corals, and starfish. Crabs, lobsters, clams, and oysters are important sources of food.

Seaweeds, the red and brown algae anchored to the ocean floor, are an important resource in its own right. A brown alga called *kelp* is harvested and used to make fertilizer. Substances extracted from algae, such as algin, agar, and carrageen, are used as stabilizers in foods like ice cream, cheese spreads, and salad dressings. Agar is also used as the base on which bacteria are grown in laboratories. Seaweed is popular as a food in many countries.

○ *What are the three categories of sea life?*
○ *Which type of sea life is the most important to humans? Explain.*

Figure 20–11. Commercial harvests from the Pacific Ocean include kelp (right) and snow crabs (left).

20.5 Ocean Pollution

Humans have been dumping waste into lakes, rivers, and oceans for hundreds of years. In addition, natural pollutants, such as dead animals and fish and decaying plants, are also being added to the waters of the world. Natural chemical processes in the ocean can decompose small amounts of some types of pollution. However, large amounts of waste upset natural processes, killing ocean life.

The chief sources of ocean pollution are sewage and garbage, industrial waste, and agricultural chemicals and wastes. Sewage is untreated human waste and household water. Most large cities have sewage treatment plants, but this treatment does not remove some types of chemicals. When treated sewage is dumped into a body of water, these chemicals go with it. The body of water may be a river or a bay or a lake, but the water eventually reaches the ocean. Many coastal areas have become unsuitable for recreation and fishing because of pollution.

Industrial chemicals kill ocean life or make seafood unfit to be eaten. Many harmful chemicals have been found in the water, in fish, and in shellfish. These chemicals become more concentrated as they move up the food chain. The effects of long-term exposure to these chemicals are not yet known.

Figure 20–12. Garbage from some cities is dumped directly into the ocean (above). Some of it returns as litter on public beaches (left).

Garbage has been polluting the oceans for many years. Most food wastes decompose, but rusting auto parts and soft drink cans litter the ocean floor. Many fish and sea mammals have died from swallowing such things as plastic bags, cans, and the plastic carriers from drink cans.

In the summer of 1987, hospital trash, including used hypodermic syringes, washed up on the beaches of New Jersey. At the same time, dead dolphins were washing up on the shore, and swimmers were complaining of skin rashes and respiratory infections. Could all these things be connected?

Many industries use water for cooling and then dump the heated water into the ocean. This causes thermal pollution. The heated water causes some fish and plants to die and some plankton to multiply rapidly, upsetting the ocean's natural balance.

Radioactive wastes from nuclear industries, such as nuclear power plants, uranium mining, and weapons manufacture, are very dangerous to ocean life. The nuclear wastes must be safely stored until they lose their radioactivity. For some wastes,

DISCOVER

Producing a Newsletter

Smog, acid rain, oil spills, contaminated wells, closed beaches, and contaminated shellfish beds are all examples of the effects of pollution. Look through several newspapers and magazines and clip articles and pictures on pollution. Also clip articles on attempts to clean up and control pollution. Then fold a large sheet of paper in half and paste your articles onto it to make a newsletter or small newspaper. Give it a title and share your newsletter with your classmates, or make a class display.

Figure 20–13. These drums of radioactive wastes were recovered from a capsized ship. Other containers of hazardous wastes are purposely stored on the ocean floor.

Figure 20–14. Oil spills present one of the greatest problems for ocean cleanup (right). Sometimes birds and other animals are killed by the toxic oil (below).

this might be 500 years; for others, 50 000 years. Experts are unsure about the safest way to store waste tanks for such long periods. The bottom of the ocean is one place where radioactive wastes are being stored; there, the tanks are covered with concrete. Would the tanks survive an earthquake without polluting the ocean?

The oil industry has been responsible for much polluting of the oceans. Each year, huge tankers carrying hundreds of thousands of tons of crude oil sail the oceans. Spills from tanker accidents have produced massive oil slicks, which have coated beaches with oil and killed many sea birds.

In recent years the industry has changed loading and unloading procedures to eliminate the discharging of oil into the sea. Greater safety measures make accidents less likely, and new

clean-up techniques make spills less disastrous, but pollution from oil is still a danger.

As the human population grows, the problem of pollution grows. And as Earth becomes more crowded, humans may look to the oceans for more resources. Pollution must be controlled to protect these resources and our future.

○ *How are the oceans being polluted by sewage?*
○ *What are the hazards of industrial waste?*

Section Review

Figure 20–15. San Diego Harbor is one of the world's busiest harbors, yet the water is fairly clean.

READING CRITICALLY

1. How has modern technology changed the fishing industry?
2. Give an example of an ocean food chain.
3. How is ocean pollution dangerous to humans?

THINKING CRITICALLY

4. Why could seafood not provide the world population with much food 100 years ago?
5. What could governments do to protect the oceans from pollution?

ACTIVITY: Recycling Water

How can you demonstrate the way nature returns fresh water to the earth through the water cycle?

MATERIALS (per group of 3 or 4)

safety goggles, glass tubing, rubber tubing, plastic bag, string, ring stand and clamp, Bunsen burner, crushed ice

PROCEDURE

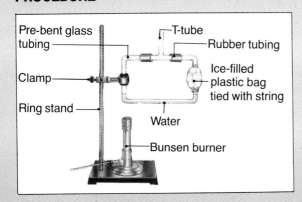

Pre-bent glass tubing
T-tube
Rubber tubing
Clamp
Ice-filled plastic bag tied with string
Ring stand
Water
Bunsen burner

CAUTION: Safety goggles must be worn at all times during this activity.

1. Set up the apparatus as shown, using the preshaped glass tubing, and then fill the bottom of the tube with water.

2. Make sure the plastic bag of ice is secured to the glass tubing with string.

3. Gently heat the water-filled part of the tube with a low Bunsen burner flame, and observe the results.

CONCLUSIONS/APPLICATIONS

1. Describe what you observed. How does this activity demonstrate the phase changes of water—solid to liquid to gas?

2. Explain how these phase changes take place in nature.

3. In what ways do machines take advantage of the phase changes of water or other fluids?

Section 2   Ocean Life   **445**

INVESTIGATION 20: Water Filtration

PURPOSE

To show how water can be purified by filtration

MATERIALS (per group of 3 or 4)

Hammer
Finishing nail
Empty soup can
Ring stand and clamp
Small beaker (150 mL)
Large beaker (500 mL)
Coarse gravel
Fine gravel
Muddy water
Crushed charcoal
Sand

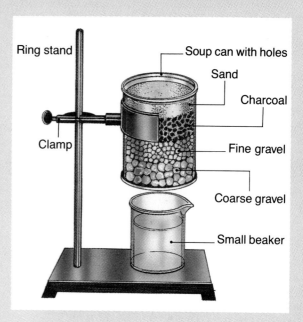

Ring stand — Soup can with holes — Sand — Charcoal — Clamp — Fine gravel — Coarse gravel — Small beaker

PROCEDURE

1. Using the hammer and finishing nail, drive about one dozen small holes into the bottom of a clean soup can.
2. Secure the can above the small beaker as shown.
3. Pour about 2 cm of coarse gravel into the bottom of the can. Then pour a 2-cm layer of fine gravel on top of the coarse gravel.
4. Pour about 100 mL of muddy water into the top of the can and collect the filtered water in the small beaker below the apparatus.
5. Describe the characteristics of the filtered water as clear, cloudy, murky, or muddy. Compare your results with the results of other groups.
6. Pour out the water in the small beaker and clean the can.
7. Repeat step 3. Then pour about 2 cm of crushed charcoal on top of the fine gravel.

Pour about 2 cm of fine sand on top of the charcoal.
8. Repeat steps 4 and 5.

ANALYSES AND CONCLUSIONS

1. Was the gravel effective in filtering the muddy water? To what extent?
2. Was the filtration process improved by the presence of charcoal and sand? To what extent?

APPLICATION

1. Why do you think fish aquariums have filters with charcoal in them?
2. How could this system be used on a large scale to clean a city's water supply?

SUMMARY

- The oceans help provide many mineral resources, such as iron, coal, and petroleum. (20.1)

- Coal was formed as the sea invaded swamps and buried organic matter under heavy sediments. (20.1)

- Ocean sediments may form deposits of petroleum. (20.1)

- Fresh water and salt may be obtained from sea water by evaporation. Salt may be mined from large deposits known as salt domes. (20.2)

- Except for manganese nodules, the mining of deep-sea sediments is not cost-effective. (20.3)

- Manganese nodules may provide an economical source of iron, manganese, nickel, and cobalt. (20.3)

- Deep-sea sediments provide clues to past geologic and climatic changes. (20.3)

- Plankton are microscopic organisms that are the basis for ocean food chains. (20.4)

- Phytoplankton are plantlike organisms that photosynthesize, while zooplankton are animal-like organisms that feed on phytoplankton. (20.4)

- Nekton are free-swimming animals. They provide much of the seafood consumed by humans. (20.4)

- Benthos, or bottom-dwelling organisms, provide many products for human use. Seaweed and many shellfish are benthos. (20.4)

- Organic waste, such as sewage, may be harmful to ocean life. (20.5)

- Industrial chemicals and nuclear wastes are dangerous to ocean life and humans. (20.5)

- Oil spills have killed many sea birds and fouled many beaches. (20.5)

Write all answers on a separate sheet of paper.

SCIENCE TERMS

Correctly use each of the following terms in a sentence.

benthos **(442)**
brine **(435)**
nekton **(441)**
plankton **(439)**
salt domes **(435)**

SCIENCE QUIZ

Modified True-False

Mark each statement *true* or *false.* If a statement is false, change the underlined word to make the statement true.

1. <u>Salt domes</u> are important for mining commercial salt.

2. Organic matter does not decay in an <u>aerobic</u> environment.

3. <u>Manganese</u> is a major mineral resource mined from the floor of the oceans.

4. Photosynthesis takes place in the <u>euphotic</u> zone.

5. Ice cream may include stabilizers extracted from <u>seaweed</u>.

Multiple Choice

Write the letter of the choice that best answers the question or completes the statement.

6. Sharks are part of the
   a) benthos.
   b) nekton.
   c) plankton.
   d) euphotic.

7. The chief sources of ocean pollution are
   a) sewage and garbage.
   b) industrial wastes.
   c) agricultural chemicals.
   d) All of the choices are correct.

continues

8. The evidence of past ages is found in
   a) records of fossil fuel consumption.
   b) ocean salinity.
   c) deep-sea sediments.
   d) salt domes.

9. Ocean sediments may be mined commercially for
   a) coal.
   b) salt.
   c) petroleum.
   d) All of the choices are correct.

10. Manganese nodules also contain
    a) gold.
    b) clay, carbonate, and quartz.
    c) salt.
    d) water.

Completion

Complete each statement by supplying the correct term or phrase.

11. Commercial amounts of iron and nickel are found in _____.

12. The iron oxide deposits of the world were formed about _____ years ago.

13. Radioactive wastes may continue to pollute the ocean for as long as _____ years.

14. The most direct way to get fresh water from sea water is by _____.

15. In a euphotic food chain, zooplankton eat _____.

16. When the sea repeatedly invaded low, swampy areas, _____ formed.

17. The extremely salty water that forms in isolated parts of the ocean is _____.

18. The only minerals, besides salt, that are mined from the oceans are found in _____.

19. Sea animals that swim about freely are _____.

20. The _____ industry is responsible for much polluting of the oceans.

Short Answer

21. Explain why marine and deep-sea sediments are useful to humans.

22. List the minerals obtained from manganese nodules. Why are these minerals important?

23. Compare and contrast plankton, nekton, and benthos.

Writing Critically

24. Explain why petroleum deposits are found on land as well as under the ocean.

25. Why are containers of radioactive waste stored on the ocean floor sometimes called "time bombs" for future generations?

EXTENSION

1. Find out how an electronic fish finder works. Then explain how it can be used to locate schools of fish.

2. Research the methods of catching tuna. Explain why porpoises are sometimes caught with tuna.

3. Locate the nearest body of water to your school and find out if it is polluted. If it is, try to locate the source and the type of pollution. Suggest ways in which this source could be eliminated or treated.

4. Read and report to the class on the productivity of one of the major estuaries along the North American coast.

APPLICATIONS/CRITICAL THINKING

1. Why do you think that deep, clear areas of ocean waters are sometimes referred to as deserts?

2. Explain this statement: The earth could probably support 20 to 30 times its present population if the ocean could be farmed as extensively as the land.

3. If you were diving with a scuba tank containing 30 m³ of air, and you were using 2.5 m³ of air per minute at 30 m depth and 1.5 m³ at 15 m depth, how long could you dive at 20 m or 25 m?

FOR FURTHER READING

Ballard, R. "How We Found the Titanic." *National Geographic* 168(December 1985): 696. This article gives an exciting minute-by-minute account of the discovery of the lost luxury liner *Titanic.*

Golden, F. "Scientist of the Year." *Discover* 8(January 1987): 50. This article is a biography of Robert Duane Ballard, the person who discovered the lost wreckage of the luxury liner *Titanic.*

Nelson, C. and K. R. Johnson. "Whales and Walruses as Tillers of the Sea Floor." *Scientific American* (February 1987) 256:112. This article describes side-scan sonar studies of the Bering Sea that resulted in some surprising discoveries.

Sibbald, J. *Homes in the Sea: From the Shore to the Deep.* Minneapolis: Dillion, 1985. This text describes the different ocean habitats and how organisms adapt to these environments.

Whitehead, H. "The Unknown Giants." *National Geographic* 166(December 1984): 774. For centuries the great sperm and blue whales were hunted for their valuable oils. Today, they are in danger of becoming extinct. This article explains how human relationships with these gentle giants are proving to be a lesson in conservation of the ocean's varied resources.

Challenge Your Thinking

This is an underwater hotel in the Florida Keys, south of Miami. Here people vacation in the quiet solitude of the coral reef. Although this hotel is rather unusual now, it may one day be much more common. How do you think the ocean might be used in the future? Why do you think the use of ocean resources is so important for the future of life on Earth?

UNIT 6 THE SCIENCE CONNECTION:

Power from the Tides

Soon a 15-story dam will be built across Canada's Bay of Fundy. Inside the dam there will be 100 turbines, turned by the flow of the world's highest tides.

This giant tidal power plant will take advantage of the supertides that flow in and out of the Bay of Fundy. You may recall that the gravity of the sun and the moon cause the surface of the ocean to rise and fall. A supertide occurs where the tidal flow enters a long, narrow, funnel-shaped inlet. The tidal wave sloshes up to the neck of the funnel and back. If the bay is the right shape, the tidewater arrives back at the entrance just in time to get a push from the next high tide. The wave is thus intensified, resulting in higher high tides and lower low tides. Inside the Bay of Fundy, tides rise and fall by as much as 15 meters.

This intense tidal flow makes the bay ideal for developing tidal energy. The Bay of Fundy project, across the Cobequid Bay inlet, is the largest attempt yet at harnessing the power of tides. Tidal plants are already generating electricity in France and the Soviet Union. On the Bay of Fundy, a small pilot project across the Annapolis River has been operating since 1983. However, none of these power plants matches the scale of the giant Cobequid project. When completed and operating at full capacity, the plant will generate 4800 MW (megawatts) of power. By comparison, the Indian Point nuclear plant in New York produces only 873 MW.

The engineering principles of tidal power have been known for many years. Early tidal mills trapped the inflowing tide behind a dam, then allowed the water to flow out over a waterwheel. Today's tidal plant has turbines that are turned by the tidal flow.

The Rance River tidal power plant

Tidal power turbine

450

Tidal power plant control room

The French tidal plant has reversible turbines, so it operates on both incoming and outgoing tides. The Cobequid turbines will turn only when the tide is flowing out. Tidal waters will flow through the turbine at a rate of 13 000 m^3/s, about three times the flow rate of Canada's largest river, the St. Lawrence. The water's speed will be controlled by a set of moveable vanes, which will open and close like venetian blinds. The turbines will spin at 50 revolutions per minute.

Tidal power plants create clean electric power without consuming fossil fuels. There is no air pollution, as there is with fossil-fuel plants, and there are no safety and waste disposal problems, as there are with nuclear plants. However, tidal power does have problems. Tidal plants can work only when the tide is rising and falling. Those times may not be the same as the times when the demand for electricity is highest. Few people need to take a hot shower or bake a cake at 3 A.M. Also, tidal plants may be located hundreds of miles from the large cities that use the most electricity.

Scientists also warn that the Bay of Fundy plant may cause environmental damage. The giant dam might affect tides as far south as Cape Cod in Massachusetts. If the dam causes tides to change by only 15 cm, it could disturb plants and animals living in tidal waters. A 15 cm rise in the tide would mean that up to 2 m of coast would be covered by ocean water. That change could endanger beachfront property.

The Bay of Fundy tidal power plant is an attempt to harness one of Earth's greatest untapped resources. Similar projects could save oil, coal, and other non-renewable resources. Yet, it also raises questions about how to balance the need for electrical power with the need to protect the natural environment.

Low tide on the Bay of Fundy

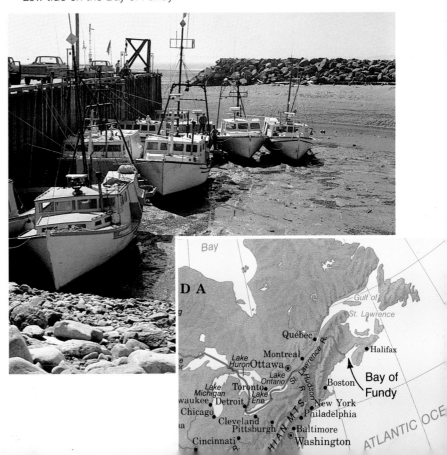

ASTRONOMY

For centuries scientists have wondered if life exists on other planets. Seeing the planet Mars dimly through telescopes, some astronomers thought that intelligent Martians lived in a world full of water and vegetation, and that they had built a system of canals. Some of the first interplanetary spacecraft traveled to Mars to investigate the possibility of life on the red planet. However, by 1976 scientists could more confidently answer the question, "Are we alone in the solar system?"

- What other bodies are there in the solar system?
- What instruments are used to study the universe?
- What are the theories concerning the origin of the universe?
- Before the Viking space probes, what did scientists know about Mars?

By reading the chapters in this unit, you will learn the answers to these questions. You will also develop an understanding of concepts that will allow you to answer many of your own questions about the universe.

"The view from Mars"

453

The Solar System

Earth and its moon are part of a system of bodies in orbit around a rather ordinary star. This solar system, named for its central feature, the sun, is probably only one of many billions of similar systems in the universe. However, much scientific study is confined to these neighbors in space.

A solar eclipse

SECTION

1 Motions of Earth and the Moon

SECTION OBJECTIVES

After completing this section, you should be able to:

- **Describe** the earth's rotation in terms of direction of motion and time.
- **Define** the term *revolution* as it relates to the earth's path through space.
- **Explain** how the tilt of the earth's axis affects the seasons.
- **Relate** the cause of solar and lunar eclipses.

NEW SCIENCE TERMS

axis
rotation
revolution
orbit
lunar eclipse
umbra
penumbra
solar eclipse

21.1 **Rotation of Earth**

Imagine that you are an astronaut looking at Earth from a stationary space station. You would see Earth turning around its axis. An **axis** is an imaginary line that passes through the center of a planet, from pole to pole. The turning of Earth on its axis is called **rotation** (roh TAY shuhn). The earth rotates much like a spinning toy top, making one complete rotation every 24 hours. This time is called the *period of rotation.* The direction of Earth's rotation is counterclockwise as seen from above the North Pole.

If your space station were positioned so that you could see North America as Earth turned, you could watch the sun's light move across the continent. The first part of the continent to receive the sun's light is the east coast of Canada. At this time, all of the United States is still in darkness. As the earth rotates, the area of darkness moves westward, until all of North America is in sunlight. This process takes about five hours.

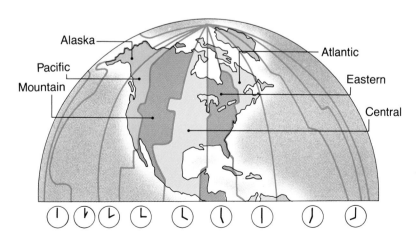

Alaska
Pacific
Mountain
Atlantic
Eastern
Central

Figure 21–1. The rotation of the earth is responsible for the formation of different time zones. The time zones of North America are shown here.

Section 1   Motions of Earth and the Moon   **455**

In order for the time of sunrise to be approximately the same all across North America, five different time zones have been established. These time zones change roughly every 15 degrees of longitude. However, time zones are not straight lines as are lines of longitude. The zones are established so that they do not divide cities. Imagine living in one time zone and going to school in another. You would have to get up an hour earlier or later than the other students just to arrive at school on time.

○ *What is an axis of a planet?*
○ *What is the earth's period of rotation?*

21.2 Revolution of Earth

From your space station, you would also observe a second motion of the earth. In addition to rotating on its axis, Earth moves around the sun. Movement of Earth around the sun is known as **revolution** (rehv uh LOO shuhn). Since it takes Earth one year, or $365\frac{1}{4}$ days, to complete one revolution, its period of revolution is one year. The path that the earth follows as it revolves around the sun is known as its **orbit.** The orbit of the earth does not change; Earth's speed of revolution remains constant as well.

Figure 21–2. The earth revolves around the sun every 365¼ days. This period of revolution determines the length of an earth year.

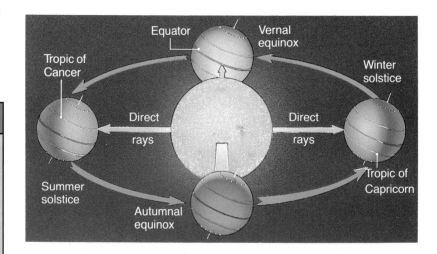

As the earth revolves around the sun, its axis remains in a fixed position, but it is tilted. The earth's tilt, or *inclination*, is always 23.5° in relation to a line perpendicular to the plane of its orbit. Imagine a table tennis ball with a toothpick as its axis. If it is tilted about 23° from vertical, its inclination would be similar to Earth's. Because this inclination is fixed, first the Northern and then the Southern Hemisphere leans toward the sun. This creates seasonal changes in climate over most of the earth.

When the Northern or Southern Hemisphere is tilted toward the sun, that hemisphere receives very direct rays from the sun. Also, the amount of time sunlight is received during the day is longer. This is summer. When the same hemisphere is tilted away from the sun, that hemisphere receives less direct sunlight. There are also fewer hours of sunlight during the day. It is winter.

On June 20 or 21 the tilt of the Northern Hemisphere toward the sun is greatest. The sun is directly over 23.5°N parallel, which is called the *Tropic of Cancer.* In the Northern Hemisphere this date is the *summer solstice.* The summer solstice is the day of the year with the most hours of daylight.

As the earth's revolution around the sun continues, the sun is directly over the equator on September 22 or 23. In the Northern Hemisphere this is the *autumnal equinox.* On this date the periods of daylight and darkness are about equal.

As the earth continues its revolution, the daylight hours in the Northern Hemisphere grow shorter until the *winter solstice* on December 21 or 22. On this date the sun is directly over the *Tropic of Capricorn* at latitude 23.5°S. This is the day with the least daylight in the Northern Hemisphere.

On March 20 or 21, the sun is directly over the equator again, and the hours of daylight and darkness are nearly equal. In the Northern Hemisphere this is known as the *vernal equinox.* As the earth completes its revolution around the sun, the hours of daylight continue to lengthen in the Northern Hemisphere until the summer solstice. In the Southern Hemisphere the summer and winter solstices are opposite to those in the Northern Hemisphere. Explain why this is so.

○ *What is the earth's period of revolution?*
○ *How does the earth's revolution around the sun affect the seasons in the Northern Hemisphere?*

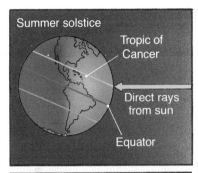
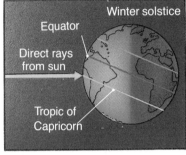

Figure 21–3. The summer solstice (top) occurs when the sun is directly over the Tropic of Cancer. The winter solstice (bottom) occurs when the sun is directly over the Tropic of Capricorn.

Figure 21–4. The seasons of the year are determined by the position of the earth with respect to the sun. In temperate climates, such as the one shown here, four seasons occur every year.

457

How can you make an instrument that can determine the distance of an object above the horizon?

MATERIALS (per person)

large index card; protractor; string, 30 cm; large metal weight; tape; plastic drinking straw

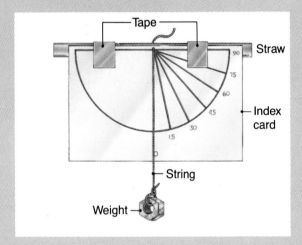

1. With a protractor, draw a line and semicircle across the index card and label it as shown.
2. Tie the weight to one end of the string. Make a small hole in the index card at the center of the straight line and thread the loose end of the string through the hole.
3. Tie the loose end of the string around the plastic straw. Tape the straw to the index card.
4. Look through the straw at a tall object or something in the sky, but do not look at the sun. Let the weight hang freely and observe where the string crosses the degree markings. That is the altitude of the object above the horizon.

CONCLUSIONS/APPLICATIONS

1. Do you think it is possible to calculate the height of a structure in meters just by knowing its angle above the horizon? Explain.
2. What other scientific instrument is similar to the astrolabe? Describe the similarities and differences.

21.3 Motions of the Moon

Studying the moon from Earth is like studying the earth from a stationary space station. The earth and the moon form a system that jointly revolves around the sun. The earth appears to be stationary in respect to any other movements of the moon.

The moon has movements that are similar to the rotation and revolution of the earth. The moon rotates on its axis as it revolves around the earth. However, since the moon's periods of rotation and revolution are the same, the same side of the moon always faces the earth.

Although the moon appears to shine on its own, moonlight is really reflected sunlight. When the entire side of the moon that faces the earth is lighted, you see a *full moon*. When the side of the moon facing the earth is not lighted, it is called a *new moon*. The series of changes in the appearance of the moon as it revolves around the earth, shown in Figure 21–5, are called the *phases of the moon*.

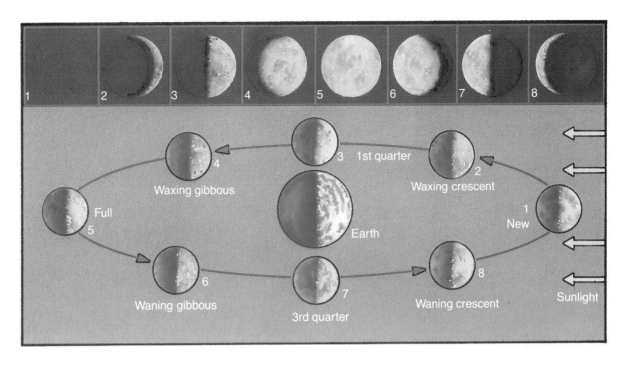

Figure 21–5. The moon goes through a series of phases every month. These phases, shown here, are due to the position of the moon and the earth in relation to the sun.

Occasionally, the moon moves into a position where the earth is directly between the moon and the sun. Then the sun's rays that would ordinarily fall on the moon's surface are temporarily blocked by the earth. When the earth's shadow falls on all or part of the moon's surface, a **lunar eclipse** (ih KLIHPS) occurs. Because of the relative sizes of the earth and the sun, two shadow zones form during a lunar eclipse. The zone of complete shadow is called the **umbra** (UHM bruh). The umbra is surrounded by a zone of partial shadow called the **penumbra** (pih NUHM bruh). During a total lunar eclipse, the moon passes through the penumbra and into the umbra. You can watch the moon pass through a region of dim light until it crosses a distinct boundary into darkness. The moon lies entirely within the umbra for almost two hours before it finally reappears in the penumbra. In a partial lunar eclipse, part of the moon passes through the penumbra but not the umbra, so the moon never completely disappears.

Figure 21–6. A lunar eclipse (below) occurs when the earth casts a shadow on the surface of the moon (left).

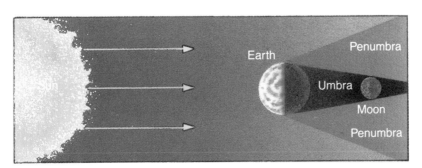

Section 1   Motions of Earth and the Moon   **459**

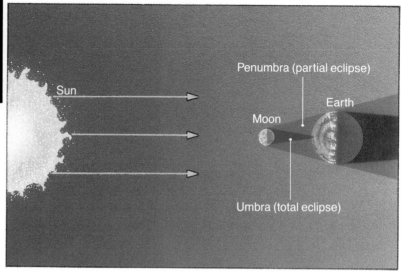

Penumbra (partial eclipse)

Sun

Earth

Moon

Umbra (total eclipse)

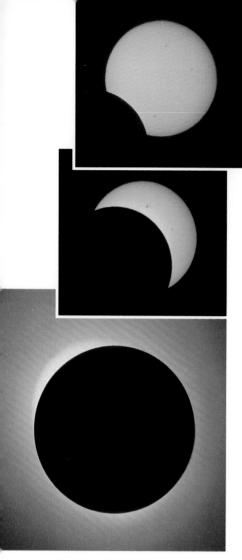

Figure 21–7. A solar eclipse (above) occurs when the moon passes between the earth and the sun (right).

Sometimes the shadow of the moon passes across the lighted side of the earth, forming a **solar eclipse.** Since the moon is much smaller than the sun, the moon forms a narrow shadow, about 258 km wide, as it moves across the face of the earth. A partial eclipse, where only part of the sun is blocked by the moon, is visible on either side of the area experiencing total solar eclipse. The moon's shadow moves rapidly across the surface of the earth. In fact, a total solar eclipse can be seen from any one position on Earth for only about 7.5 minutes. **CAUTION: Never look directly at the sun, even during an eclipse. The sun's rays may cause permanent damage to your eyes.**

○ *What are the phases of the moon?*
○ *What is a lunar eclipse? What is a solar eclipse?*

Section Review

READING CRITICALLY

1. Compare and contrast the revolution and rotation of Earth.
2. Describe the relative positions of the sun, the moon, and the earth during a full moon.
3. Explain why a solar eclipse is not observed at all points on the earth at the same time.

THINKING CRITICALLY

4. How would the seasons be different if the earth's axis were not inclined?
5. Why does a solar eclipse only last about 7.5 minutes?
6. Why is the far side of the moon never seen from Earth?

SECTION
2 The Inner Planets and the Asteroids

SECTION OBJECTIVES

After completing this section, you should be able to:

- **Describe** the arrangement and orbits of the inner planets in the planetary system.
- **Explain** why the surfaces of Mercury and the moon are heavily marked with craters while those of Venus and Earth are not.
- **Discuss** the ways in which Venus differs from Earth.
- **Tell** why Mars is red.

NEW SCIENCE TERMS

planets
elliptical
astronomical unit
satellite
asteroids
meteorites

21.4 The Solar System

The earth, moon, and sun are only part of a complex system of orbiting bodies that make up the solar system. The solar system, named for the sun, is comprised of nine planets. **Planets** are the main bodies that revolve around the sun. The nine planets of the solar system are: Mercury, Venus, Earth, Mars, Jupiter, Saturn, Uranus, Neptune, and Pluto. All the planets revolve around the sun in the same direction. Viewed from the sun's north pole, this direction, called *prograde*, is counterclockwise.

Figure 21–8. The nine planets in our solar system are shown here in their relative positions.

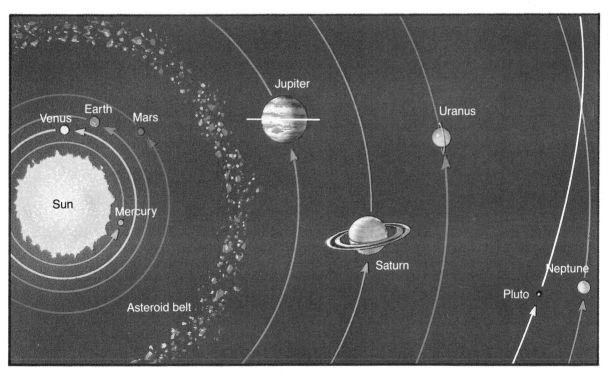

Accurate mathematical observations about the planets were made in the late 1500s by a Danish astronomer, Tycho Brahe (TEE koh BRAH hee). For many years Brahe made careful observations about the position of the planets. Upon his death he left his notes to his assistant, Johannes Kepler. Kepler continued to study the planets, using Brahe's data, and plotted the orbits of the planets.

Figure 21–9. The orbit of Earth is not circular. Rather, the path that Earth takes around the sun forms an ellipse.

Perihelion   Sun   Aphelion   Earth

The orbits of the planets as they revolve around the sun are not exactly circular. Instead, the orbits are **elliptical,** or flattened circles with two fixed points called *foci.* The sun is at one focus while the other is empty. The point along an orbit where the planet is closest to the sun is called *perihelion* (pehr uh HEEL yuhn); the point in the orbit that is farthest from the sun is called *aphelion* (uh FEE lee uhn). Planets with large orbits take longer to revolve around the sun than planets with smaller orbits. The period is the time it takes a planet to make one revolution around the sun.

A planet's distance from the sun is measured in astronomical units. An **astronomical unit** (AU) is the average distance between the earth and the sun, or about 150 000 000 km. Table 21–1 shows the most recent photograph of each planet in the solar system and gives current information about each planet.

○ *What is an elliptical orbit?*
○ *What are astronomical units?*

JOHANNES KEPLER (1571–1630)

Johannes Kepler was born on December 27, 1571. In his youth, he studied theology, mathematics, and philosophy at the University of Tübingen, in Germany. In the sixteenth century many people believed that the earth was the center of the solar system. In that model, the sun and all the planets revolved around the earth. Kepler became a strong defender of another theory, which had been proposed by Nicolas Copernicus in 1540. That theory states that the sun, and not the earth, is the center of the solar system. In his mid-twenties Kepler published a work entitled *Mysterium Cosmographicum*. In the book, he strongly put forth this theory.

He later worked with the accomplished astronomer Tycho Brahe who invited him to become an assistant. When Tycho Brahe died a year later, Kepler took Brahe's position at court and inherited all of the famed astronomer's notes and records. Brahe's notes included much important information about the position and movement of the planets.

Kepler made many of his own observations and also used Brahe's work to discover several laws about the motion of the planets. In 1609 he wrote *Astronomia Nova* (New Astronomy). In that book he proposed that the planets did not move in circles as was previously thought but instead moved in elliptical orbits around the sun. Kepler's laws of planetary motion served as the basis for Sir Isaac Newton's laws of gravity.

IRENE DUHART LONG (1951–)

Irene Duhart Long was born in Cleveland, Ohio, in 1951. She became interested in flight at a very young age, when she accompanied her father on his flying lessons. By the age of nine, she had decided that she wanted to work for NASA.

She began her college education at Northwestern University, in Evanston, Illinois, and received her medical degree from the St. Louis University School of Medicine, specializing in aerospace medicine. She completed her medical residency at the Kennedy Space Center.

In 1982 Dr. Long was named the Chief of the Medical and Environmental Health Office at NASA. She is currently Chief of Medical Operations. Dr. Long and her team of scientists study how the free-fall environment in outer space affects the health of astronauts. Her group provides medical care for the astronauts in the event of an emergency and collects data on the reactions of the human body to space travel. The group is particularly concerned with the effects of space travel on the heart and on the endocrine system. The information gathered is being used to develop methods to minimize the impact of space travel on the human body.

One of the highest-ranking women at the Kennedy Space Center, Dr. Long is the first black to be named Chief of Medical Operations. Her next goal is to travel in space herself, as the medical officer on a spaceflight. Then she could collect in-flight information about the effects of space travel first hand.

TABLE 21–1: THE SOLAR SYSTEM

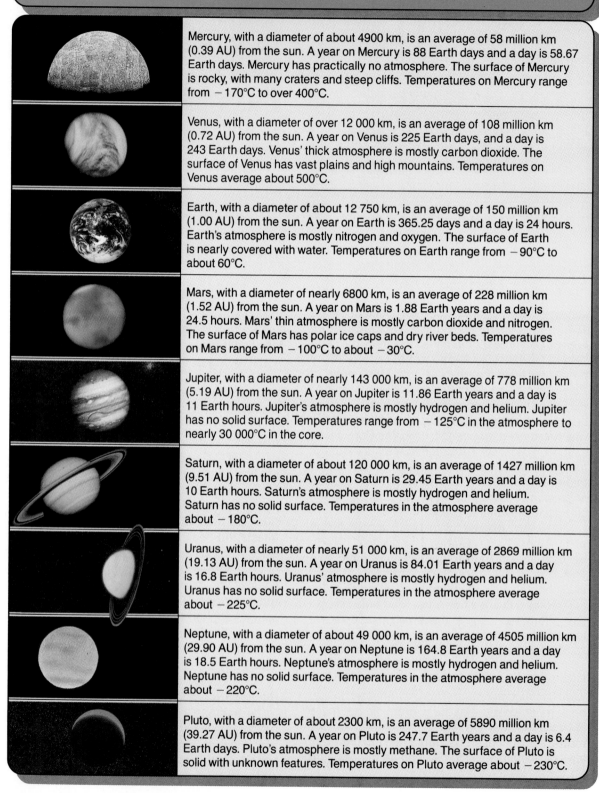

Mercury, with a diameter of about 4900 km, is an average of 58 million km (0.39 AU) from the sun. A year on Mercury is 88 Earth days and a day is 58.67 Earth days. Mercury has practically no atmosphere. The surface of Mercury is rocky, with many craters and steep cliffs. Temperatures on Mercury range from −170°C to over 400°C.

Venus, with a diameter of over 12 000 km, is an average of 108 million km (0.72 AU) from the sun. A year on Venus is 225 Earth days, and a day is 243 Earth days. Venus' thick atmosphere is mostly carbon dioxide. The surface of Venus has vast plains and high mountains. Temperatures on Venus average about 500°C.

Earth, with a diameter of about 12 750 km, is an average of 150 million km (1.00 AU) from the sun. A year on Earth is 365.25 days and a day is 24 hours. Earth's atmosphere is mostly nitrogen and oxygen. The surface of Earth is nearly covered with water. Temperatures on Earth range from −90°C to about 60°C.

Mars, with a diameter of nearly 6800 km, is an average of 228 million km (1.52 AU) from the sun. A year on Mars is 1.88 Earth years and a day is 24.5 hours. Mars' thin atmosphere is mostly carbon dioxide and nitrogen. The surface of Mars has polar ice caps and dry river beds. Temperatures on Mars range from −100°C to about −30°C.

Jupiter, with a diameter of nearly 143 000 km, is an average of 778 million km (5.19 AU) from the sun. A year on Jupiter is 11.86 Earth years and a day is 11 Earth hours. Jupiter's atmosphere is mostly hydrogen and helium. Jupiter has no solid surface. Temperatures range from −125°C in the atmosphere to nearly 30 000°C in the core.

Saturn, with a diameter of about 120 000 km, is an average of 1427 million km (9.51 AU) from the sun. A year on Saturn is 29.45 Earth years and a day is 10 Earth hours. Saturn's atmosphere is mostly hydrogen and helium. Saturn has no solid surface. Temperatures in the atmosphere average about −180°C.

Uranus, with a diameter of nearly 51 000 km, is an average of 2869 million km (19.13 AU) from the sun. A year on Uranus is 84.01 Earth years and a day is 16.8 Earth hours. Uranus' atmosphere is mostly hydrogen and helium. Uranus has no solid surface. Temperatures in the atmosphere average about −225°C.

Neptune, with a diameter of about 49 000 km, is an average of 4505 million km (29.90 AU) from the sun. A year on Neptune is 164.8 Earth years and a day is 18.5 Earth hours. Neptune's atmosphere is mostly hydrogen and helium. Neptune has no solid surface. Temperatures in the atmosphere average about −220°C.

Pluto, with a diameter of about 2300 km, is an average of 5890 million km (39.27 AU) from the sun. A year on Pluto is 247.7 Earth years and a day is 6.4 Earth days. Pluto's atmosphere is mostly methane. The surface of Pluto is solid with unknown features. Temperatures on Pluto average about −230°C.

21.5 Mercury and Venus

Mercury, the planet closest to the sun, is also the smallest of the inner planets. Even through a telescope Mercury appears as an indistinct object just above the horizon in the morning and evening sky.

Pictures of Mercury's surface show it to be similar to that of Earth's moon. There are many impact craters and plains of dark rock, which could be basalt. Mercury's surface has long ridges that seem to have originated as the crust and the interior cooled and contracted.

Mercury has little gravity because it is so small. Therefore, Mercury has virtually no atmosphere. Most of the gases that reach the planet's surface from its interior escape into space.

The density of Mercury has been calculated to be similar to that of Earth. This indicates that it may have an iron-nickel core as Earth has. However, the magnetic field of Mercury is very weak. Scientists believe this may be because the planet rotates too slowly for electrical currents to develop in its core.

Figure 21–10. Mercury is the closest planet to the sun. The surface of Mercury is similar to that of the moon. This planet was named after the fleet-footed messenger of the Roman gods. The name was chosen because Mercury has such a rapid period of revolution.

The second planet from the sun is Venus. Venus is often thought of as Earth's sister planet because of their similar masses, densities, and sizes. Often visible in the evening or in the morning, Venus appears to have phases similar to those of the moon. These phases occur because from Earth Venus is seen with different angles of solar light.

Venus rotates very slowly, taking 243.01 Earth days to rotate once on its axis. Unlike Earth and most of the other planets, Venus rotates east to west, or *retrograde*. Venus takes 224.7 Earth days to orbit the sun, so a day on Venus is longer than a Venusian year.

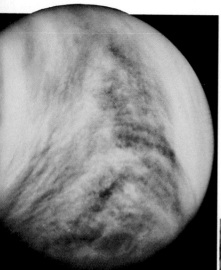

Space probes have determined that the atmosphere of Venus is very dense. The Venusian atmosphere consists of 96.0 percent carbon dioxide, 3.5 percent nitrogen, and 0.5 percent sulfur dioxide, argon, and water vapor. This dense atmosphere produces a strong greenhouse effect, which allows the surface temperature to reach 500°C.

Venus has a calculated density of 5.24 g/cm^3, probably with an iron-nickel core almost as large as Earth's. Venus has no magnetic field. The reason may be that Venus, like Mercury, rotates too slowly for electrical currents to develop in the core.

○ *Describe the surface of Mercury.*
○ *Why is Venus called Earth's sister planet?*

Figure 21–11. Venus is sometimes called the "cloud planet" because its atmosphere looks like the clouds of Earth. However, the atmosphere of Venus is made up of carbon dioxide rather than of nitrogen and oxygen as our atmosphere is. Venus was named after the Roman goddess of love and beauty.

21.6 **Earth and the Moon**

Earth is the third planet from the sun. Because of its location in the solar system, Earth's characteristics are very different from the other inner planets. Earth has many deep oceans, the atmosphere is moderately dense, and the surface temperature is moderate. These conditions probably contributed to the evolution of life on Earth.

The earth, with its iron-nickel core extending almost halfway to the surface, generates a magnetic field. The core is surrounded by a thick mantle of molten rock, which is covered with a solid crust.

Earth is the first planet from the sun to have a satellite. A **satellite** is a small body, or moon, orbiting a planet. The moon is nearly 385 000 km from Earth. Density calculations show that the moon may have a small iron-nickel core; however, it has no magnetic field. Data collected from instruments left on the moon by Apollo astronauts indicates the presence of a mantle and a 60-km-thick crust.

Figure 21–12. The earth-moon system is shown here (left). The surface of the moon is barren due to the lack of an atmosphere (right).

The moon has two major terrains: the older highlands composed of feldspar rocks, and the younger maria, or seas, formed of basalt. These seas are not filled with water; they only resemble seas when viewed from Earth. The surface of the moon is marked with craters and covered with fine rock debris up to 20 m deep.

There are two major theories about the origin of the moon. One is that the moon was a separate body revolving around the sun, and it was captured by the earth's gravity. A second theory is that the moon is a piece of the earth that was broken off by a collision with another body. Neither theory has gained general support.

○ *What conditions are unique to the earth?*
○ *What is a satellite?*

21.7 **Mars and the Asteroids**

Mars, the fourth planet from the sun, orbits the sun in 687 Earth days. Mars is often called the red planet because its surface is reddish in color when viewed from Earth. The red color is probably the result of the oxidation of iron in the surface rocks. Mars has a radius a little more than half that of Earth. It has a significant, but not large, iron-nickel core and a weak magnetic field.

The surface of Mars includes deserts, dry riverbeds and flood plains, mountains, and the highest volcano in the solar system, Olympus Mons. Olympus Mons, which is now inactive, is 600 km across and rises to a height of 26 km. There are many surface craters, especially in the southern hemisphere.

> **DISCOVER**
>
> *Traveling to Mars*
>
> Using the table of distances on page 464, determine how long it would take to get to Mars, traveling by car at a safe 85 km per hour.

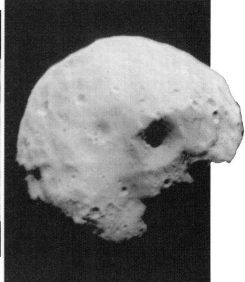

Figure 21–13. Mars is often called the "red planet" because it looks red as viewed from Earth. The surface of Mars is covered with red dust (right). Ice caps of frozen carbon dioxide are found at the poles.

The existence of dry riverbeds and flood plains may indicate that water was once present on Mars. Today, however, Mars is essentially dry, the water having been lost to space. Mars has two thin polar caps that expand and contract as the seasons change, but they are made of frozen carbon dioxide, not water. Space probes have shown that the atmosphere of Mars is very thin, consisting of 95.7 percent carbon dioxide, 2.7 percent nitrogen, and traces of argon, oxygen, carbon monoxide, and water.

Mars has two small satellites, Phobos and Deimos. Both have irregular shapes and may be captured asteroids. **Asteroids** are small planetary bodies that circle the sun in a belt between the orbits of Mars and Jupiter. The largest asteroid is Ceres, with a diameter of 1025 km. Large asteroids have a spherical shape; smaller asteroids have irregular shapes, like the two satellites of Mars.

Figure 21–14. Mars has two satellites: Phobos and Deimos. This planet was named after the Roman god of war. He had two attendants: Phobos, the bringer of fear, and Deimos, the bringer of panic.

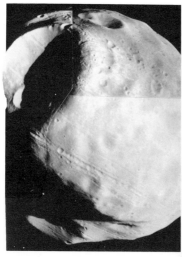

Some scientists believe that the asteroids represent bodies that failed to combine into a planet. The gravity of the neighboring planet Jupiter was probably too strong for the asteroids to form a single body. The asteroids revolve around the sun in individual orbits, often colliding with each other. These collisions break up the asteroids and disturb their orbits.

Meteorites are asteroid fragments that leave orbit. Meteorites captured by Earth's gravity fall to Earth, occasionally forming a crater. Traces of these craters are visible on Earth. One of the largest ones, Meteor Crater, is in Arizona. Most of the craters on the moon are from meteorites.

○ *Why is Mars called the red planet?*
○ *What is an asteroid?*

Figure 21–15. Ceres (top) is the largest asteroid in the asteroid belt. Meteorites (bottom) are formed when pieces of asteroids break off and leave orbit. Occasionally, a meteorite survives entry into Earth's atmosphere and crashes onto the surface of the planet (left).

Section Review

READING CRITICALLY

1. What is the probable reason why the asteroids did not form a single planet?
2. How might the moon have formed?

THINKING CRITICALLY

3. Why are the surfaces of Mercury and the moon heavily marked with impact craters while those of Venus and Earth are not?
4. What are the possibilities that there might have been life on Mars at one time? Explain.

SKILL ACTIVITY: Measuring the Diameter of the Sun

BACKGROUND

The diameter of the sun has been known since ancient times. All you need to make the same calculations that ancient astronomers made are a little mathematics and a simple knowledge of how light behaves.

PROCEDURE

In order to make a measurement of this sort, you must first construct the necessary instrument.

1. Cut a small index card as shown. Then, using a T-pin or the tip of a ballpoint pen, make a hole about 2 cm from the top of the middle of the cut card.
2. Tape another index card to the end of a meter stick as shown.
3. Slip the card with the hole onto the meter stick, so that you can slide it to any position along the stick.
4. With your back to the sun, place the meter stick over your shoulder so that you are looking at the card. Aim the back of the stick at the sun.
5. Beginning at the end closest to the sun, move the sliding card toward the end card. An image of the sun will appear on the end card.

APPLICATION

Once you have mastered the use of the instrument, you can begin to make your measurement.

1. Position the sliding card so that the image of the sun is in focus.
2. Circle the sun's image with a pencil as accurately as you can and measure the diameter in millimeters.
3. Note the position of the sliding card (in millimeters) from the end card.
4. Calculate the diameter of the sun according to the following formula:

$$D_s = \frac{D_i \times C_s}{C_i}$$

D_s is the diameter of the sun; D_i is the diameter of the sun's image in millimeters; C_s is the distance from the earth to the sun (use 1.49×10^{14} millimeters); and C_i is the distance in millimeters from the end card to the sliding card when the sun's image is in focus.

5. Record your results on a sheet of paper and compare them with your classmates' results. Explain any differences.

USING WHAT YOU HAVE LEARNED

Explain how this procedure could be used to measure the diameter of the moon.

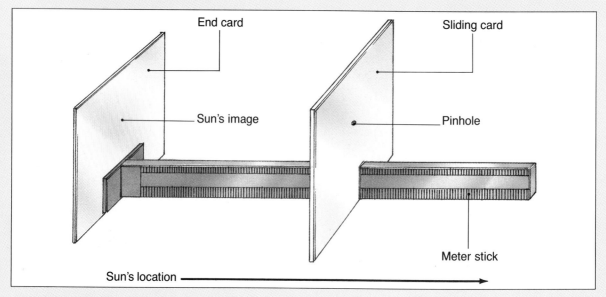

End card — Sun's image — Sliding card — Pinhole — Meter stick — Sun's location

3 The Outer Planets and the Comets

SECTION OBJECTIVES

After completing this section, you should be able to:

- **Compare** the structure of the outer planets with that of the inner planets.
- **List** the peculiarities of the Galilean satellites of Jupiter.
- **Discuss** the structure and composition of comets.

NEW SCIENCE TERM

comets

21.8 Jupiter

The outer planets include the giant planets Jupiter, Saturn, Uranus, and Neptune, and the smallest planet, Pluto. The giant planets have many features in common. They all consist mainly of frozen gases and have atmospheres that are 90 percent hydrogen and 10 percent helium. Jupiter, Saturn, and Uranus have orbiting ring systems and numerous satellites. Neptune has only two satellites that are visible from Earth and no apparent rings.

With a mass 2.5 times that of all the other planets combined, Jupiter is by far the largest planet in the solar system. Jupiter is called a *gaseous giant* because it has no solid crust as Earth has. Beneath the thick atmosphere lies Jupiter's mantle, which extends all the way to its core. The core may have a radius of 10 000 km, larger than Earth's radius.

Jupiter's atmosphere is 17 000 km thick, and although it is not warmed much by the sun, there is convection. This may be because Jupiter itself is very warm—in fact Jupiter radiates about twice as much energy into space as it receives from the sun. This energy probably comes from radioactivity or convection in the core and mantle.

Figure 21–16. Jupiter (right), the largest planet, was named for the Roman king of the gods. This planet is characterized by a huge red spot (left), which is a swirling mass of hot gases.

The light-colored bands of Jupiter's atmosphere are hot, rising gases; the dark-colored ones are cooler, sinking gases. These colored bands rotate around the planet at different speeds. A prominent feature of Jupiter's atmosphere is the Great Red Spot south of the equator. The red spot is a swirling mass of gases, with winds probably in excess of 1000 km per hour. Circulation in the red spot is counterclockwise because of Jupiter's Coriolis effect. How do you think the Coriolis effects of Jupiter and Earth compare?

The four largest of Jupiter's 16 satellites were discovered in 1610 by the Italian astronomer Galileo. In his honor they are called *Galilean satellites.* In order of distance from the planet, the satellites observed and named by Galileo are Io, Europa, Ganymede, and Callisto.

Io, which is slightly larger than Earth's moon, has more volcanic activity than any other body in the solar system. Unlike Earth, however, the energy for the volcanic activity of Io is tidal, caused by the gravity of neighboring Jupiter.

Europa is smaller than Earth's moon, and its surface is covered with a thick layer of ice. There are very few impact craters on Europa, suggesting that the surface reseals itself after a meteorite impact.

Ganymede is larger than Mercury. In fact, it is the largest satellite in the solar system. Ganymede's surface has some impact craters, but it also has features, such as surface grooves, that may indicate more recent structural changes. These grooves may be caused by tension in the crust, like the fractures caused by tension in Earth's crust. Callisto, on the other hand, is heavily cratered, indicating that its surface has not changed much since its formation.

Figure 21–17. The four largest satellites of Jupiter are shown here: Io (top), Europa (bottom left), Gannymede (bottom right), and Callisto (bottom center).

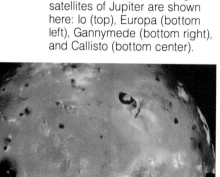

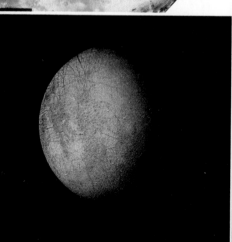

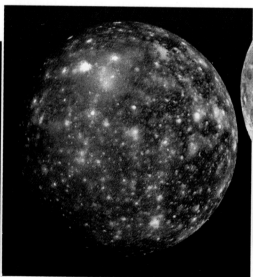

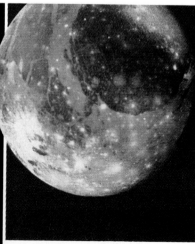

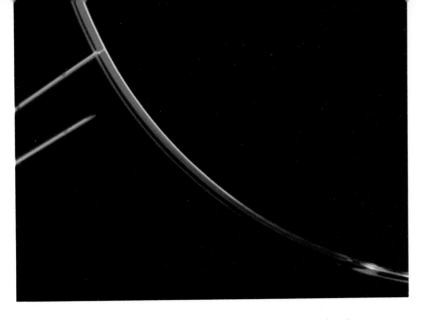

Figure 21–18. The rings of Jupiter can be seen in this Voyager photograph.

Jupiter has a ring system. The rings consist of micrometer-sized particles of metals, silicates, and ice. The ring system is not visible from Earth; it was discovered in 1976 when the United States spacecraft *Pioneer 11* photographed Jupiter.

○ *Why is Jupiter called a gaseous planet?*
○ *Name and give several characteristics of Jupiter's large satellites.*

21.9 Saturn, Uranus, Neptune, and Pluto

Saturn, shown in Figure 21–19, has the lowest density of all the planets because it consists mainly of ice. The rocky core of Saturn is small, and the structure of its mantle and atmosphere is similar to that of Jupiter. Like Jupiter, Saturn gives off more energy than it receives from the sun.

Titan, Saturn's largest satellite, is larger than Mercury but smaller than Ganymede. Titan consists of a rocky core surrounded by a thick mantle of ice. The atmosphere of Titan consists mostly of nitrogen, and there is evidence of active volcanoes on Titan. The remaining satellites of Saturn are small, heavily cratered bodies of silicate rocks and ice.

A MATTER OF FACT

Both Pluto (2300 km in diameter) and Mercury (4880 km in diameter) are smaller than Saturn's satellite Titan (5150 km in diameter) and Jupiter's satellite Ganymede (5276 km in diameter).

Figure 21–19. Saturn can be identified by its massive ring system (left). This planet was named for the Roman god of agriculture.

Figure 21–20. The ring system of Saturn (right) has been studied by many scientists since the time of Galileo. The study continues today at places such as the Jet Propulsion Laboratory in Pasadena, California (left).

Saturn has a complex ring structure, first observed by Galileo in 1610. The ring system lies on Saturn's equatorial plane and consists of hundreds of concentric rings separated by gaps of varying width. The rings are made of fragments of rock and ice ranging in size from less than 1 cm to over 10 m.

Uranus and Neptune are gaseous planets with rocky cores. Uranus has at least 15 satellites, while Neptune has only 2 that are visible from Earth. Uranus is unusual in that its axis is inclined 98° from the plane of its orbit, and it rotates retrograde. The ring system of Uranus is thin, with particle sizes similar to those of Saturn's rings. Neptune has at least a partial ring; scientists will know more after analyzing photographs taken by the *Voyager 2* spacecraft.

Figure 21–21. Of the three outer planets, Uranus (left) and Neptune (center left) are gaseous; Pluto (right) is solid. Scientists are just now learning about these planets from space probes that are passing the planets and sending back computer images and information.

Pluto is the outermost planet. Pluto is also the smallest planet in the solar system and the only outer planet without a thick atmosphere. Little is known about Pluto except that it has one known satellite, Charon, discovered in 1978.

○ *What characteristics do Saturn and Uranus share with Jupiter?*

○ *What is unusual about Uranus?*

21.10 The Comets

Beyond Pluto are bodies that orbit the sun over extremely long periods of time. **Comets** consist of silicate rock and metal particles embedded in ices. The ices are frozen gases such as carbon dioxide, methane, ammonia, and water. Comets may be condensations of planetary matter, formed when the solar system was beginning to take shape.

Comets are frequently captured by the giant planets—mainly Jupiter—and forced into orbits that take them close to the sun. When a comet approaches the sun, gases escaping from the comet are blown by the solar wind. These gases form a tail, as much as 1 million km thick and 10 million km long, pointing away from the sun. Comets form an envelope around the sun called the *Oört cloud.*

The diameter of a typical comet is about 1 km. The body of the comet loses a layer several meters thick each time the comet moves around the sun. Scientists believe that most comets do not last more than a dozen passages around the sun. Halley's comet, however, has been observed every 76 years since 240 B.C. How many passages of Halley's comet have been seen from Earth?

○ *What are comets?*

Figure 21–22. Perhaps the most famous comet is Halley's comet (left), which comes near the earth every 76 years. The core of this comet (right) is composed of rocks and ice.

DISCOVER

Looking for Halley

From the information on the period of revolution of Halley's comet, determine how old you will be the next time Halley's comet returns.

Section Review

READING CRITICALLY

1. How do the four Galilean satellites differ from each other?
2. Of what materials are the rings of Saturn made?
3. How are comets' tails formed?

THINKING CRITICALLY

4. Why are scientists so interested in the composition of comets?
5. Do you think that the ring systems of the outer planets are like unformed moons? Explain.

INVESTIGATION 21: Drawing Planet Distances to Scale

PURPOSE

To make a scale model of the sun and inner planets

MATERIALS (per group of 3 or 4)

String, 30 m
Metric ruler
Construction paper
Scissors
Tape
Markers

PROCEDURE

1. Table 1 shows the diameter of the sun and each planet as well as each planet's mean distance from the sun.
2. Using a scale in which 1 mm = 10^4 km, draw the sun and planets on a piece of construction paper. The size of each planet does not have to be exact. At this scale, you will find that they are all small and that exact dimensions are difficult to represent.
3. Draw a square around the sun and each planet, making sure to label each square for identification purposes.
4. Cut out the squares.
5. Tape the square of the sun to the end of the string.
6. According to the same scale used in step 2, determine the distance of each planet from the sun and tape it to the string.
7. Stretch out the string.

ANALYSES AND CONCLUSIONS

1. How do the sizes of the planets compare to that of the sun?
2. What difficulty did you experience in trying to draw the planets to scale?
3. Using the same scale of measurement, how far from Earth is Earth's moon (384.4×10^3 km) in centimeters?

APPLICATION

Explain how you would use astronomical units instead of kilometers in this investigation.

TABLE 1: THE SUN AND THE INNER PLANETS					
	Sun	**Mercury**	**Venus**	**Earth**	**Mars**
Diameter (km)	1.4×10^6	4900	12 000	12 750	6800
Distance from the Sun (km)		5.8×10^7	1.08×10^8	1.50×10^8	2.28×10^8

SUMMARY

- Earth rotates counterclockwise on its axis once every 24 hours. (21.1)

- Earth revolves around the sun in a specific orbit every 365-1/4 days. (21.2)

- The tilt of the axis from a line perpendicular to the plane of its orbit is about 23.5°. (21.2)

- Phases of the moon are caused by the varying illumination of the moon's surface. (21.3)

- The solar system contains nine planets, each of which revolves around the sun. (21.4)

- Mercury, the planet closest to the sun, has a cratered surface and practically no atmosphere. (21.5)

- The atmosphere of Venus, the second planet from the sun, is rich in carbon dioxide. Venus is sometimes referred to as Earth's sister planet because of its similar size, mass, and density. (21.5)

- Earth is the third planet from the sun, the first planet with a satellite, and the only planet known to have life. (21.6)

- Mars, the fourth planet from the sun, is called the red planet. Although it may have had water at one time, none presently exists. Polar ice caps of frozen carbon dioxide can be seen on its surface. (21.7)

- Asteroids are planetary fragments that did not join together to form a planet. Asteroids orbit between Mars and Jupiter. (21.7)

- Jupiter is the largest planet in the solar system. It emits twice as much energy as it receives from the sun. (21.8)

- The rings of Saturn consist of fragments of rock and ice from less than 1 cm to more than 10 m in diameter. (21.9)

- Uranus is unusual because of its nearly horizontal tilt and retrograde rotation. (21.9)

- Comets consist of metal and silicate particles embedded in ices. (21.10)

Write all answers on a separate sheet of paper.

SCIENCE TERMS

Correctly use each of the following terms in a sentence.

asteroids **(468)**

astronomical unit **(462)**

axis **(455)**

comets **(475)**

elliptical **(462)**

lunar eclipse **(459)**

meteorites **(469)**

orbit **(456)**

penumbra **(459)**

planets **(461)**

revolution **(456)**

rotation **(455)**

satellite **(466)**

solar eclipse **(460)**

umbra **(459)**

SCIENCE QUIZ

Modified True-False

Mark each statement *true* or *false*. If a statement is false, change the underlined term to make the statement true.

1. Jupiter gives off <u>less</u> energy than it receives.

2. The thick atmosphere of Venus makes its surface <u>cooler</u> than the surface of Mercury.

3. <u>Asteroids</u> falling to Earth usually burn up in Earth's atmosphere before they hit the ground.

4. The four largest moons of Jupiter were named by <u>Kepler</u>.

5. Between the orbits of Mars and Jupiter lies a belt of <u>comets</u>.

continues

Multiple Choice

Write the letter of the choice that best answers the question or completes the statement.

6. Which of the following is NOT an inner planet?
   a) Mercury   b) Venus
   c) Mars   d) Jupiter

7. The only outer planet that is NOT a gas giant is
   a) Jupiter.   b) Saturn.
   c) Pluto.   d) Neptune.

8. The planet without rings is
   a) Pluto.   b) Jupiter.
   c) Saturn.   d) Uranus.

9. What happens to the length of a planet's revolution the nearer it is to the sun?
   a) It increases.
   b) It decreases.
   c) It remains the same.
   d) It becomes retrograde.

10. Which of the following planets is solid?
    a) Jupiter   b) Saturn
    c) Uranus   d) Mars

Completion

Complete each statement by supplying the correct term or phrase.

11. The order of the inner planets from the sun is _____, Venus, Earth, and Mars.

12. The order of the outer planets from the sun is _____, Saturn, Uranus, Neptune, and Pluto.

13. The atmosphere of _____ is heated from the planet itself.

14. The planet with the shortest day is _____.

15. Earth's sister planet is _____.

Short Answer

16. List the planets that have solid surfaces.

17. What bodies besides Earth seem to have active volcanoes? What is the evidence?

18. Describe the shape of the orbits of the nine planets.

19. Why are the four largest satellites of Jupiter known as Galilean satellites?

20. Which planets rotate prograde and which rotate retrograde?

Writing Critically

21. Explain why a comet's tail always points away from the sun.

22. Explain why the surfaces of Mercury, Mars, and Earth's moon are pockmarked with craters, while the surfaces of the inner planets, Venus and Earth, are not.

23. All of the outer planets except Pluto are gas giants. What might be a possible reason why Pluto is not gaseous?

24. Why do all the planets revolve around the sun in the same direction?

EXTENSION

1. Some scientists believe that unpiloted space probes can accomplish more than space missions with humans. Collect and report information about the current "piloted versus unpiloted" spacecraft controversy.

2. Review the information sent back from the United States' Viking probes to Mars. Report to the class on what scientists did to test for the presence of life on that planet and what the results of those experiments showed.

3. Review the conditions that exist on another planet. According to the conditions of the planet, describe an imaginary creature that might live there. For example, a Venusian would have to breathe carbon dioxide and be nearly impervious to scalding heat and acid rain.

APPLICATION/CRITICAL THINKING

1. The mean distance of the planet Jupiter from the sun is 7.783×10^8 km. If Earth and Jupiter are aligned on the same side of the sun, how long will it take a radio transmission from a space probe in orbit around the giant planet to reach Earth?

2. Explain possible reasons why the asteroids did not form a planet.

3. As the engineers working with the United States space program were preparing to send space probes to the moon, they would estimate where the moon would be several days after launch and then try to hit that spot with the probe. Why could they not aim for the location of the moon on the day of the probe's launch?

FOR FURTHER READING

Gallant, R. A. *National Geographic Picture Atlas of Our Universe.* Washington, D.C.: National Geographic Society, 1986. This beautifully done volume summarizes the wonders of the solar system. The work includes photographs taken by space probes as well as illustrations of bodies that share the neighborhood of the sun.

Weinhouse, B. "Leading Ladies of the '80s: Space Age Mother." *Ladies Home Journal* 102 (May 1985): 139. This brief article describes the life of astronaut Anna L. Fisher.

Challenge Your Thinking

You may recall from reading the chapter that Jupiter gives off about twice as much energy as it receives. Scientists believe that this energy comes from nuclear reactions within the planet. With this information, hypothesize about the possibility of Jupiter being a second sun with its own solar system.

CHAPTER

The Universe

On any clear night you can see thousands of stars in the sky. People have been looking at and studying stars for thousands of years. How many stars are there? How far away are they? Could you travel to one? Are stars just lights in the sky? Many of these questions can be answered by studying the night sky.

The Orion Nebula

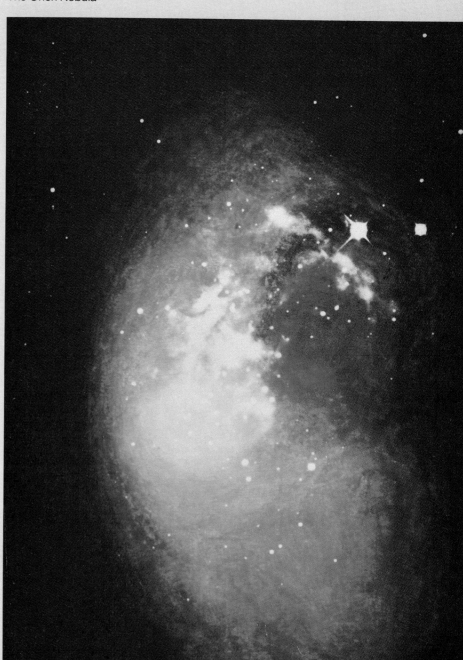

1 Stars

SECTION OBJECTIVES

After completing this section, you should be able to:

- **Describe** how a star forms.
- **Discuss** the meaning of magnitude for a star.
- **Identify** the different types of variable stars.
- **Tell** how the color and the temperature of a star are related.

NEW SCIENCE TERMS

universe
magnitude
variable stars
constellations
zodiac
neutron star
pulsar
black holes

22.1 Stars

When you observe the sky on a clear night, you can see many, many stars. What are stars, and where did they come from? Stars are huge masses of glowing gases, mainly hydrogen and helium.

Scientists believe stars form from clouds of dust and gases in the universe. The **universe** is all the matter and energy that exists: all the stars, planets, dust, gases, and energy in space. Gravity causes the clouds in space to contract. The pressure at the center increases, and the temperature of the center becomes hotter and hotter. When the temperature reaches about 1 million°C, nuclear fusion begins. Hydrogen atoms fuse, creating helium and vast amounts of energy. The mass begins to glow, and the star shines.

The brightness of a star can be measured by an instrument called a *photometer.* The photometer gives a number that is the star's magnitude. **Magnitude** indicates the star's brightness as it appears from Earth—not its actual, or absolute, brightness. The magnitude of a star depends on two things: how much light it emits and how close it is to Earth. A dim star that is nearer Earth may appear more brilliant than a bright star that is farther away.

Figure 22–1. On a clear night, many stars in the Milky Way Galaxy can be seen from Earth.

Figure 22–2. The magnitude, or brightness, of stars varies. The bright star shown here has a lower magnitude than the stars around it.

Figure 22–3. Shown here are a variable star (left) and a binary star system (right).

The ancient Greeks classified stars into six magnitudes. First-magnitude stars appear the brightest. The second, third, fourth, and fifth magnitudes include stars of decreasing brightness. Stars of the sixth magnitude are barely visible to the unaided eye. Modern telescopes have made it possible to see stars even as faint as those of the twenty-fifth magnitude. Negative magnitudes are used for very bright objects. The sun, for example, has a magnitude of –27.

Some stars are called **variable stars** because they change magnitudes. One type of variable star is the *pulsating variable.* Energy from below the surface of the star heats the gases of the visible, shining surface. The surface becomes hotter and brighter, and it expands. As the surface expands, the gases cool and the star becomes dimmer. The cooler gases contract, and a new cycle begins as the gases again become hotter and the surface expands.

The first pulsating variable discovered was *Mira,* whose magnitude changes every 331 days. The time from bright to dim and back to bright again is the star's period. Mira is a long-period variable. The period of most of the thousands of other pulsating variables that have been discovered is 100 days or less. Astronomers call these short-period variables *cepheid* (SEF ee ihd) *variables.* The North Star, for instance, is a cepheid variable with a period of about 4 days.

Other variable stars are *eclipsing binaries* and *exploding stars.* Eclipsing binaries are pairs of stars that move around each other. Their brightness appears to vary as one star periodically blocks the light from the other star. Exploding variables are stars that have bursts of energy that make them appear many thousands of times brighter for days or even years. An exploding star is called a *nova* or a *supernova* because in ancient times it was thought to be a new star; *nova* is Latin for "new." A nova eventually returns to its original brightness, but a supernova is such a huge explosion that the star blows itself apart.

○ *What is a star?*
○ *What is the system that measures a star's apparent brightness as seen from Earth?*
○ *Name three kinds of variable stars.*

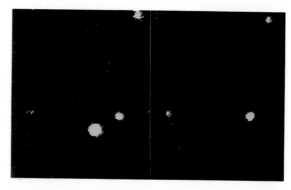

22.2 Constellations

Have you ever looked at clouds floating in a blue sky and imagined you saw pictures in the cloud shapes? Ancient astronomers looked at the stars scattered across the sky and saw pictures of animals, humans, and other objects in certain groups of stars. Rather like the way you might make a dot-to-dot picture, ancient astronomers filled in shapes around groups of stars. The groups of stars that form images are the **constellations.** The ancient Greeks recognized 48 constellations. Today 88 are recognized, and different areas of the sky are named for the constellation located in that area.

Imagine that the constellations and their stars are painted on a glass sphere that surrounds the earth. This is called a celestial sphere. Just as a globe shows Earth with an equator midway between its poles, the sphere has a midline called the *celestial equator.* The sphere also has a line that represents the earth's orbit, called the *ecliptic.* The twelve constellations along the ecliptic form the signs of the **zodiac.** These twelve constellations roughly parallel the twelve months of the year.

DISCOVER

Observing Constellations

Using the star chart on page 584 of the Reference Section, locate as many constellations of the zodiac as you can. Draw the constellation that corresponds to your zodiac sign and write a short report on its features.

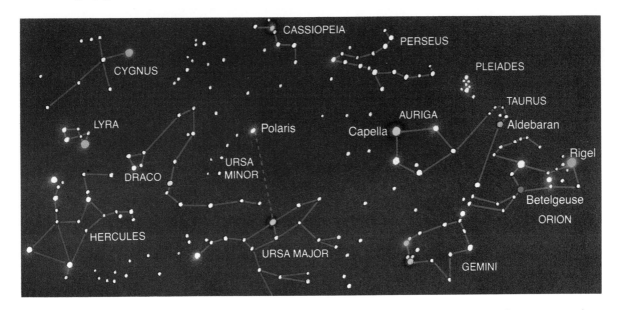

Figure 22–4. Some groups of stars form identifiable figures in the sky.

Figure 22–4 shows several of the constellations of the zodiac. You can use it to identify the major stars of the night sky. Star color can also aid in identification. Some stars, like Betelgeuse, are reddish in color, and others, like Sirius and Rigel, are bluish. You can locate these stars on the star chart and verify their colors by observing the night sky.

○ *What is a constellation?*
○ *What is the celestial equator?*
○ *What is the zodiac?*

ACTIVITY: Making a Chart of the Horizon

How can you make a chart of the local horizon to aid in locating the constellations?

MATERIALS (per student)

magnetic compass, drawing compass, construction paper, colored pencils, protractor, metric ruler

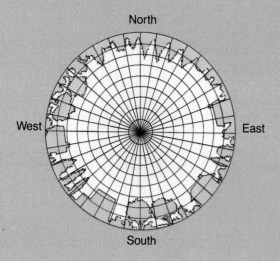

PROCEDURE

1. Use the magnetic compass to locate north, south, east, and west.
2. Use the drawing compass to draw a dark circle 18 cm in diameter on the construction paper and label the circle as shown.
3. Using a pencil, sketch the horizon so that natural landmarks correspond with their true locations on the horizon.
4. In the evening, locate familiar stars and constellations and mark their positions on the chart. You may use the astrolabe you constructed for Chapter 21, or a protractor, to help you determine the position of the constellations above the horizon.
5. Note the date and time that you made your chart.

CONCLUSIONS/APPLICATIONS

1. Why is it important to note the date and time that you made your chart?
2. What information could you gather from charts drawn at the same time every evening during the course of a year?

22.3 Red Giants, White Dwarfs, and Novas

The energy output of stars varies tremendously. Some stars emit only one one-thousandth as much energy as the sun; other stars emit 100 000 times more energy than the sun. The more energy a star emits, the hotter it is. The color of a star and its temperature are closely related. The hottest stars are bluish in color, and the cooler ones are reddish.

In 1913 American astronomer Henry Norris Russell and Danish astronomer Ejnar Hertzsprung published a graph showing that the energy emitted by a star is related to its color. This graph is known as the *Hertzsprung-Russell diagram.*

Most of the stars that have been plotted so far form a diagonal line on the diagram called the *main sequence.* Hot, blue stars are at the top left; cooler, red stars are at the bottom right. Yellow stars, like the sun, are plotted near the middle.

Figure 22–5. A star's color is based on the amount of energy the star emits.

Besides color differences, the Hertzsprung-Russell diagram shows star masses. Star mass increases from the bottom to the top of the main sequence. The stars at the bottom right have masses about one-tenth that of the sun. At the top left, there are stars with masses more than ten times greater than that of the sun. The largest stars are called *supergiants*. The smallest stars are called *dwarfs*. Although small in size, dwarf stars are extremely dense. Supergiants are large, but they may be even less dense than the earth's outer atmosphere.

Figure 22–6. This graph, called a *Hertzsprung-Russell diagram*, shows the relationship between a star's color and its energy.

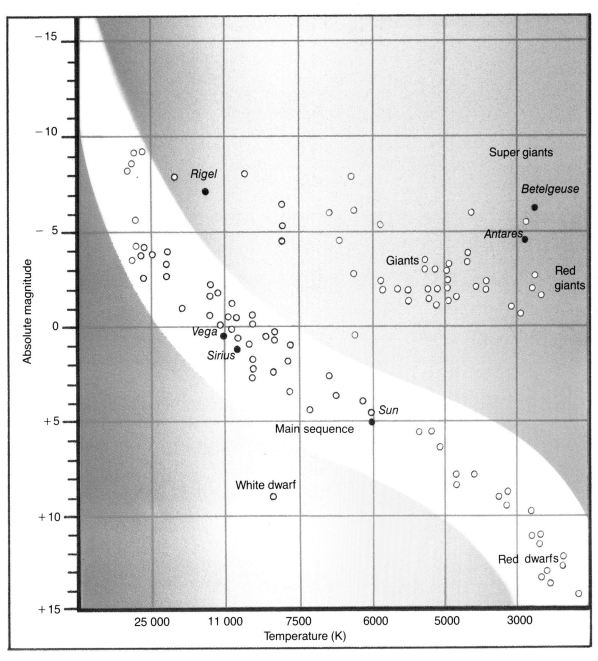

The Hertzsprung-Russell diagram can also be used to show the life cycle of stars. All stars produce energy by nuclear reactions. A star with a mass similar to the mass of the sun produces energy for about 10 billion years. The nuclear reaction in smaller stars is slower, so smaller stars last longer than larger stars.

Scientists believe that the sun is about 4.6 billion years old. It probably has enough hydrogen left in its core to continue shining as it does now for at least another 4.6 billion years. According to the most accepted theory, when the sun is about 10 billion years old, all the hydrogen in its core will be gone. Then the core will contract, its temperature will rise, and its surface will expand outward. After another 100 million years, the sun will become a red giant, its surface reaching halfway to Venus. Then the center of the sun will get hotter, its outer layer will blow off, and the sun will become a white dwarf. Eventually, the nuclear reactions in the center will stop. The sun will be a small star, no more than 10 000 km in radius, with a very high density. It will continue shining for billions of years, slowly cooling. When completely burned out, it will be a cold cinder called a *black dwarf*.

Figure 22–7. Shown here is the life cycle of a main-sequence star.

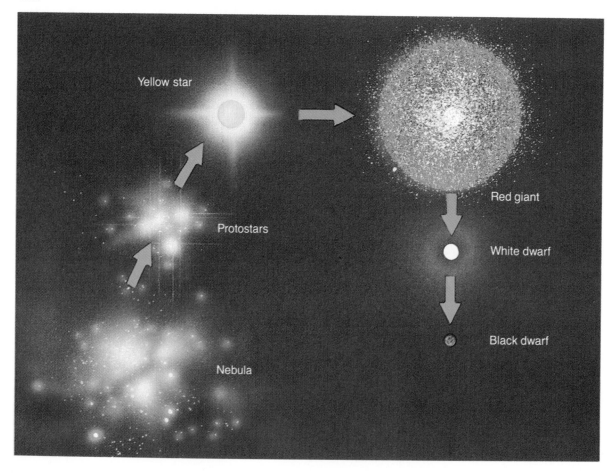

If a white dwarf is part of a binary system, it may undergo a different change. Instead of slowly cooling, it may capture hydrogen from its companion star. When the temperature of the captured hydrogen reaches 10 million°C, a nuclear explosion occurs. The star increases its brightness about 10 000 times and appears as a nova. A nova reaches its maximum brightness in a day or two and then gradually returns to its original intensity.

○ *What does the Hertzsprung-Russell diagram show?*
○ *How will a yellow star, like the sun, change when its supply of hydrogen is gone?*

22.4 Neutron Stars and Pulsars

A star that has a large mass has a different life cycle from that of a smaller star like the sun. When the hydrogen in the core of a large star is depleted, the core can no longer support the weight of the outer layers. It contracts with such violence that the entire star blows up. Actually it is a two-way explosion; the core implodes, or is blown inward, and the rest of the star explodes outward. An exploding star is called a *supernova*, and it produces more energy than the sun will produce in its entire lifetime. Astronomers can see part of this energy as light that is as bright as 500 million suns.

Figure 22–8. Novas occur when white dwarfs explode. Novas are like supernovas, except that the total mass of a nova is much less than that of a supernova.

Figure 22–9. A nebula, such as the Crab Nebula shown here, can be created by the explosion of a supernova.

During a supernova explosion, 90 percent of the star's mass scatters into space, becoming the matter from which new stars may be born. The other 10 percent, the core of the star, is blown inward, becoming a neutron star. A **neutron star** is a very small star, as small as 15 km in radius. Neutron stars rotate at high speed and have strong magnetic fields. They are also very dense. If the Great Pyramid of Cheops in Egypt, which weighs 3 million tons, had the density of a neutron star, it would be the size of a pinhead.

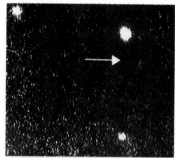

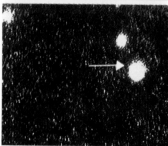

Figure 22–10. Pulsars (above) are stars that seem to blink on and off. Pulsars form during the life cycle of giant stars (right).

A neutron star may capture gas from space, or from a companion star, or nearby star. Perhaps the captured gas becomes ionized and, as the neutron star rotates, pulses of light are emitted, similar to the beacon of a lighthouse. The first rapidly pulsating neutron star, or **pulsar,** was discovered in 1967. Scientists are not certain what causes pulsars. Over 350 pulsars have been discovered since then.

If the core mass of a supernova is more than three times greater than the sun's mass, the force of the implosion collapses the core into itself so that no energy, including light, can be emitted. Collapsed stars that emit no energy are called **black holes.**

○ *What is a supernova?*
○ *What is a neutron star?*
○ *What is a black hole?*

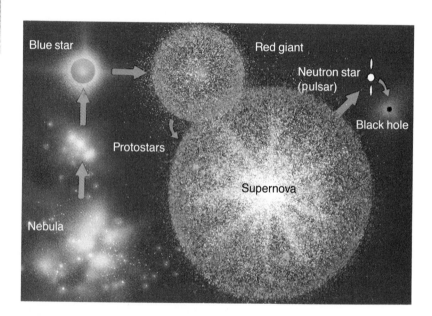

Section Review

READING CRITICALLY

1. How does a star form?
2. What are some of the differences among stars?
3. How does a star of great mass die?

THINKING CRITICALLY

4. What does it mean to say that "looking at stars is looking at the past"?
5. What can astronomers learn about the sun from studying other stars?

SKILL ACTIVITY: Using a Star Chart

BACKGROUND

The star chart below displays a map of the most familiar constellations visible in the Northern Hemisphere. However, not all of these constellations can be seen on a single night, since the view of the celestial sphere from Earth changes from day to day as our planet revolves around the sun. In this activity, you can create your own star chart to help you become familiar with the relative positions of these constellations.

PROCEDURE

1. Measure the radius of the star chart with a compass and draw a circle of the same size on construction paper. Cut out the circle with scissors.
2. Trace the star chart on tracing paper and tape the tracing paper neatly to the circular piece of construction paper.
3. Draw another, slightly larger circle (with a radius that is larger by 1 cm) and cut it out. Drawing through the center of that circle, use a ruler to divide the circle evenly into 12 slices. At the rim of the circle, label the slices, numbering 1 through 12 to represent the months in a year.
4. Place the tracing of the star chart on the large circle and pin both pieces of construction paper, at their centers, to a piece of cardboard.
5. Study the positions of the constellations by comparing their locations relative to the North Star, or polestar.

APPLICATION

Use the star chart to find some of the constellations by positioning your chart in front of you so that the constellations on the celestial sphere match the direction you are facing. Locate the constellations Hercules and Orion.

1. What are the positions of the constellations Hercules and Orion relative to the *North Star*?
2. Explain why not all constellations in the northern celestial sphere are visible at once.

USING WHAT YOU HAVE LEARNED

Use your star chart to locate as many of the constellations as you can.

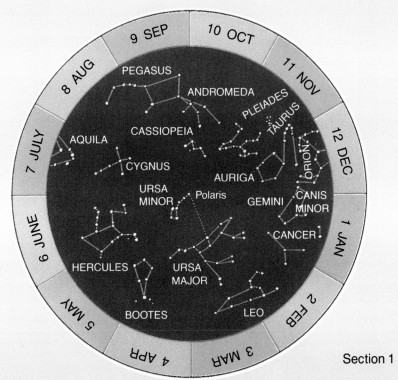

SECTION OBJECTIVES

After completing this section, you should be able to:

■ **Name** the different layers of the sun.

■ **Describe** the activity on the sun.

22.5 The Composition and Structure of the Sun

Astronomers say that the sun is just an average star. However, to the inhabitants of Earth it is the most important star in the universe, for life on Earth depends on the sun's energy.

The sun's radius is 696 000 km, 109 times greater than the radius of the earth, while its mass is over 330 000 times greater than the mass of Earth. In fact, the sun contains about 99.85 percent of all the mass of the solar system.

At the center of the sun is a dense core of hydrogen and helium, which is at a temperature of about 15 million°C. Above the core is a thick *radiative layer.* The energy produced in the core warms this layer just as heat from a radiator warms a room. In this layer, the temperature averages about 3 million°C. Next is the *convective layer,* where energy is transferred by convection. The temperature at the top of this layer is only 8000°C.

Figure 22–11. You may think of the sun simply as a source of light and heat. The complex structure of the sun produces this energy (right).

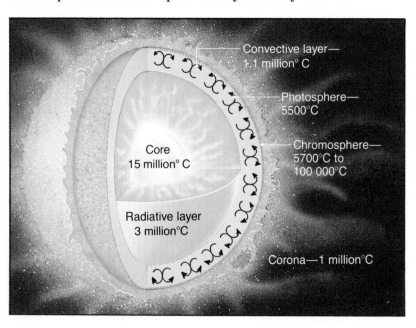

Convective layer—
1.1 million° C

Photosphere—
5500°C

Core
15 million° C

Chromosphere—
5700°C to
100 000°C

Radiative layer
3 million°C

Corona—1 million°C

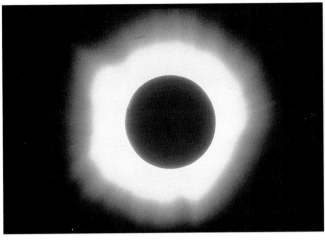
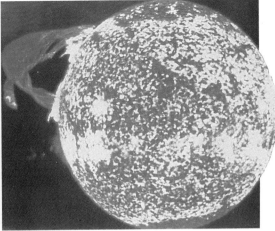

Above the convective layer is the sun's atmosphere, which has three layers. The *photosphere* is the inner layer of the sun's atmosphere, which forms the visible surface of the sun. The photosphere is similar to the surface of a pan of sauce; it boils and bubbles in much the same way.

The temperature of the photosphere is about 5500°C. The surface of the photosphere has many small, bright areas called *granules*. The granules, shown in Figure 22–12, are actually the tops of rising columns of hot gases. The dark areas between granules are sinking areas of cooler gases.

Above the photosphere is the *chromosphere*, a 2500-km-thick layer, where the temperature increases to 100 000°C. Above the chromosphere is the *corona*, a layer of thin solar gases that merge with outer space. The temperature in the corona is approximately 1 million°C.

○ *Name the layers of the sun and its atmosphere, starting from the core.*
○ *What are granules?*

Figure 22–12. During a solar eclipse (left), the corona of the sun is sometimes visible. At other times, it is possible to observe the surface of the sun (right), using a telescope equipped with a special filter.

22.6 **Motions and Activity of the Sun**

The sun rotates on its axis in the same direction as the planets revolve, that is, counterclockwise as seen from above the sun's north pole. Since the sun is a gaseous body, not all parts rotate at the same speed; the rotational speed decreases toward the sun's north and south poles. At the solar equator, one sun day is equal to about 25 Earth days.

Solar energy is produced by the fusion of hydrogen nuclei into helium nuclei. The helium nucleus that is the end product of the fusion reaction has two protons and two neutrons. Scientists calculate that it takes about 2 million years for energy from the sun's core to reach its surface and 8.3 minutes to reach Earth.

DISCOVER

Calculating Time

Using the table on page 464, calculate how long it takes for the sun's energy to reach each of the planets.

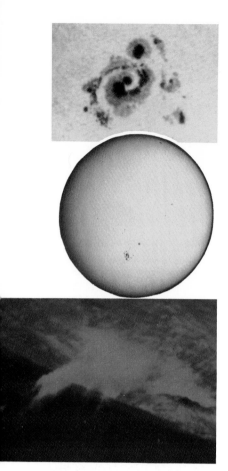

Figure 22–13. Sunspots (top and center) are the result of magnetic activity in the sun. Another form of solar activity is a solar prominence (bottom).

The convective layer and the photosphere of the sun consist of ionized atoms and free electrons. Turbulence in these layers moves the charged particles around and creates local magnetic fields. These fields tend to combine with each other as magnets do, forming larger fields. When a magnetic field grows to about 2000 km across, convection stops. The top of the field radiates energy away and becomes cooler. These cooler areas appear dark from Earth. They combine, forming larger dark areas called *sunspots*.

Sunspot formation follows an 11-year cycle. During a sunspot minimum, the surface of the sun may be completely free of sunspots, while at a sunspot maximum, several major groups may dot the surface of the sun at any given time. In sunspots, charged particles such as protons, electrons, and ions speed up until they escape from the sun's atmosphere. These particles cause an increase in strength of the solar wind. *Solar flares*, spectacular eruptions that occur near sunspots, cause periodic increases in the solar wind as well.

You may recall from Chapter 8 that the solar wind particles, which reach Earth in about 2.5 days, are responsible for disturbances in the earth's ionosphere. A particularly strong solar wind can disturb the earth's magnetic field, causing interference in radio and television transmissions and in microwave communications.

Another impressive display of solar activity is the solar prominence, shown in Figure 22–13. **Solar prominences** are streams of gases that explode above the sun's surface. They eventually fall back to the sun's surface under the pull of gravity.

○ *What are sunspots?*
○ *What is the solar wind?*
○ *What are solar flares and solar prominences?*

Section Review

READING CRITICALLY

1. How does sunlight reach the earth?
2. How do activities on the sun affect the earth?

THINKING CRITICALLY

3. Why is Earth the only planet in the solar system known to have life?
4. Why would it take 2 million years for energy from the sun's core to reach its surface but only 8.3 minutes for it to reach the earth?

3 Galaxies

22.7 Measuring Distances in the Universe

Stars generally are very far from each other. If the sun were a marble 1 cm in diameter, the nearest marble, or star, would be over 300 km away. As modern instruments make it possible to discover stars at greater distances from Earth, a unit was needed to measure these distances. The unit scientists have chosen to use is called the **light-year**—the distance light travels in one year. At a speed of over 300 000 km/sec, this is about 9.5×10^{12} km.

In 1838 German astronomer Wilhelm Bessel determined that the star known as 61 Cygni was 11 light-years away from the earth. This means that it has taken 11 years for the light from 61 Cygni to reach Earth. There are stars closer to Earth than 61 Cygni. The closest star, Proxima Centauri, is 4.26 light-years away.

○ *What is a light-year?*
○ *What star is closest to the earth?*

DISCOVER

Traveling Fast

Determine the relationship between astronomical units and the speed of light. Then determine how long it would take to travel from Earth to Pluto. Use the table on page 464 to find the astronomical units that you need to use.

Figure 22–14. Proxima Centauri, part of the bright star system shown here, is the closest star (other than the sun) to Earth.

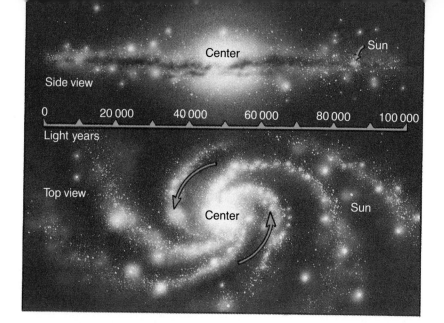

Figure 22–15. The earth is part of the Milky Way Galaxy, shown here in relation to the sun.

Figure 22–16. There are many types of star clusters. Shown here are two examples, the globular cluster (top) and the open cluster (bottom).

22.8 The Milky Way

Have you ever noticed, on a very clear night, that there is a ribbon of stars so dense that it looks like a starlit cloud? This ribbon, called the *Milky Way Galaxy*, consists of billions of stars. A **galaxy** is a large grouping of stars. All the stars you see with your unaided eye belong to the Milky Way Galaxy. The Milky Way Galaxy contains about 180 billion stars and a large amount of interstellar gases and dust. The Milky Way Galaxy is a spiral galaxy; it is shaped like a disk that bulges in the middle and has curved arms.

The Milky Way Galaxy is 100 000 light-years across and 10 000 light-years thick at the center, decreasing to 1000 light-years at the edge. The sun is on the inner rim of one arm of the galaxy, about 30 000 light-years from the center and midway between the upper and lower edges of the galaxy. As the Milky Way Galaxy rotates, the sun travels around the center of the galaxy at a speed of 250 km/s, making one complete turn every 250 million years.

The average distance between stars in the Milky Way Galaxy is several light-years. Near the sun, the average distance between stars is four to five light-years. All around the Milky Way Galaxy, there are groups of stars that are close together. These stars form star clusters. The stars in a cluster seem to be arranged in a spherical mass. They are probably the galaxy's oldest stars, estimated to be about 10 to 15 billion years old. Each cluster contains over 100 000 stars and has little or no interstellar dust.

○ *What is a galaxy?*
○ *In what galaxy is the sun?*
○ *What is a star cluster?*

22.9 Galaxies, Quasars, and the Big Bang

The Milky Way Galaxy is just one of billions of galaxies that exist in the universe. One of these galaxies is visible without a telescope and appears as a hazy star or cloud, called a *nebula*. A **nebula** is a cloud of dust and gases in space, which sometimes glows by reflected light. When astronomers began using telescopes, they found that some of these clouds were actually galaxies. Other nebulas are not galaxies, but the remains of exploded stars. For instance, the Crab Nebula, in the constellation Taurus, is the result of the explosion of a supernova in 1054.

Figure 22–17. The Magellanic Clouds are the closest galaxies to the Milky Way Galaxy.

The two galaxies closest to Earth, known as the Large Magellanic Cloud and the Small Magellanic Cloud, are about 170 000 light-years away. The Great Spiral Galaxy, visible in the constellation Andromeda, is about 2 million light-years away.

Galaxies are named for their shapes as they appear from Earth. Spiral and irregular galaxies contain a lot of dust and gases, in addition to stars. Elliptical galaxies contain very little dust and gas. Some scientists believe that spiral and irregular galaxies have a black hole at their center around which the galaxy rotates. Elliptical galaxies either do not rotate or rotate very slowly.

Figure 22–18. Galaxies have several shapes (left). The Southern Pinwheel Galaxy (right), in the constellation Hydra, is an example of a barred galaxy.

| Elliptical | Spiral | Barred | Irregular |

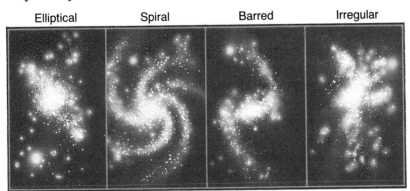

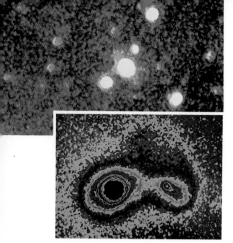

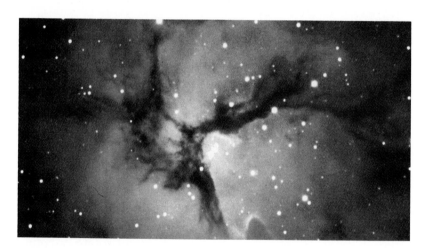

In the 1960s astronomers located what appear to be the largest energy sources in the universe. **Quasars** emit radio waves and light. Although they are several billion light-years away, they may be the brightest objects in the universe, emitting the energy of hundreds of galaxies. However, scientists still do not fully understand the nature of quasars.

How did the universe come to be? Physicists claim that the universe began about 15 billion years ago. Currently, the most accepted theory for the beginning of the universe is the big-bang theory. According to the **big-bang theory,** a tremendous explosion created the universe.

If, as American astronomer Edwin Hubble observed in 1929, the universe is expanding, it must have been smaller at one time. If the universe were run in reverse, it would contract to a densely packed mass. Probably about 15 billion years ago, the mass exploded into a giant cloud of gases and dust that has been expanding and cooling ever since.

○ *What are the three kinds of galaxies?*
○ *What is a quasar?*
○ *What is the big-bang theory of the origin of the universe?*

Figure 22–19. Quasars are most easily observed using radiotelescopes. The "pictures" that astronomers receive look like those above.

A MATTER OF FACT

If time since the big bang were compressed into 24 hours, Earth would not have formed until sunset, and humans would have been on Earth for only one second.

Figure 22–20. The big-bang theory explains one way in which the universe may have originated. Shown here is an artist's conception of how the initial explosion may have looked.

Section Review

READING CRITICALLY

1. Why are some galaxies originally thought to be nebulas?
2. Why is the universe expanding?

THINKING CRITICALLY

3. Why has the study of astronomy changed so much over the years?
4. Why would the density of the universe relate to its future?

CAREERS

ASTRONOMER

The study of the movements of stars, planets, and galaxies is called *astronomy*. An *astronomer* is trained to observe, catalog, and draw conclusions about the patterns of movement of stars and other heavenly bodies.

Professional astronomers attend colleges and universities and usually have a master's degree or a doctorate. However, many amateur astronomers with only a high-school education have also made important contributions to the study of the universe.

The universe is vast, and the catalog of objects continues to grow with the work of both professionals and amateurs.

For Additional Information
Amateur Astronomers, Inc.
Sperry Observatory
1033 Springfield Avenue
Cranford, NJ 07016

ASTROPHYSICIST

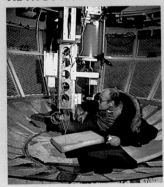

An *astrophysicist* studies the physical characteristics of the universe. Such a scientist needs a great deal of training in the physical sciences, especially physics and mathematics. Most astrophysicists have doctorates, and they teach and conduct research at universities. They must have a clear understanding of current theories regarding the relationship between the matter and energy that comprise the universe.

The major divisions of astrophysics include the following studies: solar system, stellar atmospheres, nebulae and interstellar matter, and stellar evolution.

For Additional Information
American Astronomical Society
c/o Dr. Peter B. Boyce
2000 Florida Ave., N.W. #300
Washington, DC 20009

ASTRONAUT

The word *astronaut* means "star voyager." Since exploration of the heavens is both challenging and dangerous, astronauts must be highly trained before undertaking missions above Earth's atmosphere. They must be well educated, confident, and resourceful.

The first American astronauts were military test pilots. However, since that time research scientists have joined the space program.

In the 1960s only 43 men were assigned as astronauts on the Mercury, Gemini, Apollo, and Skylab projects. Since 1978 many candidates, both males and females, have been selected for assignments aboard the space shuttle.

For Additional Information
Young Astronauts Council
P.O. Box 65432
1211 Connecticut Ave., N.W.
Washington, DC 20036

INVESTIGATION 22: Determining Distance by Parallax

PURPOSE

To learn how parallax is used to determine the distance of stars

MATERIALS (per student)

Drawing compass
Construction paper
Scissors
Tape
Insect pin
Modeling clay
Metric ruler
Protractor
Table of Tangents

PROCEDURE

1. Use the drawing compass to make a half-circle 25 to 30 cm in diameter on the construction paper.
2. Cut out a long strip of construction paper that will represent the "celestial sphere" as shown. Perpendicular to the diameter of the half-circle, draw a line to the middle of the celestial sphere.
3. Tape the celestial sphere to the half-circle and place it at the edge of a flat surface.
4. Draw a small image of a star, and attach it to the insect pin. Stand it up, using the modeling clay as a base. Place the star anywhere along the midline of the celestial sphere.
5. Kneel down and, with the bridge of your nose at the position indicated, sight the distant star with one eye. Mark the position of your eye on the diameter line and draw a line from that point to the star.
6. Measure the distance from your eye position to the position of the bridge of your nose when you made the sighting.
7. Using the protractor, measure the angle made by the line to the star and the line between the sighting eye and nose positions.
8. Use a Table of Tangents to find the tangent of the angle.
9. Calculate the distance to the star, using the formula

$$D_S = T_A \times D_E$$

where D_S is the distance to the star, T_A is the tangent of the angle, and D_E is the distance from the sighting eye to the bridge of the nose.
10. Repeat steps 5 to 9 several times, moving your eye position each time.

ANALYSES AND CONCLUSIONS

1. After taking several measurements, what did you discover about the tangent of the angle as the distance of the star from your nose decreased?
2. If you try closing first one eye and then the other during a sighting, the position of the star shifts back and forth. This shifting is called *parallax* and is useful in measuring the distance to objects. What factors would limit this type of procedure in measuring the distance to many stars?

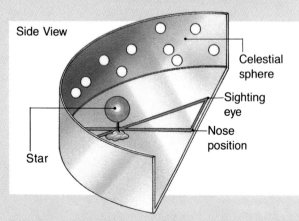

Side View

Celestial sphere

Sighting eye

Nose position

Star

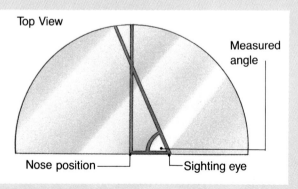

Top View

Measured angle

Nose position

Sighting eye

SUMMARY

- A star is a huge mass of glowing gases, mainly hydrogen and helium, which formed from a swirling cloud of dust and gases. (22.1)

- The magnitude of a star is a number indicating its apparent brightness as seen from the earth. (22.1)

- A constellation is a group of stars in a pattern that seems to form an image. (22.2)

- The Hertzsprung-Russell diagram shows how the energy emitted by a star is related to its color. (22.3)

- A nova is a star that becomes suddenly brighter due to an explosion and then returns to its original brightness. (22.3)

- A neutron star is the small, extremely dense core of a supernova. (22.4)

- A black hole is the core of a supernova that completely collapses into itself. (22.4)

- The sun has a layered structure consisting of the core, the radiative layer, the convective layer, and the atmosphere. The atmosphere has three layers: the photosphere, the chromosphere, and the corona. (22.5)

- Sunspots are cooler areas on the sun's surface that appear dark from Earth. (22.6)

- Solar flares are eruptions of the photosphere near sunspots. Solar prominences are streams of gases that form in the corona and are pulled back to the sun. (22.6)

- A light-year is the distance light travels in a year—it is equal to 9.5 trillion km. (22.7)

- A galaxy is a grouping of stars. The sun is in the Milky Way Galaxy. (22.8)

- A quasar is a starlike source of radio waves billions of light-years away from Earth. (22.9)

- The most widely accepted theory of the beginning of the universe is the big-bang theory. (22.9)

Write all answers on a separate sheet of paper.

SCIENCE TERMS

Correctly use each of the following terms in a sentence.

big-bang theory **(496)**
black holes **(488)**
constellations **(483)**
galaxy **(494)**
light-year **(493)**
magnitude **(481)**
nebula **(495)**
neutron star **(487)**
pulsar **(488)**
quasars **(496)**
solar prominences **(492)**
universe **(481)**
variable stars **(482)**
zodiac **(483)**

SCIENCE QUIZ

Modified True-False

Mark each statement *true* or *false.* If the statement is false, change the underlined word to make the statement true.

1. Most galaxies contain <u>thousands</u> of stars.

2. Star groupings that form a pattern as seen from Earth are called <u>galaxies</u>.

3. Cepheid-variable stars change magnitude in periods of 100 days or <u>less</u>.

4. According to the big-bang theory, the universe may continue to <u>expand</u>.

5. Rapidly rotating neutron stars are called <u>quasars</u>.

6. The theory that explains how the universe may have formed is called the <u>big-bang</u> theory.

continues

Multiple Choice

Write the letter of the choice that best answers the question or completes the statement.

7. What elements fuse in the reactions of the sun?
   a) hydrogen and oxygen
   b) hydrogen
   c) hydrogen and carbon
   d) all heavy elements

8. How far away is the nearest star?
   a) 4 260 000 kilometers
   b) 4.26 astronomical units
   c) 4.26 light-years
   d) 10 light-years

9. The most distant objects in the sky are known as
   a) quasars.
   b) galaxies.
   c) supernovas.
   d) spiral galaxies.

10. How old is the universe according to the big-bang theory?
    a) 4.5 million years
    b) 4.5 billion years
    c) 15 million years
    d) 15 billion years

11. The leftover core of a supernova may be any of the following except a
    a) black hole.   b) pulsar.
    c) quasar.   d) neutron star.

12. Most stars of the main sequence, such as the sun, will eventually end up as
    a) black holes.
    b) black dwarfs.
    c) white dwarfs.
    d) supernovas.

Completion

Complete each statement by supplying the correct term or phrase.

13. The _____ theory states that an explosion created the universe.

14. A _____ magnitude star is the brightest.

15. The life cycle of stars is represented on the _____ diagram.

16. The center of a rotating spiral galaxy may contain a _____.

17. The universe will continue to expand if the theory of _____ is correct.

Short Answer

18. Describe several types of galaxies.

19. Explain how energy from the sun's core reaches its surface.

20. Describe the differences between the types of star clusters.

21. Name four different types of stars.

22. List the phases of the sun's evolution from its birth to its death.

23. Draw the arrangement of stars in five constellations.

Writing Critically

24. Explain why the energy of the sun's core takes so long to reach the sun's surface.

25. Compare and contrast opposing theories about the future of the universe.

EXTENSION

1. Research the steady-state theory of the universe; then write a report contrasting this theory with the currently accepted big-bang theory.

2. Research and report on the methods astronomers use to approximate the distance to the stars. Be sure to include the advantages and limitations of each method.

3. Make a table showing the distances to several close stars in kilometers, astronomical units, and light-years.

APPLICATION/CRITICAL THINKING

1. Calculate the distance to Proxima Centauri in kilometers.

2. Describe how variable stars might be used to calculate stellar distances.

3. Since black holes give off no energy, how can they be detected?

4. Most scientists believe that the universe is still expanding from the big bang. Some scientists believe the universe will eventually start to contract, and that it will continue contracting until another big bang occurs. Describe the events that might take place during this "big crunch."

FOR FURTHER READING

Davies, O. "Comet Collectors." *Omni* 8 (July 1986): 24. This article is a brief essay describing the successful attempts by scientists to use existing satellites for gathering information on the comets Giacobini-Zinner and Halley.

Heidmann, J. *Extragalactic Adventure: Our Strange Universe.* New York: Cambridge University Press, 1982. This book describes the evolution of the universe, space, time, black holes, and the possible existence of extraterrestrials.

Mitton, S. and J. Mitton. *Invitation to Astronomy.* New York: Basil Blackwell, 1986. The authors explain what astronomy is all about and what astronomers do. It is an excellent book for anyone who is considering a career in astronomy.

Wolkomir, R. "SS-433, What Are You?" *Omni* 8 (September 1986): 30. This article describes the star designated SS-433, which is now known to be a binary star. Before an in-depth analysis was performed on the object, no one was quite sure what it was.

Challenge Your Thinking

This photograph is an artist's idea of a black hole. In some movies the crew of spacecraft travel through a black hole to return home safely. Using your knowledge of black holes, explain why this would not be possible.

Observing the Universe

Stars, gas clouds, novae, and neutron stars all emit energy. Cosmic rays, gamma rays, X rays, ultraviolet rays, visible light rays, infrared rays, and radio waves all travel through space at the speed of light. How are these rays detected? What instruments do astronomers use?

Kitt Peak Observatory, in Arizona

SECTION
1 Optical Astronomy

SECTION OBJECTIVES

After completing this section, you should be able to:
- **Discuss** the importance of telescopes.
- **Explain** what refraction and reflection are.
- **Define** what is meant by *focus* and *focal length*.
- **Describe** the difference between refractors and reflectors.

NEW SCIENCE TERMS

refraction
reflection
lens
focal length
telescope
magnification
spectrograph

23.1 Lenses and Light

Have you ever looked at the millions of stars in the night sky? The stars shine because of the energy they give off. The energy you see is in the form of light. You may recall from Chapter 8 that light travels through space as particles called *photons*.

Visible light is a very small part of all the energy in the universe. Some other forms of energy include radio waves, X rays, infrared radiation, and ultraviolet radiation. Visible light, however, is the easiest for people to detect. Light travels through space at the constant speed of 299 792 458 m/s (3.0×10^8 m/s). Light also travels through transparent matter but at slower rates of speed. Table 23–1 shows the speed of light in various substances. Notice that the speed of light in air is similar to the speed of light in space.

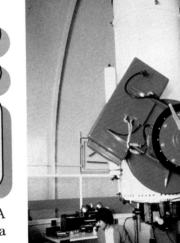

Figure 23–1. Observing the stars is not simply a matter of looking through a telescope. Modern astronomical telescopes are controlled by computers.

TABLE 23–1: THE SPEED OF LIGHT IN DIFFERENT MEDIA	
Substance	**Speed of Light (m/s)**
Empty space	299 792 458 (3.0×10^8)
Air	299 702 547 (3.0×10^8)
Water	225 407 863 (2.2×10^8)
Glass	197 231 880 (2.0×10^8)
Diamond	123 881 181 (1.2×10^8)

Refraction Light behaves in certain predictable ways. A beam of light traveling through air will slow down if it strikes a glass pane. If the beam hits the glass at a right angle, even though it slows down, it will come out of the glass going in the same direction as when it went in. If a light beam strikes a glass pane at an angle other than 90 degrees, it not only slows down but also changes direction.

Consider a drill team marching in formation. As they come to a corner, the person on the end of a line turns the corner first and then marches in place. The rest of the people in line continue marching until they reach the corner, and then they also turn and march in place. When the entire line has turned, they again begin marching in a straight line.

Figure 23–2. The sun's energy easily passes through clear glass into this room, and is trapped inside as heat.

If the person who turned the corner first did not march in place, then the line would bend as a beam of light does. One side of the light beam hits the glass first and is slowed before the other side even reaches the glass. This causes the beam to bend. As the beam emerges from the other side of the glass, the side of the beam that reached the glass first leaves the glass first and moves faster than the other side. The beam then changes direction again. The change in direction of a light beam as it passes from one type of matter to another is called **refraction.**

Figure 23–3. Light striking a shiny surface is reflected (left). Light that strikes a pane of glass at a 90° angle passes through the glass without refracting (center). If light strikes a pane of glass at a lesser angle, the light is refracted (right).

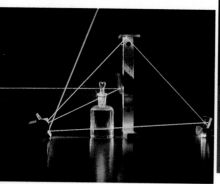

Reflection Light is bounced back from a smooth, shiny surface. This bouncing back of light is called **reflection.** If light hits a shiny surface at any angle other than 90°, it bounces off at an angle. The angle at which the light is reflected is equal to the angle at which the light hit the surface. This property of reflected light enables anyone with a mirror to bounce light in practically any direction.

Lenses The properties of refraction and reflection of light are used in astronomy. Understanding refraction allows scientists to use a series of lenses to focus the light given off by distant stars. A piece of curved glass that can be used to focus light is called a **lens.** If the surfaces are curved outward as shown in Figure 23–4, all light rays will converge, or focus, at a single point. Lenses that cause light rays to converge are called *convex lenses.* Lenses that cause light rays to diverge, or separate, are called *concave lenses.* These lenses curve inward as shown in Figure 23–4. The distance between the center of the lens and the point of focus is called the **focal length.**

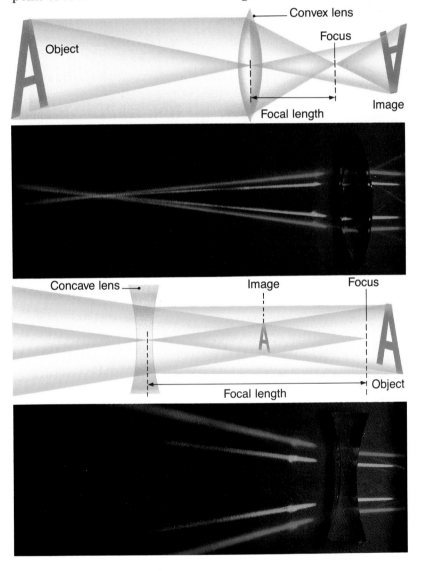

Figure 23–4. Light rays passing through a convex lens will converge at a focus (top). Light rays passing through a concave lens will diverge (bottom).

○ *What is refraction of light?*
○ *What is reflection of light?*
○ *What is a lens?*

Figure 23–5. A refracting telescope, such as the one that Galileo used (above), has convex lenses (right).

23.2 **Refracting Telescopes**

A **telescope** is an instrument used to enlarge the image of a distant object. In 1609 Galileo Galilei, an Italian astronomer and physicist, made a telescope based on the principle of refraction. Although this instrument was primitive, Galileo made many discoveries, such as the craters and mountains on the moon, the phases of Venus, and the four large satellites of Jupiter.

A *refracting telescope* is usually made with two tubes that fit together snugly. Light from the object being viewed passes through the *objective lens.* The light comes to focus inside the tube and then passes through a second lens, called the *ocular lens,* to the observer's eye. The tubes can be moved to get the best focus for the observer.

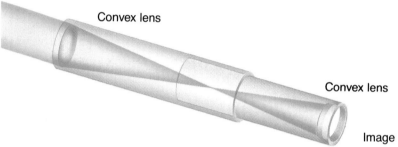

Convex lens

Convex lens

Image

The **magnification** of a telescope is equal to the focal length of the objective lens divided by the focal length of the ocular lens. The combination of a long objective-lens focal length and a short ocular-lens focal length produces a telescope with great magnification. Galileo's first telescope had a magnification of 3. The brightness of the image depends on the size of the surface of the objective lens. The larger the surface area of the objective lens, the greater the amount of light captured. Increasing the amount of light entering a telescope increases the brightness of the image.

Figure 23–6. A modern refracting telescope has basically the same structure as the one that Galileo used in the early 1600s.

In order to improve the quality of refracting telescopes, larger lenses were made. The first large lenses were 38 cm in diameter and were made in Germany in the 1830s. The Yerkes Observatory, founded in 1895 in Williams Bay, Wisconsin, has the largest refracting telescope ever made. The objective lens of the Yerkes telescope is 102 cm in diameter. The glass used to make refracting lenses must be perfectly smooth. For this reason it is very difficult to make large lenses. Lenses larger than that of the Yerkes refractor will probably not be made because the glass in lenses larger than 100 cm in diameter tends to change shape, causing distortion.

Until the end of the last century, most astronomical observations were made with the human eye. Now most observations are made using photographic plates, which astronomers then study. Computers align the telescope, and motors on the telescope keep it focused on an object for hours or even days. This tracking system allows enough time for even very faint objects to leave an image on the photographic plates.

Photographic plates taken of the same area of the universe at different times can be studied in order to see changes in the stars. When a supernova was observed in 1987, studying photographic plates made it possible to identify which star had exploded, since the star simply was no longer there.

Telescopes have also been connected to spectrographs to study the composition of stars. A **spectrograph** is an optical instrument used for observing the spectrum of light given off by a star. A spectrograph can be used to determine the composition of stars because light from each element gives off a unique spectrum. When the spectrum from a star is compared with the spectrum of elements known on Earth, astronomers can identify the elements present in the star.

○ *What is a refracting telescope?*
○ *What is a spectrograph?*

Figure 23–7. Many astronomical observations are made by studying film exposed through a large telescope (left). In some cases, film has been replaced by an electronic plate that records light more precisely than film does (below).

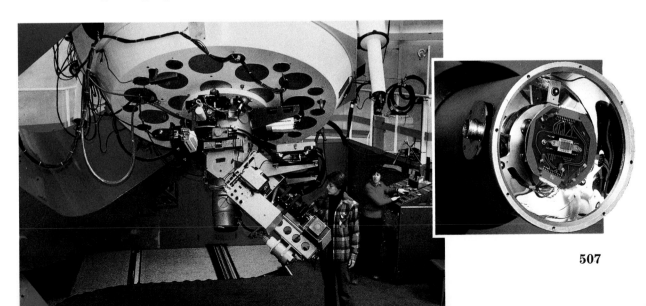

How can you show that white light is refracted as it passes through glass?

MATERIALS (per group of 3 or 4)

glass blocks (2), flashlight, white construction paper

PROCEDURE

1. On a flat surface, stand one glass block on end, point the flashlight at the block, and turn the flashlight on.
2. Hold the white construction paper as shown, and move it around the glass block until the paper reflects an image of the light coming through the glass block.
3. Position the second glass block between the paper and the first block, and note the angle of the light as it passes through both blocks.

CONCLUSIONS/APPLICATIONS

1. How is the white light changed as it passes through the first block? Give a specific description of the nature of the change.
2. What happens to the light passed through the first block when it is intercepted and passed through the second block?

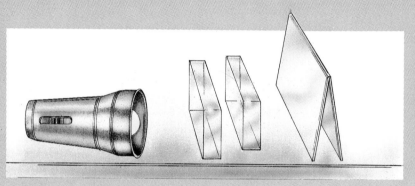

Figure 23–8. This telescope is a replica of the reflecting telescope made by Isaac Newton.

23.3 Reflecting Telescopes

Sir Isaac Newton, an English scientist, constructed the first *reflecting telescope*. Newton realized that concave mirrors have three advantages over lenses. First, they reflect light instead of refracting it and therefore produce no distortion. Second, a good mirror absorbs less light than even the most transparent glass, so the image will be brighter. Third, a mirror is not as heavy as a lens of the same size. Therefore, a mirror is easier to support, and much larger telescopes can be constructed.

Reflecting telescopes have two mirrors. The primary mirror is a large concave mirror that receives the image and reflects it to a secondary mirror. The secondary mirror, a flat mirror placed in front of the focus, reflects the light through an ocular lens to the observer. The secondary mirror is so small in comparison to the primary mirror that the light it intercepts does not affect the quality of the image.

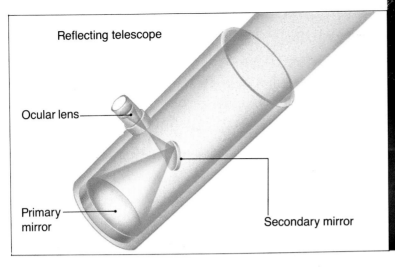

Reflecting telescope

Ocular lens

Primary mirror

Secondary mirror

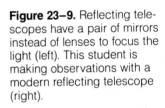

The largest telescopes used today are reflectors. The largest reflector in the world, located on Mt. Semirodriki in the Soviet Union, has a primary mirror 6 m across. The second-largest reflector is located at the Palomar Observatory, near San Diego, California.

Even though the atmosphere lets in visible light, the light is distorted by the atmosphere and seems to flicker. This is the reason why stars seem to twinkle. As a result of this flickering, images produced by ground-based telescopes tend to be a bit fuzzy. There are plans to place into Earth orbit a large reflector for the observation of distant space. A telescope operating outside the atmosphere will produce sharper images and allow astronomers to study objects in space in greater detail.

Figure 23–9. Reflecting telescopes have a pair of mirrors instead of lenses to focus the light (left). This student is making observations with a modern reflecting telescope (right).

Figure 23–10. The primary mirror of a reflecting telescope can be quite large (right). Astronomical observatories are often placed on mountain tops to eliminate much of the distortion caused by the earth's atmosphere (left).

Infrared radiation, or heat energy, from stars is studied, as is visible light, with reflecting telescopes. One problem is that surrounding buildings and the telescope are warm and, therefore, produce their own infrared radiation. A photographic plate at the focus of an infrared reflector receives more infrared radiation from the mirror itself than it does from space. To minimize this problem, infrared telescopes are placed on top of remote mountains and cooled with liquid nitrogen or liquid helium.

○ *How does a reflecting telescope focus light?*

Figure 23–11. Shown here are infrared photographs of a nova (above) and of the edge of the Milky Way Galaxy (right).

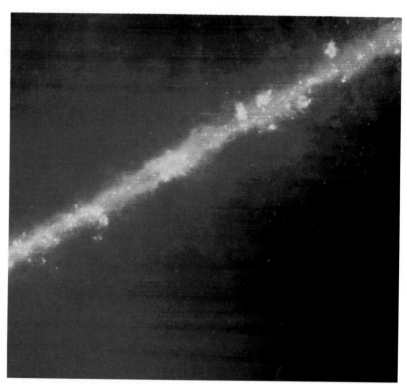

Section Review

READING CRITICALLY

1. What is the difference between a reflecting telescope and a refracting telescope?
2. What advantages do reflecting telescopes have over refracting telescopes?
3. Why is the spectrograph a useful instrument in astronomy?

THINKING CRITICALLY

4. If the city of San Diego expands, what will happen to the effectiveness of the Palomar Observatory?
5. How has the telescope advanced the knowledge of astronomy?

SKILL ACTIVITY: Using a Telescope

BACKGROUND

The word *telescope* comes from the Greek word *teleskopos,* meaning "to see from a distance." A telescope is a device used to make distant objects, such as stars, appear brighter than the unaided eye perceives them. The first refracting telescopes combined a convex and concave lens as shown. In this activity, you will construct and use a telescope similar to the one used by the great astronomer Galileo Galilei.

PROCEDURE

1. Fit together two short paper tubes made of rolled construction paper. One tube should fit snugly into the other and slide back and forth with ease.
2. Into the end of the smaller-diameter tube (the eyepiece), insert a double concave lens. Hold the lens in place with modeling clay.

3. Into the end of the larger-diameter tube (the objective), insert a double convex lens, using clay to secure it.
4. Slide the tubes back and forth to focus and enlarge distant objects.

APPLICATION

1. Use the telescope to view objects in your classroom.
2. Draw to scale several objects as they appear both with the unaided eye and through the telescope.

USING WHAT YOU HAVE LEARNED

1. Use your telescope to view the moon. What features can you see with the telescope that you could not see before?
2. When you have mastered the use of this telescope, read about reflecting telescopes and try making and using a simple one.

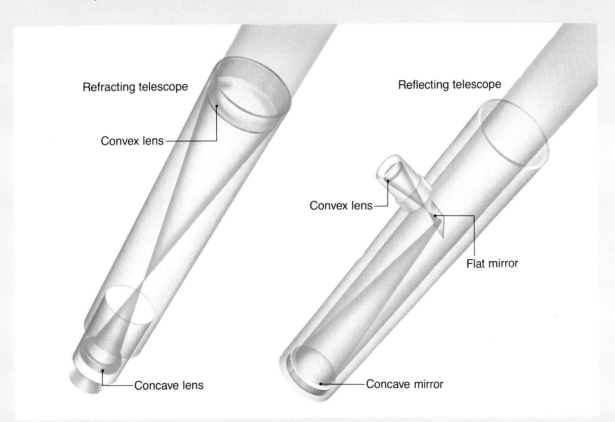

Refracting telescope

Convex lens

Concave lens

Reflecting telescope

Convex lens

Flat mirror

Concave mirror

② Radio Astronomy

SECTION OBJECTIVES

After completing this section, you should be able to:

- **Explain** the difference between radio waves and visible light.
- **Describe** how radiotelescopes work.
- **List** the major discoveries made with radiotelescopes.

23.4 Radio Waves

Objects in the universe that cannot be seen can still be studied. Other forms of energy, especially radio waves, can be detected from Earth by special telescopes. **Radio waves** are the long-wavelength electromagnetic waves. Light rays can have a wavelength as short as 0.000 000 5 m. Radio waves, on the other hand, can have wavelengths that range from 10 m to several kilometers. Any time you listen to a radio, you are hearing sounds that are transmitted as radio waves. Your radio is tuned to receive a particular frequency, or wavelength, of radio waves. In 1931 an American engineer named Karl Jansky wondered why he was getting static on his radio receiver. He hypothesized that the static was radio waves coming from space.

Figure 23–12. Radio waves have very long wavelengths.

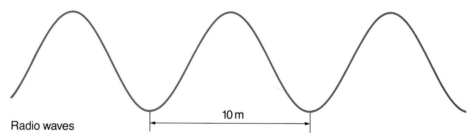

.000001m
Light waves

Radio waves

10 m

Now scientists know that radio waves of many wavelengths are produced in outer space. Stars, planets as big as Jupiter, and the cores of many galaxies emit radio waves. Quasars emit powerful radio waves. Radio waves are not absorbed by matter as visible light waves are. Radio waves from outer space may cross the entire universe with little absorption.

Astronomers capture radio waves from outer space with a special instrument called a **radiotelescope.** These waves provide astronomers with much information about the structure of the universe and the distribution of matter.

○ *What are radio waves?*
○ *What is a radiotelescope?*

23.5 Radiotelescopes

Since the 1940s, radiotelescopes have been used to study space. Some radiotelescopes are like large TV antennas. Most, however, are metallic dishes like the one shown in Figure 23–13.

Radio waves are reflected by metallic surfaces just as light is reflected by a mirror. The wavelength of radio waves is very long; therefore, it is possible to build very large dishes for receiving distant signals. Like optical telescopes, the dishes can be rotated in different directions. The largest movable radiotelescope is 100 m across.

The largest radiotelescope is near Arecibo, Puerto Rico. A wire mesh acts as a reflecting surface. This telescope cannot be moved because it is built in a natural depression, but the rotation of the earth enables the telescope to point in different directions.

At the focus of a radiotelescope is a radio antenna. Electrons in the metal of the antenna are excited by the radio waves focused onto it. The resulting current is fed through a tuner and an amplifier, much as in a common radio. The radio astronomer uses the tuner to select the radio frequency he or she wishes to study. The strength and wavelengths of the signals are recorded, and the source of their direction is plotted.

Figure 23–13. The shape of this large radiotelescope antenna in Puerto Rico focuses radio waves to the receiver above the dish.

A MATTER OF FACT

The first radiotelescope was built in the 1930s.

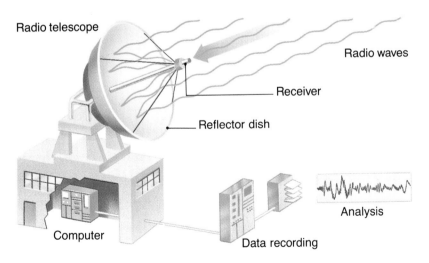

Radio telescope

Radio waves

Receiver

Reflector dish

Computer

Data recording

Analysis

Figure 23–14. This diagram illustrates the operation of a radiotelescope.

A single radiotelescope provides an image of the sky that is less precise than the image provided by an optical telescope. Radiotelescopes can be linked together, however, so that the image received by one can be compared and mixed with the image received by another. Astronomers can use this type of information to measure astronomical distances and wavelengths of energy from space. Combining images from several radiotelescopes can produce a much clearer radio image of the sky. Near Socorro, New Mexico, an array of 27 dishes has been built in the desert. This group of radiotelescope dishes provides an image of the sky that compares favorably with the image provided by the best optical telescopes.

Radio astronomers have made major discoveries about the universe. The first discovery, made in 1931, was that the center of the Milky Way Galaxy emits radio waves. It was later discovered that all galaxies emit radio waves and that some galaxies emit more intense waves than others. Even stronger is the radio-wave emission of quasars.

The most important discovery made by radio astronomers was the detection of background radiation resulting from the big bang. The detection of this radiation in 1964 provided strong evidence to support the big-bang theory.

○ *How does a radiotelescope collect information?*
○ *What are the advantages of radiotelescopes over optical telescopes?*

Figure 23–15. This array of 27 radiotelescopes provides a much clearer image of the universe than a single dish does.

Section Review

READING CRITICALLY

1. How are radio waves different from visible light?
2. Why are radio waves from outer space important in astronomical studies?

THINKING CRITICALLY

3. Why are radiotelescopes curved?
4. Why is the big bang still considered a theory, even with the information provided by radio astronomy?
5. Do you think the spectrograph could help explain the beginnings of the solar system? Why?

TECHNOLOGY: Radiotelescopes

Modern optical telescopes are capable of revealing many startling cosmic objects that would not ordinarily be visible to the unaided eye. These large refracting, and even larger reflecting, telescopes allow astronomers to view in detail the wonders of objects never seen before. Nevertheless, despite their advanced capabilities, optical telescopes are "blind" to many other structures that exist in our universe. They cannot "see" many of the events that alter and transform the size and shape of the cosmos.

The visible spectrum is only a small fraction of the entire range of electromagnetic radiation that floods the vacuum of outer space. Gamma rays, X rays, ultraviolet and infrared rays, radar, microwaves, shortwaves, and radio waves are given off by cosmic bodies. Telescopes of many different designs have been engineered to capture these other important forms of energy.

Radiotelescopes must have very powerful amplifiers be-cause the radio waves that reach Earth are extremely weak.

A radiotelescope consists of three components: (1) a large reflecting surface that collects and focuses the incident radiation, (2) an electronic receiver that detects and amplifies cosmic signals, and (3) a data display device used by astronomy laboratory technicians to view and record the information.

The principle of operation of the radiotelescope is very similar to that of the reflecting telescope. The electromagnetic energy strikes a reflecting, concave surface and is focused on the receiver, just as visible light is reflected by a concave mirror and focused through the eyepiece of an optical telescope.

However, the reflecting surface of a radiotelescope must be precisely measured to take into account the frequency and wavelength of the energy it is designed to capture. Radiotelescopes have a parabolic shape that must focus all radiation to the same, exact point; otherwise, the image will be blurred.

A Radiotelescope at Stanford, California

A recent innovation in radiotelescope technology has revolutionized the study of radio waves from space. Astronomers prefer to construct large radiotelescopes that can detect large fields of radiation. However, because of the accuracy that is required in the construction of these giants, larger and larger structures are more difficult to engineer. Therefore, astronomers have worked out a method of combining the efforts of a series of individual radiotelescopes to form what is called a very large array, or VLA. One such array in New Mexico includes a series of 27 radio dishes that can achieve a resolution equal to that of a single dish 40 km wide.

A VLA requires a large complex of computers to integrate the information coming in from the separate radiotelescopes. However, astrophysicists believe that the effort is worth it. With such an array, they can study the most distant objects of the cosmos.

Radio picture of the Andromeda Galaxy

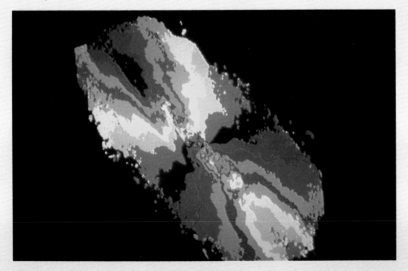

515

INVESTIGATION 23: Measuring Focal Length

PURPOSE

To measure the focal length of different convex lenses

MATERIALS (per group of 3 or 4)

Short candle
Matches
Petri dish
Meter stick
Scissors
White construction paper
Modeling clay
Convex lenses of varying thickness

PROCEDURE

1. **CAUTION: An open flame is used in this investigation; be careful of hair and clothing.** Carefully warm the bottom of the candle with a lighted match and allow the melting wax to drip into the center of the petri dish. Blow out the match and put the candle in the hot wax so that as the wax cools the candle is held in place.
2. Place the meter stick several meters behind the dish.
3. Cut the white construction paper as shown so that it can stand vertically without support. Then place the paper near the end of the meter stick.
4. Mold the modeling clay as a base for the lenses and mount each lens in the clay as shown.
5. Light the candle and put one of the lenses between the candle and the construction paper. Darken the room. Focus the image of the candle on the paper.
6. When the image is sharp, measure and record the distance from the lens to the construction paper. This is the focal length of the lens.
7. Repeat the procedure for the other lenses, keeping a record of each focal length.

ANALYSES AND CONCLUSIONS

1. Prepare a chart to compare and contrast the focal lengths of the lenses. Can you see any relationships between the two quantities? Explain.
2. Does the actual candle flame differ from its image on the construction paper? If yes, how? Be specific.
3. What happens to the focal length of a lens as its thickness increases?

APPLICATION

How could this procedure be used to find the focal length of telescopes and other optical instruments?

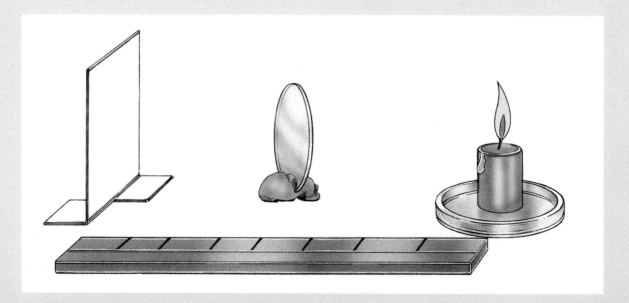

SUMMARY

- Refraction occurs when a beam of light hits the surface of a substance through which light travels at a slower speed. (23.1)

- A lens that causes light to converge on one point is called a convex lens. A lens that causes light to diverge is called a concave lens. (23.1)

- Focal length is the distance between the center of a lens and its focus. (23.1)

- Refracting telescopes have two lenses that focus the light and produce an upside-down image. (23.2)

- The size of refractors is limited by the problems associated with making large lenses. (23.2)

- The use of photographic plates can greatly extend the range of telescopes. (23.2)

- Reflectors are telescopes that focus light with mirrors. (23.3)

- Reflecting telescopes have no distortion, absorb less light than lenses, and can be made larger than refractors. (23.3)

- Radio waves are long-wavelength energy. Many objects in space produce radio waves. (23.4)

- Radiotelescopes are used to detect long-wavelength radiation from outer space. (23.4)

- Radiotelescopes consist of movable dishes that focus radio waves on an antenna. (23.5)

- Many important discoveries about the universe have been made by radio astronomers. (23.5)

Write all answers on a separate sheet of paper.

SCIENCE TERMS

Correctly use each of the following terms in a sentence.

focal length **(505)**
lens **(505)**
magnification **(506)**
radio waves **(512)**
radiotelescope **(512)**
reflection **(504)**
refraction **(504)**
spectrograph **(507)**
telescope **(506)**

SCIENCE QUIZ

Modified True-False

Mark each statement *true* or *false*. If a statement is false, change the underlined term to make the statement true.

1. Energy is emitted from quasars as <u>light</u> waves.

2. The speed of light in air is nearly the same as it is in <u>space</u>.

3. Light bouncing off an object is <u>reflected</u>.

4. Distortions occur in <u>reflecting</u> telescopes.

5. The refracting telescope was used by the Italian astronomer <u>Galileo</u>.

Multiple Choice

Write the letter of the choice that best answers the question or completes the statement.

6. Which of the following emit radio waves?
   a) stars
   b) the moon
   c) asteroids
   d) All of the choices are correct.

continues

7. What will increasing the surface area of a lens in a refracting telescope do?
   a) brighten the image
   b) darken the image
   c) sharpen the image
   d) nothing

8. Which of the following have longer wavelengths than light?
   a) radio waves
   b) X rays
   c) ultraviolet rays
   d) gamma rays

9. What happens to a light wave as it hits glass at a 90° angle?
   a) It bends.
   b) It increases in wavelength.
   c) It speeds up.
   d) It slows down.

10. Who first used a reflecting telescope?
    a) Galileo
    b) Newton
    c) Kepler
    d) Brahe

Completion

Complete each statement by supplying the correct term or phrase.

11. Lenses _____ light.

12. A telescope with an objective lens and an ocular lens is called a _____ telescope.

13. The magnification of a refracting telescope is related to the _____ of the lenses.

14. Refracting telescopes usually produce an _____ image.

15. Spectrographs are used to determine the _____ of a star.

Short Answer

16. Compare and contrast the operation of refracting and reflecting telescopes.

17. Why can reflecting telescopes be larger than refracting telescopes?

18. Explain how a spectrograph can be used to determine the composition of a star.

Writing Critically

19. Explain how the radiotelescope has added to the catalog of known objects in the universe.

20. Discuss the reasons why astronomers would like to launch optical telescopes, infrared telescopes, and radiotelescopes into orbit around the earth.

EXTENSION

1. Write a report on the efforts of NASA to design, launch, and maintain a fleet of observatories in space.

2. Research and report on the uses of orbiting optical and infrared telescopes that are currently pointed at Earth.

3. Find out how a body, such as Pluto, appears through different telescopes. Then make a drawing of the different views and describe them to your classmates.

4. Research and write a report about the Hubble space telescope.

APPLICATION/CRITICAL THINKING

1. The Hubble space telescope is to be placed into orbit above the earth's atmosphere by the space shuttle. What advantage is there to placing a telescope above the earth's atmosphere? Why are scientists so interested in viewing the distant parts of the universe?

2. What is the magnification of a telescope with an ocular focal length of 2.5 cm and an objective focal length of 25 cm?

3. If the frequency of an electromagnetic wave traveling through a vacuum is about 24 000 vibrations per second, what is its frequency in air? Explain.

FOR FURTHER READING

Bartusiak, M. "Megascope." *Omni* 8(July 1986): 16. This article is a discussion of how improved technology may once again make arrays of reflecting telescopes valuable to astronomers.

Bartusiak, M. "Ultimate Catalog." *Omni* 8(February 1986): 26. This article is a summary of how scientists use complex computers to catalog over 20 million celestial objects.

Berry, R. *Build Your Own Telescope: Complete Plans for Five High-Quality Telescopes That Anyone Can Build.* New York: Scribner's, 1985. This is a step-by-step manual that shows anyone with basic carpentry skills how to build a telescope.

Kerrod, R. *The All Color Book of Space.* New York: Arco, 1985. This is a series of essays that relates astronomy to space exploration. The author describes early telescopes, space exploration technology, and space pioneers.

Renner-Smith, S. "New Satellite Antennas." *Popular Science* 227(December 1985): 69. This article gives pointers on how to select inexpensive, state-of-the-art television receivers.

Challenge Your Thinking

This array of radiotelescopes is "eavesdropping" on the universe, listening for signals that might indicate the presence of other life forms. If other life forms are listening to us, what kinds of things do you think they have learned about humans in the last 10 years?

Exploring Space

"Space, the final frontier": These are familiar words to even the most casual "Trekkie," but they are a reality in your lifetime. Humans first went into space nearly 30 years ago, and it is likely that there are humans in space as you read this. Humans have been to the moon and are now ready to explore the planets in person. Are you ready for the adventure?

A space-shuttle launch

1 The Road to Space

SECTION OBJECTIVES

After completing this section, you should be able to:

- **State** the principle on which rocket propulsion is based.
- **Discuss** the contributions to rocketry of Tsiolkovsky and Goddard.
- **Name** the first Russian and the first American to orbit the earth.

NEW SCIENCE TERM

momentum

24.1 Rocket Propulsion

Have you ever blown up a toy balloon and then released it to fly around a room? If so, you have demonstrated the basic principle of rocket propulsion. Rocket propulsion is based on Newton's third law of motion; that is, for each action there is an equal and opposite reaction. The air rushes out of the back of the balloon, and the balloon is pushed forward.

This law establishes the principle of conservation of momentum. Mathematically, **momentum** is equal to the mass of an object times its velocity, but a simple example provides a more useful description. A loaded cannon has no momentum. When the cannon is fired, however, the forward momentum of the cannonball is equal to the backward momentum of the cannon. The sum of the forward momentum and backward momentum is zero because they are equal and in opposite directions. The original momentum, which was zero, is conserved.

Fireworks, jet engines, and rockets all work on the same principle. Fuel is burned, and the gases produced by the burning are expelled through a nozzle facing the rear. The momentum of the gases toward the rear propels the object forward.

Figure 24–1. This jet has momentum because of the exhaust gases. The momentum for the runner is provided by leg muscles.

Figure 24–2. Momentum for launching a rocket is provided by the burning of a fuel. The momentum of the balloon is from escaping air.

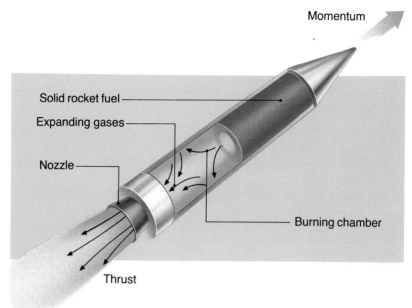

Momentum

Solid rocket fuel

Expanding gases

Nozzle

Burning chamber

Thrust

Fireworks, jets, and rockets are launched by igniting fuel. Air, oxygen, or a chemical oxidant allows the fuel to burn. Fireworks fuel consists of gunpowder; solid fuel for rockets contains mixtures of powdered metals and plastics or rubber. Jet engines and some rockets use kerosene as fuel. The fuel is compressed, ignited, and expelled.

○ *What is Newton's third law of motion?*
○ *What force propels jet engines and rockets forward?*

24.2 Rocket Pioneers

The first person to scientifically study the use of rockets for space travel was Konstantin Tsiolkovsky (KAHN stuhn teen tseel KOHF skee), a Russian inventor. In the 1920s, Tsiolkovsky designed a spaceship with a bullet-shaped passenger cabin in

Figure 24–3. Konstantin Tsiolkovsky is shown here with a model rocket.

the nose and fuel tanks filled with liquid hydrogen and liquid oxygen in the tail. He planned for the fuel and oxygen to burn in a cone-shaped chamber that faced the rear. Tsiolkovsky designed support systems for his astronauts and pressurized suits to enable them to leave their cabins and construct space stations. Although Tsiolkovsky never launched a rocket himself, his thinking was fundamental to the development of rocket flight.

At about the same time, an American physicist, Dr. Robert Goddard, was actually experimenting with rockets. Goddard contributed to the development of liquid fuels for rockets. In 1926 Goddard launched the world's first liquid-fueled rocket, shown in Figure 24–4. The fuel was gasoline, and the oxidant was liquid oxygen. The rocket was 1.2 m long, and it rose to a height of 56 m at an average speed of 103 km/h.

During World War II, the Germans constructed many rockets for delivering bombs. On October 3, 1944, a rocket called the *V-2* was used to bomb London. The rocket was self-steering and was launched from a land-based platform.

Figure 24–4. Robert Goddard (right) launched a gasoline-fueled rocket in 1926. After World War II, the United States conducted experiments with captured German *V-2* rockets (left).

At the end of World War II, scientists in the United States used captured German rockets to scientifically explore the upper atmosphere. An American two-stage rocket, with a captured German *V-2* rocket as its first stage, reached a record altitude of 393 km in 1949.

○ *Who was the first person to study the use of rockets for space travel?*
○ *For what purpose did the Americans use captured V-2 rockets?*

Figure 24–5. *Sputnik* (right) was the first satellite placed into Earth orbit. *Explorer* (left) was the first satellite placed into orbit by the United States.

24.3 **Humans in Space**

The development and use of rockets proceeded slowly until October 4, 1957, when the Soviet Union surprised the world by placing into orbit an artificial satellite weighing 83.6 kg. Called *Sputnik 1*, the satellite carried instruments to measure the density and temperature of the upper atmosphere. A month later, on November 3, 1957, the Soviet Union put another satellite, weighing 508 kg, into orbit. This second satellite carried a passenger—a dog named Laika. The purpose for orbiting the animal was to study the effects of space travel on a living organism.

The first United States satellite, *Explorer 1*, was launched on January 31, 1958. The rocket was 2.05 m long and carried a scientific package for measuring temperature, cosmic rays, and meteoroids. *Explorer 1* discovered the Van Allen belts of radiation described in Chapter 8.

Figure 24–6. Yuri Gagarin (right) was the first human to be launched into space. The United States' first astronaut to orbit the earth was John Glenn (left).

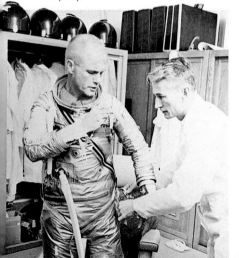

On April 12, 1961, Yuri Gagarin (UHR ee gah GAHR ihn), a 27-year-old lieutenant in the Soviet Air Force, was launched into space. Gagarin orbited the earth once, reaching an altitude of 327 km.

Lieutenant Colonel John Glenn of the United States Marine Corps was launched into space on February 20, 1962, and made three Earth orbits. Glenn's flight followed two suborbital flights, each of which lasted about 15 minutes.

SKILL ACTIVITY: Making a Star Chart for the Seasons

BACKGROUND

The earth's equator is inclined 23.5° from its orbital plane. Because of this, our view of the sky, seen from either the Northern or Southern Hemisphere, alters during the course of the year. In the Skill Activity in Chapter 22, you made a star chart that helped you become familiar with the most visible constellations. In this activity, you will complete your star chart so that it gives a more accurate view of these constellations as they appear during the changing seasons.

PROCEDURE

1. Use a compass to measure the radius of your star chart and draw a circle of equal size on a piece of construction paper.
2. On tracing paper, trace the circle shown in Figure A, cut it out (leaving the tab intact), and use it as a guide to draw a similar ellipse onto the circle made of construction paper.

3. Use a ruler to divide the circle into 24 sections, representing the hours in a day.
4. Put a pin through the tab and attach the circle to the star chart as shown in Figure B.

APPLICATION

When viewing stars at night, align the time and approximate date before viewing the window of the celestial sphere.

1. Why would you need to make an adjustment for the time of night?
2. Why is the date important in using the sphere correctly?

USING WHAT YOU HAVE LEARNED

What modifications would have to be made to the star chart if you lived near the equator or near the Arctic Circle? Explain.

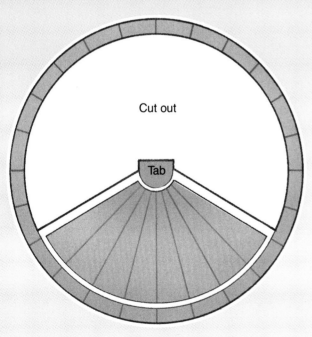

Figure A

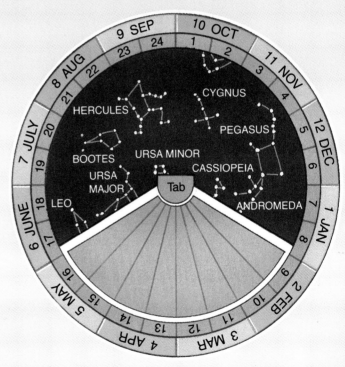

Figure B

The first woman to go into space was Valentina Tereshkova (tehr uhsh KOH vah), launched in June 1963. In 1965 Soviet Lieutenant Colonel Alexei Leonov (LEE uhn uhf) became the first human to "walk" in space. Leonov, dressed in a protective space-suit, left his pressurized spacecraft cabin and floated for 10 minutes in space. All these space firsts had been accomplished in less than 10 years after the launch of *Sputnik 1;* the next 10 years would be even more dramatic.

○ *When did the United States put its first satellite into orbit? What was it called?*

○ *Who was the first person to "walk" in space?*

Figure 24–7. Former space-shuttle astronaut Sally Ride was the first woman astronaut for the United States.

Section Review

READING CRITICALLY

1. What kinds of fuel did Goddard use to launch his early rockets?
2. Who were the first Soviet and American humans launched into Earth orbit?

THINKING CRITICALLY

3. Why do you think that the Soviet Union is so far ahead of the United States in some areas of space exploration?
4. What effects of space travel did the Soviets wish to study by sending Laika into orbit?

ACTIVITY: Observing Newton's Third Law

How can you apply Newton's third law of Motion?

MATERIALS (per group of 3 or 4)

string (10 m), balloon, plastic straw, tape, ring stands (2)

PROCEDURE

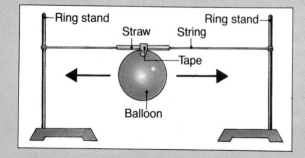

1. Place the ring stands on desks on opposite sides of the room.
2. Tie one end of the string to the top of one of the ring stands.
3. Thread the other end of the string through the plastic straw and tie that end to the other ring stand. Make sure that the string is tightly stretched.
4. Inflate the balloon but do not tie it closed. Hold the balloon closed and tape it to the straw.
5. On the count of "5, 4, 3, 2, 1 launch!" release the balloon at one end of the string and observe it as it flies across the room.

CONCLUSIONS/APPLICATIONS

1. Why is this a demonstration of Newton's third law of motion?
2. What forces drove the balloon forward?
3. What forces tended to slow the balloon down?

SECTION

② Space Travel

SECTION OBJECTIVES

After completing this section, you should be able to:
- **Discuss** why it was important for humans to go to the moon.
- **Explain** the importance of space stations.
- **Describe** what the space shuttle can do.

NEW SCIENCE TERMS

There are no new terms in this section.

24.4 Humans on the Moon

As a challenge to the Soviet Union's leadership in space, President John F. Kennedy committed the United States to placing an astronaut on the moon before the end of the 1960s. In preparation for this journey to the moon, many additional Earth-orbiting missions, called *Mercury* and *Gemini,* were launched. American astronauts learned to live and work in space.

In October 1968, the Apollo system, which was designed to send three astronauts to the moon, was tested in Earth orbit. *Apollo 7* consisted of a three-stage rocket, called *Saturn V,* and a command module for the flight crew. *Apollo 8,* led by Colonel Frank Borman, made the first flight to the moon in December 1968, but it did not land. The spacecraft orbited the moon 10 times and sent televised views of the lunar surface back to Earth.

Landing humans on the lunar surface involved launching a spaceship with three separate units: a command module, a service module, and a lunar module. The command module was the capsule in which the astronauts traveled to the moon and

Figure 24–8. The *Saturn V* rocket (right) is the most powerful rocket ever built. It was used to launch the Apollo missions to the moon (left).

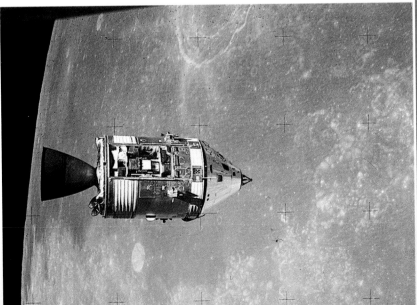

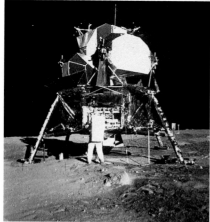

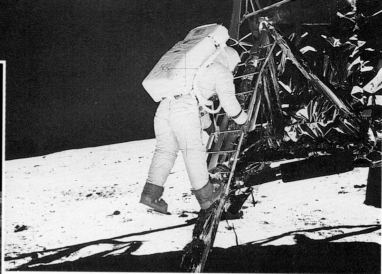

Figure 24–9. The LEM *Eagle* (left) landed on the moon July 20, 1969. Neil Armstrong and "Buzz" Aldrin (right) spent 2-1/2 hours walking on the moon.

returned to Earth. The service module supplied oxygen, power, and fuel to the command module. The lunar excursion module (LEM) landed two astronauts on the moon's surface. The LEM included an engine to slow the craft for descent and four legs for a landing platform. The lunar-module engine was used for the liftoff from the moon and the return to the command module.

On July 16, 1969, the crew of *Apollo 11* were launched from Cape Canaveral, Florida. Neil Armstrong and Edwin "Buzz" Aldrin were to land on the moon's surface. The third astronaut, Michael Collins, was to remain in orbit around the moon.

The lunar module landed on the surface of the moon on July 20 at 8:17 P.M. Greenwich Mean Time (GMT). Armstrong and Aldrin left the lunar module at 2:56 A.M. GMT and spent 2-1/2 hours on the surface of the moon. They collected rocks, set up experiments, and took many pictures. Their mission accomplished, the astronauts left the moon and returned to Earth, landing in the Pacific Ocean 195 hours and 18 minutes after launch.

The last lunar mission, *Apollo 17*, was completed on December 19, 1972. No humans have been to the moon since then, but there are preliminary plans for a permanent station on the moon early in the twenty-first century.

○ *Describe the three parts of the spaceship used to land on the lunar surface.*
○ *Name the astronauts aboard the first mission to land on the moon.*

Figure 24–10. The "Lunar Rover" was used by astronauts to explore wide areas of the moon's surface.

24.5 **Space Stations**

In addition to the information obtained from the Apollo moon missions, astronaut-scientists have conducted many experiments from space stations. The first space station was the Soviet Union's *Salyut 1*, launched in April 1971. *Salyut 1* was cylindrical in shape; it was 15 m long, over 4 m in diameter, and weighed 18.9 metric tons. Three Soviet cosmonauts made a variety of scientific observations from this station.

In May 1973 the United States launched *Skylab*, shown in Figure 24–11. *Skylab* was much bigger than *Salyut 1*; it had a length of 25.7 m and a diameter of 6 m. *Skylab* was joined three times by Apollo-type spacecraft carrying teams of three astronauts. The three *Skylab* missions, all in 1973, lasted 28, 59, and 84 days, respectively.

Figure 24–11. The *Skylab* space station (left) was used three times in 1973. Inside *Skylab* (center and right), astronauts conducted many experiments in the reduced gravity of space.

On July 15, 1975, an American Apollo spacecraft joined with a Soviet Soyuz spacecraft 225 km above Earth. The three American astronauts and two Soviet cosmonauts visited each other through a passage connecting the two spacecraft and jointly conducted several small experiments.

Figure 24–12. Astronauts from the United States and the Soviet Union (left) met in space for the first time on July 15, 1975. The Soyuz and Apollo spacecraft were linked by a common passage (right).

529

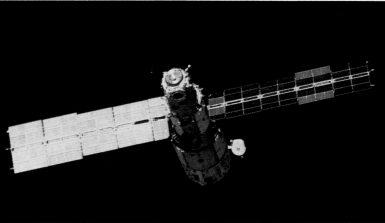

Figure 24–13. Scientists from the Soviet Union have occupied various space stations almost continuously since 1971.

In September 1977 a new space station, *Salyut 6*, was launched by the Soviet Union. *Salyut 6* had two docking ports. On December 11, 1977, *Soyuz 26* docked at the rear port, and on January 11, 1978, *Soyuz 27* docked at the front port. For the first time, three spacecraft were linked in space. There was a total of six people on board.

On January 20, 1978, the first unpiloted ferry craft was launched by the Soviets to resupply *Salyut 6*. Resupplying by crewless flights meant that the crews on *Salyut 6* could stay in space indefinitely. *Salyut 6* was in operation until the summer of 1982, when it was replaced by *Salyut 7*, or *Mir*. In 1984 a two-person crew remained aboard *Mir* for 238 days. The Soviet Union is now developing a shuttle to ferry crews and supplies to and from future space stations.

The Soviet Union is clearly ahead of the United States in establishing a permanent space station. The only American space station to date was *Skylab*, which operated from May 1973 to February 1974. Although plans are being considered for an American space station, it will not become operational for several more years.

○ *Name the Soviet and American space stations.*
○ *Why did the Soviets wish to resupply space stations by unpiloted flights?*

A MATTER OF FACT

Although a single one of the space shuttle's main rocket engines has a mass of only 3000 kg, it has a power output equal to seven Hoover Dams.

24.6 The Space Shuttle

A space vehicle that could be lifted into orbit by rockets and returned to Earth like an airplane seemed necessary for the construction of an American space station. The final design was the space shuttle, a system consisting of four parts: two rocket boosters with solid fuel, one disposable liquid-fuel tank, and an orbiter.

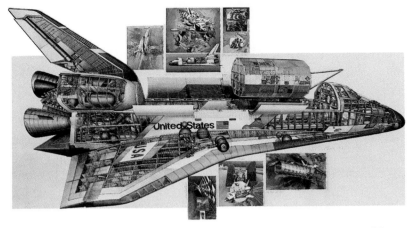

Figure 24–14. The United States' space shuttles are very complex spacecraft.

Columbia, the first orbiter to be used in space, looks like a snub-nosed airplane with short wings. The orbiter is 37.24 m long, 17.27 m high, and has a wingspan of 23.79 m. *Columbia* has three main engines, each with a thrust of nearly 170 metric tons. The two smaller maneuvering engines have a thrust of nearly 3 tons. The cargo deck, which is 18 m long, is used for scientific equipment or to carry satellites into orbit.

The solid rocket boosters contain rubber and aluminum powder mixed with an oxidant as fuel. The thrust of each booster engine is over 1200 metric tons. The external fuel tank contains liquid oxygen and liquid hydrogen, which are mixed and burned in the orbiter's engines.

Figure 24–15. Space shuttles are prepared in the huge Vehicle Assembly Building (left) at the Kennedy Space Center in Florida. After landing in California (below), space shuttles are flown back to Florida on top of a "jumbo jet."

Figure 24–16. The crew of a space shuttle can launch satellites (left) or conduct scientific experiments inside the Spacelab module (right).

The shuttle system is lifted from its launch pad at Kennedy Space Center in Florida by the rocket engines of the boosters and the orbiter. At an altitude of 45 km, the boosters are detached and returned to Earth by parachutes for recovery and reuse. The external tank, however, is dropped into the Atlantic Ocean. The orbiter is placed into orbit by its maneuvering engines, which are also fired to slow the craft for reentry and landing. The orbiter lands at about 350 km/h on a long aircraft runway.

Three similar orbiters, *Challenger, Discovery,* and *Atlantis,* were added to the space shuttle fleet between 1980 and 1985. However, the January 1986 flight of *Challenger* ended in tragedy as one of the two boosters developed a leak and the assembly exploded 75 seconds into the flight. The seven astronauts died.

A consortium of 11 European nations has built a small laboratory, called *Spacelab,* that fits into the cargo bay of a shuttle orbiter. The first Spacelab mission was in November 1983. Since that time, there have been additional Spacelab missions, all of which produced much scientific information.

○ *What was the name of the first American space shuttle to be used in space?*
○ *Which parts of the space shuttle system are recovered and reused, and which parts are not reused?*

Section Review

READING CRITICALLY

1. Of what components did the Apollo launch system consist?
2. What dimensions did *Skylab* have?
3. How is the space shuttle launched, and how does it return to Earth?

THINKING CRITICALLY

4. Why did the United States decide to send astronauts to the moon?
5. Explain the importance of Americans and Soviets meeting in space in 1975.

TECHNOLOGY: The Space Shuttle

The exploration of the solar system presents one of the most exciting challenges ever faced by a generation of pioneers. While piloted missions to the moon, planets, and possibly even the stars will continue to excite the imagination, unpiloted satellites are already hard at work for the benefit of all people.

At this very moment, orbiting research satellites search the sky and conduct experiments in the nearly perfect vacuum hundreds of kilometers above the earth's atmosphere. Weather satellites monitor cloud patterns, communications satellites carry phone conversations, and land satellites watch crops grow. Yet, as valuable as they are, the design, launch, and maintenance of this army of orbital machinery is expensive in time and materials.

The space shuttle represents an innovation in the exciting task of working and living in space. Before the development of the shuttle, payloads were launched into orbit with rockets that were discarded after a single launch.

Space shuttle launch

Repairing a satellite

The boosters of these multiple-stage giants were dropped into the ocean and lost forever. Such a procedure is, obviously, wasteful because new boosters must be designed and built for every satellite put into orbit. The space shuttle, on the other hand, is designed as a recyclable vehicle.

In April 1981, with astronauts John Young and Robert Crippen at the controls, the first operational orbiter, *Columbia,* was launched into space from Kennedy Space Center in Florida. With a wingspan of about 24 m, the orbiter rode on the back of an external fuel tank attached to two solid-rocket boosters, each having a thrust of nearly 12 900 000 N. On the launch pad the entire assembly had a mass of 1 996 000 kg and stood more than 56 m high.

After lifting *Columbia* to the fringe of the atmosphere, the external fuel tank and solid-

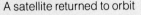

A satellite returned to orbit

rocket boosters separated from the orbiter and dropped into the ocean. The external fuel tank is expendable, but the solid rocket-boosters parachute safely to the sea and are recovered by United State Navy vessels. These boosters can be repaired and prepared for another launch.

The orbiter's three main engines, each having a thrust of 1 668 000 N, gave the craft its last needed push into orbit.

The first flight of *Columbia* lasted 54 hours, and after 36 orbits it reentered Earth's atmosphere and glided to a successful landing at Edwards Air Force Base in California.

Since that time, the space shuttle fleet has placed many satellites into orbit and has even retrieved a few.

Before this and the next generation of pioneers can venture to the planets, the space shuttle will assist NASA personnel in constructing a space station that will serve as the threshold to the planets. In the next few decades, the space shuttle will be the workhorse of the space program. It remains the most efficient vehicle for getting our astronauts and their valuable payloads into space and back.

3 Space Science

SECTION OBJECTIVES

After completing this section, you should be able to:
- **Explain** what communications satellites do.
- **Tell** why weather satellites are important.
- **Summarize** the major scientific advances produced by space exploration.

24.7 Communications Satellites

Why do you think countries and corporations spend millions of dollars every year sending communications satellites into space? What can be achieved in space that cannot be done on Earth?

Communications satellites are important for relaying information between points on Earth. Without communications satellites, it would not be possible to watch live television from around the world or to talk inexpensively to a relative or friend across the country.

English scientist Arthur Clarke proposed that three satellites equally spaced in orbits 35 870 km above the equator and moving with the earth could be used to form a worldwide communications network. Today there are many satellites in these orbits, called *Clarke orbits*. A satellite in a Clarke orbit is **geostationary** because from Earth it does not appear to move from its position.

Figure 24–17. Communications satellites in geostationary orbits (right) allow people to view live television programs broadcast anywhere in the world (left).

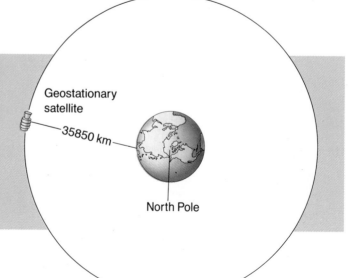

Geostationary
satellite

35850 km

North Pole

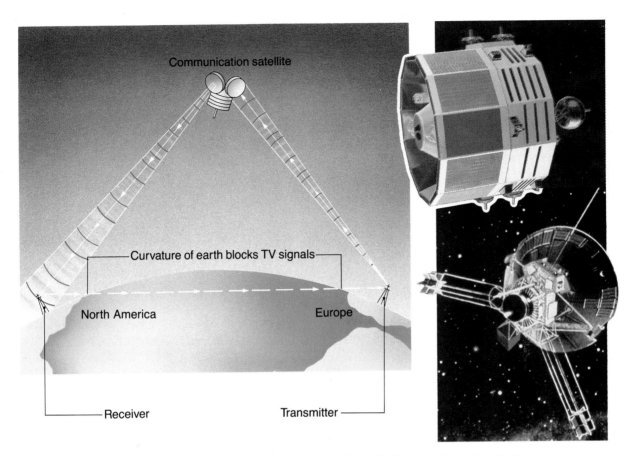

Labels in figure: Communication satellite; Curvature of earth blocks TV signals; North America; Europe; Receiver; Transmitter

Figure 24–18. Communications data is sent from an Earth station to a satellite (left), and then to another Earth station. Data can also be sent from one satellite to another satellite to complete a transmission to the other side of the earth.

Communications satellites are also important for relaying information from other satellites and spacecraft. Before there can be wide-range exploration of space, it is necessary to have many communications satellites in place.

In 1960 the United States launched the first communications satellite, *Echo 1.* Echo was simply an aluminum-coated, helium-filled balloon that expanded once it was in orbit. Voice and television communications sent from one land station bounced off Echo's metal skin and were received by another land station.

In 1962 the United States placed into orbit a more sophisticated communications satellite called *Telstar 1.* Telstar satellites had a capacity of 600 voice channels and one television channel and could receive and transmit signals, not just bounce them.

Several large nations, such as the United States, the Soviet Union, India, Indonesia, Australia, and Brazil, have developed independent domestic satellite communications systems. These systems serve the purpose of linking nations internally and of providing a sense of unity for their people.

○ *Why are communications satellites important?*
○ *Name and describe the first communications satellite launched by the United States.*

A MATTER OF FACT

It is now possible to build satellite dishes that are small enough to fit onto the roof of a motor home.

Figure 24–19. Before they are launched, many satellites are tested in chambers that simulate space conditions.

Figure 24–20. Weather satellites can be used to track and predict the movement of blizzards (right) and hurricanes (left).

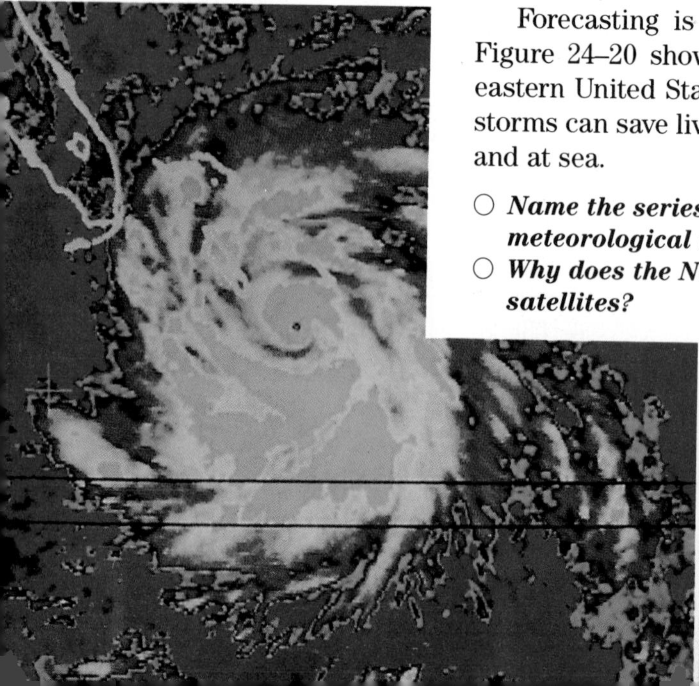

24.8 Weather Satellites

Weather has a direct impact on human welfare. Many people can be killed in hurricanes or blizzards if no warning is given. In order to alert populations to severe weather, as well as to generally improve the forecasting of weather, satellites were developed that could photograph storms and relay those photographs back to Earth.

Nine early weather satellites, each called *Tiros*, were launched by the United States between 1960 and 1965. These satellites observed the earth, using visible light and infrared radiation. Between 1966 and 1969, nine Environmental Science Services Administration (ESSA) satellites, equipped with large television cameras, were launched.

One Improved Tiros Operational Satellite (ITOS), launched in 1970, did the work of two ESSA satellites and measured the vertical temperature profile of the atmosphere as well. A series of five more ITOSs, National Oceanic and Atmospheric Administration (NOAA) numbers 1 to 5, were launched between 1970 and 1976. The Tiros-ESSA-ITOS-NOAA series of satellites were placed in polar orbits. Satellites that circle in polar orbits can observe the entire earth as it rotates. Their observations were supplemented by information from geostationary satellites. These satellites provide the bulk of meteorological information used by the United States and other countries.

Today weather satellites monitor the earth continuously. The cloud cover is photographed constantly, and the pictures are transmitted to Earth stations. The pictures that the meteorologist puts on your television screen each night come from these satellites. With the aid of these pictures, the National Weather Service can forecast the weather with a fair degree of accuracy several days in advance.

Forecasting is also important to air and sea navigation. Figure 24–20 shows a recent hurricane as it approached the eastern United States. Advance notice of hurricanes and other storms can save lives and reduce property damage both on land and at sea.

○ *Name the series of satellites that first provided meteorological information for the United States.*
○ *Why does the National Weather Service depend on weather satellites?*

24.9 Observing the Earth

The value of observing the earth from space became clear when the first piloted flights began. Photographs taken by early astronauts and cosmonauts showed features of the earth that were not clear from the ground. The first satellite specifically designed to photograph the earth was *ERTS 1*, launched into a polar orbit on July 23, 1972. *ERTS 1*, which stands for *Earth Resources Technology Satellite 1*, was later renamed *Landsat 1*. Five Landsats were launched between 1972 and 1984.

The first application of Landsats was in cartography—the process of making maps. In addition, Landsat pictures of crop areas have made it possible to forecast production and to establish a worldwide food watch. Landsat photos are also important in monitoring dust storms, forest fires, and air pollution.

Landsats have also been used to locate mineral resources. The structure of the upper crust of the earth is studied as a means of identifying subsurface mineral deposits. The color of bare rocks and the pattern of vegetation where bedrock is covered often reveal the structure and mineral composition of the rocks. The changing snow cover in middle and high latitudes is also monitored by Landsats as a way of forecasting spring floods. The ice cover in Antarctica, Greenland, and many mountainous areas is monitored as a way of predicting possible sea-level changes. Several Landsat photographs, similar to the ones in Figure 24–21, have been used in this textbook to illustrate Earth features that standard aerial photography could not show nearly as well.

Figure 24–21. Landsat photographs of New York (left), Chicago (center), and Los Angeles (right) are shown here.

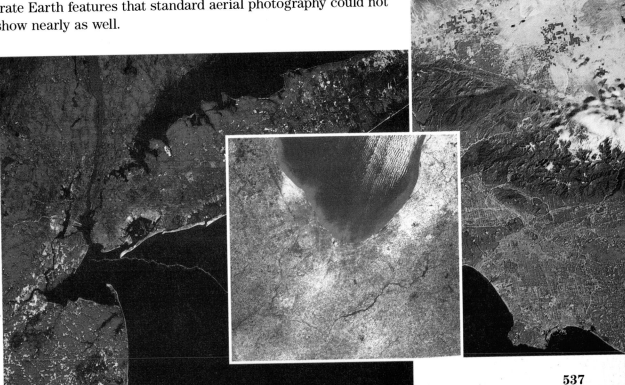

A satellite called *Seasat 1* was launched in 1978 for the specific purpose of studying the oceans. *Seasat 1* measures wave heights, the speed and direction of currents, and the surface temperature of the ocean. The satellite also records the world distribution of sea ice.

○ *Name four uses for pictures of the earth taken by Landsats.*
○ *Name the satellite used to study the oceans of the world.*

Figure 24–22. Seasat photographs of the oceans are similar to Landsat photographs of the continents.

24.10 Exploring the Solar System

The sun has been the object of intense study for hundreds of years. The United States launched eight *Pioneer* space probes between 1959 and 1973 to study the sun. **Space probes** are spacecraft that are launched into the solar system rather than into Earth orbits as are satellites.

The exploration of the planets started with *Pioneer 5*, launched on March 11, 1960. *Pioneer 5* returned data from a distance of 36.5 million km from Earth. Since that time the United States and the Soviet Union have sent many probes into the solar system. Some of the discoveries made by these probes are shown in Table 24–1.

The space probes that have been launched during the past 30 years have provided scientists with a wealth of data and information about the earth and the solar system. This information has been used by scientists to study the earth and the origin and evolution of the solar system.

○ *What was the first space probe to be launched into the solar system?*
○ *What new information about the solar system has been discovered by these probes?*

DISCOVER

Learning About Mars

Research the missions of the Viking probes that landed on Mars. Find out what kind of experiments were used in the search for life on Mars.

TABLE 24–1: SPACE PROBES

Pioneer   The United States launched eight Pioneer probes between 1959 and 1973. Studies of the sun were conducted by many of the Pioneers. *Pioneer 5,* launched in 1960, was the first probe sent to explore interplanetary space. *Pioneer 10* and *Pioneer 11* sent back to Earth the first close-up photographs of Jupiter in 1972 and 1973. *Pioneer 11* sent back close-up photographs of Saturn in 1974.

Mariner   The United States launched a series of Mariner probes beginning in 1962. *Mariner 2* passed close to Venus in 1962, and *Mariner 4* passed close to Mars in 1964. Both made many important new discoveries. *Mariner 9* orbited Mars in 1971, taking many photographs, and *Mariner 10* passed close to Mercury in 1973. The photographs of Mercury showed it to be heavily cratered like the moon.

Venera   A series of Venera probes was launched by the Soviet Union. *Venera 9* and *Venera 10* landed on Venus in 1975, taking one photograph of the surface before being destroyed. *Venera 13* and *Venera 14* landed on Venus in 1982. Together they took four photographs, including the first color photograph of the surface of Venus. Many other measurements of the planet were made as well.

Viking   *Viking 1* and *Viking 2,* launched by the United States, orbited Mars in 1976, taking over 50 000 photographs. Landers were sent to the surface to make direct observations of the surface. The Viking landers took photographs of the surface, recorded quakes and weather conditions, analyzed Martian soil, and conducted experiments to search for life on Mars.

Voyager   Two Voyager probes were launched by the United States to explore the outer planets. In 1979 and 1980 *Voyager 1* and *Voyager 2* passed close to Jupiter, discovering many new satellites and a system of rings. In 1980 and 1981 the Voyager probes passed close to Saturn, again making many discoveries. Voyager continued on to explore Uranus in 1986 and Neptune in 1989.

Section Review

READING CRITICALLY

1. What is a geostationary satellite?
2. What is the advantage of a polar orbit?

THINKING CRITICALLY

3. Describe three ways in which satellites affect your life.
4. Why do you think certain nations have spent so much money and effort exploring other planets in the solar system?

INVESTIGATION 24: Measuring Balloon Rocket Thrust

PURPOSE

To measure the force required to accelerate a balloon rocket

MATERIALS (per group of 3 or 4)

Laboratory balance
Balloons (2)
String (10 m)
Plastic straw
Tape
Ring stands (2)
Metric ruler
Stopwatch

PROCEDURE

1. Use a balance to find the mass in grams of an inflated balloon. Record that mass.
2. Turn to the Procedure of the Activity on page 526, and follow steps 1 through 4. This time, measure and record the distance between the ring stands to the nearest centimeter.
3. Set the stopwatch to zero and begin the countdown.
4. Measure the time it takes for the balloon to race to the end of the string.
5. Calculate the balloon's average velocity during its brief trip, using the formula

$$v = \frac{d}{t}$$

where v is the average velocity, d is the distance between the ring stands, and t is the time.
6. Calculate the momentum of the balloon when it reached the end of the string, using the formula

$$p = m \times v$$

where p is the momentum, m is the mass of the similarly inflated balloon on the balance, and v is the average velocity of the balloon.
7. Repeat the procedure several times, taking an average measure for momentum. Since the velocity, and therefore the momentum, of the balloon became zero at the end of the string, p represents a measure of the *force* required to stop the balloon's forward motion.

ANALYSES AND CONCLUSIONS

1. What was the momentum of the balloon at the instant prior to its collision with the ring stand?
2. How much force was required to stop the balloon? What is the unit of measure for this force?

APPLICATION

How could this procedure be used to measure the momentum of a small rocket motor that could be purchased at a hobby shop?

SUMMARY

- The momentum of burned gases expelled from a rocket propels the rocket forward. (24.1)

- Solid rocket fuels consist of mixtures of powdered metals, plastics, or rubber. (24.1)

- The liquid fuel used by Robert Goddard for his first rocket was gasoline. (24.2)

- The *V-2* was a rocket adapted to warfare by the Germans. (24.2)

- The Soviets put *Sputnik 1,* the first artificial satellite, into orbit in 1957. (24.3)

- The first American in orbit was Lieutenant Colonel John Glenn. Glenn orbited the earth three times in 1962. (24.3)

- *Apollo 11,* launched in 1969, put the first humans on the moon. (24.4)

- The Soviets have orbited a series of seven space stations called *Salyut.* (24.5)

- *Skylab* was a large American space station that operated during 1973–1974. (24.5)

- The space shuttle is a spacecraft that is launched by rockets but is capable of landing on an airstrip as an airplane does. (24.6)

- Geostationary orbits are high above the equator, while polar orbits pass above the poles. (24.7–24.8)

- Weather satellites continuously monitor the earth and allow the forecasting of weather. (24.8)

- Landsats are designed to observe the earth's surface. (24.9)

- The sun and the solar system have been studied by a series of space probes launched by the United States and the Soviet Union. (24.10)

- *Voyager 1* and *Voyager 2* discovered a total of 20 new moons orbiting Jupiter, Saturn, and Uranus. (24.10)

Write all answers on a separate sheet of paper.

SCIENCE TERMS

Correctly use each of the following terms in a sentence.

geostationary **(534)**
momentum **(521)**
space probes **(538)**

SCIENCE QUIZ

Modified True-False

Mark each statement *true* or *false*. If the statement is false, change the underlined word to make the statement true.

1. Rocketry is an application of Newton's <u>first</u> law of motion.

2. The first human to orbit Earth was <u>John Glenn</u>.

3. The fact that <u>momentum</u> is conserved makes rocket travel possible.

4. <u>John F. Kennedy</u> committed the United States to landing humans on the moon.

5. Geostationary orbits are also called <u>Clarke</u> orbits.

Multiple Choice

Write the letter of the choice that best answers the question or completes the statement.

6. One of the first American rocket pioneers was
   a) Konstantin Tsiolkovsky.
   b) Robert Goddard.
   c) John Glenn.
   d) Yuri Gagarin.

7. In order to burn their fuel, most rockets must carry a chemical
   a) propellant.
   b) catalyst.
   c) oxidant.
   d) repellent.

continues

8. Who was the first person to set foot on the moon?
   a) Neil Armstrong   b) John Glenn
   c) Yuri Gagarin   d) Frank Borman

9. Which of the following is a good reason to launch satellites?
   a) monitor weather
   b) scientific research
   c) communications
   d) All of the choices are correct.

10. Which of the following is a function of *Landsat 1?*
    a) monitor ice sheets
    b) monitor forest fires
    c) Both choices are correct.
    d) Both choices are incorrect.

11. Which space probe was the first to land on Mars?
    a) Viking
    b) Venera
    c) Pioneer
    d) Mariner

12. What was the United States space program that landed the first humans on the moon?
    a) Mercury
    b) Gemini
    c) Apollo
    d) Columbia

13. Which space shuttle was destroyed shortly after launch on January 28, 1986?
    a) Columbia
    b) Challenger
    c) Discovery
    d) Atlantis

14. What was the first United States space station called?
    a) *Skylab*
    b) *Mir*
    c) *Salyut*
    d) *Spacelab*

15. Which of the following is not a part of a space shuttle?
    a) orbiter
    b) external fuel tank
    c) solid rocket boosters
    d) LEM

Completion

Complete each statement by supplying the correct term or phrase.

16. The Mercury and Gemini space missions increased the ability of humans to _____.

17. *Viking 1* soft-landed on the planet _____.

18. The outer planets, Jupiter, Saturn, and Uranus, have been photographed in detail by the space probes _____.

19. A satellite that remains over a single spot on Earth is in a _____ orbit.

20. During World War II, Germany used a rocket called the _____ to bomb London, England.

Short Answer

21. List the names of two early rocket pioneers and describe their accomplishments.

22. Describe three early probes into space and tell what each did.

23. Describe the mission of *Apollo 11.*

Writing Critically

24. Why did the United States speed up its efforts to develop a productive space program after the launch of *Sputnik 1?*

25. Why is an operational space shuttle necessary to the development of a permanant United States space station?

EXTENSION

1. Do library research and write a report on the Mercury, Gemini, Apollo, Skylab, or Apollo-Soyuz space projects.

2. Debate the need for the United States to have a permanant space station.

3. Read about the causes of the *Challenger* explosion, and report on the corrections that have been made to the space shuttle system to avoid a similar tragedy in the future.

4. Read about the history of NASA, the National Aeronautics and Space Administration, and write a summary on the accomplishments of the United States space program.

APPLICATION/CRITICAL THINKING

1. The space shuttle orbiters are covered with hundreds of heat-absorbing tiles that protect the craft and the crew during reentry into Earth's atmosphere. During the first few shuttle flights, several of the tiles fell off. What kinds of problems might occur if a large number of these tiles were lost during a mission?

2. Why was there such a long delay in launching a space shuttle after the *Challenger* explosion?

FOR FURTHER READING

Black, R. "Shuttle Hybrid." *Omni* 8(September 1986): 20. Because of the *Challenger* explosion, the author of this article considers alternative methods of getting large payloads into space without the shuttle.

Joels, K. *The Mars One Crew Manual.* NY: Ballantine Books, 1985. The author shows how modern technology will be used to make the first trip to Mars. The book is your crew member's manual, describing ship systems, crew's duties, flight time, and experiments to be conducted.

Kerrod, R. *The Illustrated History of NASA.* New York: Gallery Books, 1986. This book is a photographic history of the National Aeronautics and Space Administration.

Wolkomir, R. "Alien Worlds: The Search Heats Up." *Discovery* 8(October 1987): 66. This article explains that wobbling stars provide evidence that this is not the only solar system in the Milky Way Galaxy.

Challenge Your Thinking

In January 1986, seven American astronauts died in the fiery explosion of the space shuttle *Challenger.* Yet, the families of every one of the astronauts have encouraged the continued human exploration of space. Why is it necessary for humans to fly into space when robot probes, such as Viking, have sent so much data back from space?

UNIT 7 THE SCIENCE CONNECTION:

Life on Mars

In the summer of 1976, two Viking spacecraft landed on Mars and sampled its soil and atmosphere. Some scientists hoped that the Mars landing would yield clues about the existence of life on the red planet. However, the Viking landers' tests demonstrated that no life, as we know it, presently exists on Mars. With the Viking landings, science fiction's long fantasy of great civilizations beyond Earth finally ended.

Launch of a Viking aboard a Titan rocket

The fantasy began in the early 1900s with an amateur astronomer named Percival Lowell. Lowell built an observatory in Arizona just to look at Mars. He claimed that an intelligent race of Martians had built irrigation canals on the dry planet. Lowell also believed that polar ice caps proved the existence of water on Mars. Color changes on the planet's surface seemed to indicate the seasonal growth of plants.

The hopes for finding life on Mars were dashed in 1965 when a Mariner spacecraft sent back pictures of the Martian surface. The photos revealed a dry, cratered planet, more like the moon than like the earth. Mariner also revealed that Mars had virtually no atmosphere. The atmospheric pressure

A Viking lander

on Mars was only one two-hundredth that of Earth. The pressure was too low to hold liquid water; it would boil away into space.

A second Mariner renewed hopes for finding life on Mars. It photographed dry channels that might have been the result of water runoff. Perhaps, scientists said, life on Mars had begun during a wet period and had evolved to survive the recent dry conditions.

Enough scientists still believed in the possibility of life on Mars that biological experiments were built into the Viking landers. Two Viking spacecraft landed on the Martian surface. Each lander analyzed Martian soil. One biological experiment involved mixing Martian soil with a nutrient-rich liquid nicknamed

A Viking lander on Mars

The Viking probes provided much evidence as to why Mars has no life. Mars has almost no water vapor in its atmosphere. The polar ice caps are made of frozen carbon dioxide, or dry ice, rather than water. Earth has oceans full of life-sustaining water; Mars has only a few bucketfuls at most. Besides having no water, Mars has virtually no atmosphere. This allows its surface to be bombarded by ultraviolet rays. These rays, which destroy many known life-forms, are filtered out by the ozone in Earth's atmosphere. Finally, the surface of Mars is covered with metal oxides, similar to rust. Oxides are poisonous to most primitive forms of life.

Even though no life was found on Mars, scientists did learn something about life on Earth. Life as we know it can survive only with very special conditions. Those conditions exist on Earth but nowhere else in the solar system. Therefore, Earth's life-forms are all closely related. All life shares a basic chemical and genetic makeup.

The surface of Mars

"chicken soup." Scientists hoped that Martian micro-organisms would reveal themselves by digesting some of the liquid and giving off detectable gases.

Other scientists scanned the surface with Viking's television camera, searching for signs of movement. After days of watching the television screen, project scientists admitted that "no rock has gotten up and moved away." A reaction had taken place in the "chicken soup"; however, it was almost certainly a chemical process, since it provided no evidence of any life form. The surface of Mars, it seems, was entirely dead.

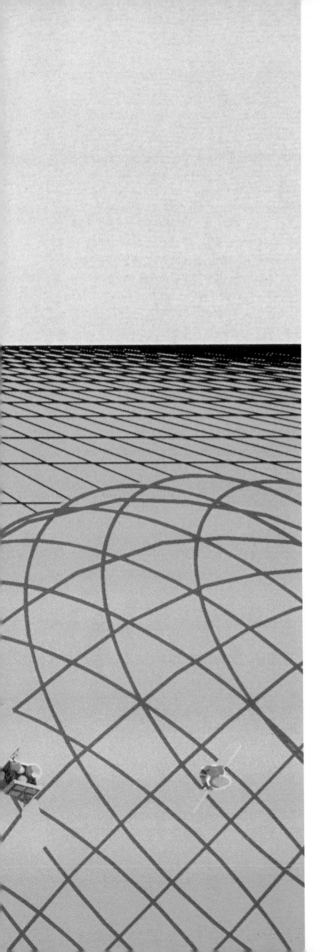

REFERENCE SECTION

Safety Guidelines

Participating in laboratory investigations should be an enjoyable experience as well as a learning experience. You can ensure both learning and enjoyment from the experience by making the laboratory a safe place in which to work. Carelessness, lack of attention, and showing off are the major causes of laboratory accidents. It is, therefore, important that you follow safety guidelines at all times. If an accident should occur, you should know exactly where to locate emergency equipment. Good safety practice means being responsible for your fellow students' safety as well as your own.

You will be expected to practice the following safety guidelines whenever you are in the laboratory.

1. **Preparation** Study your laboratory assignment in advance. Before beginning your investigation, ask your teacher to explain any procedures you do not understand.

2. **Eye Safety** Wear goggles when handling acids or bases, using an open flame, or performing any activity that could harm the eyes. If a solution is splashed into the eyes, wash the eyes with plenty of water and notify your teacher at once. Never use reflected sunlight to illuminate a microscope. This practice is dangerous to the eyes.

3. **Safety Equipment** Know the location of all safety equipment. This includes fire extinguishers, fire blankets, first-aid kits, eyewash fountains, and emergency showers. Note the location of the nearest telephone. Take responsibility for your fellow students and report all accidents and emergencies to your teacher immediately.

4. **Neatness** Keep work areas free of all unnecessary books and papers. Tie back long, loose hair and button or roll up loose sleeves when working with chemicals or near an open fire.

5. **Chemicals and Other Dangerous Substances** Chemicals can be dangerous if they are handled carelessly.

   Never taste chemicals or place them near your eyes. Do not use mouth suction when using a pipette to transfer chemicals. Use a suction bulb instead.

   Never pour water into a strong acid or base. The mixture produces heat. Sometimes the heat causes splattering. To keep the mixture cool, pour the acid or base slowly into the water.

   If any solution is spilled on a work surface, wash it off at once with plenty of water. When noting the odor of chemical substances, wave the fumes toward your nose with your hand rather than putting your nose close to the source of the odor.

   Never eat in the laboratory. Counters and glassware may contain substances that can contaminate food. Handle toxic substances in a well-ventilated area or under a ventilation hood.

   Do not use flammable substances near flame. Do not place flammable chemicals in a refrigerator. Sparks from the refrigeration unit can ignite these substances or their fumes.

6. **Heat** Whenever possible, use an electric hot plate instead of an open flame. If you must use an open flame, shield it with a wire screen that has a ceramic center. When heating chemicals in a test tube, do not point the test tube toward anyone.

Keep combustible materials away from heat sources.

7. **Electricity** Be cautious around electrical wiring. When using a microscope with a lamp, do not place its cord where it can cause someone to trip and fall. Do not let cords hang loose over a table edge in a way that permits equipment to fall if the cord is tugged. Do not use equipment with frayed cords.

8. **Knives** Use knives, razor blades, and other sharp instruments with extreme care. Do not use double-edged razor blades in the laboratory.

9. **Glassware** Examine all glassware before heating. Glass containers for heating should be made of boro-silicate glass or some other heat-resistant material. Never use cracked or chipped glassware. Never force glass tubing into rubber stoppers. Broken glassware should be swept up immediately, never picked up with the fingers. Broken glassware should be discarded in a special container, never into a sink.

10. **First Aid** In case of severe bleeding, apply pressure or a compress directly to the wounded area and see that the injured student reports immediately to the school nurse or a physician.

   Minor burns caused by heat should be treated by applying ice. Immerse the burn in cold water if ice is not available. Treat acid burns by applying sodium bicarbonate (baking soda). Use boric acid to treat burns caused by bases. Any burn, regardless of cause, should be reported to your teacher and referred to the school nurse or a physician.

   In case of fainting, place the person's head lower than the rest of the body and see that the person has fresh air. Report to your teacher immediately.

   In case of poisoning, report to your teacher at once. Try to determine the poisoning agent if possible.

11. **Unauthorized Experiments** Do not perform any experiment that has not been assigned by your teacher. Never work alone in the laboratory.

12. **Cleanup** Wash your hands immediately after handling hazardous materials. Before leaving the laboratory, clean up all work areas. Put away all equipment and supplies. Make sure water, gas, burners, and electric hot plates are turned off.

Remember at all times that a laboratory is a safe place only if you regard laboratory work as serious work.

The instructions for your laboratory investigations will include cautionary statements where necessary. In addition, you will find that the following safety symbols appear whenever a procedure requires extra caution:

 Wear safety goggles

 Flame/heat

 Wear laboratory apron

 Dangerous chemical/poison

 Sharp/pointed object

 Electrical hazard

 Biohazard/disease-causing organisms

 Radioactive material

Laboratory Procedures

Reading a Metric Ruler

1. Examine your metric ruler. The numbers on it represent lengths in centimeters. The usual metric ruler is about 30 cm long. There are 10 marked spaces within each centimeter, which represent tenths of centimeters (0.1 cm).
2. To measure the width of a piece of paper, place the ruler on the paper. The zero end of the ruler must line up exactly with one edge of the paper. Look at the other edge of the paper to see which of the marks on the ruler is closest to that edge. In Figure A, for example, the edge of the paper is nearest to the second line beyond the 7. Therefore, the width of the paper is 7.2 cm.

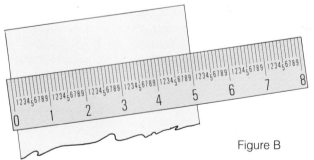

Figure A

3. The edge of the paper might fall exactly on one of the centimeter marks. In Figure B, the edge is just on the 5-cm mark. The width of this paper is 5.0 cm. You must write in the .0 to indicate that the measurement is accurate to the nearest tenth of a centimeter; that is, it is more than 4.95 cm and less than 5.05 cm.

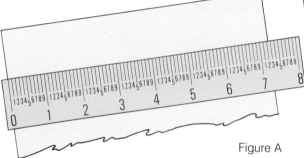

Figure B

4. Sometimes you may want to make a reading with more accuracy. It is possible to estimate readings to the nearest hundredth of a cen-timeter, but you must be very careful. Look at Figure A again. You can guess the number of tenths in the distance between the marks. The edge of the paper is about 3 tenths of the space between 7.2 and 7.3. The best estimate, then, is that the width of the paper is 7.23 cm.

5. In Figure C, the edge of the paper falls exactly on the 8.6 mark. If you are taking careful read-ings, accurate to the nearest hundredth of a centimeter, you must record the width as 8.60 cm.

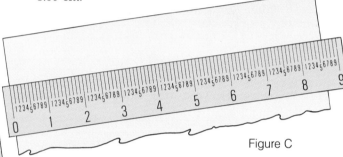

Figure C

6. Note the general rule: You can estimate scale readings to the nearest tenth of a scale divi-sion. If the scale is marked in tenths, you can estimate the hundredths place, but never more than that.

Metric Conversion Table

In SI, it is easy to convert from unit to unit. To convert from a larger unit to a smaller unit, move the decimal to the left. To convert from a smaller unit to a larger unit, move the decimal to the right. Figure D shows you how to move the deci-mals to convert in SI.

Figure D

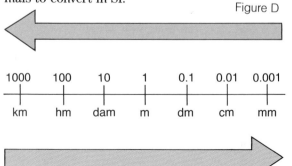

METRIC CONVERSION TABLE

Metric Units		Converting Metric to English		Converting English to Metric	
Length					
kilometer (km)	= 1000 m	1 km	= 0.62 mile	1 mile	= 1.609 km
meter (m)	= 100 cm	1 m	= 1.09 yards	1 yard	= 0.914 m
			= 3.28 feet	1 foot	= 0.305 m
centimeter (cm)	= 0.01 m	1 cm	= 0.394 inch		= 30.5 cm
millimeter (mm)	= 0.001 m	1 mm	= 0.039 inch	1 inch	= 2.54 cm
micrometer (μm)	= 0.000 001 m				
nanometer (nm)	= 0.000 000 001 m				
Area					
square kilometer (km²)	= 100 hectares	1 km²	= 0.3861 square mile	1 square mile	= 2.590 km²
hectare (ha)	= 10 000 m²	1 ha	= 2.471 acres	1 acre	= 0.4047 ha
square meter (m²)	= 10 000 cm²	1 m²	= 1.1960 square yards	1 square yard	= 0.8361 m²
				1 square foot	= 0.0929 m²
square centimeter (cm²)	= 100 mm²	1 cm²	= 0.155 square inch	1 square inch	= 6.4516 cm²
Mass					
kilogram (kg)	= 1000 g	1 kg	= 2.205 pounds	1 pound	= 0.4536 kg
gram (g)	= 1000 mg	1 g	= 0.0353 ounce	1 ounce	= 28.35 g
milligram (mg)	= 0.001 g				
microgram (μg)	= 0.000 001 g				
Volume of Solids					
1 cubic meter (m³)	= 1 000 000 cm³	1 m³	= 1.3080 cubic yards	1 cubic yard	= 0.7646 m³
			= 35.315 cubic feet	1 cubic foot	= 0.0283 m³
1 cubic centimeter (cm³)	= 1000 mm³	1 cm³	= 0.0610 cubic inch	1 cubic inch	= 16.387 cm³
Volume of Liquids					
kiloliter (kL)	= 1000 L	1 kL	= 264.17 gallons	1 gallon	= 3.785 L
liter (L)	= 1000 mL	1 L	= 1.06 quarts	1 quart	= 0.94 L
milliliter (mL)	= 0.001 L	1 mL	= 0.034 fluid ounce	1 pint	= 0.47 L
microliter (μL)	= 0.000 001 L			1 fluid ounce	= 29.57 mL

Reading a Graduate

1. Examine the graduate and note how the scale is marked. The units are milliliters (mL). A milliliter is a thousandth of a liter, and is equal to a cubic centimeter. Note carefully how many milliliters are represented by each scale division on the graduate.

2. Pour some liquid into the graduate and set the cylinder on a level surface. Notice that the upper surface of the liquid is flat in the center, and curved at the edges. This curve is called the *meniscus* and may be either upward or downward. In reading the volume, you must ignore the curvature and read the scale at the flat part of the surface.

3. Bring your eye to the level of the surface and read the scale at the level of the flat surface of the liquid.

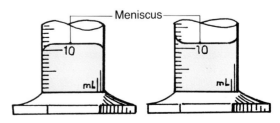

Using a Laboratory Balance

1. Make sure the balance is on a level surface. Use the leveling screws at the bottom of the balance to make any necessary adjustments.
2. Place all the countermasses at zero. The pointer should be at zero. If it is not, adjust the balancing knob until the pointer rests at zero.
3. Place the object you wish to measure on the pan. **CAUTION: Do not place hot objects or chemicals directly on the balance pan as they can damage its surface.**
4. Move the largest countermass along the beam to the right until it is at the last notch that does not tip the balance. Follow the same procedure with the next largest countermass. Then move the smallest countermass until the pointer rests at zero.
5. Determine the readings on all beams and add them together to determine the mass of the object.
6. When massing crystals or powders, use a piece of filter paper. First, mass the paper; then add the crystals or powders and remass. The actual mass is the total minus the mass of the paper. When massing liquids, first mass the empty container, then mass the liquid and container. Finally, subtract the mass of the container from the mass of the liquid and the container to get the mass of the liquid.

Triple-beam balance

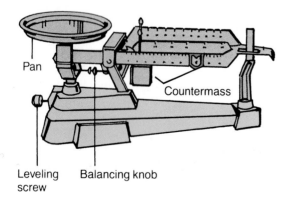

Pan

Countermass

Leveling screw

Balancing knob

Using a Bunsen Burner

1. Before lighting the burner, observe the locations of fire extinguishers, fire blankets, and sand buckets. Wear safety goggles, gloves, and an apron. Tie back long hair and roll up long sleeves.
2. **CAUTION: If the burner is not operating properly, the flame may burn inside the base of the barrel. Carbon monoxide, an odorless gas, is released from this type of flame. Should this situation occur, immediately turn off the gas at the laboratory gas valve. Do not touch the barrel of the burner.** After the barrel has cooled, partially close the air ports before relighting the burner.
3. Close the air ports of the burner and turn the gas full on by using the valve at the laboratory outlet.

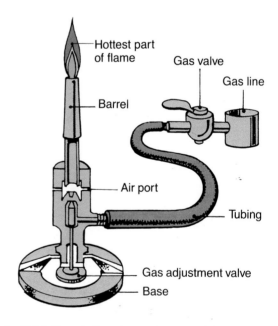

Hottest part of flame

Gas valve

Gas line

Barrel

Air port

Tubing

Gas adjustment valve

Base

4. Hold the striker in such a position that the spark will be placed just above the rim of the burner. Strike a spark.
5. Open the air ports until you can see a blue cone inside the flame. If you hear a roaring sound, the ports are open too wide.

6. Adjust the gas-flow valve and the air ports on the burner until you get a flame of the desired size—not roaring—with a blue cone. The hottest part of the flame is just above the tip of the blue cone.

Filtering Techniques

1. To separate a precipitate from a solution, pass the mixture through filter paper. To do this, first obtain a glass funnel and a piece of filter paper.
2. Fold the filter paper in fourths. Then open up one fourth of the folded paper, as shown in Figure A. Put the paper, pointed end down, into the funnel.

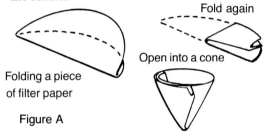

Fold again

Open into a cone

Folding a piece
of filter paper

Figure A

3. Support the funnel in a ring stand, with its stem inserted into a beaker.
4. Stir the mixture well and pour it quickly into the filter paper within the funnel. Wait until all the liquid has flowed through.

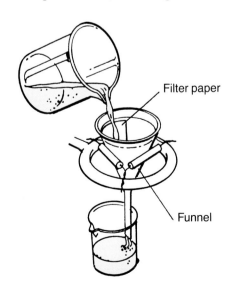

Filter paper

Funnel

5. You may wish to wash the solid that is left in the filter paper. If this is the case, pour some distilled water into the filter paper. Your teacher will tell you how much water to use.

Using Reagents

1. For safety reasons, it is important to learn how to pour a reagent from a bottle into a flask or beaker. Begin with the reagent bottle on the table.
2. While holding the bottle steady with your left hand, grasp the stopper of the bottle between the first and second fingers of your right hand. Remove the stopper from the bottle. DO NOT put the stopper down on the table top.

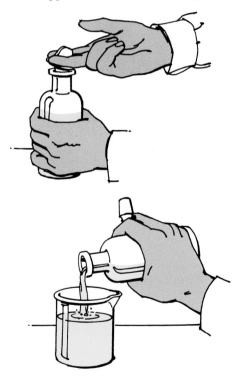

3. While still holding the stopper, lift the bottle with your right hand and pour the reagent into your container.
4. Replace the bottle on the table top and replace the stopper.
(If you are left-handed, these instructions may be reversed.)

Common Laboratory Equipment

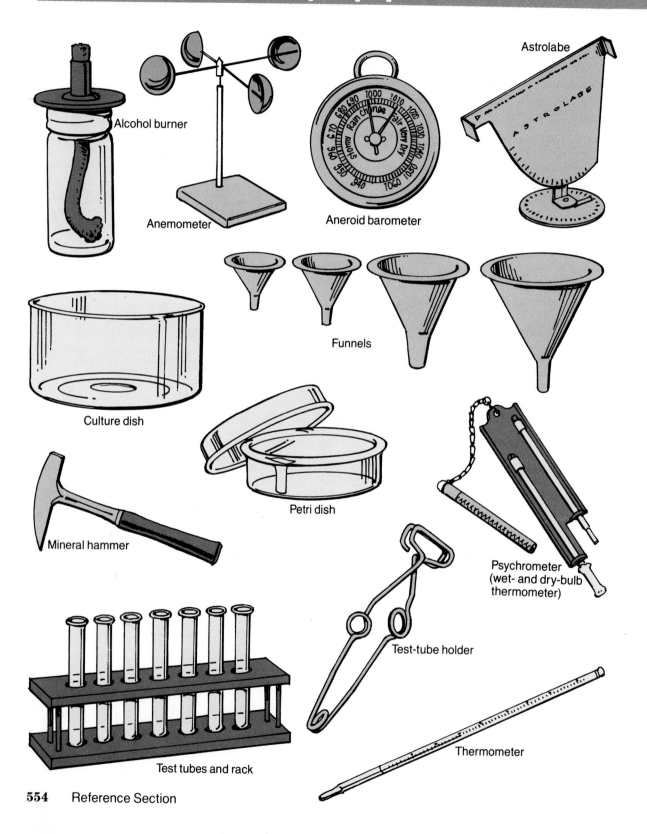

Alcohol burner

Anemometer

Aneroid barometer

Astrolabe

Funnels

Culture dish

Petri dish

Mineral hammer

Psychrometer
(wet- and dry-bulb
thermometer)

Test-tube holder

Test tubes and rack

Thermometer

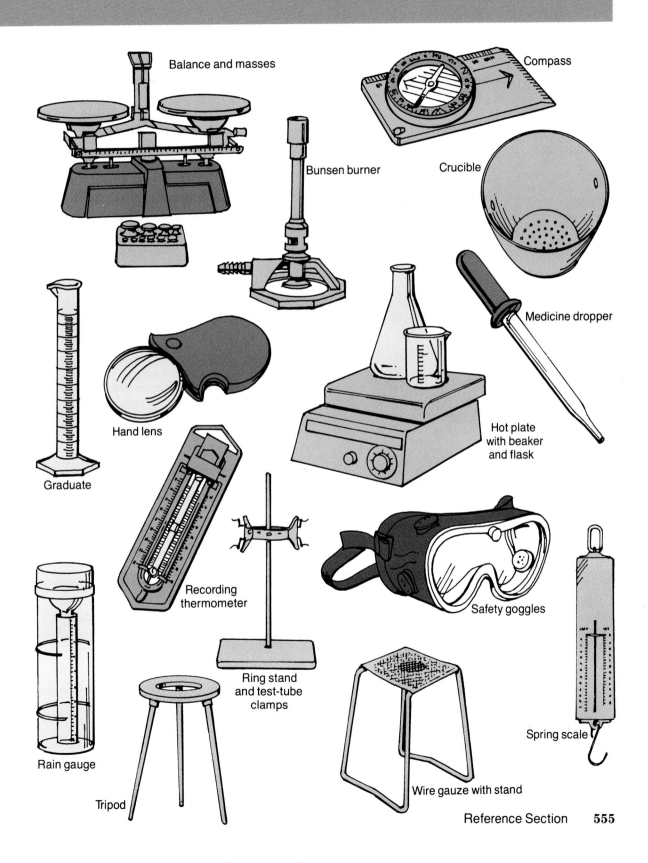

Balance and masses

Compass

Bunsen burner

Crucible

Medicine dropper

Graduate

Hand lens

Hot plate
with beaker
and flask

Recording
thermometer

Safety goggles

Spring scale

Rain gauge

Ring stand
and test-tube
clamps

Tripod

Wire gauze with stand

Key Discoveries in Earth Science

3000 B.C.–501 B.C.

3000 B.C. The Sumerians of the Tigris-Euphrates valley reach high levels in the use of bronze. They are aware that copper can be obtained from mineral ores by using fire, mixed with tin, and cast into bronze instruments.

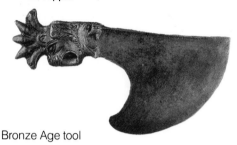

Bronze Age tool

500 B.C.–799 A.D.

500 B.C. Greek philosophers Thales and Anaximander recognize that fossil fish are preserved evidence of once-living fish.

Fossil fish

1600s

1608 Hans Lippershey invents the first telescope by accidentally combining two lenses. In 1609 Galileo refines the telescope and discovers four satellites orbiting Jupiter, rings around Saturn, and mountains and craters on the moon.

1643 Torricelli, an Italian inventor, designs the first barometer, using a column of water in a tube to measure air pressure.

Galileo's telescope

1700s

1785 James Hutton publishes *The Theory of the Earth,* which stresses the geological activity of the internal heat of the earth.

1799 William Smith publishes a usable method of rock classification.

1900s-1930s

1908 Anschutz-Kampfe, a German engineer, develops a gyrocompass. A gyrocompass points to true north rather than to magnetic north as a magnetic compass does.

1940s

1947 John Bardeen, Walter Brattain, and William Shockley invent a transistor out of germanium. These tiny transistors replace vacuum tubes in many common electronic instruments.

1950s

1957 *Sputnik* is launched into an orbit around the earth by USSR. This is followed in 1958 by the launch of *Explorer,* the first United States satellite placed into orbit.

1961 Yuri Gagarin of the Soviet Union becomes the first human to see the earth from space.

Explorer

The key discoveries given here are only a limited selection over 2500 years and are not meant to be a definitive list. The number of women scientists has increased dramatically over the past 10 years.

A.D. 829 Al-Mamum founds an astronomical observatory in Baghdad. Observations of the precession of the equinox and the orbit of the earth are made.

Early cartographer

1250 Roger Bacon of Oxford University states that the true student should learn natural science by experimentation, not from scholars who base their opinions on fallible authority or the weight of custom.

1569 Gerardus Mercator publishes a new map, based on a projection that shows meridians and parallels as straight lines that cross each other at right angles. This is a great aid to navigation.

1824 Joseph Aspdin of Leeds, England, invents cement by pulverizing limestone, mixing it with clay and water, and then heating it in a kiln.

1830 Charles Lyell publishes *The Principles of Geology* as an attempt to explain that changes in the earth's surface were caused by processes still in operation. This is basically a summary of the earlier theories of James Hutton.

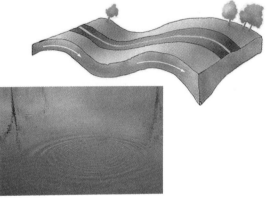

Surface waves

1969 Neil Armstrong of the United States becomes the first human to walk on the moon.

"One small step"

1976 The United States places two *Viking* spacecraft in orbit around Mars. Each orbiter sends a lander to the Martian surface, where they take photographs and conduct experiments on the Martian soil, searching for signs of life.

1986 Archaeological dig near Titusville, Florida unearth preserved brain tissue of humans that lived over 5000 years ago.

Archaeological dig

Periodic Table

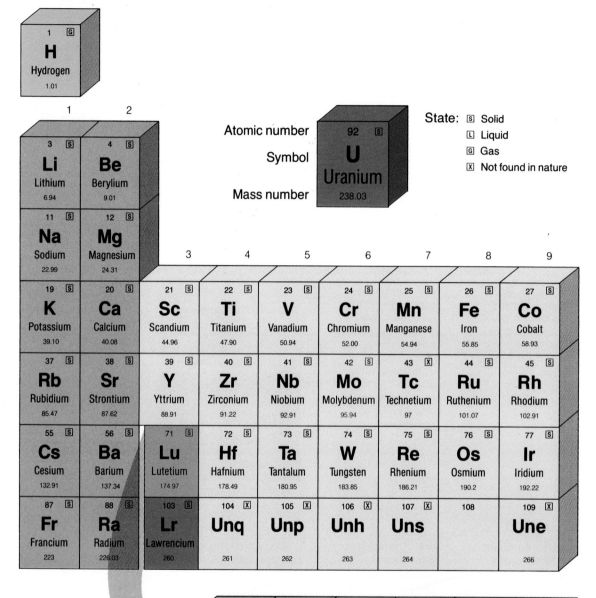

State:
- Ⓢ Solid
- Ⓛ Liquid
- Ⓖ Gas
- Ⓧ Not found in nature

Atomic number	92 Ⓢ
Symbol	**U** Uranium
Mass number	238.03

1 Ⓖ											
H Hydrogen 1.01											

1	**2**	**3**	**4**	**5**	**6**	**7**	**8**	**9**
3 Ⓢ **Li** Lithium 6.94	4 Ⓢ **Be** Berylium 9.01							
11 Ⓢ **Na** Sodium 22.99	12 Ⓢ **Mg** Magnesium 24.31							
19 Ⓢ **K** Potassium 39.10	20 Ⓢ **Ca** Calcium 40.08	21 Ⓢ **Sc** Scandium 44.96	22 Ⓢ **Ti** Titanium 47.90	23 Ⓢ **V** Vanadium 50.94	24 Ⓢ **Cr** Chromium 52.00	25 Ⓢ **Mn** Manganese 54.94	26 Ⓢ **Fe** Iron 55.85	27 Ⓢ **Co** Cobalt 58.93
37 Ⓢ **Rb** Rubidium 85.47	38 Ⓢ **Sr** Strontium 87.62	39 Ⓢ **Y** Yttrium 88.91	40 Ⓢ **Zr** Zirconium 91.22	41 Ⓢ **Nb** Niobium 92.91	42 Ⓢ **Mo** Molybdenum 95.94	43 Ⓧ **Tc** Technetium 97	44 Ⓢ **Ru** Ruthenium 101.07	45 Ⓢ **Rh** Rhodium 102.91
55 Ⓢ **Cs** Cesium 132.91	56 Ⓢ **Ba** Barium 137.34	71 Ⓢ **Lu** Lutetium 174.97	72 Ⓢ **Hf** Hafnium 178.49	73 Ⓢ **Ta** Tantalum 180.95	74 Ⓢ **W** Tungsten 183.85	75 Ⓢ **Re** Rhenium 186.21	76 Ⓢ **Os** Osmium 190.2	77 Ⓢ **Ir** Iridium 192.22
87 Ⓢ **Fr** Francium 223	88 Ⓢ **Ra** Radium 226.03	103 Ⓢ **Lr** Lawrencium 260	104 Ⓧ **Unq** 261	105 Ⓧ **Unp** 262	106 Ⓧ **Unh** 263	107 Ⓧ **Uns** 264	108	109 Ⓧ **Une** 266

57 Ⓢ **La** Lanthanum 138.91	58 Ⓢ **Ce** Cerium 140.12	59 Ⓢ **Pr** Praseodymium 140.91	60 Ⓢ **Nd** Neodymium 144.24	61 Ⓧ **Pm** Promethium 145	62 Ⓢ **Sm** Samarium 150.4
89 Ⓢ **Ac** Actinium 227	90 Ⓢ **Th** Thorium 232.04	91 Ⓢ **Pa** Protactinium 231.04	92 Ⓢ **U** Uranium 238.03	93 Ⓧ **Np** Neptunium 237.05	94 Ⓧ **Pu** Plutonium 244

Metals

Transition metals

Nonmetals

Noble gases

Lanthanide series

Actinide series

				13	14	15	16	17	18
									2 G **He** Helium 4.00
				5 S **B** Boron 10.81	6 S **C** Carbon 12.01	7 G **N** Nitrogen 14.01	8 G **O** Oxygen 16.00	9 G **F** Fluorine 19.00	10 G **Ne** Neon 20.18
	10	11	12	13 S **Al** Aluminum 26.98	14 S **Si** Silicon 28.09	15 S **P** Phosphorus 30.97	16 S **S** Sulfur 32.06	17 G **Cl** Chlorine 35.45	18 G **Ar** Argon 39.95
	28 S **Ni** Nickel 58.71	29 S **Cu** Copper 63.55	30 S **Zn** Zinc 65.38	31 S **Ga** Gallium 69.72	32 S **Ge** Germanium 72.59	33 S **As** Arsenic 74.92	34 S **Se** Selenium 78.96	35 L **Br** Bromine 79.90	36 G **Kr** Krypton 83.80
	46 S **Pd** Palladium 106.4	47 S **Ag** Silver 107.87	48 S **Cd** Cadmium 112.40	49 S **In** Indium 114.82	50 S **Sn** Tin 118.69	51 S **Sb** Antimony 121.75	52 S **Te** Tellurium 127.60	53 S **I** Iodine 126.90	54 G **Xe** Xenon 131.30
	78 S **Pt** Platinum 195.09	79 S **Au** Gold 196.97	80 L **Hg** Mercury 200.59	81 S **Tl** Thallium 204.37	82 S **Pb** Lead 207.2	83 S **Bi** Bismuth 208.98	84 S **Po** Polonium 209	85 S **At** Astatine 210	86 G **Rn** Radon 222

63 S **Eu** Europium 151.96	64 S **Gd** Gadolinium 157.25	65 S **Tb** Terbium 158.93	66 S **Dy** Dysprosium 162.50	67 S **Ho** Holmium 164.93	68 S **Er** Erbium 167.26	69 S **Tm** Thulium 168.93	70 S **Yb** Ytterbium 173.04
95 X **Am** Americium 243	96 S **Cm** Curium 247	97 X **Bk** Berkelium 247	98 X **Cf** Californium 251	99 X **Es** Einsteinium 254	100 X **Fm** Fermium 257	101 X **Md** Mendelevium 258	102 X **No** Nobelium 259

Scientific Notation

The distance from the earth to the sun is 150 000 000 000 m. The diameter of a hydrogen atom is 0.000 000 000 053 m. You could get dizzy counting all those zeroes, so scientists have developed a much neater method, called *scientific notation,* to express very large and very small numbers.

People often give approximate values in round numbers, such as "a thousand" or "about a million." In scientific notation, this sort of approximation is stated by writing the number 10 raised to some power. For example:

$100 = 10^2$

a hundred thousand $= 100\ 000 = 10^5$

a billion $= 1\ 000\ 000\ 000 = 10^9$

In each case, *the number of zeroes after the 1 is the exponent of 10 in scientific notation.*

To write a more exact number in scientific notation, use a technique like this:

$450\ 000 = 4.5 \times 100\ 000 = 4.5 \times 10^5$

The trick is to place the decimal point just after the first digit. Then count the number of places from there to the end of the number. This gives you the power of 10.

To express the distance from the earth to the sun (150 000 000 000 m) in scientific notation, put the decimal point after the 1. From that point, there are 11 digits to the end of the number. Therefore, the distance from the earth to the sun is 1.5×10^{11} m.

For numbers less than 1, the exponent of 10 is negative. Place the decimal point after the first nonzero digit. Then count the number of places from there, going left to the decimal point. That number is the negative exponent of 10.

To express the diameter of a hydrogen atom (0.000 000 000 053 m) in scientific notation, place the decimal point after the 5. Counting left from the decimal point, there are 11 digits. Thus, the diameter of a hydrogen atom is 5.3×10^{-11} m.

Significant Digits

In measuring, there is an important difference between an answer of 9.2 cm and 9.20 cm. If you write 9.2 cm, you are saying that you know the value to the nearest 0.1 cm. That is, the correct value lies between 9.15 cm and 9.25 cm. A measurement of 9.20 cm says that you know the value correct to the nearest hundredth of a cm. The correct value lies between 9.195 cm and 9.205 cm.

The accuracy of a measurement is expressed by the number of significant digits it contains. A reading of 9.20 cm has three digits. It is a more accurate measurement than is expressed by the two digits of 9.2 cm.

Accuracy has nothing to do with the location of the decimal point. Thus, if the length of a road is given as 4935 m, the reading has four significant digits. It is a more accurate measurement than a statement that the width of the road is 20.7 m, which has only three digits.

The zeroes in a number are not always significant. Thus, a measurement made in meters might come out to be 0.058 m. This is no different from 5.8 cm—both numbers have only two significant

digits. Zeroes before the first nonzero digit are needed only to give the location of a decimal point, and are never significant.

Final zeroes may or may not be significant. If you guess the length of a field as "about 350 m," the final zero is not even a guess. It is put in only to locate the decimal point. The 5 is a guess, so the number has two significant digits. However, if you measure the field carefully and it comes to 350 m, the zero is significant. The reading has three significant digits.

Sometimes, a decimal point is put after a zero to indicate that the final zero is significant. Using this system, 350 m has two significant digits, and 350. m has three significant digits.

Zeroes after the decimal point are not needed to locate the decimal point and are never written unless they are significant. A measurement of 9.300 has four significant digits.

In scientific notation, all digits are significant. The guess about the length of the field would be written 3.5×10^2 m, and the measurement would be 3.50×10^2 m.

Suggested Science Projects

1. Make a table showing the metric measurement of some common objects. For example, a doorknob is about 1 m off the floor.

2. Make a topographic map of your neighborhood, showing all the houses and streets; also, use contour lines to show changes in elevation.

3. Make a collection of mineral crystals. Start your collection with some large crystals of salt.

4. Make a list of some different kinds of rock and ways in which they are used in your community. For example, your school may be built of limestone or sandstone.

5. Under the supervision of your science teacher, try smelting a piece of copper ore to recover the copper metal.

6. Make a model of several layers of sedimentary rocks with different colors of modeling clay; then try folding and faulting the layers to produce various landforms.

7. Make a model of a cinder-cone volcano out of plaster or papier-maché. With the supervision of your science teacher, create an eruption in the volcano by igniting potassium permanganate.

8. Make a model showing the Coriolis effect, using a marble and a record turntable. Vary the turntable speed to see what it does to the Coriolis effect.

9. Try making a cloud in a laboratory flask by first adding some condensation nuclei in the form of dust or smoke, and then reducing the air pressure in the flask with a laboratory pump.

10. Set up a weather station with a barometer, hygrometer, recording thermometer, and anemometer, and try making weather forecasts. Start by predicting the weather for the next 12 hours. Then try forecasting for longer periods with the aid of weather maps and satellite photographs from a newspaper.

11. Try making several different types of soil from varying percentages of sand, humus, and clay. Plant some seeds in each of the soils to determine the best soil mixture.

12. Make a model of stream erosion, using a flat pan, some soil, and a garden hose. Try creating different stream patterns and different types of landforms to be eroded.

13. Using topographic maps of Wisconsin, Pennsylvania, or New York, try to find the southern limit of the last glaciation of these areas by looking for glaciated landforms, such as drumlins, kettle lakes, and moraines.

14. Visit a national, state, or local park near your home. Ask the attendants for as much information as they can provide for you on the topography, geology, and hydrology of the area.

15. Make a collection of fossils from the area in which you live, or from some place with which you are familiar. Look for fossils in layers of sedimentary rock, particularly shale.

16. Make a diorama of one of the periods of Earth's history. Paint the background scenery with typical plants and animals of the period, then make models of the organisms for the center of the display.

17. Compare the heat value of various fossil fuels by burning equal amounts of each in a simple calorimeter. A calorimeter can be made from a soda can, a test tube, a one-holed stopper, a thermometer, and a large beaker.

18. Make a working model of one of the locks that raise and lower ships along the St. Lawrence Seaway.

19. If you live near the ocean or a bay, determine the salinity of the streams or rivers that flow into the bay or ocean. Determine the salinity at 1 km intervals from the ocean or bay.

20. Make a list of the products that you use daily that come directly or indirectly from the ocean. For example, shrimp come directly from the ocean, while oil comes indirectly from the fossils of organisms that once lived in the ocean.

21. Make a model of the solar system, showing all of the planets and their known satellites. Try making each planet and satellite its correct relative size. As an alternative, try making a model of the orbits of the inner planets, making the orbits of each the proper relative size.

22. Determine what constellations are visible, and their location on different nights, or at different times during the same night.

23. Make a telescope out of two paper-towel rolls and a pair of convex lenses.

24. Make a model of the Apollo/Saturn system that carried astronauts to the moon. Using the correct modules for each phase, model a typical mission.

25. Make a relief model of the surface of the earth, showing mountains, lakes, oceans, and seas. Show the elevations of the land and the depths of the oceans as close as possible to relative scale.

Building a Science Vocabulary

Are you confused and intimidated by all those big words used in science books? Some students are so overwhelmed by them that they just give up, which is a pity, because scientific words are not really hard once you know a trick about them.

The thing to realize about words used in science is that they are made up of little "pieces," or word elements, from Greek and Latin. Those elements are parts of words that you probably already know. For example, *helicopter* is made up of the word elements *helix* and *pter. Helix* means "spiral," and *pter* means "wing." Do you recall that a pterodactyl is a "wing-fingered" dinosaur? Now, look at the science word *orthoptera.* You know that *pter* means "wing," and you know that an orthodontist straightens crooked teeth, so orthoptera means "something with straight wings." It is the scientific way to say *dragonfly.* When you see a difficult word, look for those parts whose meanings you know; then try to figure out the rest.

Abrasion is one of the first words you encounter in the glossary of this book. You probably already know the word, but here is what the elements mean. *Ab* means "away" (as in *absence*), and *raso* means "rub" (as in *erase*). Therefore, *abrasion* means "rubbing away."

Alluvial is a word that might cause some difficulty unless you know that the first part, *al,* means "to" or "at" (as in *ally*) and the last part, *luvia,* means "wash" (as in *lavatory*). *Alluvial* comes from "to wash," and an alluvial fan is a fan-shaped pile of debris that has been washed over the ground by a river's flow.

Anticline is easy if you know that *anti* means "against" and *cline* means "to lean" (as in "That wall is inclining so much that it may soon recline on the floor"). An anticline is a place where layers of rocks have been pushed upward and lean against each other.

Asthenosphere means "a weak part of a sphere." The word element *a* (or *an*) at the beginning of a science word means "no" or "not" (as in *asexual,* or *asymmetrical*). The Greek element *sthenos* means "strength," and *sphere* means "ball." Of course, you know that the earth is a sphere, but the outer part of that sphere is the only part that is solid. The inside is still in the molten state because it has not cooled off enough to be solid. That is why we call the interior part of the earth (the lower mantle, the inner core, and the outer core) the *asthenosphere,* or the "no-strength part of the ball."

Atmosphere is another word that incorporates the word *sphere.* The Greek word for vapor is *atmos,* and the atmosphere is that part of the spherical earth that contains air, water, and other vapors.

Barometer can better be remembered if you know what its word elements mean. *Baros* means "heavy," and *metr* means "to measure." Just as a speedometer measures the speed of a car, a barometer measures how heavy the atmosphere is.

Benthos is a very deep subject—deep in the sea, that is, because *benthic* creatures live deep in the sea. *Benthos* is close to the word *bathos*, which really means "deep." A bathysphere is a round vehicle that people travel in to the bottom of the sea.

Caldera is an interesting word because we all know that a caldron is a large, open-topped kettle. If you get too close to a boiling caldron, it might scald you. The center of a volcanic cone, the caldera, looks rather like a giant boiling kettle.

Cartography might seem, at first glance, to be a funny name for mapmaking. However, the Greek word *charte* means "a piece of paper," and you may have read about sailing uncharted seas. The word *card* also means "paper," and *graphy* means "description," so *cartography* refers to descriptions on pieces of paper.

Convection is one of the ways in which heat is transferred, but you might have difficulty remembering that. The first element, *con*, means "with" (as in *convoy* and *conjunction*). The word element *vec* means "to carry" (as in *convey*). It should now be easy to understand that when heat is moved or transferred by convection, it is carried with something, such as a liquid or a gas.

Delta is not part of a word; it is a letter in the Greek alphabet, and it is shaped like a triangle. *Delta* has two meanings in science. The symbol means "change"; the word means "the triangular-shaped pile of silt deposited at the mouth of a muddy river."

Exfoliation will be easy to remember if you know that *ex* means "off." Many small leaves make up foliage, and *exfoliation* means "the falling away of small flakes of rock from boulders," similar to the falling of autumn leaves from trees.

Foliated rocks are rocks that are in layers. This textbook has many layers as well, but we call them *leaves*, or in Latin, *foli*.

Galaxy In the sky at night, you can see a galaxy that is called the *Milky Way*. The Greeks called it that also, because *galax* means "milk." They thought that the ribbon of stars they saw looked as white as milk. At parties, the Greeks would wear their whitest clothes. That is why, even now, a dress-up party is called a *gala*.

Geologic and **geomorphic** both contain the word element *geo*, which means "earth." The other elements in these words are *logos*, which means "a study of," and *morphic*, which means "form" or "shape."

Igneous is a hot item in earth science classes because it means "fire" (as in *ignite*). Whenever you think of igneous rocks, think about the fact that they form in the earth's hot mantle.

Lithosphere *Lithos* is a Greek word that means "stone." *Lithosphere* means "rock sphere" and refers to the rock-solid outer layer of the earth.

Magma is a strange name for the molten rock inside the earth. It becomes more understandable when you think about the consistency of molten lava—about the same as that of the dough from which bread is made. *Magma* is a Latin word that means "dough."

Metamorphic contains the word element *morphic*, which means "form." The element *meta* means "among" or "changing," so metamorphic rocks are rocks that have changed to form new rocks.

Nekton is a Greek word meaning "swim near the surface of the sea." This word is a little like the word *nata*, which just means "to swim." However, Nekto was a mortal who angered a god and was forced to swim at the surface of the sea forever.

Nucleus is a nutty-sounding word—and it should be—because when scientists first looked inside a cell through a microscope, they saw what looked like a nut (a walnut, to be exact). The Latin name for walnut is *nucleus*. We use the same term for the inner core of an atom.

Ozone is the term used for a molecule of three oxygen atoms. The word element *ozo* means "smell." Pure ozone has a smell that is hard to forget. It smells a little like acetylene gas or the odor from a spark generator. Its toxic fumes will quickly give you a headache.

Penumbra refers to a kind of shadow. The first part of the word, *penu*, means "almost" (as in *penultimate*, or "next to last"). *Umbra* means "shade" (as in *umbrella*), so *penumbra* means "an area of partial eclipse."

Permeable has the word element *per*, which means "through" (as in *perjury* and *perspective*). Think of permeable things as being able to let fluids pass though with ease.

Playa is neither Greek nor Latin. It is a Spanish word that means "beach." The word came from the early Spanish conquistadors who saw shallow pools of water in the desert. The area around the pools looked like a beach, so the name stuck.

Refraction is usually a hard word to remember. Think of the word element *fract*, which is part of *fracture*. *Fract* means "broken." Refracted light is light that is broken, or bent. A fraction is a "broken" whole number.

Saltation is not a salty term. It is more of a bouncy, dancing little word because *salta* actually means "to bounce" or "to dance." Think of little pieces of sand and gravel being blown so hard that they bounce when they strike a surface.

A **seismograph** is a device for measuring earthquakes. You know what a graph is; would it surprise you to learn that *seism* means "to shake"?

Smog is a fairly new word. It was first used in the early 1950s. The word is a combination of *smoke* and *fog*, describing what smog looks like.

A **syncline** is the opposite of an anticline. Since *cline* means "to lean," and *syn* means "within," you can figure out that a syncline must be a formation in which the rocks lean toward one another.

Tectonics (as in "the theory of plate tectonics") has the word element *techton*. *Techton* is a Greek word meaning "carpenter." Plate tectonics is a theory that describes how huge continental plates drift into different patterns over the ages, much as large boards can be put together by a carpenter into different shapes.

Zodiac is a term that is not merely a part of astrology. The word element *zo* means "animal" (as in *zoo* and *protozoa*). The term *zodiac* arose because ancient astronomers thought they could see the shapes of animals in certain patterns of stars. They called those patterns the *zodiac*.

Element	Meaning	Examples
A	No, Not	Atheist, Anarchy, Asthenosphere
AB	From, Away	Abrasion, Absorb, Abscond
ANTI	Against	Antidote, Antiseptic, Anticline
AQUA	Water	Aquatic, Aqueduct, Aquifer
BARI	Heavy	Barometer, Barium, Baritone
BENTHO	Depths (of the sea)	Benthic, Benthal, Benthos
CALD	Hot	Caldera, Caldron, Scald
CART	Piece of paper	Cartography, Chart, Magna Charta
CLAST	To break	Iconoclast, Clastic, Orthoclase
CLIMA	Region, Slope	Climate, Climax
CLINO	Lean	Anticline, Incline, Recline
CO	With	Cooperate, Covalent bond, Collaborate
COM	Together	Composite, Complex, Complicate
CON	With, Together	Convection, Conclude, Convey
DELTA	Triangle shape	River delta, Deltoid, Deltoidal
EPI	Upon	Epicenter, Epidemic, Epitaph
EX	Out of	Extrusive, Exodus, Exclude
FER	To bear, Ferry	Coniferous, Ferry, Aquifer
FLATU	Wind	Deflation, Inflate
FOLI	Plant leaves	Foliage, Folio, Foliated
FRACT	Broken	Fraction, Refraction, Fracture
GALA	Milky white	Galaxy, Galactic, Gala party
GEO	Earth	Geology, Geomorphic, Geographical
GRAPHY	Written description	Cartography, Geography, Photography
HYDRO	Water	Hydrophobia, Hydraulic, Hydrogen
HYPO	Under	Hypothesis, Hypodermic, Hypotonic
IGNIS	Fire	Ignite, Igneous rock, Ignition
ION	Violet color	Cation, Ionize, Iodine
ISO	Equal	Isobar, Isosceles, Isomer
LAVO	To wash	Alluvial, Lavatory, Laundry
LITH	Stone, Rock	Lithosphere, Monolith, Neolithic
LOGO	Speech, Thought	Analog, Dialogue, Logical
MAGMA	Dough, To knead	Magma, Magmatic
META	Changing	Metaphase, Metastasis, Metamorphic
METR	To measure	Barometer, Centimeter, Thermometer
MORPH	Form, Shape	Mesomorph, Metamorphic, Morphology
NEKTON	Swim	Nekton, Nekteric, Nektonic
NUCLE	Nut, Walnut	Nuclear, Nucleoplasm, Nucleus
OZO	Smell	Ozone, Ozonosphere
PENU	Almost	Penultimate, Penumbra, Penumbrous
PER	Through	Perjury, Perspective, Permeable
POSI	Place	Position, Composite, Deposit
RAS	Scrape, Rub away	Rasp, Abrasion, Razor
SALTA	Jump, Dance	Saltation, Saltatory, Salticidae
SEDI	Sit	Sediment, Sedentary, Sedimentary rock
SEISM	Shake	Seismic, Seismograph, Seismology
STHENOS	Strength	Asthenosphere, Asthenia, Asthenolith
TECTON	Carpenter	Plate tectonics, Tectology, Tectological
TOMO	Cut	Atom, Appendectomy, Epitome
TOPO	Surface	Topography, Topectomy, Topographical
TRUDO	Thrust	Intrusive, Protrude, Extrude
UMBRA	Shade	Umbrella, Penumbra, Umbrage
VALE	Value, Worth	Covalent bond, Equivalent, Valence
VEC	To carry	Convection, Vector, Convey
VERG	To incline	Converge, Diverge, Divergent
ZO	Animal	Zodiac, Zoo, Protozoa

Mineral Classification Key

DIRECTIONS: Examine your mineral sample. First, determine whether the sample has a metallic luster, which places it in Category I, or a nonmetallic luster, which places it in Category II. If the luster is metallic, determine the mineral's streak. If the streak is black (Choice A), determine the mineral's hardness. If the streak is yellow or red, move to the Key of Mineral Groups.

If the luster is nonmetallic, determine whether the mineral is light colored (Choice A) or dark colored (Choice B). With each choice, determine the mineral's hardness, then move to the proper group in the Key of Mineral Groups.

Key for Determining Mineral Groups

I. Metallic luster

   A. Streak black, greenish-black, or dark gray
      1. Hardness 3 or less . . . . Group 1
      2. Hardness between 3 and 6 . . . . Group 2
      3. Hardness 6 or over . . . . Group 3

   B. Streak yellow to brown . . . . Group 4
   C. Streak red . . . . Group 5

II. Nonmetallic Luster

   A. Mineral white or light colored
      1. Hardness 3 or less . . . . Group 6
      2. Hardness between 3 and 6 . . . . Group 7
      3. Hardness 6 or over . . . . Group 8

   B. Mineral dark colored, black, or green
      1. Hardness 3 or less . . . . Group 9
      2. Hardness between 3 and 6 . . . . Group 10
      3. Hardness 6 or over . . . . Group 11

Key of Mineral Groups

DEFINITIONS: The following terms are used in the Key of Mineral Groups.

   Cleavage—
      basal—one direction
      cubic—three directions, 90° angles
      diamond—two directions, less than 90° angles
      rhombohedral—three directions, less than 90° angles
      square—two directions, 90° angles
      octahedral—four directions
      dodecahedron—six directions
   Concretionary—looks like chunks of different-colored minerals glued together; luster is earthy or dull, color of samples
   Saline—tastes like salt
   Malleable—will bend without breaking

Group 1
Galena—cubic cleavage, very high specific gravity
Graphite—basal cleavage, greasy feeling

Group 2
Chalcopyrite—dark brassy color

Group 3
Magnetite—magnetic, black color
Pyrite—light yellow color

Group 4
Limonite—no cleavage, rarely has metallic luster

Group 5
Copper—malleable, reddish color
Hematite—brittle, black to reddish in color, can also have appearances of glitter

Group 6
Bauxite—no cleavage, earthy, concretionary, tan, white color
Calcite—rhombohedral cleavage, hardness 3
Gypsum—nonelastic, hardness 2, basal cleavage not often evident
Halite—cubic cleavage, saline taste
Kaolinite—no cleavage, white color, earthy, smells like clay when moist
Muscovite—perfect basal cleavage
Serpentine—variety of asbestos, separates into silky fibers, although this may not always be evident
Talc—may show basal cleavage, nonelastic, greasy feeling

Group 7
Amphibole—hornblende, variety of tremolite, diamond-shaped cleavage, white to gray color
Apatite—no cleavage, brown or green color
Barite—three cleavage planes, two at right angles, high specific gravity
Dolomite—rhombohedral cleavage
Fluorite—octahedral cleavage, hardness 4
Malachite—no cleavage, green color
Sphalerite—dodecahedral cleavage, resinous luster

Group 8
Corundum—no cleavage, may show parting, hardness 9

Group 8 (continued)
Orthoclase—feldspar, two cleavages at right angles, hardness 5, tan or pink color
Plagioclase—feldspar, two cleavages at 86° and 94°, striations, blue-gray to white color
Quartz—no cleavage, hardness 7, conchoidal fracture
Tourmaline—no cleavage, pink, blue, or green color

Group 9
Bauxite—no cleavage, earthy, concretionary
Biotite—perfect basal cleavage, elastic, black color
Chlorite—perfect cleavage, nonelastic, green color
Graphite—basal cleavage, greasy feeling
Hematite—red streak, high specific gravity
Limonite—no cleavage, brown streak
Muscovite—perfect basal cleavage, elastic
Talc—basal cleavage, nonelastic, greasy feeling

Group 10
Amphibole—hornblende, diamond-shaped cleavage, two planes not equal to 90°
Apatite—no cleavage, brown or green color, hardness 5
Azurite—no cleavage, blue color
Fluorite—octahedral cleavage, hardness 4
Hematite—red streak, high specific gravity
Limonite—no cleavage, brown streak
Malachite—no cleavage, green color, green streak
Pyroxene—resembles hornblende, square cleavage, two planes at 92°
Serpentine—no cleavage, green, waxy luster
Sphalerite—dodecahedral cleavage, resinous luster

Group 11
Corundum—no cleavage, hardness 9
Garnet—no cleavage, red color
Olivine—olive-green color, never occurs with quartz
Orthoclase—feldspar, two cleavages at right angles, hardness 6, white, tan, or pink color
Plagioclase—feldspar, two cleavages at 86° and 94°, striations, white to blue-gray color, hardness 6
Quartz—no cleavage, hardness 7
Tourmaline—no cleavage, usually black color

Rock Classification Key

DIRECTIONS: Examine your rock sample. First, determine if it belongs in Category I by asking yourself the question, "Does the rock have layers or clastic particles?" If you decide the answer is yes, then determine whether choice A or B applies to your rock sample.

If you decide that the rock does not belong in Category I, then go to Category II. Continue through the Categories until you find one that fits your rock sample. Then read the detailed choices to determine the name of your rock sample.

I. Does the sample have layers, or clastic particles glued together? If it does, go to choice A. If it does not, go to Category II.

   A. Does the rock react to HCl? If it does, go to Example 1. If it does not, go to Choice B.
      1. Limestone—not shiny; whole surface reacts
      2. Calcareous sandstone—sand grains present
      3. Limestone conglomerate—pebble-sized particles; HCl reacts with matrix
      4. Marble—shiny

   B. If the rock does not react to HCl and it is not foliated, go to Example 1. If the rock is foliated, go to Category II.
      1. Conglomerate—rounded pebbles
      2. Breccia—sharp-edged pebbles
      3. Sandstone—sand-sized grains
      4. Shale—clay-sized grains; frequently shows thin layers
      5. Bituminous coal—black color which rubs off on your hands

II. Is the rock foliated or banded? If the sample is foliated or banded, go to Choice A. If the sample is not foliated or banded, go to Category III.

   A. Banded
      1. Gneiss—shows layers of different-colored minerals

   B. Foliated
      1. Schist—distinct mica-type layering with wavy surfaces
      2. Phyllite—shows some "mica shine" but looks less foliated than schist

III. Does the rock contain holes? If it does, go to Choice A. If it does not, go to Category IV.

   A. Scoria—dark color, dense and heavy

   B. Pumice—light color, tan, brown, or red, lightweight

IV. Does the rock have luster like glass? If it does, it is obsidian or anthracite. If it does not, go to Category V.

   A. Obsidian—hard, shows conchoidal fracture, black to gray color

   B. Anthracite coal—very lightweight, black color, glassy luster

V. Can mineral crystals be identified? If yes, the rock is granite. If not, go to Choice A.

   A. Rhyolite—light color, dull luster

   B. Basalt—dark, dull luster

   C. Obsidian—shiny luster like glass

Granite

Basalt

Pumice

Rhyolite

Obsidian

Scoria

Limestone

Conglomerate

Breccia

Sandstone

Shale

Siltstone

Coquina

Chert

Rock salt

Marble

Quartzite

Slate

RELATIVE HUMIDITY TABLE: CELSIUS

Wet-Bulb Depression (°C)
(Dry-bulb temperature minus wet-bulb temperature)

Dry-Bulb Temperature (°C)

Dry-Bulb Temp (°C)	1	2	3	4	5	6	7	8	9	10	11	12	13	14	15	16	17	18	19	20
-10	67	35																		
-8	71	43	15																	
-6	74	49	25																	
-4	77	55	33	12																
-2	79	60	40	22																
0	81	64	46	29	13															
2	84	68	52	37	22	7														
4	85	71	57	43	29	16														
6	86	73	60	48	35	24	11													
8	87	75	66	51	40	29	19	8												
10	88	77	66	55	44	34	24	15	6											
12	88	78	68	58	48	39	29	21	12											
14	90	79	70	60	51	42	34	26	18	10										
16	90	81	71	63	54	46	38	30	23	15	8									
18	91	82	73	65	57	49	41	34	27	20	14	7								
20	91	83	74	66	59	51	44	37	31	24	18	12	6							
22	92	83	76	68	61	54	47	40	34	28	22	17	11	6						
24	92	84	77	69	62	56	49	43	37	31	26	20	15	10	5					
26	92	85	78	71	64	58	51	46	40	34	29	24	19	14	10					
28	93	85	78	72	65	59	53	48	42	37	32	27	22	18	13	9	5			
30	93	86	79	73	67	61	55	50	44	39	35	30	25	20	16	13	9	5		
32	93	86	80	74	68	62	57	51	46	41	37	32	28	24	20	16	12	8	5	
34	93	87	81	75	69	63	58	53	48	43	39	35	30	26	23	19	15	12	8	5
36	94	87	81	75	70	64	59	54	50	45	41	37	33	29	25	21	18	15	11	8
38	94	88	82	76	71	66	61	56	51	47	43	39	35	31	27	24	20	17	14	11
40	94	88	82	77	72	67	62	57	53	48	44	40	36	33	29	26	23	20	16	14

DEW-POINT TEMPERATURE TABLE: CELSIUS

Wet-Bulb Depression (°C)
(Dry-bulb temperature minus wet-bulb temperature)

Dry-Bulb Temperature (°C)

Dry-Bulb Temp (°C)	1	2	3	4	5	6	7	8	9	10	11	12	13	14	15	16	17	18	19	20
-10	-15	-22																		
-8	-12	-18	-30																	
-6	-9	-14	-23																	
-4	-7	-11	-17	-30																
-2	-5	-8	-13	-20																
0	-2.5	-6	-10	-15	-25															
2	-0.5	-3	-7	-11	-18	-30														
4	2	-1	-4	-7.5	-12	-17														
6	4	1.5	-1	-4	-8	-14	-22													
8	6	4	1	-1.7	-4.5	-9	-15	-20												
10	8	6	4	1	-1.5	-5	-9.5	-15	-28											
12	10	8	6	4	1	-2	-5.5	-10	-16	-30										
14	12	11	8	6	4	1	-2	-6	-10	-17.5										
16	14	12.5	11	8.5	6	4	1	-2	-6	-10	-18									
18	16	14.5	13	11	9	6.5	4	1	-2	-4.5	-10	-18								
20	18	16.7	15	13	10.5	9.5	7	4.5	2	-1	-5	-10	-18							
22	20	18.7	17	16	13.5	11.5	10	7.5	5	2	-1.5	-5	-10	-18						
24	22	20.7	19	17.5	16	14	12	10	8	5	2.5	-1	-5	-10	-18					
26	24	22.7	21	19.5	18	16.5	15	13	10.5	8	6	2.5	-1	-5	-10	-18				
28	26	24.7	23	22	20	19	17	15	13	11	9	6	3	-1	-5	-10	-18			
30	28	26.7	25	24	22	21.5	20	18	16	14	12	10	6	3	-1	-5	-10	-18		
32	30	28.7	27	26	24	23	22	20	18	17	15	13	10	6	3	-1	-5	-10	-18	
34	32	30.7	29	28	26	25	24	22	20	19	17	15	13	10	6	3	-1	-5	-10	-18
36	34	32.7	31	30	28	27	26	24	22	21	19	17	15	13	10	6	3	-1	-5	-10
38	36	34.7	33	32.5	30	29	28	26	24	23	21	19	17	15	13	10	6	3	-1	-5
40	38	36.9	35	34	32	31	30	28	26	25	23	21	19	17	15	13	10	6	3	-1

HEAT INDEX CHART (APPARENT TEMPERATURE)
RELATIVE HUMIDITY (%)

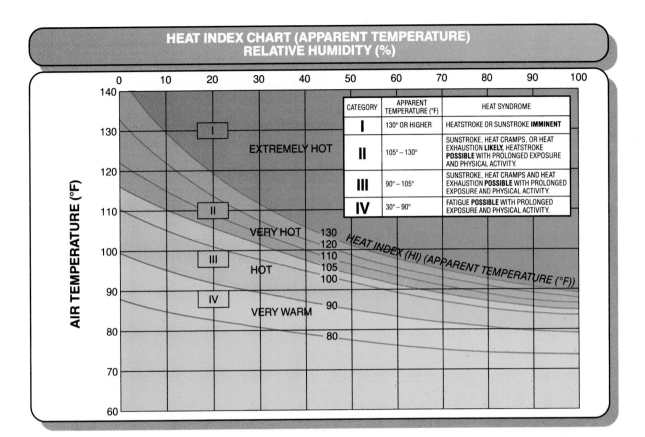

CATEGORY	APPARENT TEMPERATURE (°F)	HEAT SYNDROME
I	130° OR HIGHER	HEATSTROKE OR SUNSTROKE **IMMINENT**
II	105° – 130°	SUNSTROKE, HEAT CRAMPS, OR HEAT EXHAUSTION **LIKELY,** HEATSTROKE **POSSIBLE** WITH PROLONGED EXPOSURE AND PHYSICAL ACTIVITY.
III	90° – 105°	SUNSTROKE, HEAT CRAMPS AND HEAT EXHAUSTION **POSSIBLE** WITH PROLONGED EXPOSURE AND PHYSICAL ACTIVITY.
IV	30° – 90°	FATIGUE **POSSIBLE** WITH PROLONGED EXPOSURE AND PHYSICAL ACTIVITY.

WIND CHILL CHART (APPARENT TEMPERATURE)

Wind Speed		Cooling Power of Wind Expressed as "Equivalent Chill Temperature"																			
Knots	**MPH**	**Temperature (°F)**																			
Calm	**Calm**	40	35	30	25	20	15	10	5	0	−5	−10	−15	−20	−25	−30	−35	−40	−45	−50	−55	−60
		Equivalent Chill Temperature																			
3	6	5	35	30	25	20	15	10	5	0	−5	−10	−15	−20	−25	−30	−35	−40	−45	−50	−55	−65	−70
7	10	10	30	20	15	10	5	0	−10	−15	−20	−25	−35	−40	−45	−50	−60	−65	−70	−75	−80	−90	−95
11	15	15	25	15	10	0	−5	−10	−20	−25	−30	−40	−45	−50	−60	−65	−70	−80	−85	−90	−100	−105	−110
16	19	20	20	10	5	0	−10	−15	−25	−30	−35	−45	−50	−60	−65	−75	−80	−85	−95	−100	−110	−115	−120
20	23	25	15	10	0	−5	−15	−20	−30	−35	−45	−50	−60	−65	−75	−80	−90	−95	−105	−110	−120	−125	−135
24	28	30	10	5	0	−10	−20	−25	−30	−40	−50	−55	−65	−70	−80	−85	−95	−100	−110	−115	−125	−130	−140
29	32	35	10	5	−5	−10	−20	−30	−35	−40	−50	−60	−65	−75	−80	−90	−100	−105	−115	−120	−130	−135	−145
33	36	40	10	0	−5	−15	−20	−30	−35	−45	−55	−60	−70	−75	−85	−95	−100	−110	−115	−125	−130	−140	−150

Winds above 40 have little additional effect	Little Danger	Increasing Danger (Flesh may freeze within 1 min.)	Great Danger (Flesh may freeze within 30 sec.)
	Danger of freezing exposed flesh for properly clothed persons		

CLIMATIC SUMMARY

January

January	Avg Max	Avg Min	Rec Max	Rec Min	Mean Days of Precip.	Total Avg	Total Max	Snowfall Avg	Snowfall Max	% Possible Sunshine	Mean Clear Days	Mean Cloudy Days	Rel. Humidity (%)	Avg. Wind Speed (MPH)
Albuquerque, NM	47.2	22.3	69	-17	3.8	.41	1.32	2.5	9.5	72	13.0	10.2	40	8.1
Anchorage, AK	20.0	6.0	50	-34	7.0	.80	2.09	10.1	21.1	40	8.3	18.2	70	6.1
Atlanta, GA	51.2	32.6	79	-5	11.6	4.91	10.82	0.8	8.3	49	8.3	16.4	60	10.6
Bismarck, ND	17.5	-4.2	62	-44	7.9	.51	1.29	7.2	25.0	54	6.8	16.5	67	10.0
Boise, ID	37.1	22.6	63	-17	12.4	1.64	3.87	7.2	21.4	39	4.4	21.8	70	8.2
Boston, MA	36.4	22.8	63	-12	11.7	3.99	10.55	12.6	35.9	53	9.3	15.0	57	14.0
Buffalo, NY	30.0	17.0	72	-16	20.2	3.02	6.88	23.9	66.3	32	1.4	23.4	73	14.3
Casper, WY	32.5	11.9	60	-40	7.4	.50	1.19	10.1	22.1	NA	6.5	16.8	60	16.5
Charleston, SC	58.8	36.9	83	10	9.8	3.33	6.68	0.1	1.0	58	8.9	15.6	55	9.2
Charlotte, NC	50.3	30.7	78	-3	10.2	3.80	7.44	2.0	11.7	55	9.1	15.5	56	7.9
Chicago, IL	29.2	13.6	61	-26	11.3	1.60	4.11	11.1	34.3	42	6.9	18.2	67	11.6
Cincinnati, OH	37.3	20.4	69	-25	12.1	3.13	9.43	7.5	31.5	41	5.1	19.7	68	10.7
Dallas, TX	54.0	33.9	88	4	7.0	1.65	3.60	1.4	12.1	53	9.8	15.5	59	11.1
Denver, CO	43.1	15.9	73	-25	5.8	.51	1.44	7.7	23.7	71	10.0	11.8	46	8.8
Detroit, MI	30.6	16.1	62	-21	13.0	1.86	3.63	9.6	29.6	40	4.3	20.0	70	11.7
Helena, MT	28.1	8.1	62	-42	8.2	.66	2.78	9.1	35.6	46	4.6	20.5	61	6.9
Honolulu, HI	79.9	65.3	87	53	8.2	3.79	14.74	0.0	–	62	9.1	8.9	63	9.8
Houston, TX	61.9	40.8	84	12	10.3	3.21	7.68	0.2	2.0	43	7.3	18.3	65	8.2
Kansas City, MO	34.5	17.2	68	-17	8.2	1.08	2.66	5.8	14.2	59	9.6	15.4	64	11.0
Las Vegas, NV	56.0	33.0	77	8	3.0	.50	2.41	1.2	16.7	77	13.9	10.9	30	11.2
Lexington, KY	39.8	23.1	76	-21	12.6	3.57	16.65	6.2	21.9	NA	5.8	19.3	69	11.2
Los Angeles, CA	64.6	47.3	88	23	6.1	3.06	9.60	T	T	69	12.0	10.8	56	6.7
Memphis, TN	48.3	30.9	78	-4	10.1	4.61	12.21	2.3	12.2	50	7.9	17.2	63	10.3
Miami, FL	75.0	59.2	88	31	6.5	2.08	6.66	0.0	–	67	9.6	8.5	60	9.4
Milwaukee, WI	26.0	11.3	62	-26	11.1	1.64	4.04	13.0	33.6	44	7.2	17.7	68	12.8
Minneapolis, MN	19.9	2.4	58	-34	8.8	.82	3.63	9.9	46.4	52	8.4	15.3	66	10.4
New Orleans, LA	61.8	43.0	83	14	10.1	4.97	13.63	0.0	0.1	50	6.7	16.9	66	9.4
New York, NY	38.0	25.6	72	-6	11.2	3.21	10.52	7.6	27.4	50	8.1	13.7	60	10.7
Philadelphia, PA	38.6	23.8	74	-7	11.0	3.18	8.86	6.4	23.4	50	7.4	16.1	59	10.3
Phoenix, AZ	65.2	39.4	88	17	3.8	.73	2.41	T	T	78	13.8	10.3	32	5.3
Pittsburgh, PA	34.1	19.2	68	-18	16.4	2.86	6.25	12.1	40.2	33	3.0	21.9	65	10.7
Portland, OR	44.3	33.5	62	-2	18.6	6.16	12.83	3.9	41.4	27	2.8	25.0	76	10.0
Sacramento, CA	52.6	37.9	70	23	10.2	4.03	9.14	T	T	45	6.3	18.7	71	7.6
Salt Lake City, UT	37.4	19.7	62	-22	9.9	1.35	3.14	13.2	32.3	46	5.5	18.9	68	7.7
San Antonio, TX	61.7	39.0	89	0	8.1	1.55	8.52	0.2	4.7	48	8.9	15.9	57	9.2
San Francisco, CA	55.5	41.5	72	24	10.8	4.65	10.43	0.0	1.5	56	8.6	14.7	66	7.1
San Diego, CA	65.2	48.4	88	29	6.9	2.11	6.26	T	T	72	12.4	11.2	56	5.8
Seattle, WA	43.9	34.3	64	0	19.0	6.04	12.92	6.0	57.2	24	2.6	24.5	74	9.9
Tampa, FL	70.0	49.5	84	22	6.3	2.17	8.02	0.0	0.2	64	9.5	11.6	59	8.8
Washington, D.C.	42.9	27.5	79	-5	10.3	2.76	7.11	5.1	21.3	48	7.6	16.2	54	10.0

March

March	Avg Max	Avg Min	Rec Max	Rec Min	Mean Days of Precip.	Total Avg	Total Max	Snowfall Avg	Snowfall Max	% Possible Sunshine	Mean Clear Days	Mean Cloudy Days	Rel. Humidity (%)	Avg. Wind Speed (MPH)
Albuquerque, NM	60.7	31.7	85	8	4.5	.52	2.18	1.9	13.9	73	11.3	9.6	24	10.2
Anchorage, AK	31.7	15.7	51	-24	7.7	.69	2.76	9.1	31.0	53	7.4	17.9	56	6.7
Atlanta, GA	63.2	41.7	85	10	11.7	5.91	11.66	0.4	7.9	58	8.9	14.9	51	10.9
Bismarck, ND	36.4	15.6	81	-31	8.2	.70	3.19	8.4	31.1	59	5.4	17.2	63	11.0
Boise, ID	51.8	41.4	81	6	9.7	1.03	2.76	1.9	11.9	62	6.0	17.8	45	10.1
Boston, MA	45.0	31.8	81	6	11.9	4.13	11.0	7.8	31.2	57	7.8	15.3	57	13.8
Buffalo, NY	40.4	25.6	81	-7	16.2	2.97	5.59	11.8	29.2	45	3.6	19.9	67	13.4
Casper, WY	43.4	20.2	73	-21	9.6	.99	2.43	15.2	36.2	NA	5.8	16.5	48	14.0
Charleston, SC	68.0	45.3	90	15	10.4	4.38	11.11	0.1	2.0	67	9.1	13.7	51	10.1
Charlotte, NC	61.6	39.1	90	4	11.4	4.83	8.76	1.5	19.3	63	9.1	13.8	51	8.9
Chicago, IL	44.3	27.6	87	-8	12.7	2.59	5.91	7.6	24.7	50	4.7	17.6	61	11.8
Cincinnati, OH	51.5	32.0	83	-11	13.2	3.95	12.8	4.5	13.0	28	5.0	19.3	61	11.1
Dallas, TX	67.2	44.9	96	15	7.3	2.42	6.39	0.2	2.5	59	9.6	13.7	51	13.0
Denver, CO	51.2	24.7	84	-11	8.7	1.21	4.56	13.1	30.5	70	7.8	12.9	41	9.8
Detroit, MI	43.4	26.5	77	-4	13.2	2.54	4.48	6.8	16.1	51	5.3	18.3	62	11.7
Helena, MT	42.5	20.6	77	-30	8.6	.69	1.62	7.4	21.6	60	3.6	19.1	46	8.5
Honolulu, HI	81.4	67.3	88	55	9.0	3.48	20.79	0.0	–	68	7.3	9.6	58	11.5
Houston, TX	72.1	49.8	90	22	9.8	2.68	8.52	0.0	–	47	6.5	17.9	60	9.5
Kansas City, MO	51.3	31.7	82	-10	10.7	2.41	9.08	3.9	11.4	58	6.3	16.9	59	12.5
Las Vegas, NV	68.3	42.3	91	23	2.9	.41	1.83	0.0	0.1	83	13.8	8.2	22	10.1
Lexington, KY	53.7	34.1	83	-2	13.1	4.83	10.38	2.8	17.7	NA	5.5	18.0	58	11.5
Los Angeles, CA	65.1	49.7	93	34	5.8	1.76	6.37	0.0	–	73	11.6	10.7	61	8.1
Memphis, TN	61.4	41.9	85	12	11.0	5.44	12.08	1.0	17.3	56	8.1	16.3	56	11.1
Miami, FL	79.3	64.1	92	32	5.7	1.89	7.22	0.0	–	77	8.7	8.1	56	10.4
Milwaukee, WI	39.2	24.9	81	-10	11.9	2.58	6.93	9.2	26.7	50	6.0	17.2	65	13.0
Minneapolis, MN	37.5	20.8	83	-32	10.3	1.71	4.75	10.6	40.0	55	6.9	16.4	62	11.3
New Orleans, LA	71.2	51.6	89	25	9.4	4.73	19.09	T	T	57	7.7	15.1	60	9.9
New York, NY	48.6	34.1	86	3	11.5	4.22	10.41	5.1	30.5	56	8.8	12.1	55	11.0
Philadelphia, PA	50.5	33.1	87	7	11.1	3.86	7.01	4.0	13.4	55	7.5	15.3	53	11.4
Phoenix, AZ	74.5	46.7	95	25	3.5	.81	4.16	T	T	83	14.6	8.6	24	6.7
Pittsburgh, PA	47.6	29.4	80	-1	16.1	3.58	6.10	8.7	21.3	44	4.0	20.1	58	10.9
Portland, OR	54.5	37.4	80	19	17.2	3.61	7.52	0.5	12.9	46	3.1	23.6	60	8.3
Sacramento, CA	64.1	42.4	86	26	8.4	2.06	7.12	T	T	72	10.4	12.5	53	8.8
Salt Lake City, UT	51.5	29.9	78	2	11.5	1.72	3.97	10.4	41.9	63	7.0	15.7	47	9.3
San Antonio, TX	73.7	49.8	100	19	7.0	1.33	4.19	T	T	57	8.5	15.2	47	10.6
San Francisco, CA	60.6	44.9	85	30	9.7	2.64	9.01	T	T	69	9.7	12.6	63	10.4
San Diego, CA	65.9	52.1	93	39	7.2	1.60	6.57	0.0	–	71	11.2	10.5	59	7.3
Seattle, WA	51.1	37.2	72	11	17.1	3.59	8.40	1.4	18.2	49	3.1	22.1	62	9.9
Tampa, FL	76.2	56.1	91	29	6.8	3.46	12.64	T	T	71	10.2	10.8	55	9.6
Washington, D.C.	55.0	36.6	89	11	11.0	3.46	7.43	2.3	17.1	55	7.3	15.1	50	10.9

CLIMATIC SUMMARY

July

City	Avg Max	Avg Min	Rec Max	Rec Min	Mean Days of Precip.	Total Avg	Total Max	Snow Avg	Snow Max	% Possible Sunshine	Mean Clear Days	Mean Cloudy Days	Relative Humidity (%)	Avg. Wind Speed (MPH)
Albuquerque, NM	92.8	64.7	105	54	8.8	1.30	3.33	0.0	–	76	12.1	4.7	27	9.1
Anchorage, AK	65.1	51.1	81	38	11.6	1.97	4.44	0.0	–	43	3.4	22.0	61	7.1
Atlanta, GA	87.9	69.2	105	53	12.0	4.73	11.26	0.0	–	63	5.6	12.0	61	7.5
Bismarck, ND	84.4	56.4	109	35	8.9	2.05	5.24	0.0	–	75	11.5	6.7	42	9.2
Boise, ID	90.6	58.5	111	40	2.3	.26	1.62	T		87	20.9	3.1	22	8.4
Boston, MA	81.8	65.1	102	54	9.1	2.68	8.12	0.0	–	67	7.0	11.6	55	10.9
Buffalo, NY	80.2	61.2	94	43	9.8	2.96	6.43	0.0	–	68	7.0	10.9	54	10.4
Casper, WY	87.1	54.7	104	30	7.8	1.06	3.05	0.0	–	NA	14.1	5.9	26	10.1
Charleston, SC	89.4	71.6	101	53	13.7	7.33	18.46	0.0	–	68	4.9	13.8	63	7.9
Charlotte, NC	88.3	68.7	103	53	11.3	3.92	9.12	0.0	–	68	6.5	13.0	58	6.6
Chicago, IL	83.3	62.7	102	40	9.8	3.63	8.33	0.0	–	69	8.8	9.2	57	8.1
Cincinnati, OH	85.8	64.9	101	47	10.0	4.28	8.36	0.0	–	70	7.4	11.9	57	7.1
Dallas, TX	97.8	74.7	110	59	4.9	2.59	11.13	0.0	–	81	15.3	6.4	44	9.5
Denver, CO	88.0	58.7	104	43	9.1	1.93	6.41	0.0	–	72	9.2	6.0	34	8.4
Detroit, MI	83.1	60.7	102	41	9.7	3.10	6.02	0.0	–	70	9.3	9.5	53	8.3
Helena, MT	83.6	52.2	102	36	7.4	1.04	3.89	T		78	14.8	5.3	29	7.8
Honolulu, HI	87.1	73.1	91	67	7.5	.54	2.01	0.0	–	73	7.5	5.4	52	13.5
Houston, TX	93.6	72.5	104	62	9.7	3.33	8.10	0.0	–	66	7.7	8.1	58	6.8
Kansas City, MO	88.5	68.5	107	52	7.1	4.35	8.71	0.0	–	74	14.5	7.3	53	8.9
Las Vegas, NV	104.5	75.9	116	61	2.6	.45	2.48	0.0	–	87	19.6	3.4	15	10.1
Lexington, KY	85.9	65.9	103	47	11.2	4.95	10.64	0.0	–	82	7.8	10.5	59	7.4
Los Angeles, CA	75.3	62.6	97	49	0.7	.01	.15	0.0	–	82	12.7	5.4	67	7.7
Memphis, TN	91.5	72.6	108	52	8.7	4.03	8.84	0.0	–	74	10.4	8.8	57	7.5
Miami, FL	88.7	76.2	98	69	15.8	5.98	13.51	0.0	–	79	2.5	11.3	63	7.8
Milwaukee, WI	79.8	61.1	101	40	9.5	3.54	7.66	0.0	–	70	10.3	9.5	61	9.6
Minneapolis, MN	83.4	62.7	104	43	9.7	3.51	7.10	0.0	–	71	10.4	9.1	53	9.3
New Orleans, LA	90.7	73.5	101	60	14.8	6.73	13.07	0.0	–	62	5.1	11.5	66	6.1
New York, NY	85.3	68.2	106	52	10.4	3.77	11.89	0.0	–	65	8.5	9.5	55	7.6
Philadelphia, PA	86.1	66.8	104	51	9.1	3.88	8.33	0.0	–	62	7.1	12.0	54	8.0
Phoenix, AZ	105.0	79.5	118	61	4.4	.74	5.15	0.0	–	85	16.3	4.4	20	7.1
Pittsburgh, PA	82.7	61.3	99	42	10.8	3.83	7.43	0.0	–	59	5.3	12.5	53	7.3
Portland, OR	79.5	55.8	107	43	3.8	.46	2.68	0.0	–	70	13.0	9.3	45	7.6
Sacramento, CA	93.3	57.9	114	48	0.3	.05	.79	0.0	–	97	27.1	0.9	28	9.0
Salt Lake City, UT	93.2	61.8	107	40	4.5	.72	2.57	0.0	–	83	16.9	4.3	21	9.5
San Antonio, TX	94.9	74.3	106	62	4.3	1.92	8.19	0.0	–	74	9.2	6.6	45	9.3
San Francisco, CA	71.0	53.3	104	43	0.3	.03	.35	0.0	–	66	20.6	2.9	59	13.5
San Diego, CA	75.6	64.9	93	55	0.3	.01	.19	0.0	–	69	13.3	4.8	65	7.3
Seattle, WA	75.2	54.3	98	43	5.1	.74	2.39	T		65	10.4	10.4	49	8.3
Tampa, FL	90.0	74.2	97	63	15.7	7.35	20.59	0.0	–	62	2.2	12.8	63	7.4
Washington, D.C.	87.9	69.9	103	55	9.7	3.88	11.06	0.0	–	64	7.8	11.4	53	8.2

September

City	Avg Max	Avg Min	Rec Max	Rec Min	Mean Days of Precip.	Total Avg	Total Max	Snow Avg	Snow Max	% Possible Sunshine	Mean Clear Days	Mean Cloudy Days	Relative Humidity (%)	Avg. Wind Speed (MPH)
Albuquerque, NM	83.0	54.9	100	37	5.7	.85	1.99	T		79	16.9	5.3	31	8.6
Anchorage, AK	55.2	41.1	73	20	13.8	2.45	5.43	0.2	4.6	41	4.0	20.7	63	6.2
Atlanta, GA	82.3	63.6	98	36	7.5	3.17	7.52	0.0	–	64	9.7	10.5	60	8.0
Bismarck, ND	71.4	43.2	105	11	7.0	1.38	6.93	0.3	5.0	66	10.4	10.8	44	10.0
Boise, ID	77.6	48.7	102	23	3.7	.58	2.54	0.0	–	81	16.9	6.0	30	8.3
Boston, MA	72.3	56.9	100	38	8.7	3.41	8.31	0.0	–	64	10.5	11.3	60	11.1
Buffalo, NY	71.4	52.7	98	32	10.8	3.37	8.99	T		59	6.5	13.6	60	10.4
Casper, WY	74.2	42.5	96	16	6.3	.76	3.40	1.2	11.5	NA	13.4	8.0	30	11.0
Charleston, SC	84.6	66.7	99	42	9.4	4.94	17.31	0.0	–	63	6.6	12.8	62	7.9
Charlotte, NC	81.7	62.3	104	39	7.2	3.59	10.89	0.0	–	67	9.2	11.4	57	6.7
Chicago, IL	75.5	53.9	99	28	9.6	3.35	11.44	T		58	9.0	11.3	58	8.8
Cincinnati, OH	78.7	56.3	98	33	8.0	2.91	8.61	0.0	–	64	9.5	11.6	58	7.4
Dallas, TX	89.7	67.5	105	40	6.8	3.31	9.52	0.0	–	74	13.1	8.4	54	9.4
Denver, CO	77.5	47.7	97	20	6.1	1.23	4.67	1.6	21.3	75	13.5	7.5	34	8.0
Detroit, MI	74.4	52.2	98	29	9.7	2.25	5.83	0.0	–	61	8.3	12.2	57	8.7
Helena, MT	70.3	40.8	99	18	6.6	.83	3.37	1.6	13.7	67	10.2	11.1	35	7.4
Honolulu, HI	88.2	72.9	93	66	7.0	.62	2.74	0.0	–	75	8.2	5.9	52	11.6
Houston, TX	88.7	68.1	100	48	9.5	4.93	11.35	0.0	–	62	8.9	10.6	62	6.8
Kansas City, MO	78.6	58.1	98	33	7.6	4.14	11.34	0.0	–	65	11.8	10.0	59	9.3
Las Vegas, NV	94.7	65.6	113	46	1.8	.32	1.58	0.0	–	91	22.7	2.3	17	8.8
Lexington, KY	79.3	58.1	103	35	8.1	3.28	9.69	0.0	–	NA	10.6	10.9	59	7.7
Los Angeles, CA	76.4	62.5	110	47	1.0	.15	4.39	0.0	–	79	12.9	6.6	65	7.2
Memphis, TN	84.3	64.1	103	36	7.0	3.62	7.61	0.0	–	70	12.4	10.0	56	7.5
Miami, FL	87.8	75.7	95	68	17.6	8.07	24.40	0.0	–	72	2.3	12.9	67	8.2
Milwaukee, WI	71.2	52.5	98	28	9.0	2.88	9.87	T		59	9.7	11.1	63	10.5
Minneapolis, MN	71.0	50.2	98	26	9.4	2.50	7.53	0.0	1.7	61	10.1	11.3	59	9.9
New Orleans, LA	86.8	70.1	101	42	9.8	5.87	16.74	0.0	–	63	9.9	9.8	65	7.3
New York, NY	76.4	60.1	102	35	8.3	3.66	16.85	0.0	–	63	10.6	9.4	57	8.1
Philadelphia, PA	77.8	58.6	100	39	7.9	3.42	8.78	0.0	–	60	9.8	11.3	55	8.2
Phoenix, AZ	98.2	70.9	118	47	3.0	.64	4.23	0.0	–	89	22.0	3.0	23	6.3
Pittsburgh, PA	74.8	53.3	97	31	9.3	2.80	5.42	0.0	–	58	7.6	12.1	56	7.4
Portland, OR	74.2	51.1	101	34	7.8	1.61	3.98	T		61	10.1	11.9	49	6.5
Sacramento, CA	87.6	55.8	108	43	1.2	.27	1.81	0.0	–	93	23.7	2.2	31	7.7
Salt Lake City, UT	80.0	50.0	100	27	5.3	.89	7.04	0.1	4.0	83	16.9	5.1	29	9.1
San Antonio, TX	89.3	69.4	102	41	7.1	3.75	15.78	0.0	–	67	9.6	8.3	52	8.6
San Francisco, CA	73.4	54.3	103	38	0.9	.19	2.30	0.0	–	72	18.4	3.6	57	11.0
San Diego, CA	76.8	65.1	111	51	1.1	.19	1.90	0.0	–	68	15.0	5.8	65	6.9
Seattle, WA	68.7	51.2	94	35	9.4	2.02	5.95	T		59	7.6	13.6	58	8.1
Tampa, FL	88.9	72.8	96	57	13.1	6.23	13.98	0.0	–	61	4.5	12.0	62	8.0
Washington, D.C.	80.1	62.0	101	39	7.6	3.22	12.36	0.0	–	62	10.1	11.7	55	8.2

Geologic Time Line

Cryptozoic Eon	Phanerozoic										
	Paleozoic Era										
Precambrian or Hadean, Archean, and Proterozoic Eras	Cambrian Period			Ordovician Period		Silurian Period	Devonian Period			Carboniferous Period	
	* 570			500		430	395			345	
	Early	Middle	Late	Early	Late		Early	Middle	Late	Mississippian	Pennsylvanian

* Time is in millions of years before the present.

Eon																
		Mesozoic Era								Cenozoic Era						
Permian Period		Triassic Period			Jurassic Period			Cretaceous Period		Tertiary Period					Quaternary Period	
280		225			190			136		65					2.5	
Early	Late	Early	Middle	Late	Early	Middle	Late	Early	Late	Paleocene	Eocene	Oligocene	Miocene	Pliocene	Pleistocene	Holocene

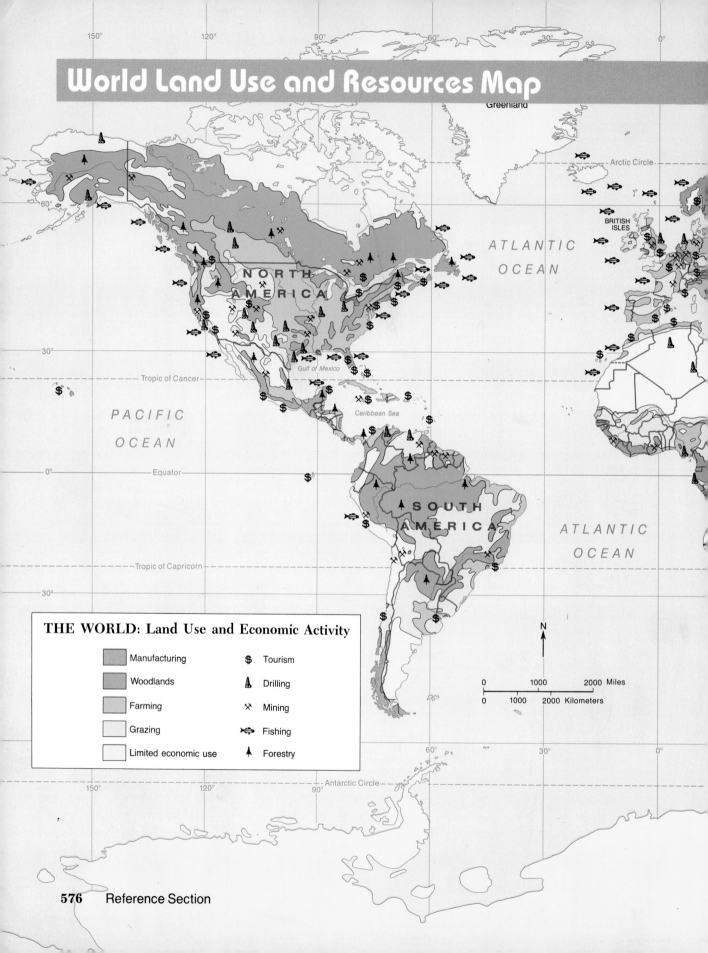

World Land Use and Resources Map

Greenland

Arctic Circle

ATLANTIC
OCEAN

BRITISH
ISLES

NORTH
AMERICA

60°

30°

Tropic of Cancer

Gulf of Mexico

Caribbean Sea

PACIFIC
OCEAN

Equator

0°

SOUTH
AMERICA

ATLANTIC
OCEAN

Tropic of Capricorn

30°

THE WORLD: Land Use and Economic Activity

▨	Manufacturing	$	Tourism
▨	Woodlands	⚒	Drilling
▨	Farming	✗	Mining
▨	Grazing	🐟	Fishing
▢	Limited economic use	♠	Forestry

N

0	1000	2000 Miles
0	1000	2000 Kilometers

150°   120°   90°   60°   30°   0°

Arctic Circle

Map of World Climates

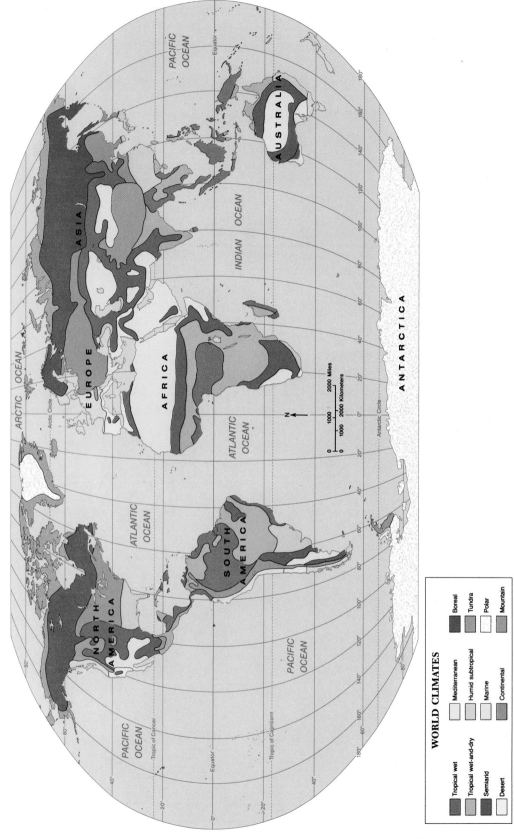

WORLD CLIMATES

- Tropical wet
- Tropical wet-and-dry
- Semiarid
- Desert
- Mediterranean
- Humid subtropical
- Marine
- Continental
- Boreal
- Tundra
- Polar
- Mountain

Map of Ocean Currents

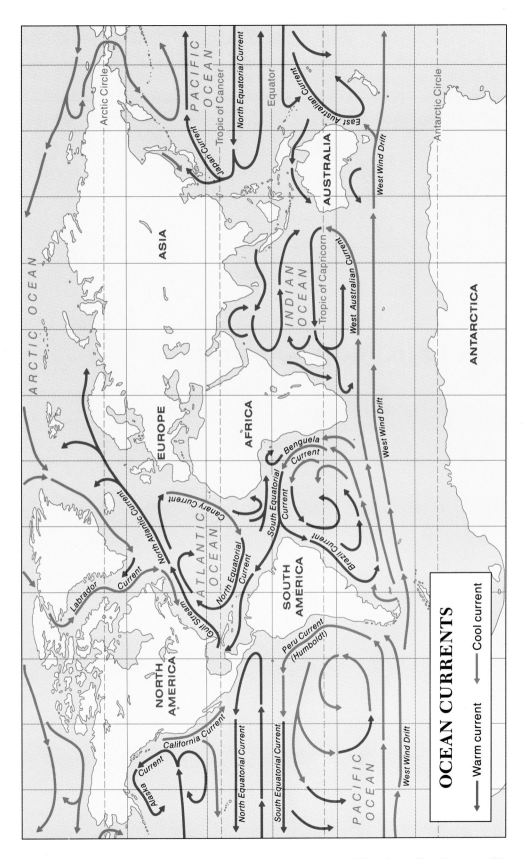

OCEAN CURRENTS

→ Warm current   → Cool current

ARCTIC OCEAN

PACIFIC OCEAN

Arctic Circle

Tropic of Cancer

Equator

North Equatorial Current

Japan Current

ASIA

AUSTRALIA

East Australian Current

West Wind Drift

Antarctic Circle

INDIAN OCEAN

Tropic of Capricorn

West Australian Current

West Wind Drift

ANTARCTICA

EUROPE

AFRICA

Benguela Current

South Equatorial Current

Brazil Current

Canary Current

North Atlantic Current

ATLANTIC OCEAN

North Equatorial Current

Gulf Stream

Labrador Current

NORTH AMERICA

SOUTH AMERICA

Peru Current (Humboldt)

California Current

Alaska Current

North Equatorial Current

South Equatorial Current

West Wind Drift

PACIFIC OCEAN

West Wind Drift

Topographic Maps and Map Symbols

Symbol	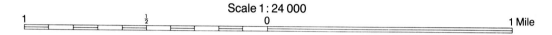								
Color	Black	Black	Red or black	Black	Blue	Blue	Blue	Brown	Blue or green
Meaning	Buildings	School	Road	Railroad	Stream	Intermittent Stream	Lake or pond	Depression	Marsh or swamp

Scale 1 : 24 000

1   ½   0   1 Mile

Geologic Map Symbols

Symbols indicating structure

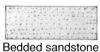 Strike and dip symbol showing beds striking N30E, dipping 40° to the southeast

 Strike symbol for vertical beds

 Axis of anticline

Strike and dip symbol indicating overturned beds

 Symbol for horizontal beds

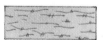 Axis of syncline

Geological ages

Q   Quaternary

T   Tertiary

K   Cretaceous

J   Jurassic

TR   Triassic

P   Permian

P   Pennsylvanian

M   Mississippian

D   Devonian

S   Silurian

O   Ordovician

Є   Cambrian

PЄ   Precambrian

Symbols indicating rock types

Conglomerate

Basic lava flows

Bedded sandstone

Other lava flows

Shale

Granitic rock

Massive limestone

Folded schist

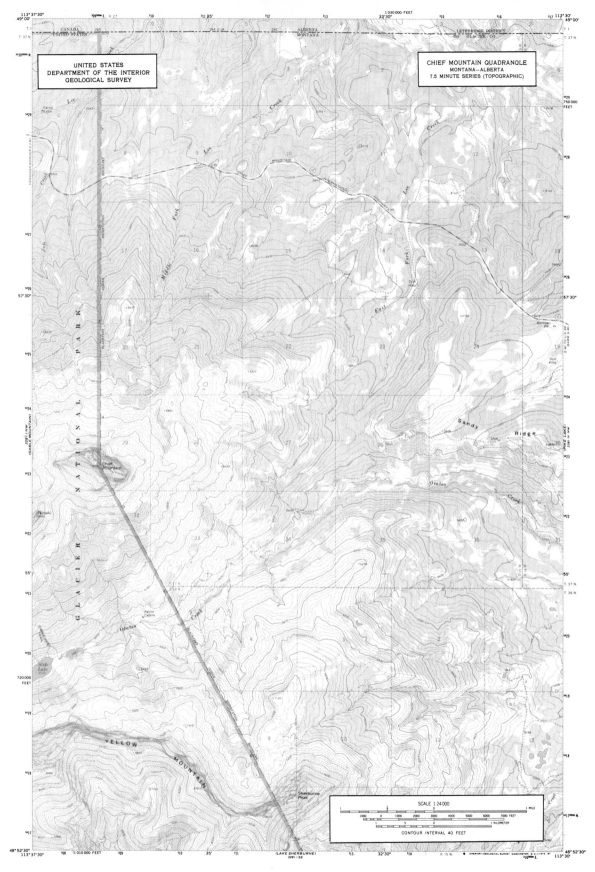

Use with Investigation 13 on page 294.

UNITED STATES
DEPARTMENT OF THE INTERIOR
GEOLOGICAL SURVEY

AYER QUADRANGLE
MASSACHUSETTS
7.5 MINUTE SERIES (TOPOGRAPHIC)

SCALE 1:25 000

CONTOUR INTERVAL 10 FEET
NATIONAL GEODETIC VERTICAL DATUM OF 1929

Use with Investigation 13 on page 294.

Weather Map Symbols

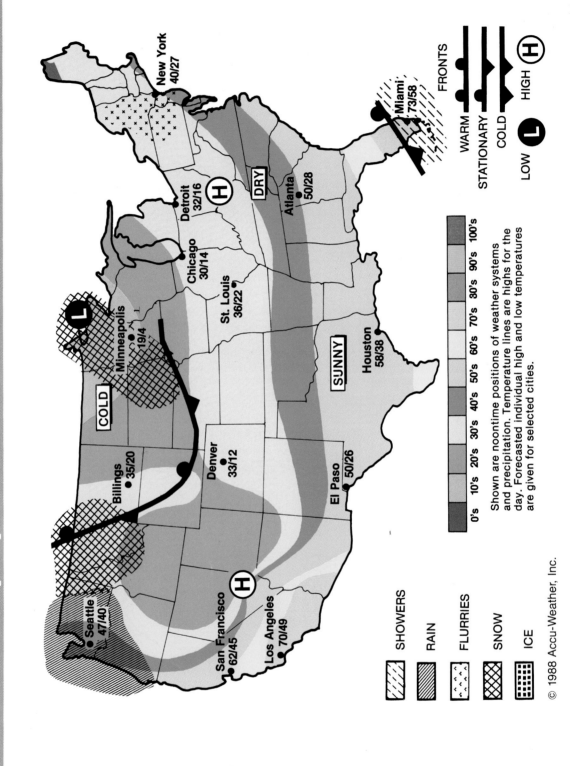

New York 40/27

Miami 73/58

Detroit 32/16

DRY

Atlanta 50/28

Chicago 30/14

St. Louis 36/22

COLD

Minneapolis 19/4

Houston 58/38

SUNNY

Billings 35/20

Denver 33/12

El Paso 50/26

Seattle 47/40

San Francisco 62/45

Los Angeles 70/49

FRONTS

WARM

STATIONARY

COLD

LOW **L**   HIGH **H**

0's 10's 20's 30's 40's 50's 60's 70's 80's 90's 100's

Shown are noontime positions of weather systems and precipitation. Temperature lines are highs for the day. Forecasted individual high and low temperatures are given for selected cities.

SHOWERS

RAIN

FLURRIES

SNOW

ICE

© 1988 Accu-Weather, Inc.

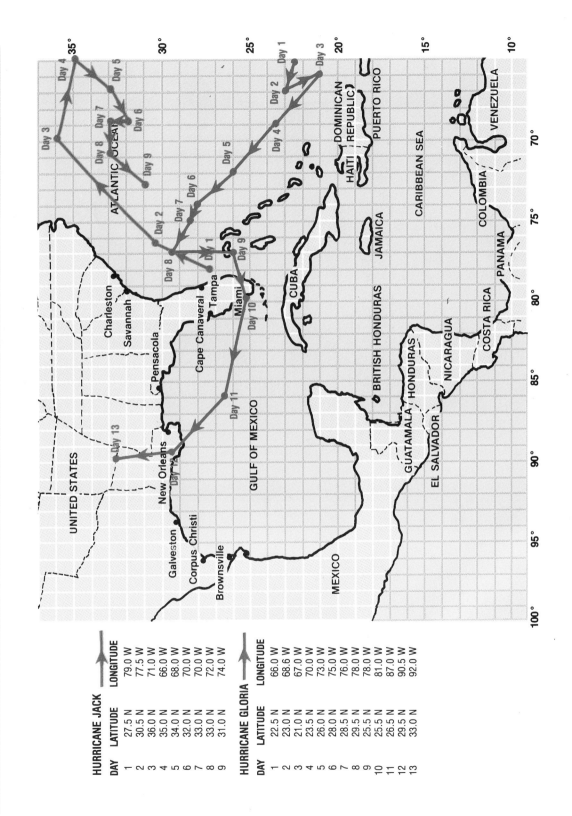

Hurricane Tracking Map

HURRICANE JACK

DAY	LATITUDE	LONGITUDE
1	27.5 N	79.0 W
2	30.5 N	77.5 W
3	36.0 N	71.0 W
4	35.0 N	66.0 W
5	34.0 N	68.0 W
6	32.0 N	70.0 W
7	33.0 N	70.0 W
8	33.0 N	72.0 W
9	31.0 N	74.0 W

HURRICANE GLORIA

DAY	LATITUDE	LONGITUDE
1	22.5 N	66.0 W
2	23.0 N	68.6 W
3	21.0 N	67.0 W
4	23.5 N	70.0 W
5	26.0 N	73.0 W
6	28.0 N	75.0 W
7	28.5 N	76.0 W
8	29.5 N	78.0 W
9	25.5 N	78.0 W
10	25.5 N	81.0 W
11	26.5 N	87.0 W
12	29.5 N	90.5 W
13	33.0 N	92.0 W

Weather Maps with Satellite Photos

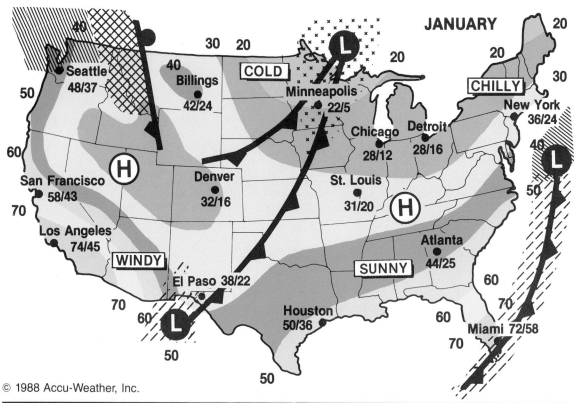

JANUARY

COLD

CHILLY

WINDY

SUNNY

Seattle 48/37
Billings 42/24
Minneapolis 22/5
Chicago 28/12
Detroit 28/16
New York 36/24
San Francisco 58/43
Denver 32/16
St. Louis 31/20
Los Angeles 74/45
El Paso 38/22
Atlanta 44/25
Houston 50/36
Miami 72/58

© 1988 Accu-Weather, Inc.

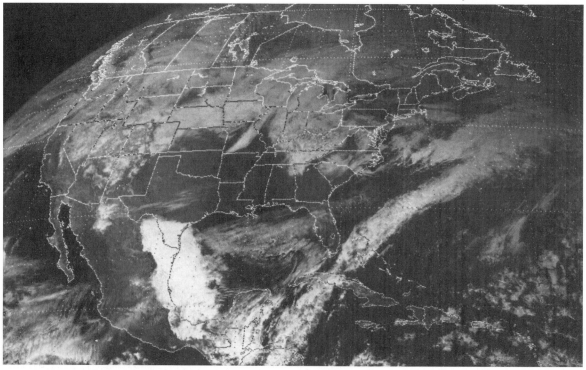

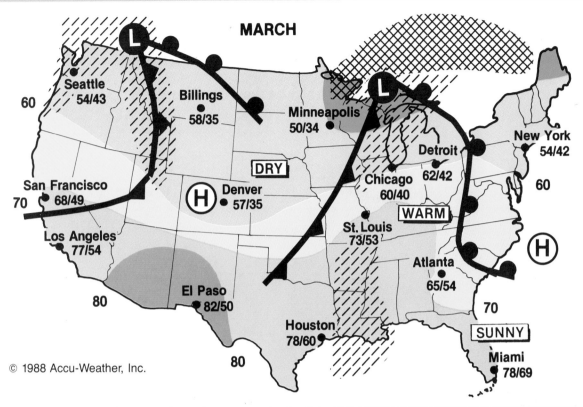

MARCH

Seattle
54/43

60

Billings
58/35

Minneapolis
50/34

DRY

San Francisco
70 • 68/49

Denver
H • 57/35

Detroit
62/42

New York
54/42

60

Chicago
60/40

WARM

H

Los Angeles
77/54

St. Louis
73/53

Atlanta
65/54

El Paso
80 • 82/50

70

Houston
78/60

SUNNY

© 1988 Accu-Weather, Inc.

80

Miami
78/69

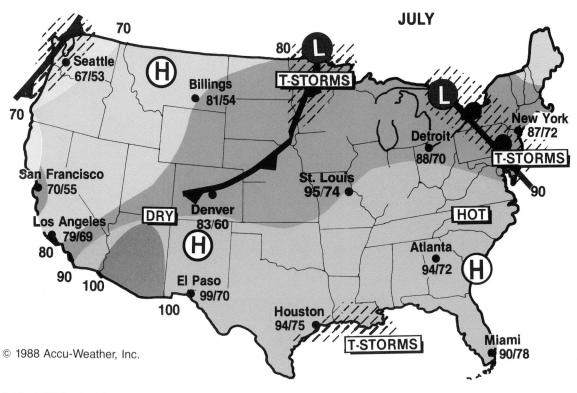

JULY

Seattle 67/53
Billings 81/54
San Francisco 70/55
Los Angeles 79/69
Denver 83/60
El Paso 99/70
St. Louis 95/74
Detroit 88/70
New York 87/72
Atlanta 94/72
Houston 94/75
Miami 90/78

70
80
70
70
80
90
100
100
90

H
L
L
H
H

T-STORMS
DRY
HOT
T-STORMS
T-STORMS

© 1988 Accu-Weather, Inc.

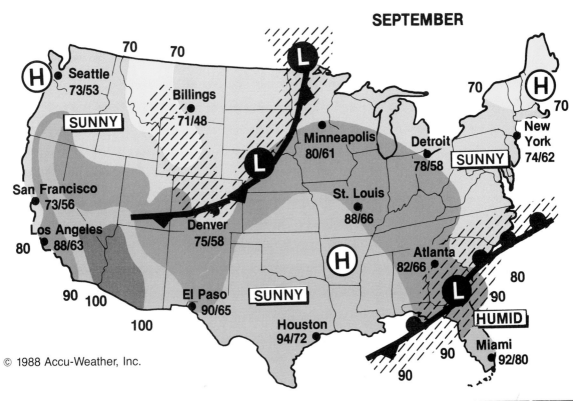

SEPTEMBER

70 70

(H) Seattle 73/53

SUNNY

Billings 71/48

L

L

Minneapolis 80/61

Detroit 78/58

70

(H) 70

New York 74/62

SUNNY

San Francisco 73/56

Los Angeles 80 88/63

90 100

100

El Paso 90/65

Denver 75/58

SUNNY

St. Louis 88/66

(H)

Atlanta 82/66

L

Houston 94/72

90

80

90

HUMID

Miami 92/80

90

© 1988 Accu-Weather, Inc.

Star Chart

Physical Map of the World

Ice and snow
High barren areas
Tundra and alpine vegetation
Needleleaf trees
Broadleaf deciduous trees
Tropical rainforest
Grassland
Desert

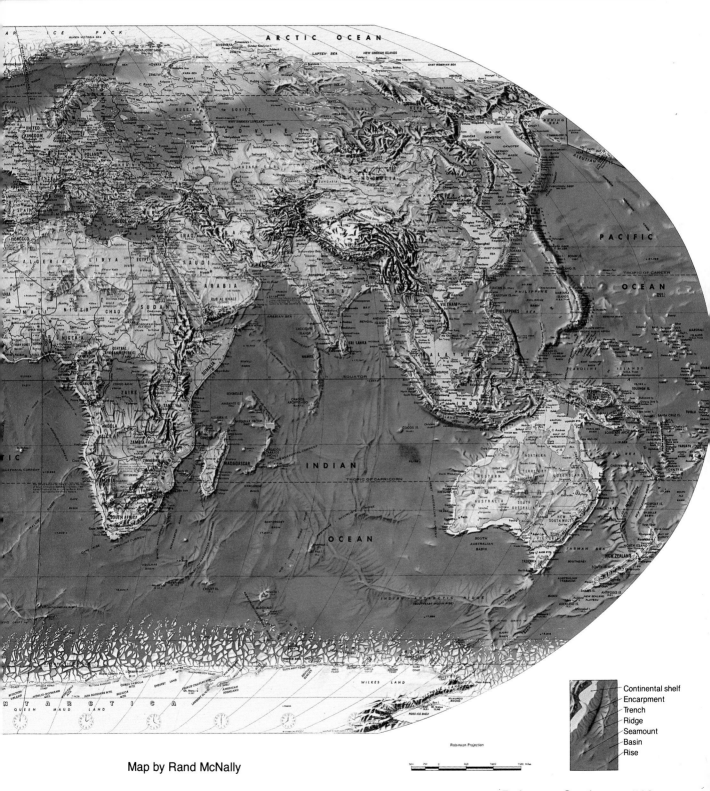

Continental shelf
Encarpment
Trench
Ridge
Seamount
Basin
Rise

Map by Rand McNally

Robinson Projection

GLOSSARY

A

abrasion weathering by physical contact, causing the rounding of rocks **(234)**

absolute dating process of determining the age of a sample by using the radioactive isotopes in the sample **(329)**

aerosol solid in the air that acts as a nucleus for condensation **(195)**

air mass large body of air with uniform temperature and humidity **(207)**

alluvial fan

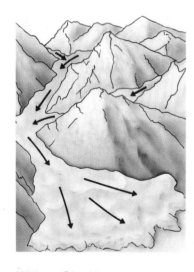

alloy mixture of two or more metals, or a mixture of metals and nonmetals **(100)**

alluvial fan fan-shaped deposit at the base of a mountain **(263)**

anticline structure formed when rock layers are folded upward **(123)**

aquifer permeable rock layer that is between two nonpermeable layers and transports water **(267)**

arid dry; usually regarding climate **(275)**

asteroid small planetary body that circles the sun in a belt between the orbits of Mars and Jupiter **(468)**

asthenosphere (as THEHN uh sfihr) 100-km-thick layer of the earth, near the top of the mantle **(119)**

astronomical unit average distance between the earth and the sun, or about 150 000 000 km **(462)**

astronomy study of the characteristics of the planets, the sun, and the stars **(5)**

atmosphere   layer of gases that surrounds the earth **(165)**

atmospheric pressure   weight of the atmosphere **(175)**

atom   building block of matter **(51)**

atomic number   number of protons an atom contains **(56)**

axis   imaginary line that passes through the center of a planet from pole to pole **(455)**

barchan (BAHR kahn) **dune**   crescent-shaped dune with the rounded side facing into the wind **(280)**

barometer (buh RAHM uh tuhr)   instrument used to measure atmospheric pressure **(175)**

base level   low point of a stream or river **(259)**

batholith   body of magma that cools beneath the surface, forming an intrusion **(143)**

bay   indentation of the sea, partially enclosed by land **(402)**

bench mark   bronze marker indicating location and elevation of a specific surveyed place in the United States **(27)**

benthos   organisms that live in or on the bottom of the ocean **(442)**

big-bang theory   theory stating that a tremendous explosion created the universe; the most accepted theory for the beginning of the universe **(496)**

biomass   once-living matter **(379)**

black hole   collapsed star that emits no energy **(488)**

brine   extremely salty water that collects in areas totally or partially isolated from the open ocean **(435)**

bench mark

caldera   wide basin in the center of a volcanic cone **(146)**

carbonation   chemical weathering process in which carbonic acid reacts with the minerals in rocks such as limestone, dissolving them **(236)**

cartography (kahr TAHG ruh fee)   science of mapmaking **(26)**

cellular respiration   breakdown of carbohydrates in the presence of oxygen **(346)**

cementation   process in which water that carries natural cements with it soaks through the earth, coating sediments and binding them together **(84)**

chemical sediment   sediment that comes from minerals dissolved in water **(85)**

cinder-cone volcano   steep-sided volcano that consists of alternating layers of ash, tuff, and lava **(145)**

cirrus cloud   highest cloud; made of ice crystals **(197)**

clastic sediments   pieces of other rocks **(85)**

cleavage   tendency of minerals to break along flat surfaces **(66)**

climate   all the weather that characterizes a region **(211)**

cinder-cone volcano

comet

cloud   collection of water droplets in the atmosphere   **(195)**

comet   object consisting of silicate rock and metal particles embedded in ices   **(475)**

composite volcano   volcano that consists of layers produced by alternating eruptions of different lavas   **(145)**

compound   combination of two or more different kinds of elements   **(54)**

compression   pushing together of the earth's crust   **(126)**

conclusion   statement about an initial hypothesis, based on all the information that was gathered   **(8)**

condensation   change from a vapor to a liquid   **(194)**

conservation   wise use of materials   **(372)**

constellation   group of stars that forms an image   **(483)**

continental drift   slow movement of large portions of the earth's crust   **(129)**

continental shelf   extension of the continents, forming a gently sloping ocean floor   **(394)**

continental slope   steep ocean floor farther offshore than the continental shelf   **(394)**

control   in an experiment, the part that does not change   **(8)**

convection   transfer of heat by the circulation or movement of solids, liquids, or gases   **(120)**

convergent boundary   boundary where two plates of land press together   **(140)**

core   innermost layer of the earth   **(119)**

Coriolis (cawr ee OH lihs) **effect**   tendency of objects to move to the right in the Northern Hemisphere   **(181)**

covalent bond   bond formed when atoms share electrons   **(55)**

creep   slow, downhill movement of rock materials saturated with water   **(237)**

crossbedding   design in rocks that shows directional changes in wind patterns   **(325)**

crust   thin, rocky layer on the surface of the earth   **(117)**

crystal   three-dimensional structure with each face, or surface, having a definite shape and orientation   **(69)**

cumulus (KYOO myoo luhs) **cloud**   cloud that forms as hot, humid air rises   **(197)**

current   stream of water that moves like a river through the ocean   **(419)**

cyclone   any weather system that circulates around a low-pressure area   **(209)**

cumulus cloud

D

data   information gathered by observation or investigation   **(7)**

deflation   removal of all the soil of an area by wind erosion   **(279)**

delta   area of sand and soil deposits, formed by a stream that slows down as it enters a large body of water   **(262)**

density   measure of the mass contained in a certain volume of matter   **(18)**

dew point   temperature at which dew forms   **(194)**

differential weathering   difference in weathering of rocks that are in the same formation and are exposed to the same environment   **(240)**

dip   angle at which a rock layer is tilted from the horizontal   **(123)**

divergent boundary   area of the earth where the crust is spreading apart   **(139)**

drainage basin   area drained by a river and its tributaries   **(255)**

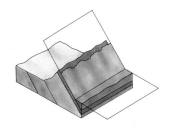

dip

earthquake   violent shaking of the earth's interior   **(139)**

electromagnetic wave   energy wave that travels at the speed of light   **(170)**

electron   subatomic particle that has a negative electrical charge   **(52)**

element   substance made of only one kind of atom   **(51)**

elliptical   describes a flattened circle   **(462)**

epicenter   area on the earth's surface directly above the focus of an earthquake   **(151)**

erosion   process that moves soil and reshapes landscapes   **(254)**

evaporation   process of changing a liquid to a vapor   **(191)**

evaporite (ih VAP uh ryte)   chemical sedimentary rock that forms when minerals dissolved in water are left as the water evaporates   **(87)**

exfoliation (ehks foh lee AY shuhn)   weathering process in which the surface of rocks flakes off in sheets   **(233)**

experiment   test designed to give a scientist information under carefully controlled conditions   **(8)**

extrusive (ehk STROOS ihv) **rocks**   fine-grained igneous rocks formed from lava   **(80)**

exfoliation

fault   rock fracture along which there is movement   **(125)**

fermentation   partial breakdown of sugar to release energy   **(344)**

fiord (fee AWRD)   long, deep, narrow, U-shaped sea channel that was originally formed by ice   **(402)**

firn   snow pile in which half of the air spaces are filled with ice   **(283)**

fission   breakdown of atomic nuclei, with the release of energy as a result   **(379)**

flood plain   area over which a stream spreads in times of high-volume runoff   **(260)**

flow   process in which water-saturated rock material moves downhill at a noticeable rate   **(238)**

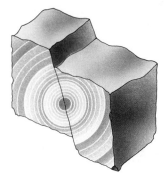

focus

focal length distance between the center of a lens and the focal point **(505)**

focus source of an earthquake **(151)**

fold in a rock layer, a bend caused by force **(122)**

foliated rock metamorphic rock having obvious layers, formed from more than one mineral type **(92)**

forge furnace through which a stream of air is forced **(102)**

fossil remains or traces of a once-living organism, preserved in the earth **(5)**

fracture break in minerals that occurs along curved or irregular surfaces **(66)**

front surface between different air masses **(207)**

frost action weathering process that involves the freezing and cracking of rocks **(233)**

fusion energy energy produced when small atomic nuclei fuse to form new atoms **(379)**

G

galaxy large grouping of stars **(494)**

gem mineral that has beauty, durability, and value **(69)**

geologic map map that shows different layers and types of rock by the use of different colors **(29)**

geology study of the solid earth **(4)**

geomorphic province area with a distinctive pattern of landforms **(301)**

geomorphology study of landforms of the earth **(299)**

geostationary describes a satellite that is in a Clarke orbit and does not appear to move from its position relative to the earth **(534)**

geostrophic current ocean current that flows from a high water level to a lower level **(421)**

geothermal power steam power from the earth **(377)**

glaciation period in which great continental ice sheets advance and retreat repeatedly **(354)**

glacier sheet of ice that covers mountains or a large part of a continent **(282)**

greenhouse effect temperature rise caused by the addition of carbon dioxide to the atmosphere **(174)**

H

half-life time it takes for half of the atoms of a radioactive isotope to decay **(328)**

headward erosion process by which streams increase their length **(258)**

horizon development of soil in horizontal layers **(246)**

hot spot area of concentrated heat deep within the earth **(147)**

humidity amount of water vapor in the air **(193)**

humus (HYOO muhs) organic part of the soil **(244)**

hurricane large storm that develops over a tropical ocean **(219)**

horizon

hydration   chemical weathering process in which water is combined with certain minerals to form new compounds   **(235)**

hydrocarbon   compound containing only hydrogen and carbon   **(107)**

hydroelectric power   power produced from the energy of falling water   **(375)**

hydrologic map   map that shows where surface water and underground water are located, as well as special landforms produced by water   **(29)**

hydrology   study of all the water on land   **(4)**

hypothesis (hy PAHTH uh sihs)   educated guess made by scientists about what they think will occur   **(8)**

iceberg   great mass of freshwater ice freed from a glacier by a process called *calving*   **(283)**

igneous (IHG nee uhs) **rock**   rock that forms from magma or lava   **(76)**

intrusive (ihn TROO sihv) **rock**   igneous rock produced when magma cools slowly and forms large crystals   **(79)**

ion   atom that has gained or lost one or more electrons and therefore has an electrical charge   **(53)**

ionic bond   bond that forms when electrons move from one atom to another   **(54)**

isotope   atom of an element that has a different number of neutrons   **(53)**

iceberg

jet stream   high-speed wind circling both polar areas at an altitude of 12 km   **(183)**

lake   large body of fresh or salt water completely surrounded by land   **(398)**

landslide   mass movement that occurs when dry soil or rock moves rapidly and unexpectedly down a steep slope   **(238)**

latitude (LAT uh tood)   distance from the earth's equator measured by a parallel imaginary line that runs east and west   **(30)**

lava   molten rock that reaches the surface of the earth   **(76)**

law   description of what happens in a given situation   **(8)**

leaching   movement of minerals from an upper horizon to a lower horizon   **(247)**

lens   piece of curved glass that can be used to focus light   **(505)**

lightning   stroke of electrons traveling at speeds of about 1000 km/s   **(214)**

light-year   distance that light travels in one year   **(493)**

lithosphere (LIHTH uh sfihr)   solid layer of the earth, composed of the crust and upper mantle   **(119)**

load   amount of eroded material carried by a stream or a river   **(254)**

lightning

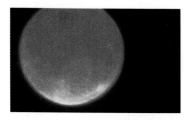

lunar eclipse

loess (LEHS)   thick layers of fine glacial till   **(287)**

longitude (LAHN juh tood)   distance measured by a vertical imaginary line that runs around the earth through the North and South Poles   **(30)**

lunar eclipse (ih KLIHPS)   occurs when Earth's shadow falls on all or part of the moon's surface   **(459)**

luster   appearance of a mineral's surface in reflected light   **(63)**

magma   molten rock within the earth   **(76)**

magnification   mathematically, the focal length of the objective lens of a telescope divided by the focal length of the ocular lens   **(506)**

magnitude   star's brightness as it appears from Earth—not its actual, or absolute, brightness   **(481)**

mantle   layer of the earth immediately beneath the Moho   **(118)**

mass   amount of matter an object contains   **(17)**

mass movement   downhill movement of rocks and soil due to the force of gravity   **(237)**

meander

meander   broad curve formed by the erosion of a mature stream   **(261)**

metamorphic (meht uh MAWR fihk) **rock**   rock produced from an existing rock as heat and pressure cause the minerals to form new combinations or new crystals   **(77)**

meteorite   asteroid fragment that leaves orbit and enters the earth's atmosphere   **(469)**

meteorology   study of the earth's atmosphere   **(4)**

mineral   naturally occurring substance consisting of a single element or compound   **(59)**

model   something scientists create to help them understand how things work   **(11)**

molecule   group of two or more atoms   **(54)**

momentum   mathematically, the mass of an object multiplied by the object's velocity   **(521)**

moraine   mound of unsorted rock materials that builds up along the edge of a glacier   **(286)**

nebula   cloud of dust and gas in space; sometimes glows by reflected light   **(495)**

nekton   sea animals that swim about freely   **(441)**

neutron   subatomic particle that has no electrical charge   **(52)**

neutron star   very small, dense star, as small as 15 km in radius   **(487)**

nonfoliated rock   metamorphic rock having no visible layers   **(92)**

nucleus   core of an atom   **(52)**

ocean major expanse of salt water with a surface area larger than 10 million km² **(391)**

oceanography study of the waters of the oceans **(4)**

orbit path that the earth follows as it revolves around the sun **(456)**

ore economically important mineral compound of a metal **(99)**

organic sediment hard remains of a once-living organism **(85)**

orogeny (oh RAHJ uh nee) process of mountain building **(122)**

outwash plain thick blanket of sediments formed as melt-water deposits material in front of a stationary glacier **(287)**

oxidation chemical weathering process in which oxygen combines directly with an element **(235)**

ozone gas molecule consisting of three oxygen atoms **(168)**

outwash plain

penumbra in a lunar eclipse, a zone of partial shadow surrounding the umbra **(459)**

permeability ability of rocks to allow water to pass through **(266)**

pesticide chemical used to kill crop-destroying insects **(370)**

photon (FOH tahn) particle of light, which can travel from the sun to the earth in 8.3 minutes **(170)**

photosynthesis (foht oh SIHN thuh sihs) process of making food from water and carbon dioxide **(166)**

planet main body that revolves around the sun **(461)**

planetesimal (plan uh TEHS uh muhl) small body of gases and solids that formed a planet **(342)**

plankton microscopic plantlike and animal-like organisms that drift with the tides and currents **(439)**

playa desert plain where runoff collects temporarily in muddy pools **(263)**

plucking process in which glacial ice freezes around rocks and plucks them out of the ground as the glacier moves **(284)**

pollution unwanted dirt and waste that fouls water, air, and land **(364)**

porosity ratio of the volume of air space in a rock to the total volume of the rock **(266)**

precipitate (pruh SIHP uh tayt) chemical sedimentary rock that forms when dissolved minerals precipitate, or fall out of solution **(87)**

precipitation solid or liquid water that falls to the earth's surface **(199)**

primary coast coast resulting from erosional processes of the land **(402)**

playa

quasar

rip current

principle of superposition principle stating that younger rocks are deposited on top of older rocks **(324)**

principle of uniformitarianism principle that similar processes produce similar results in different times **(323)**

proton subatomic particle that has a positive electrical charge **(52)**

pulsar rapidly pulsating neutron star **(488)**

P wave primary wave; longitudinal wave produced during an earthquake **(151)**

quasar body that emits radio waves and light in a starlike fashion; largest energy source in the universe **(496)**

radiation balance balance between solar radiation coming into Earth's atmosphere and radiation going out from Earth **(172)**

radioactive decay process in which particles and energy are given off by unstable nuclei **(328)**

radiotelescope instrument for capturing radio waves from outer space **(512)**

radio wave long-wavelength electromagnetic wave **(512)**

rain shadow area shielded from precipitation by a mountain or other topographic feature **(276)**

recycle to use something again **(371)**

reflection light bounced back from a smooth, shiny surface **(504)**

refraction change in direction of a light beam as it passes from one type of matter to another **(504)**

relative dating dating an igneous formation absolutely, and then estimating the age of the neighboring formation by its position relative to the igneous formation **(332)**

revolution movement of Earth around the sun **(456)**

Richter (RIHK tur) **scale** special scale used to report the size, or magnitude, of an earthquake **(154)**

rip current returning water flowing at right angles to the beach **(425)**

rock naturally occurring combination of minerals **(75)**

rock cycle continuous process of change in which new rocks are formed from old rock material **(78)**

rotation (roh TAY shuhn) turning of the earth on its axis **(455)**

runoff movement of water on the surface of the earth toward a lower level **(253)**

salinity measure of the amount of salts dissolved in ocean water **(412)**

saltation bouncing of sand that is blown by the wind **(279)**

salt dome mound of salt 2 km to 3 km across and 5 km to 10 km high **(435)**

satellite   small body orbiting a larger body or planet   **(466)**

saturated air   air that is filled with as much water as it can hold   **(192)**

scale   relationship between a distance on a map and a distance on the earth's surface   **(36)**

science   organized body of knowledge that has developed through observation and experimentation   **(5)**

scientific method   steps that scientists use to answer questions   **(9)**

sea   expanse of salt water less than 3 million km^2 in area   **(392)**

seamount   underwater volcano   **(147)**

secondary coast   coast shaped by erosion caused by the sea   **(403)**

sedimentary (sehd uh MEHN tuhr ee) **rock**   rock that forms from compressed sediments   **(76)**

seismograph (SYZ muh graf)   instrument used to record earthquake waves   **(152)**

shear   action of pieces of the earth's crust dragging against each other in opposite directions   **(127)**

shield volcano   volcano with a gentle slope and a wide base formed by quiet eruptions of lava   **(144)**

smelting   process of separating metals from their ores by using heat   **(99)**

smog   chemical fog produced by the reaction between sunlight and pollutants   **(367)**

soil   combination of small rock fragments and organic material in which plants can grow   **(231)**

solar cell   device that produces electricity from sunlight   **(378)**

solar eclipse   occurs when the moon's shadow passes across the lighted side of the earth   **(460)**

solar power   power produced by energy from the sun   **(377)**

solar prominence   stream of gases that explodes above the sun's surface   **(492)**

solar wind   high-energy radiation from the sun   **(165)**

sonar   device that bounces sound waves off a solid surface to determine its distance   **(394)**

space probe   spacecraft that is sent out into the solar system but is not sent into Earth orbit   **(538)**

spectrograph   optical instrument used for observing the colors of light given off by a star   **(507)**

stratus cloud   extended cloud layer that may cover the entire sky   **(197)**

streak   color of the powder that is left when a mineral is rubbed on a rough surface   **(63)**

strike   straight line along the earth's surface, formed where a rock layer dips   **(124)**

secondary coast

space probe

subsidence (suhb sy duhns)   sinking of rock material to a lower level without its sliding down a hillside   **(238)**

sunspot   enormous whorl of gases on the surface of the sun   **(168)**

surface wave   earthquake wave that is produced by the interaction of P waves and S waves and that causes the earth's surface to roll   **(152)**

S wave   secondary wave; transverse wave produced during an earthquake   **(151)**

syncline   structure formed when rock layers are folded downward   **(123)**

telescope

telescope   instrument used to enlarge the image of a distant object   **(506)**

tension   force that causes the earth's crust to pull apart   **(125)**

tetrahedron   network of atoms that forms a four-sided structure   **(59)**

texture   size of the crystals in igneous rocks   **(79)**

theory   statement that explains why things happen the way they do   **(8)**

theory of plate tectonics   theory stating that the lithosphere is divided into plates, each moving independently of the others **(138)**

thermocline   separation between warm surface water and colder deep water   **(415)**

thermohaline   ocean circulation due to temperature and salinity differences   **(415)**

thunderstorm   violent storm that develops from cumulonimbus clouds   **(214)**

tidal power   energy of changing tides   **(376)**

tide   crest of ocean water pulled up by the gravity of the moon as the earth turns on its axis   **(425)**

till   unsorted rock debris deposited directly by glaciers   **(285)**

topographic map

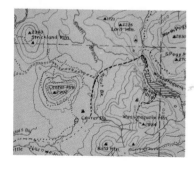

topographic map   map that shows the shape of the land, using fine lines drawn in patterns to indicate features such as mountains, valleys, and rivers   **(29)**

topography   contour or shape of the land surface   **(39)**

tornado   small, extremely violent storm   **(216)**

transform fault   type of lateral fault that forms when two plates slide past each other in parallel but opposite directions   **(142)**

umbra (UHM bruh)   zone of complete shadow in a lunar eclipse   **(459)**

universe   all the matter and energy that exists: all the stars, planets, dust, gas, and energy in space   **(481)**

Van Allen belts   two atmospheric belts that trap incoming protons and electrons and protect the earth from high-energy particles; discovered in 1958 by an American physicist, James Van Allen **(180)**

variable   in a scientific experiment, something that can be changed **(8)**

variable star   star that changes magnitude **(482)**

viscosity   liquid's resistance to flow **(144)**

volcano   opening in the earth's crust that allows magma to reach the earth's surface **(139)**

volume   total space that an object occupies **(15)**

waterspout   vortex that occurs over open water **(218)**

water table   zone of saturation, where all the air spaces in the soil are filled with water **(265)**

wave   surface motion of water, produced by a local wind **(423)**

weathering   processes that break rocks into smaller fragments, eventually producing soil **(231)**

weight   measure of the force, or pull, of gravity on the mass of an object **(17)**

zodiac   signs of the 12 constellations along the ecliptic **(483)**

waterspout

INDEX

Boldface numbers refers to an illustration on that page.

A

Aa lava, 144, **144**
Abrasion (weathering), 234
Absolute dating, 329
Absolute humidity, 193
Acadia National Park, 313
Accelerated mass spectrometry, 335
Acid rain, 167, 368
   damage of, **368**
   effects of, **236**
Aeration, **365**
Aerosols, 195
Agar, 442
Air, 367–368
Air mass, 207
Air pollution, **367, 368**
Air scrubbers, 368
Alaskan Province, 307–308, **308**
Alaska Range, 308, **308**
Aldrin, Edwin "Buzz," 528, **528**
Aleutians, 307, 396
Algin, 442
Allahabad, 195
Alloy, 100
Alluvial fans, 263, **263**
Alpine glaciers, 289–290
Aluminum, 371
Alvarez, Luis, biography of, 19, **19**
Alvarez, Walter, 321, 386–387, **387**
Amazon River, 419
Amber, 332, **332**
Amino acids, 344, **357**
Amphibians
   of Devonian Period, 351
Amplitude (wave), 170
Anchorage, **211**
Ancient map, **25**
Andes Mountains, 141
Andromeda, 495
Aneroid barometer, 175, **175**
Angel Falls, 259
Annapolis River (Canada), 450
Antarctica, 282
Antarctic Circumpolar Current, 420
Antarctic Ocean, 392
Anthracite, 92, 93, 110
Anticlines, 123, **123, 135**
Anticyclones, 209, **209**

Apatite, **65**
Aphelion, 462
Apollo/Soyuz linkup, 529
Apollo spacecraft, **529**
Apollo spaceflights, 527, 528
Appalachian Mountain range, 129
Appalachian Plateau, 312, **312**
Appalachian provinces, 310–313
Apprenticeships, 58
Aquifers, 267, **267**
Arabian Sea, **140**
Aral Sea, 398
Archean Era, 344–345
Arctic Ocean, 229, 308, **414,** 431, **431**
Arctic Slope, 308, **308**
Arecibo, 513
Arid environments, 275–277
Armstrong, Neil, 528, **528**
Artesian wells, 267, **267**
Asbestos, **66**
Asia, central, **212**
Asian plate, 141
Asphalts, 108
Asteroids, 468–469
Asthenosphere, 119, **120,** 138
Astronauts, 17. *See also*
      Individual astronauts
   career as, 497, **497**
Astronomer, **5**
   career as, 497, **497**
Astronomia Nova (Kepler), 463
Astronomical unit, 462
Astronomy
   defined, 5
   optical, 503–511
   radio, 512–515
Astrophysicist, career as, 497, **497**
Aswan Dam, 260
Atacama Desert, 199, **199**
Atlantic Ocean, 229, 394–395
Atlantis, 532
Atlas, earth science, 576–593
Atmosphere (earth's), **178**
   composition of, 165–168
   defined, 165
   gases of, 165–166
   layers of, 177–180
   motions of, 181–185
   present, 167–168
   primitive, 165–167
   primitive and present
      compared, *table* 166
   structure of, 175–180
   temperature of, **179**

Atmosphere chemist, career as, 176, **176**
Atmospheric diving suits (ADS), 438
Atmospheric pressure, 175–177
Atolls, 308
Atomic number, 56
Atomic structure, 51–53
   atoms of, 51–52
   and ions, 53
   and isotopes, 53
Atoms, 51–52
   computer model of, **52**
   and radioactive decay, 328
Aurora, 179
Aurora australis, 164
Aurora borealis, 164, **164**
Automotive technician, career as, 176, **176**
Autumnal equinox, 457
Axis
   of crystal shapes, *table* 68
   of the earth, 455
   of the moon, 458
   of Uranus, 474

B

Bacteria, 166
Bagnold, R.A., 296, 297
Balance, **17, 18**
Bald eagle, **370**
Ballard, Robert Duane, biography of, 418, **418;** 438
Baltic Sea, 393
Baraboo Dome, 309
Barchan dune, 280, 281
Barometer, 175
Barrier islands, 404, **404**
Barrow, **225**
Basalt, 79, **79,** 80
Base level, **259**
   of streams, **259**
   temporary, **259**
Basin (geologic feature), **135**
Basin and Range Province, 305
Batholith, 143
Bay, 402
Bay of Fundy, 389, 426, 450–451, **451**
Bench mark, 27, **27,** 35
Benioff, Victor Hugo, 155
Benthos, 442
Beryl, 69

N

Credits

Kabilinsky/Peter Arnold, Inc.; 176(b), Grapes Michaud/Photo Researchers; 177, Phil Degginger; 179, Yoshikazu Shirakawa/The Image Bank; 180, NASA; 184, Paolo Koch/Photo Researchers; 185(l), Stephen Krasemann/Photo Researchers; 185(r), Jules Bucher/Photo Researchers; 189, AP/Wide World Photos; 190, Eric Meola/The Image Bank; 191(l), Harvey Lloyd/The Stock Market; 191(r), NASA; 192, HBJ Photo/Earl Kogler; 193(t), Thermometer Corporation of America; 193(b), HBJ Photo/Rodney Jones; 194(t), Nicholas Foster/The Image Bank; 194(bl), Bob Silverman/The Stock Market; 194(br), Chris Jones/The Stock Market; 195(l), Michael Schneps/The Image Bank; 195(r), Frank Whitney/The Image Bank; 196(t to b), Leonard Lee Rue III/Photo Researchers; 196, Gregory K. Scott/Photo Researchers; 196, Henry Lansford/Photo Researchers; 196, Joyce Photographics/Photo Researchers; 196, Mary Fuller/Taurus Photos; 197(t to b), Richard Kolar/Earth Scenes; 197, Joe Devenney/The Image Bank; 197, H. Wendler/The Image Bank; 197, Edna Bennett/Photo Researchers; 197, Eric Meola/The Image Bank; 197, Howie Bluestein/Photo Researchers; 199(tl), Frans Lanting/Photo Researchres; 199(tr), Francois Gohier/Photo Researchers; 199(b), Science Source/Photo Researchers; 200(l), NOAA; 200(r), E. R. Degginger; 201(l), Universitetsbiblioteket 1 Oslo, Hovedbiblioteket; 202, HBJ Photo/Earl Kogler; 204, Larry Dale Gordon/The Image Bank; 205(l), John Gerlach/Earth Scenes; 205(r), NCAR; 206, Bruce Coleman, Inc.; 207, Walter Bibikow/The Image Bank; 211(l), Carleton Ray/Photo Researchers; 211(c), Margot Granitsas/Photo Researchers; 211(r), Mathias Oppersdorff/Photo Researchers; 212(tl), G. Whiteley/Photo Researchers; 212(bl), William Bacon III/Photo Researchers; 212(r), E. R. Degginger; 213(tl), Hank Morgan/Photo Researchers; 213(tr), Hank Morgan/Photo Researchers; 213(b), Hank Morgan/Photo Researchers; 214, Larry Dale Gordon/The Image Bank; 215(l), Lionel Brown/The Image Bank; 215(r), E. R. Degginger; 216(tl), Howard B. Bluestein/Science Source/Photo Researchers; 216(tc), Howard B. Bluestein/Science Source/Photo Researchers; 216(tr), Howard B. Bluestein/Science Source/Photo Researchers; 216(bl), Howard B. Bluestein/Science Source/Photo Researchers; 216(bc), Howard B. Bluestein/Science Source/Photo Researchers; 217, E. R. Degginger; 218, C. C. Lockwood/Earth Scenes; 220, NOAA/Science Photo Library/Photo Researchers; 221(l), UPI/Bettmann Newsphotos; 221(r), AP/Wide World Photos; 225, Jim Cartier/Photo Researchers; 226(l), National Severe Storms Lab, Norman, OK.; 226(r), Photri; 227(t), Howard Bluestein/Photo Researchers; 227(b), E. R. Degginger;

UNIT 4: 228—229, National Park Service; 230, E. R. Degginger; 231(l), Sepp Seitz/Woodfin Camp & Assoc.; 231(r), HBJ Photo/Sam Joosten; 232(l), Townsend P. Dickinson/Photo Researchers; 232(r), Townsend P. Dickinson/Photo Researchers; 233, John Eastcott/Yva Momatiuk/Woodfin Camp & Assoc.; 234(t), Miriam Reinhart/Photo Researchers; 234(bl), E. R. Degginger; 234(br), Townsend P. Dickinson/Photo Researchers; 235(t), Franke Keating/Photo Researchers; 235(c), C. C. Lockwood/Earth Scenes; 235(bl), Joy Glenn; 235(br), HBJ Photo/Sam Joosten; 236(t), Mike Andrews/Earth Scenes; 236(b), Gregory G. Dimijian/Photo Researchers; 237, Georg Gerster/Photo Researchers; 238(tl), R. Call/Gamma-Liaison; 238(tr), E. R. Degginger/Earth Scenes; 238(b), Leif Skoogfors/Woodfin Camp & Assoc.; 239(l), E. R. Degginger; 239(r), E. R. Degginger; 240(t), Bill Ross/Woodfin Camp & Assoc.; 240(bl), E. R. Degginger; 240(br), U. S. Geological Survey, Department of the Interior; 241, Photri; 243, E. R. Degginger; 244(l), Pat & Tom Leeson/Photo Researchers; 244(r), E. R. Degginger; 245(t), Betsy Blass/Photo Researchers; 245(b), Phil Degginger; 245(inset), Lowell Georgia/Photo Researchers; 246, USDA; 247(tl), Owen Franken/Stock, Boston; 247(tr), E. R. Degginger; 247(b), E. R. Degginger; 251(l), UPI/Bettmann Newsphotos; 251(r), The Bettmann Archive; 252, E. R. Degginger; 253, Mike Maple/Woodfin Camp & Assoc.; 254(l), E. R. Degginger; 254(r), Charles R. Belinky/Photo Researchers; 255, NASA; 257(l), The Bancroft Library/University of California, Berkeley; 257(r), Fred Seidman; 258(l), Ty & Julie Hotchkiss/Photo Researchers; 258(r), E. R. Degginger; 259, Ralph A. Reinhold/Earth Scenes; 260(l), John Bova/Photo Researchers; 260(r), John Lemker/Earth Scenes; 261(l), Dan Getsug/Photo Researchers; 261(tr), C. C. Lockwood/Earth Scenes; 261(cr), Dan Guravich/Photo Researchers; 261(br), Earth Scenes; 262(l), NASA/Photri; 262(r), Dan Guravich/Photo Researchers; 265(t), William M. Partington/Photo Researchers; 265(b), Francois Gohier/Photo Researchers; 266(l), E. R. Degginger; 266(r), E. R. Degginger; 267, E. R. Degginger; 268, U. S. Geological Survey, Department of the Interior; 269(t), Breck P. Kent/Earth Scenes; 269(bl), Bruce Roberts/Photo Researchers; 269(br), Robert Knowles/Photo Researchers; 272, R. R. Donnelley & Sons Company Cartographic Services; 273, Breck P. Kent/Earth Scenes; 274, Richard Weymouth Brooks/Photo Researchers; 276(t), Zig Leszczynski/Animals Animals; 276(bl), George Holton/Photo Researchers; 276(br), Kummerly & Frey, Bern edition 1984; 277, Mike Yamashita/Woodfin Camp & Assoc.; 278(t), Larry Miller/Photo Researchers; 278(bl), Bill Bachman/Photo Researchers; 278(br), Georg Gerster/Photo Researchers; 279(tl), Charlie Ott/Photo Researchers; 279(tr), E. R. Degginger; 279(b), E. R. Degginger; 280(t), Phil Degginger; 280(b), E. R. Degginger; 282, W. Curtsinger/Photo Researchers; 283(tl), E. R. Degginger; 283(tc), E. R. Degginger; 283(tr), Farrell Grehan/Photo Researchers; 283(bl), Nuridsany et Perennou/Photo Researchers; 283(bc), Thomas W. Martin/Photo Researchers; 283(br), Nuridsany et Perennou/Photo Researchers; 284(t), Paolo Koch/Photo Researchers; 284(c), Francois Gohier/Science Source/Photo Researchers; 284(b), Leonard Lee Rue III/Earth Scenes; 285(t), Paolo Koch/Photo Researchers; 285(b), Steve Coombs/Photo Researchers; 286(tl), Jim Brandenburg/Woodfin Camp & Assoc.; 286(tr), John Lemker/Earth Scenes; 286(bl), E. R. Degginger; 286(br), Charles R. Belinky/Photo Researchers; 287(tl), E. R. Degginger; 287(tr), D. Pulleston/Photo Researchers; 287(b), Ira Block/The Image Bank; 288(t), Bob Daemmrich/Stock, Boston; 288(c), Carleton Ray/Photo Researchers; 288(b), Bonnie Freer/Photo Researchers; 289, E. R. Degginger/Bruce Coleman, Inc.; 290(tl), Breck P. Kent/Earth Scenes; 290(tr), Leo Touchet/Woodfin Camp & Assoc.; 290(bl), Sven Olof Lindblad/Photo Researchers; 290(bc), E. R. Degginger; 290(br), NASA; 291, Michael Giannechini/Photo Researchers; 292(l), The Granger Collection; 292(r), R. R. Donnelley & Sons Company Cartographic Services; 293(l), E. R. Degginger; 293(r), William W. Bacon III/Photo Researchers; 294(l), U. S. Geological Survey, Department of the Interior; 294(r), U. S. Geological Survey, Department of the Interior; 297, Dick Rowan/Photo Researchers; 298, Pat & Tom Leeson/Photo Researchers; 299(l), E. R. Degginger; 299(r), Mickey Gibson/Earth Scenes; 301(t), R. R. Donnelley & Sons Company Cartographic Services; 301(bl), Richard Kolar/Earth Scenes; 301(bc), John Bova/Photo Researchers; 301(br), E. R. Degginger; 302(l), The Granger Collection; 302(r), AP/Wide World Photos; 303(l), Doug Wechsler/Earth Scenes; 303(r), Dan Guravich/Photo Researchers; 304(t), R. R. Donnelley & Sons Company Cartographic Services; 304(bl), Kent & Donna Dannen/Photo Researchers; 304(br), E. R. Degginger; 305(l), Stouffer Enterprises/Earth Scenes; 305(r), Lillian N. Bolstad/Photo Researchers; 306(l), Steve Van Matre/National Audubon Society/Photo Researchers; 306(r), E. R. Degginger; 307(l), E. R. Degginger; 307(r), Roger Archibald/Earth Scenes; 308(t), R. R. Donnelley & Sons Company Cartographic Services; 308(c), Johnny Johnson/Earth Scenes; 308(bl), Stephen J. Krasemann/Photo Researchers; 308(br), Stephen J. Krasemann/Photo Researchers; 309(tl), E. R. Degginger; 309(tc), E. R. Degginger; 309(tr), R. R. Donnelley & Sons Company Cartographic Services; 309(b), Thomas Hollyman/Photo Researchers; 310(l), E. R. Degginger; 310(r), R. R. Donnelley & Sons Company Cartographic Services; 311(t), Bill Curtsinger/Photo Researchers; 311(bl), E. R. Degginger; 311(bc), J. H. Robinson/Photo Researchers; 311(br), John Lemker/Earth Scenes; 312(tl), Wendell D. Metzen/Bruce Coleman, Inc.; 312(tr), Photo Researchers; 312(bl), Michael Habicht/Earth Scenes; 312(br), Kent & Donna Dannen/Photo Researchers; 313(t), Stephen J. Krasemann/Photo Researchers; 313(b), E. R. Degginger; 314(l), U. S. Geological Survey, Department of the Interior; 314(r), Map courtesy of the Commonwealth of Virginia; 316, Photri; 317, McCarthy/The Image Bank; 318(l), National Park Service; 318(r), Phil Degginger/Bruce Coleman, Inc.; 319(t), National Park Service; 319(c), National Park Service; 319(b), National Park Service;

UNIT 5: 320—321, Tom McHugh/Photo Researchers; 322, Michael & Barbara Reed/Earth Scenes; 323, Paolo Koch/Photo Researchers; 324(t), E. R. Degginger/Earth Scenes; 324(bl), E. R. Degginger; 324(br), C. C. Lockwood/Earth Scenes; 325(tl), E. R. Degginger/Earth Scenes; 325(tr), E. R. Degginger; 325(b), Breck P. Kent/Earth Scenes; 326, E. R. Degginger; 327(tl), Townsend P. Dickinson/Photo Researchers; 327(tc), Townsend P. Dickinson/Photo Researchers; 327(tr), Townsend P. Dickinson/Photo Researchers; 327(bl), Thomas R. Taylor/Photo Researchers; 327(c), Louise K. Broman/Photo Researchers; 327(bc), Townsend P. Dickinson/Photo Researchers; 327(br), Townsend P. Dickinson/Photo Researchers; 328, Campbell/Sygma; 331(l), Charles Naeser/U. S. Geological Survey; 331(tr), Paolo Koch/Photo Researchers; 331(br), Milner/Sygma; 332(t), E. R. Degginger; 332(bl), J. Koivula/Photo Researchers; 332(br), Novosti from Sovfoto; 333, Richard Weiss/Peter Arnold, Inc.; 334, Michael Collier/Stock, Boston; 335(tl), Photo Researchers; 335(tr), The Stock Market; 335(b), John Zoiner/Peter Arnold, Inc.; 336, Louise K. Broman/Photo Researchers; 339, E. R. Degginger; 340, Paul Fusco/Magnum Photos; 341(l), Lauren Freudmann/Woodfin Camp & Assoc.; 341(r), Bill Bachman/Photo Researchers; 342(t), Thomas Hopker/Woodfin Camp & Assoc.; 342(b), NASA/Photo Researchers; 343(l), Paolo Koch/Photo Researchers; 343(r), Chris Alan Wilton/The Image Bank; 344(l), E. R. Degginger; 344(r), UPI/Bettmann Newsphotos; 345(tl), Michael Abbey/Photo Researchers; 345(tr), Eric Grave/Science Source/Photo Researchers; 345(b), Eugene A. Shinn; 346(t), M. I. Walker/Photo Researchers; 346(bl), Photo Researchers; 346(bc), Biophoto Assoc./Photo Researchers; 346(br), Michael Abbey/Photo Researchers; 347(l), Nancy Sefton/Photo Researchers; 347(r), Bill Curtsinger/Photo Researchers; 349, J. L. Lepore/Photo Researchers; 350(t), Tom McHugh/Photo Researchers; 350(b), E. R. Degginger; 351(t), Porterfield/Chickering/Photo Researchers; 351(b), Fred Whitehead/Animals Animals; 353(l), John Zoiner/Peter Arnold, Inc.; 353(r), Stephen J. Krasemann/Peter Arnold, Inc.; 354(l), Michael Giannechini/Photo Researchers; 354(r), Emil Muench/Photo Researchers; 355(l), Chicago Zoological Park/Tom McHugh/Photo Researchers; 355(r), Helen Williams/Photo Researchers; 356, Will & Deni McIntyre/Photo Researchers; 357(l), AP/Wide World Photos; 357(r), Boston University Photo Services; 361, Jeff Foott/Bruce Coleman, Inc.; 362, F. Stuart Westmorland/Tom Stack & Assoc.; 363(l), Grant Heilman Photography; 363(r), Grant Heilman Photography; 364(l), Bruce Coleman, Inc.; 364(r), Kees Van Den Berg/Photo Researchers; 365(tl), William Felger/Grant Heilman Photography; 365(tc), Runk-Schoenberger/Grant Heilman Photography; 365(tr), Runk-Schoenberger/Grant Heilman Photography; 365(cr), R. Perron/Nawrocki Stock Photo; 365(bl), Steve Peterson/Tom Stack & Assoc.; 265(br), Doug Lee/Tom Stack & Assoc.; 366(tl), Holt Confer/Grant Heilman Photography; 366(tr),Larry Lefever/Grant Heilman Photography; 366(b), HBJ Photo/Beverly Brosius; 367(t), Grant Heilman Photography; 367(bl), Rick McIntyre/Tom Stack & Assoc.; 367(br), Steve Solum/Bruce Coleman, Inc.; 368(tl), Runk-Schoenberger/Grant Heilman Photography; 368(tr), John Shaw/Tom Stack & Assoc.; 368(b), Jake Rajs/The Image Bank; 369(tl), Betsy Blass/Photo Researchers; 369(tr), John L. Stage/The Image Bank; 369(bl), Richard Hoffman/Sygma; 369(br), Grant Heilman Photography; 370(tl), Charles G. Summers, Jr./Tom Stack & Assoc.; 370(tr), Grant Heilman Photography; 370(bl), Grant Heilman Photography; 370(br), Adrienne T. Gibson/Tom Stack & Assoc.; 371, The Image Bank; 372(tl), Grant Heilman Photography; 372(tr), E. R. Degginger/Bruce Coleman, Inc.; 372(b), Jan Halaska/Photo Researchers; 374(t), Lowell Georgia/Photo Researchers; 374(b), Dick Durrance/Woodfin Camp & Assoc.; 375(tl), R. L. Carlton/Photo Researchers; 375(tr), Grant Heilman Photography; 375(b), Grant Heilman Photography; 376(tl), William Felger/Grant Heilman Photography; 376(tr), Georg Gerster/Photo Researchers; 376(b), A. Bollinger/The Image Bank; 377(tl), Gerald A. Corsi/Tom Stack & Assoc.; 377(tr), J. Glab/The Stock Shop; 377(bl), Tom McHugh/Photo Researchers; 377(br), John Colwell/Grant Heilman Photography; 378(tl), Nancy A. Potter/Bruce Coleman, Inc.; 378(tc), Jean Guichard/Photo Researchers; 378(tr), Jim Brandenburg/Bruce Coleman, Inc.; 378(bl), Tom McHugh/

Photo Researchers; 378(br), D. Lyons/Bruce Coleman, Inc.; 379(t), Lee Foster/Bruce Coleman, Inc.; 379(b), TASS/Sovfoto; 380(t), Mark Sherman/Bruce Coleman, Inc.; 380(b), Mark Sherman/Bruce Coleman, Inc.; 381(t), U. S. Dept. of Energy/Science Photo Library/Photo Researchers; 381(c), Ray Ellis/Photo Researchers; 381(b), Ray Ellis/Photo Researchers; 385(t), Martin Bond/Science Photo Library/Photo Researchers; 385(b), Y. Arthus Bertrand/Peter Arnold, Inc.; 386(l), Kent & Donna Dannen/Photo Researchers; 386(c), Lee Balterman/Marilyn Gartman Agency; 386(r), Prof. W. Alvarez/Science Photo Library/Photo Researchers; 387(t), M. Paternostro/Science Photo Library/Photo Researchers; 387(b), Lawrence Berkeley Laboratory/Science Photo Library/Photo Researchers;

UNIT 6: 388—389, Joseph Devenney/The Image Bank; 390, Joel Meyerowitz/The Image Bank; 391, Grant Heilman Photography; 393(l), Juergen Schmitt/The Image Bank; 393(r), Photri; 394, Runk-Schoenberger/Grant Heilman Photography; 395, NOAA/Pacific Marine Environmental Labs; 396(l), Woods Hole Oceanographic Institution; 396(r), Woods Hole Oceanographic Institution; 398(l), Gerald Brimacombe/The Image Bank; 398(r), Photri; 399(l), David J. Maenza/The Image Bank; 399(r), Gary Crall;aae/The Image Bank; 400(t), Rona Photography/Bruce Coleman, Inc.; 400(c), Frans Lanting/Photo Researchers; 400(b), M. Timothy O'Keefe/Bruce Coleman, Inc.; 401(t), George Hall/Woodfin Camp & Assoc.; 401(b), Walter Iooss, Jr./The Image Bank; 402(tl), Lowell Georgia/Photo Researchers; 402(tr), Dick Rowan/Photo Researchers; 402(bl), Lou Jacobs, Jr./Grant Heilman Photography; 402(br), Tim McCabe/Taurus Photos; 403(tl), William Felger/Grant Heilman Photography; 403(tr), S. L. Craig, Jr./Bruce Coleman, Inc.; 403(b), Alex Bartel/Science Photo Library/Photo Researchers; 404(tl), Breck P. Kent/Earth Scenes; 404(tr), Fred Whitehead/Earth Scenes; 404(c), S. E. Cornelius/Photo Researchers; 404(b), Cameron Davidson/Bruce Coleman, Inc.; 405, Chip Henderson/Woodfin Camp & Assoc.; 409(l), Grant Heilman Photography; 409(r), Dwight R. Kuhn/Bruce Coleman, Inc.; 410, Tony Arruza/Bruce Coleman, Inc.; 411, S. Dinkins/Photo Researchers; 412(t), E. R. Degginger; 412(b), Woods Hole Oceanographic Institution; 413(l), Photo Researchers; 413(r), Photo Researchers; 414(t), Rangefinders; 414(b), Harry Groom/Photo Researchers; 416(l), The Bettmann Archive; 416(r), Official U. S. Navy Photograph; 417(l), Woods Hole Oceanographic Institution; 417(r), Woods Hole Oceanographic Institution; 418(l), Culver Pictures; 418(r), Hank Morgan/Photo Researchers; 419, Dr. Richard Legeckis/Science Photo Library/Photo Researchers; 421(l), J. Messerschmidt/Bruce Coleman, Inc.; 421(r), Mark Sherman/Bruce Coleman, Inc.; 423(tl), Mark Romanelli/The Image Bank; 423(tr), Mark Romanelli/The Image Bank; 423(b), R. Rowan/Photo Researchers; 424(t), Norman Owen Tomalin/Bruce Coleman, Inc.; 424(b), Larry Dale Gordon/The Image Bank; 425(t), Tony Arruza/Bruce Coleman, Inc.; 425(b), UPI/Bettmann Newsphotos; 426(t), Scott Blackman/Tom Stack & Assoc.; 426(tc), HBJ Photo/Charlie Burton; 426(bc), HBJ Photo/Charlie Burton; 426(bl), Breck P. Kent/Earth Scenes; 426(br), Breck P. Kent/Earth Scenes; 431(l), Zig Leszczynski/Earth Scenes; 431(r), Breck P. Kent/Earth Scenes; 432, Dotte/Larsen/Bruce Coleman, Inc.; 433(l), James H. Karales/Peter Arnold, Inc.; 433(r), Charlie Ott/National Audubon Society; 434, Georg Gerster/Photo Researchers; 435(t), U. S. Geological Survey, Department of the Interior; 435(b), Porterfield/Chickering/Photo Researchers; 436(l), Steinhart Aquarium/Tom McHugh/Photo Researchers; 436(r), Manfred Kage/Peter Arnold, Inc.; 437(l), Rona Photography/Bruce Coleman, Inc.; 437(r), GECO/UK/Science Photo Library/Photo Researchers; 438(t), Bob Evans/Peter Arnold, Inc.; 438(inset), Bob Evans/Peter Arnold, Inc.; 438(b), Bill Wood/Bruce Coleman, Inc.; 439, E. R. Degginger; 440(t), E. R. Degginger; 440(b), E. R. Degginger; 441(tl), Jeff Rotman/Peter Arnold, Inc.; 441(tr), HBJ Photo/Earl Kogler; 441(bl), Robert De Cast/Photo Researchers; 441(br), Dick Rowan/Photo Researchers; 442(l), Heinz Steenmans/Wheeler Pictures; 442(r), Bob Evans/Peter Arnold, Inc.; 443(l), The Image Bank; 443(r), Gary Guisinger/Photo Researchers; 444(tl), John Launois/Black Star; 444(tr), Jean Gaumy/Magnum Photos; 444(c), Jean Gaumy/Magnum Photos; 444(bl), G. Gscheidle/The Image Bank; 444(br), TDM/Photo Researchers; 445, Julius Felcete/The Stock Market; 449, Chesher/Photo Researchers; 449(inset), R. Holland/Gamma-Liaison; 450(l), Michelangelo Durazzo/Magnum Photos; 450(r), Michelangelo Durazzo/Magnum Photos; 451(t), Stephen J. Krasemann/Photo Researchers; 451(c), A. Schoenfeld/Photo Researchers; 451(b), R. R. Donnelley & Sons Company Cartographic Services;

UNIT 7: 452—453, Don Landwenrle/The Image Bank; 454, James Carmichael/The Image Bank; 457(tl), Daniel Zirinsky/Photo Researchers; 457(tr), Daniel Zirinsky/Photo Researchers; 457(bl), Daniel Zirinsky/Photo Researchers; 457(br), Daniel Zirinsky/Photo Researchers; 459, John Sanford/Science Photo Library/Photo Researchers; 460(t), John Bova/Photo Researchers; 460(c), John Bova/Photo Researchers; 460(b), E. R. Degginger; 463(l), Dr. Jeremy Burgess/Science Photo Library/Photo Researchers; 463(r), NASA; 464(t to b), NASA/Photri; 464, NASA; 464, NASA/Omni-Photo Communications; 464, NASA; 464, NASA/Science Source/Photo Researchers; 464, NASA; 464, Science Source/Photo Researchers; 464, NASA; 464, NASA; 465(l), NASA/Hansen Planetarium; 465(r), NASA/Photri; 466(t), NASA/Photri; 466(b), Fotokhronika TASS; 467(l), Chris Bjornberg/Photo Researchers; 467(r), NASA; 468(tl), Wards SCI/Science Source/Photo Researchers; 468(tr), NASA/Omni-Photo Communications; 468(c), A.S.P./Science Source/Photo Researchers; 468(bl), Jet Propulsion Laboratory; 468(br), Jet Propulsion Laboratory; 469(l), Georg Gerster/Photo Researchers; 469(tr), Julian Baum/Science Photo Library/Photo Researchers; 469(br), E. R. Degginger; 471(l), NASA/Science Source/Photo Researchers; 471(r), NASA; 472(t), A.S.P./Science Source/Photo Researchers; 472(bl), NASA; 472(bc), NASA; 472(br), NASA; 473(t), NASA/Photri; 473(b), NASA/Science Source/Photo Researchers; 474(tl), NASA/Photri; 474(tr), NASA/Photri; 474(cl), Peter Ryan/Science Photo Library/Photo Researchers; 474(bl), NASA; 474(bc), Photo Researchers; 474(br), Science Source/Photo Researchers; 475(l), Harvard College Observatory/Science Photo Library/Photo Researchers; 475(r), David Parker/Science Photo Library/Photo Researchers; 479, NASA/Omni-Photo Communications; 480, NASA; 481, Ronald Royer/Science Photo Library/Photo Researchers; 482(t), Ronald Royer/Science Photo Library/Photo Researchers; 482(bl), Photri; 482(br), John Sanford/Science Photo Library/Photo Researchers; 484, John Sanford/Science Photo Library/Photo Researchers; 487(t), John Sanford/Science Photo Library/Photo Researchers; 487(b), Finley Holiday Film Corp.; 488, Photri; 490, Henley & Savage/The Stock Market; 491(l), National Optical Astronomy Observatories; 491(r), Science Photo Library/Photo Researchers; 492(t), National Optical Astronomy Observatories; 492(c), National Optical Astronomy Observatories; 492(b), National Optical Astronomy Observatories; 493, Ronald Royer/Science Photo Library/Photo Researchers; 494(t), Bill Iburg/Science Photo Library/Photo Researchers; 494(b), Jerry Schad/Photo Researchers; 495(t), National Optical Astronomy Observatories; 495(b), National Optical Astronomy Observatories; 496(t), National Optical Astronomy Observatories; 496(c), National Optical Astronomy Observatories; 496(b), Finley Holiday Film Corp.; 497(t), Will & Deni McIntyre/Photo Researchers; 497(c), National Optical Astronomy Observatories; 497(b), Don Klumpp/The Image Bank; 501, Finley Holiday Film Corp./Don Dixon; 502, Paul Shambroom/Photo Researchers; 503, National Optical Astronomy Observatories; 504(t), Hans Namuth/Photo Researchers; 504(bl), Photo Researchers; 504(bc), Richard Megna/Fundamental Photographs; 504(br), Richard Megna/Fundamental Photographs; 505(t), Fundamental Photographs; 505(b), Fundamental Photographs; 506(t), Art Resource; 506(b), Barbara J. Fiegles/Stock, Boston; 507(l), National Optical Astronomy Observatories; 507(r), Royal Greenwich Observatory/Science Photo Library/Photo Researchers; 508, The Granger Collection; 509(t), John Bova/Photo Researchers; 509(bl), Science Photo Library/Photo Researchers; 509(br), Science Photo Library/Photo Researchers; 510(l), NASA; 510(r), NASA/Science Source/Photo Researchers; 513, Robert Jureit/The Stock Market; 514, Peter Menzel/Wheeler Pictures; 515(t), Mark E. Gibson/The Stock Market; 515(b), E. Brinks/Leiden Observatory/Science Photo Library/Photo Researchers; 519, P. & G. Bowater/The Image Bank; 520, Mike Phillips/Peter Arnold, Inc.; 521(l), John Dommers/Photo Researchers; 521(r), Jean Marc Barey/Vandystadt/Photo Researchers; 522(t), HBJ Photo/Beverly Brosius; 522(b), Sovfoto; 523(l), The Bettmann Archive; 523(r), Culver Pictures; 524(tl), NASA; 524(tr), TASS/Sovfoto; 524(bl), NASA; 524(br), TASS/Sovfoto; 526, NASA/Peter Arnold, Inc.; 527(l), NASA; 527(r), NASA/Science Source/Photo Researchers; 528(tl), NASA; 528(tr), NASA; 528(b), NASA; 529(tl), NASA; 529(tc), NASA; 529(tr), NASA; 529(bl), TASS/Sovfoto; 529(br), NASA; 530(l), TASS/Sovfoto; 530(r), TASS/Sovfoto; 531(t), NASA; 531(bl), Frank Whitney/The Image Bank; 531(bc), NASA; 531(br), NASA; 532(l), NASA/Omni-Photo Communications; 532(r), NASA/Omni-Photo Communications; 533(t), NASA/Omni-Photo Communications; 533(c), NASA/Omni-Photo Communications; 533(b), NASA/Omni-Photo Communications; 534, HBJ Photo/Earl Kogler; 535(t), Photri; 535(b), NASA/Science Source/Photo Researchers; 536(t), Gregory Heisler/The Image Bank; 536(bl), Gamma-Liaison; 536(br), NOAA/NESDIS/NCDC; 537(l), The Image Bank; 537(c), The Image Bank; 537(r), The Image Bank; 538, NOAA/NESDIS/SDSD; 539(t), NASA; 539(ct), JPL/NASA; 539(c), JPL/NASA; 539(cb), Science Photo Library/Photo Researchers; 539(b), JPL/NASA; 540, HBJ Photo/Beverly Brosius; 543, NASA; 544(l), Photri; 544(r), Photri; 545(t), Photri; 545(bl), Photri; 545(br), Pete Turner/The Image Bank;

REFERENCE SECTION: 546-547, Michael Rochipp/The Image Bank; 556(tl), The Granger Collection; 556(tr), E. R. Degginger; 556(c), Art Resource; 556(b), NASA; 557(t), E. R. Degginger; 557(bl), NASA; 557(br), Richard Brunck/F. S. U.; 569(l-r, t-b), E. R. Degginger; Charles R. Belinky/Photo Researchers; Breck P. Kent/Earth Scenes; E. R. Degginger; Breck P. Kent/Earth Scenes; Breck P. Kent/Earth Scenes; E. R. Degginger; Breck P. Kent/Earth Scenes; E. R. Degginger; E. R. Degginger; E. R. Degginger; Breck P. Kent/Earth Scenes; Breck P. Kent/Earth Scenes; Breck P. Kent/Earth Scenes; E. R. Degginger; Breck P. Kent/Earth Scenes; Breck P. Kent/Earth Scenes; Breck P. Kent/Earth Scenes; 574(tl), Stephen J. Krasemann/DRK Photo; 574(tr), E. R. Degginger/Bruce Coleman, Inc.; 574(cl), M. Abbey/Photo Researchers; 574(cr), Porterfield/Chickering/Photo Researchers; 574(b), Michael Furman/The Stock Market; 575, Runk/Schoenberger from Grant Heilman Photography; 576-577, R. R. Donnelley & Sons Company Cartographic Services; 578, R. R. Donnelley & Sons Company Cartographic Services; 579, R. R. Donnelley & Sons Company Cartographic Services; 581, U. S. Geological Survey/Department of the Interior; 582, U. S. Geological Survey/Department of the Interior; 583, U. S. Geological Survey/Department of the Interior; 584, ©1988 Accu-Weather, Inc., State College, Pennsylvania 16801; 586(t), ©1988 Accu-Weather, Inc., State College, Pennsylvania 16801; 586(b), NOAA/National Earth Satellite Service; 587(t), ©1988 Accu-Weather, Inc., State College, Pennsylvania 16801; 587(b), NOAA/National Earth Satellite Service; 588(t), ©1988 Accu-Weather, Inc., State College, Pennsylvania 16801; 588(b), NOAA/National Earth Satellite Service; 589(t), ©1988 Accu-Weather, Inc., State College, Pennsylvania 16801; 589(b), NOAA/National Earth Satellite Service; 590-591, ©1982 by Thomas Filsinger from THE MAP OF THE UNIVERSE, published by Celestial Arts, Berkeley, CA; 592-593, from World Portrait Map created for Hubbard Scientific ©1982 by Rand McNally, R.L.88-S-95;

GLOSSARY: 595(t), Photo Researchers; 595(b), Krafft-Explorer/Photo Researchers; 596(t), Harvard College Observatory/Science Photo Library/Photo Researchers; 596(b), The Image Bank; 597, John Eastcott/Yva Momatiuk/Woodfin Camp & Assoc.; 598, USDA; 599(t), E. R. Degginger; 599(b), Lionel Brown/The Image Bank; 600(t), John Sanford/Science Photo Library/Photo Researchers; 600(b), Earth Scenes; 601, E. R. Degginger; 602(t), National Optical Astronomy Observatories; 602(b), Tony Arruza/Bruce Coleman, Inc.; 603(t), Walter Iooss, Jr./The Image Bank; 603(b), JPL/NASA; 604(t), John Bova/Photo Researchers; 604(b), HBJ Library; 605, C. C. Lockwood/Earth Scenes.

G 2
H 3
I 4
J 5